STATIC AND DYNAMIC HIGH PRESSURE MINERAL PHYSICS

High-pressure mineral physics is a field that has shaped our understanding of deep planetary interiors and revealed new material phenomena occurring at extreme conditions. Comprised of 16 chapters written by well-established experts, this book covers recent advances in static and dynamic compression techniques and enhanced diagnostic capabilities, including synchrotron X-ray and neutron diffraction, spectroscopic measurements, in situ X-ray diffraction under dynamic loading, and multigrain crystallography at megabar pressures. Applications range from measuring equations of state, elasticity, and deformation of materials at high pressure, to synthesis and thermochemistry of high-pressure phases, and new molecular compounds and superconductivity under extreme conditions. This book also introduces experimental geochemistry in the laser-heated diamond anvil cell enabled by the focused ion beam technique for sample recovery and quantitative chemical analysis at submicron scale. Each chapter ends with an insightful perspective of future directions, making it an invaluable resource for graduate students and researchers.

YINGWEI FEI is a leading high-pressure experimentalist with broad interests in Earth and planetary science and high-pressure physics and materials synthesis. He is a staff scientist in the Earth and Planets Laboratory at the Carnegie Institution for Science and is an elected Fellow of the Mineralogical Society of America, the American Geophysical Union, and the Geochemical Society.

MICHAEL J. WALTER is an experimental petrologist who investigates the chemistry, structure, and properties of melts and minerals under the extreme pressures and temperatures in deep planetary interiors. He is the Director of the Earth and Planets Laboratory at the Carnegie Institution for Science and is a Fellow of the Mineralogical Society of America and the American Geophysical Union.

STATIC AND DYNAMIC HIGH PRESSURE MINERAL PHYSICS

Edited by

YINGWEI FEI
Carnegie Institution for Science, Washington, DC

MICHAEL J. WALTER
Carnegie Institution for Science, Washington, DC

CAMBRIDGE
UNIVERSITY PRESS

CAMBRIDGE
UNIVERSITY PRESS

Shaftesbury Road, Cambridge CB2 8EA, United Kingdom

One Liberty Plaza, 20th Floor, New York, NY 10006, USA

477 Williamstown Road, Port Melbourne, VIC 3207, Australia

314–321, 3rd Floor, Plot 3, Splendor Forum, Jasola District Centre, New Delhi – 110025, India

103 Penang Road, #05–06/07, Visioncrest Commercial, Singapore 238467

Cambridge University Press is part of Cambridge University Press & Assessment,
a department of the University of Cambridge.

We share the University's mission to contribute to society through the pursuit of
education, learning and research at the highest international levels of excellence.

www.cambridge.org
Information on this title: www.cambridge.org/9781108479752

DOI: 10.1017/9781108806145

First published 2022

A catalogue record for this publication is available from the British Library.

Library of Congress Cataloging-in-Publication Data
Names: Fei, Yingwei, editor. | Walter, Michael (Michael J.), editor.
Title: Static and dynamic high pressure mineral physics / edited by Yingwei Fei (Carnegie Institution of
Washington, Washington DC), Michael J. Walter (Carnegie Institution of Washington, Washington DC).
Description: Cambridge, United Kingdom ; New York, NY : Cambridge University Press, 2022. |
Includes bibliographical references and index.
Identifiers: LCCN 2022016928 (print) | LCCN 2022016929 (ebook) | ISBN 9781108479752 (hardback) |
ISBN 9781108806145 (epub)
Subjects: LCSH: High pressure geosciences. | Mineralogy. | Geochemistry. | BISAC: SCIENCE /
Earth Sciences / Mineralogy
Classification: LCC QE33.2.H54 S73 2022 (print) | LCC QE33.2.H54 (ebook) |
DDC 550.1/5311–dc23/eng20220623
LC record available at https://lccn.loc.gov/2022016928
LC ebook record available at https://lccn.loc.gov/2022016929

ISBN 978-1-108-47975-2 Hardback

Contents

Contributors

James Badro
Université de Paris, Institut de physique du globe de Paris, CNRS, 75005 Paris, France

Reinhard Boehler
Neutron Scattering Division, Neutron Sciences Directorate, Oak Ridge National Laboratory, Oak Ridge, TN 37830, USA

Andrew J. Campbell
Department of the Geophysical Sciences, University of Chicago, Chicago, IL 60637, USA

Thomas Duffy
Department of Geosciences, Princeton University, Princeton, NJ 08544, USA

Mikhail I. Eremets
Max-Planck-Institut für Chemie, 55128 Mainz, Germany

Yingwei Fei
Earth and Planets Laboratory, Carnegie Institution for Science, Washington, DC 20015, USA

Alexander F. Goncharov
Earth and Planets Laboratory, Carnegie Institution for Science, Washington, DC 20015, USA

Malcom Guthrie
European Spallation Source, 224 84 Lund, Sweden

Bianca Haberl
Neutron Scattering Division, Neutron Sciences Directorate, Oak Ridge National Laboratory, Oak Ridge, TN 37830, USA

Kei Hirose
Department of Earth and Planetary Science, University of Tokyo, Tokyo 113-0033, Japan

Raymond Jeanloz
Departments of Earth and Planetary Science and of Astronomy, University of California, Berkeley, CA 94720, USA

Seiji Kamada
Frontier Research Institute for Interdisciplinary Sciences, and Department of Earth Science, Graduate School of Science, Tohoku University, Sendai 980-8578, Japan

Jie Li
Department of Earth and Environmental Sciences, University of Michigan, Ann Arbor, MI 48109, USA

Oliver T. Lord
School of Earth Sciences, University of Bristol, Bristol BS8 1RJ, UK

Ho-kwang Mao
Center for High Pressure Science and Technology Advanced Research (HPSTAR), Shanghai 201203, China

Wendy L. Mao
Department of Geological Sciences, Department of Geological Sciences, Stanford University, Stanford, CA 94305, USA, and Stanford Institute for Materials and Energy Sciences, SLAC National Accelerator Laboratory, Menlo Park, CA 94025, USA

Sébastien Merkel
Université de Lille, CNRS, INRAE, ENSCL, UMR 8207 - UMET– Unité Matériaux et Transformations, 59000 Lille, France

Jamie J. Molaison
Neutron Scattering Division, Neutron Sciences Directorate, Oak Ridge National Laboratory, Oak Ridge, TN 37830, USA

Alexandra Navrotsky
School of Molecular Sciences and Center for Materials of the Universe, Arizona State University, Tempe, AZ 85287, USA

Eiji Ohtani
Department of Earth Science, Graduate School of Science, Tohoku University, Sendai 980-8578, Japan

Tatsuya Sakamaki
Department of Earth Science, Graduate School of Science, Tohoku University, Sendai 980-8578, Japan

Guoyin Shen
HPCAT, X-ray Science Division, Argonne National Laboratory, IL 60439, USA

Sally J. Tracy
Earth and Planets Laboratory, Carnegie Institution for Science, Washington, DC 20015, USA

Michael J. Walter
Earth and Planets Laboratory, Carnegie Institution for Science, Washington, DC 20015, USA

Junyue Wang
Center for High Pressure Science and Technology Advanced Research (HPSTAR), Shanghai 201203, China

Takehiko Yagi
Geochemical Research Center, Graduate School of Science, University of Tokyo, Tokyo 113-0033, Japan

Li Zhang
Center for High Pressure Science and Technology Advanced Research (HPSTAR), Shanghai 201203, China

Figure 1.0 Dave Mao at the Geophysical Laboratory, Washington DC, 2018. For the color version, refer to the plate section.
(photos courtesy of the Carnegie Institution for Science).

1

Introduction to Static and Dynamic High-Pressure Mineral Physics

MICHAEL J. WALTER AND YINGWEI FEI

In October of 2018, a group of scientists gathered at the Broad Branch Road campus of the Carnegie Institution for Science to celebrate 50 years of high-pressure research by Ho-Kwang "Dave" Mao at the Geophysical Laboratory. The celebration highlighted the growth of high-pressure mineral physics over the last half century, which has matured into a vibrant discipline in the physical sciences because of its intimate connections to Earth and planetary sciences, solid-state physics, and materials science. Dave's impact in high-pressure research for over a half a century has been immense, with a history of innovation and discovery spanning from the Earth and planetary sciences to fundamental materials physics. Dave has always been an intrepid pioneer in high-pressure science, and together with his numerous colleagues and collaborators across the world he has driven the field to ever higher pressures and temperatures, guided the community in adopting and adapting a spectrum of new technologies for in situ interrogation of samples at extreme conditions, and relentlessly explored the materials that make up the deep interiors of planets. In this volume, we assemble 15 chapters from authors who have worked with, been inspired by, or mentored by Dave over his amazing career, spanning a range of subjects that covers the entire field of high-pressure mineral physics.

1.1 Introduction

High-pressure mineral physics focuses on the physical properties of materials at high pressure, a field that has shaped our understanding of deep planetary interiors and revealed new material phenomena appearing at extreme conditions. Beginning in the early 1970s, the field has made major contributions to Earth and planetary sciences, condensed matter physics, and high-pressure materials synthesis, through ever-expanding capabilities for reaching higher pressures and probing smaller samples. Ho-Kwang "Dave" Mao has been at the forefront and led the growth of the field since its beginning. Dave started his epic adventure in high-pressure research as a doctoral student at the University of Rochester working in the lab of William A. "Bill" Bassett. After graduating in 1968, Dave began as a postdoctoral fellow at the Geophysical Laboratory (GL) of the Carnegie Institution of Washington, working closely with GL staff scientist Peter Bell. Shortly thereafter, in 1972,

Dave was appointed as a staff scientist by GL Director Hatten Yoder, and he retired in 2019 after more than 50 years of discovery, innovation, impact, and influence applying mineral physics in the realms of Earth and planetary science and fundamental materials science.

While a graduate student in Rochester, Dave was first introduced to an entirely new high-pressure technology, the diamond anvil cell, which had recently been developed at the National Institute of Standards and Technology (NIST)/National Bureau of Standards (NBS) (Piermarini, 2001). Dave's graduate work, measuring lattice parameters of iron and iron oxides using in situ X-ray diffraction, working together with diamond anvil cell (DAC) pioneer Bill Bassett and fellow student Taro Takahashi, set him on his path of high-pressure discovery (Mao et al., 1967, 1974). Upon his arrival at the Geophysical Lab, Dave worked together with Peter Bell (Figure 1.1) on improvements to the original lever-arm DAC design (Piermarini, 2001), and they were the first to achieve pressures of a megabar and above (Mao and Bell, 1978), opening up the lower mantle and core to relatively routine exploration by other eager high-pressure Earth scientists. Dave recognized that measuring pressure quickly and reliably was a prerequisite for this device to take hold in the high-pressure community, and Dave again leaped on a new method, the ruby fluorescence pressure scale developed at NIST/NBS, calibrating the scale using specific volume measurements of four metals combined with their shock wave equations of state to provide a

Figure 1.1 Ho-Kwang "Dave" Mao in the early years at the Geophysical Laboratory, shown here with staff scientist Peter Bell (a,c) and holding a lever arm DAC (b) in preparation for an experiment using a first-generation laser heating system (photos courtesy of the Carnegie Institution for Science).

reliable, quasihydrostatic in situ pressure scale that is still widely used today (Mao et al., 1978, 1986).

The use of the DAC as a high-pressure device to address problems in Earth and planetary sciences, condensed matter physics, and materials sciences has drastically expanded by combining the DAC with laser heating to achieve simultaneous high pressure and temperature and coupling the DAC with synchrotron X-radiation for in situ measurements. Dave led both expansions to address a wide range of scientific questions in high-pressure science. Together with GL postdoctoral fellow Takehiko Yagi at the Geophysical Lab in the late 1970s, Dave was among the first to combine laser heating with the DAC to synthesize and investigate the structure and crystal chemistry of minerals at lower mantle conditions (Bell et al., 1979; Yagi et al., 1978, 1979), following the successful synthesis of the $MgSiO_3$-perovskite phase (bridgmanite) (Liu, 1976). The development of the double-sided laser-heating technique (Shen et al., 1996) and the symmetric DAC (Shen and Mao, 2017) further advanced the application of DAC techniques.

Dave recognized early on that the ability to probe samples in the DAC with focused energy at infrared, optical, and X-ray wavelengths permits a vast landscape of possible measurements to be made in situ at high pressure and often at high temperature. From his first forays into coupling synchrotron radiation with high-pressure experiments at the Brookhaven National Lab in the 1980s to his vision and dedication to the construction of large facilities dedicated to high-pressure science at the Advanced Photon Source (High Pressure Collaborative Access Team [HPCAT], HPSynC) and most recently the Shanghai Synchrotron Radiation Facility (SSRF), Dave and like-minded colleagues have led the community to its current state, where making measurements at megabar pressures and extreme temperatures while probing with energetic beams at scales reaching to the nanometric in scale has become commonplace and, importantly, easily accessible (Mao and Hemley, 1996; Shen et al., 1996, 2010; Mao et al., 2001a; Zhao et al., 2004; Hemley et al., 2005; Mao et al., 2016; Shen and Mao, 2017; Goncharov et al., 2019).

In the realm of Earth sciences, Dave has spent a career investigating the fundamental phase equilibria, thermodynamic, and physical properties of solid and liquid phases at pressure–temperature conditions relevant to Earth's lower mantle and core – his body of work is truly remarkable and has been tremendously impactful. From his studies on deep mantle silicates and oxides to his extensive investigations on iron and iron alloys, there can be little doubt that our current understanding of Earth's deep interior has Dave's footprints all over the territory, and it is hard to find a path Dave has not trodden (Yagi et al., 1978; Mao and Bell, 1979; Jephcoat et al., 1986; Mao et al., 1989, 1997, 1998, 1990, 2001b, 2006b; Stixrude et al., 1992; Fei and Mao, 1994; Duffy et al., 1995; Saxena et al., 1995; Shen et al., 1998; Shieh et al., 1998; Badro et al., 1999; Hirose et al., 1999; Zha et al., 2000; Merkel et al., 2002; Li et al., 2004; Lin et al., 2005). One might argue that Dave's most recently discovered path leading to new high-pressure hydrous phases and superoxide iron-rich phases, potentially tying together the Earth's surface, mantle, and core, represents a fitting destination that began with his studies on iron and iron oxide phases as a graduate student (Hu et al., 2016; Liu et al., 2017, 2019; Mao et al., 2017; Zhang et al., 2018 Lin et al., 2020).

Expanding his horizons beyond the terrestrial landscape, Dave ventured into the realm of gas giants early in his career, aiming to probe the high-pressure behavior of hydrogen and leading the elusive search for its metallicity, the "holy grail" in high-pressure physics (Sharma et al., 1980; Hemley and Mao, 1988; Mao et al., 1988a, 1988b; Badding et al., 1991; Mao and Hemley, 1994; Loubeyre et al., 1996; Gregoryanz et al., 2003). This ultimately led to numerous investigations of a vast array of molecular compounds, exploration of novel new materials, and the search for room temperature superconductivity in materials at high pressure (Mao et al., 1988a, 2006a; Vos et al., 1993; Goncharov et al., 1996; Eremets et al., 2001; Meng et al., 2004; Yoshimura et al., 2006; Gregoryanz et al., 2007; Chen et al., 2008; Somayazulu et al., 2010; Zhu et al., 2013; Zhou et al., 2016; Zeng et al., 2017; Wang et al., 2018; Ji et al., 2019).

Perhaps the greatest legacy of Dave's epic journey is the vast number of scientists (Figure 1.2) who he has trained, collaborated with, mentored, interacted with, and paved

Figure 1.2 Dave Mao and colleagues in the 1990s and 2000s at the Geophysical Laboratory.

the way for over the last 50-plus years, a tradition that he continues to this day. The operation of the dedicated high-pressure beamline (HPCAT) at the Advanced Photon Source has further expanded Dave's collaborations with researchers around the globe (Figure 1.3). This volume, *Static and Dynamic High-Pressure Mineral Physics*, represents

Figure 1.3 Dave Mao at HPCAT, the Advanced Photon Source. For the color version, refer to the plate section.

Figure 1.4 Attendees at the symposium to honor Ho Kwang "Dave" Mao and 50 years of high-pressure science at the Geophysical Laboratory, held in October 2018, at the Broad Branch Road Campus of the Carnegie Institution for Science. For the color version, refer to the plate section.
(photo courtesy of the Carnegie Institution for Science).

an outgrowth of the workshop held in Dave's honor at the Geophysical Lab in 2018 (Figure 1.4) celebrating his half-century journey in high-pressure science. Many of those who attended the workshop have contributed chapters to this book, together with collaborators and colleagues both old and new who have the pleasure of tagging along on Dave's journey of discovery in high-pressure science.

1.2 Chapter Summaries

Leading off is Professor Takehiko Yagi (University of Tokyo) who worked with Dave as a postdoc at the Geophysical Laboratory in the late 1970s and who has himself been on a lifelong journey in high-pressure research. In Chapter 2, Yagi provides a unique historical perspective on the "Developments of Static High-Pressure Techniques and the Study of the Earth's Deep Interior in the Last 50 Years and Its Future." In this chapter, Yagi describes the evolution of both large-volume multi-anvil and diamond anvil techniques and the numerous kinds of experimental and analytical techniques that have been combined with these high-pressure devices to obtain what is now a mountain of information about the properties of minerals and melts at high pressures and temperatures. Yagi discusses how advances in coupling synchrotron radiation to the DAC played a key role in our understanding of the deep Earth and ends by describing current state-of-the-art efforts to extend the pressure range of the diamond anvil cell far beyond what is routinely capable in existing high-pressure devices in order to reach the next frontier in high-pressure science.

In Chapter 3, Guoyin Shen (Argonne National Laboratory), together with Wendy Mao (Stanford University), continues the theme of the key role that high-brilliance synchrotron radiation plays in high-pressure science in their contribution "Applications of Synchrotron and FEL X-Rays in High-Pressure Research." These authors provide an extensive review of developments in synchrotron and free electron laser (FEL) technology, and provide numerous examples of the many spectroscopic techniques that have been utilized in high-pressure research. They describe double-sided laser heating and how its coupling to X-ray diffraction through the use of small X-ray beams available at synchrotron light sources revolutionized high-pressure science in the DAC, including through X-ray mapping and the ability to make single-crystal and multigrain measurements. They discuss how techniques such as absorption and emission spectroscopy, inelastic scattering, nuclear forward scattering, X-radiography, and transmission X-ray microscopes enable determination of equations of state, interrogation of the electronic state of materials, spin transitions, site occupancies, magnetic transitions, thermodynamic properties, sound velocities, microstructural evolution, and more. These authors then highlight some active areas of development in high-pressure X-ray research and provide a forward-looking perspective on future opportunities becoming available with upgrades in both synchrotron and FEL facilities worldwide.

Low-Z materials are not easily noticed by X-rays and require other methods of interrogation, and in Chapter 4, "Neutron Diamond Anvil Cell Project at ORNL," Reinhard Boehler and his colleagues Bianca Haberl and Jamie J. Molaison from the Oak Ridge National Laboratory (ORNL), together with Malcom Guthrie (European Spallation Source), report on a project to expand the pressure range of neutron diffraction in the DAC, which grew from Dave Mao's vision as director of the Department of Energy (DOE) Energy Frontier Research in Extreme Environments (EFree). The team describes efforts to develop techniques for reaching much higher pressures than previously achieved while maintaining the relatively large sample sizes required for diffraction measurements at low neutron flux. These authors first recap high-pressure advances using neutrons in high-pressure science at ORNL over the previous decade, including the Spallation Neutrons at Pressure (SNAP) diffractometer. They then describe how breakthroughs in synthesizing multicarat diamond anvils grown by chemical vapor deposition (CVD), together with the latest developments in large anvil support designs and compact multiton hydraulic diamond cells and new gasket designs, have allowed neutron diffraction experiments at pressures approaching a megabar. They finish off with an example of neutron diffraction measurements on solid D_2O (Ice VII) at 60 GPa.

While static compression experiments provide a wealth of high-pressure information, Chapters 5 and 6 highlight the importance of laboratory shock compression data for understanding the high-pressure behavior of silicates and other geological materials during planetary formation and shock metamorphism, and in deep Earth and other planetary objects. Shock experiments provide distinct, yet complementary, information to static experiments, and recently developed capabilities that allow for in situ examination of the atomic-level structure and exploration of material properties in the terapascal pressure range by laser- or magnetically driven dynamic compression have reinvigorated the field.

In Chapter 5, "Light-Source Diffraction Studies of Planetary Materials under Dynamic Loading," Sally J. Tracy (Carnegie Institution for Science) highlights the recent experimental developments that make measurements possible during the nanosecond to microsecond duration of shock experiments, especially at facilities that couple dynamic compression with high-flux X-rays for in situ X-ray diffraction under dynamic loading. Tracy provides a brief introduction into the theory of shockwave experiments and the different shock platforms and describes the ambiguity in determining features and phases formed during the compression and unloading processes due to the fast time scales of shockwave experiments. Tracy then describes developments at the Dynamic Compression Sector (DCS) at the Advanced Photon Source (APS) for in situ X-ray diffraction under plate impact shock loading that provide important information during compression and unloading, and provides examples of results in the systems SiO_2, Mg_2SiO_4, and carbon. This is followed by a discussion of the new and exciting possibilities of laser shock studies at X-Ray Free Electron Laser Sources (XFEL) that generate ultrafast bursts (fs) of X-rays with peak X-ray brightness 10 orders of magnitude higher than synchrotron sources, and reviews recent results on silicate liquids and glasses, hydrocarbons, and carbides.

Thomas Duffy (Princeton University) builds on the theme of shock compression, and in Chapter 6, "New Analysis of Shock-Compression Data for Selected Silicates," discusses issues with identification of the high-pressure phases formed under shock loading using density comparisons with static data or postmortem analysis of samples. Duffy then summarizes pressure-density shock wave data for garnet, tourmaline, nepheline, topaz, and spodumene and compares their Hugoniot compression behavior with data from static compression and theoretical studies. Duffy shows that there is good agreement with recent 300 K single-crystal X-ray diffraction data, illustrating the range of silicate behavior under shock loading. Duffy ends with a perspective on newly developed in situ X-ray experimental capabilities and the reflects on the many open questions that can be addressed with this new ability.

In Chapter 7, "Scaling Relations for Combined Static and Dynamic High-Pressure Experiments," Raymond Jeanloz (University of California at Berkeley) shows how combining static and dynamic experiments can maximize material compression by tuning the viscous dissipation that occurs especially under shock loading. In this contribution, Jeanloz summarizes scaling relations for evaluating the internal energy (E) dissipated as "waste heat" upon dynamic loading and shows that the dissipated energy increases rapidly with final compression. However, Jeanloz's analysis shows that the waste heat is significantly reduced by precompressing the target sample. For a given final density, increased precompression pushes the dynamically loaded state toward an isentrope, and in the limit of numerous steps approaches isentropic ramp compression. This means that by combining static precompression and dynamic methods, final pressure–density–temperature (P–ρ–T) states achieved in samples can be "tuned" to minimize final temperatures and maximum compression.

Accurate equations of state (EoS) for the solid phases that constitute planetary interiors are fundamental for modeling their density with pressure and temperature, and with the advent of synchrotron radiation as a primary tool in high-pressure science, the equation of state of "pressure markers" can be utilized to monitor pressure in situ in experiments. In Chapter 8, "Equations of State of Selected Solids for High-Pressure Research and Planetary Interior Density Models," Yingwei Fei (Carnegie Institution for Science) reviews the current state of the art of using thermal EoS in the Earth and planetary sciences. Fei begins by describing the experimental methods for making EOS measurements, including both static and dynamic approaches, and the importance of in situ X-ray diffraction at synchrotrons. Fei then presents a compilation of EoS parameters for a range of different solid materials that are commonly used as pressure standards in experiments, including platinum, gold, neon, MgO, and NaCl. He then discusses the internal consistency among the standards and compares their performance in extrapolation to the multimegabar pressure range. Fei then summarizes EoS data for key phases in Earth's mantle required to accurately model Earth's interior composition based on seismological measurements of density, including bridgmanite, ferropericlase, and $CaSiO_3$-perovskite, and then discusses the effect of light elements on the EoS of iron (Fe) alloys relevant to Earth's core. Fei emphasizes how future work combining both static and dynamic EoS measurements are required to improve pressure standards and our understanding of deep planetary interiors.

After nearly a century of investigations, the compositions of the liquid outer core and the solid inner core are not uniquely known. Seismically deduced density and elastic wave velocities can potentially reveal the core composition, but only if these properties are known in candidate iron alloy compositions at the extreme high pressures and temperatures of Earth's core. In Chapter 9, "Elasticity at High Pressure with Implication for the Earth's Inner Core," Seiji Kamada (Tohoku University), Tatsuya Sakamaki (Tohoku University), and Eiji Ohtani (Tohoku University) take us on a journey to the core through the lens of the elasticity of iron and iron alloys. They provide the rationale for elasticity data to probe the secrets of core composition and describe how this line of experimental inquiry has developed since the pioneering work at megabar pressures of Dave Mao and colleagues in the early 1990s. Sakamaki and coauthors provide a comprehensive and informative review of the experimental methods for measuring compressional and shear-wave velocities at high pressures and temperatures (including ultrasonic, Brillouin spectroscopy; inelastic X-ray scattering; nuclear inelastic scattering; shock waves; pulsed laser; and radial diffraction). The authors provide a careful assessment of the wave velocity data at high pressure and room temperature on a range of iron alloy systems (e.g., Fe-C; Fe-O; Fe-S; Fe-Si; Fe-H), which is followed by an evaluation of much scarcer data obtained at combined high pressure and high temperature on iron alloys using inelastic X-ray scattering. The journey ends with an evaluation of the constraints placed on inner core compositions using extrapolations of data to inner core conditions, and a call for future studies especially in obtaining shear velocity data at high temperature.

Determining the correct structure of phases synthesized in diamond anvil cell experiments is the crucial first step for investigating materials at high pressure. In Chapter 10, "Multigrain Crystallography at Megabar Pressures," Li Zhang, Junyue Wang, and Dave Mao (Center for High Pressure Science and Technology Advanced Research) provide an in-depth look at how new developments in multigrain X-ray diffraction methods can revolutionize structural characterization of samples made in the laser-heated DAC at multimegabar pressures. In situ powder X-ray diffraction is the industry standard, but it has limitations for obtaining detailed structural information. While single-crystal X-ray diffraction is challenging at megabar pressures, multigrain crystallography provides an innovative and powerful structural solution by taking advantage of separated reflection spots on diffraction images. Zhang and coauthors describe how several hundred submicron grains can be simultaneously indexed from an individual high-pressure experiment, and a nearly full convergence of the structure can be achieved when applying the multigrain method. To demonstrate the potential of this method, the authors present a schematic setup for data collection in a DAC at a synchrotron beamline and provide an example structure determination for seifertite at 129 GPa. They conclude with a discussion of future software developments that can facilitate wide and routine application of multigrain techniques.

Knowledge of how minerals deform at high pressures is key to understanding how planetary interiors evolve. In Chapter 11, "Deformation and Plasticity of Materials under Extreme Conditions," Sébastian Merkel (Université de Lille) provides a stimulating review of the tremendous experimental advances made in the last quarter century, especially the coupling of high-pressure instruments with synchrotron radiation, for addressing the deformation and plasticity of materials under extreme conditions. Merkel discusses deformation experiments in both large volume presses and diamond anvil cells, noting their differences, advantages, and limitations, as well as new advances using torsion devices combined with tomographic measurements. Following on from Zhang and colleagues' chapter, Merkel describes how studies of microstructures and plastic deformation have advanced to new heights using multigrain X-ray diffraction to study lattice-preferred orientation and microstructural elements to constrain plastic deformation mechanisms, and shows how deformation experiments are best interpreted using self-consistent methods that treat each grain of the polycrystal as an inclusion in a homogeneous yet anisotropic medium. Merkel then reviews applications of deformation experiments to deep Earth materials such as core-forming iron metal, and the lower mantle phases ferropericlase and bridgmanite, discussing how plastic deformation depends on the complex interplay between microstructure and the properties of each phase. Merkel provides an optimistic perspective of the future of deformation studies at high pressure, noting how improvements in high-pressure techniques and coupled in situ measurements will allow interrogation of samples at increasingly high pressures and over a wide range of strain rates.

The multi-anvil press (MAP) has for decades been a high-pressure experimental workhorse, and in Chapter 12, "Synthesis of High-Pressure Silicate Polymorphs Using Multi-Anvil Press," Jie Li (University of Michigan) extols the virtues of the MAP for material synthesis, structure and property investigations, and studies of phase equilibria and chemical reactions in the geological and materials sciences. Li begins by providing the basics of pressure generation and the factors that limit sample size in the MAP and delves into the realms of pressure calibration and uncertainties as well as temperature generation, measurement, and variability. The main thrust of the chapter is in describing synthesis strategies for growing large (e.g., mm-sized), pure, single crystals at high pressure that are required for investigating the key properties needed to understand deep planetary interiors. Li discusses the theory of nucleation and growth of phases from melts and fluids and in solid-state transformations, and provides examples and phase diagrams for the key transition zone and lower-mantle phases wadsleyite, ringwoodite, and bridgmanite, including a discussion of the characterization of synthesis products through combination of microanalytical techniques. Li concludes by stressing the importance of new initiatives to support construction of mammoth multi-anvil presses (e.g., ~60,000 tons) for synthesizing large single crystals and polycrystalline samples related to deep planetary interiors as well as for synthesis of superhard materials including polycrystalline nanodiamonds, carbides, and nitrides.

Recent technological advances make the diamond anvil cell an increasingly valuable petrological tool, and in Chapter 13, Yingwei Fei and Michael Walter (Carnegie Institution for Science), James Badro (Institut de Physique du Globe de Paris), Kei Hirose (Earth-Life Science Institute), Oliver T. Lord (University of Bristol), Andrew J. Campbell (University of Chicago), and Eiji Ohtani (Tohoku University) team up to describe in recent efforts in "Investigation of Chemical Interactions and Melting Using Laser-Heated Diamond Anvil Cell." In this chapter, the authors describe how technical developments, especially focused ion beam (FIB) technology, allow ever more reliable investigations of element partitioning and melting phase relations at pressures approaching the center of the Earth. The authors begin with a discussion of temperature and pressure determination in the laser-heated DAC (LHDAC) and the importance of matching the temperature distribution to textural and chemical observations in recovered samples. They describe methods for preparing homogeneous starting materials, including ultrafast quenching methods such as aerodynamic levitation and ball-mill preparation of homogeneous metal alloys, and discuss new loading techniques using FIB fabrication. FIB technology is also key for ex situ chemical and textural analyses, allowing precisely milled cross sections of the heated spot at the micron and submicron scale. Examples utilizing these cutting-edge techniques are provided, including studies of metal-silicate interaction, melting phase relations of mantle compositions, and melting of core alloys. The chapter ends by advocating for coordinated collaborative efforts to share well-characterized starting materials, and a challenge to achieve a direct determination of melting relations at the inner core boundary with sample recovery for quantitative chemical analysis to help solve the longstanding problem of the composition of Earth's core.

Knowledge of the behavior of light element molecular compounds (e.g., H, He, N) at extreme pressures is key to unlocking the secrets of gas giant planetary interiors as well as for understanding their fundamental physics. In Chapter 14, "Molecular Compounds under Extreme Conditions," Alexander Goncharov (Carnegie Institution for Science) reviews experimental studies of molecular solids at high pressures over the past several decades. Goncharov emphasizes the importance of the pressure variable and the balance between intermolecular and intramolecular bonds leading to the formation of structurally complex crystals. Goncharov relates Dave Mao's pioneering work and legacy in high-pressure research of molecular solids, especially in relation to hydrogen (metallization, clathrates), and the technical advances in experimental (e.g., gas loading) and analytical techniques needed to study these challenging materials at extreme pressures. Also described are the analytical developments that were crucial for probing the structure and behavior of low-Z molecular compounds, especially the three complementary optical spectroscopy techniques (Raman, Brouillian, and infrared), as well as X-ray scattering techniques. Goncharov describes a number of fascinating examples, including dense H_2O ices (e.g., VII and X), clathrate structures, nitrogen compounds, inclusion compounds (cocrystals), and "high-T" superconductors (e.g., H_3S). Goncharov suggests a promising future in further investigating these fascinating compounds by combining experimental static and dynamic high-pressure techniques to reach ever more extreme conditions and through further integrating experiments and theoretical approaches.

Achieving superconductivity in a solid material at room temperature has been a "holy grail" in solid-state physics for decades. In Chapter 15, "Superconductivity at High Pressure," Mikhail I. Eremets (Max Planck Institute for Chemistry) presents a fascinating narrative of the search for high-temperature superconductivity and the progress that has been made in using high pressure to investigate candidate materials. The epic begins with the search for metallic hydrogen at the Geophysical Lab driven by Dave Mao and colleagues as they entered into the megabar realm using the diamond anvil cell. Eremets describes the elusive metallization of hydrogen and the exploration of other materials such as xenon and lithium and the early successes when the highest superconducting temperatures were of the order 20 K. But metallization of hydrogen remained elusive, and still does, and Eremets relates how the search moved to hydrogen alloys. The material landscape frontier expanded with the advent of ab initio calculations that could find new structures and make predictions of superconductivity coupled with improvements in DAC techniques that allowed multimegabar experiments. Eremets regales us with the tale of H_2S, a compound that metallizes at about a megabar with a superconducting temperature of the order 80 K, and how work on this and related compounds led to superconductivity measurements at megabar pressures in the range of 200 K (e.g., H_3S). The drama continues with the recent work on lanthanum hydrides, where superconductivity at temperatures approaching room temperature (~250 K) have been achieved. Eremets ends the tale with a forward-looking and optimistic perspective on the search for room temperature superconductivity and the best candidates for future research.

In Chapter 16, "Thermochemistry of High-Pressure Phases," Alexandra Navrotsky (University of California at Davis) honors Dave Mao with a thoughtfully crafted review on the application of calorimetry to high-pressure research. The review begins with a concise treatment of the fundamental thermodynamic relationships related to the Gibbs free energy of reaction and the wide utility of thermochemical data, so long as it is made internally consistent in both values and format. Describing calorimetry as "the science and art of measuring heat effects," Navrotsky explains how knowledge of the enthalpy of reaction, ΔH, especially in combination with the entropy of reaction obtained from heat capacity measurements, provides the basis for positing the fundamental question of solid-state chemistry: "What structures can form at a given composition and why?" Navrotsky guides us through advances in calorimetric methodology, describing cryogenic heat capacity measurements in adiabatic calorimeters ranging from the large vessels holding grams of material to miniaturized "calorimeters-on-a-chip" to measure of micrograms of samples, as well as high-temperature solution and reaction calorimetry. This is followed by a systematic look at a range of materials including silicate spinels, perovskites, chalcogenides, nitrides, and carbides, as well as an enlightening discussion of water and defects in high-pressure phases and nanoscale effects. Navrotsky ends with a thoughtful perspective of the important role of calorimetry in combination with high-pressure synthesis and phase equilibria, especially when integrated with crystallographic, spectroscopic, and theoretical studies, and how together they lead to essential insights into the structure and dynamics of the interior of the Earth and other planetary bodies.

References

Badding, J. V., Hemley, R. J., Mao, H. K. (1991). High-pressure chemistry of hydrogen in metals – in situ study of iron hydride. *Science*, **253**, 421–424.

Badro, J., Struzhkin, V. V., Shu, J., et al. (1999). Magnetism in FeO at megabar pressures from X-ray emission spectroscopy. *Physical Review Letters*, **83**, 4101.

Bell, P., Yagi, T., Mao, H. (1979). Iron-magnesium distribution coefficients between spinel [(Mg, Fe) 2SiO4], magnesiowüstite [(Mg, Fe) O], and perovskite [(Mg, Fe) SiO3]. *Year Book Carnegie Institute Washington*, **78**, 618–621.

Chen, X.-J., Wang, J.-L., Struzhkin, V. V., Mao, H.-k., Hemley, R. J., Lin, H.-Q. (2008). Superconducting behavior in compressed solid SiH 4 with a layered structure. *Physical Review Letters*, **101**, 077002.

Duffy, T. S., Hemley, R. J., Mao, H. K. (1995). Equation of state and shear-strength at multimegabar pressures – magnesium-oxide to 227GPa. *Physical Review Letters*, **74**, 1371–1374.

Eremets, M. I., Struzhkin, V. W., Mao, H. K., Hemley, R. J. (2001). Superconductivity in boron. *Science*, **293**, 272–274.

Fei, Y. W., Mao, H. K. (1994). In-situ determination of the NiAs phase of FeO at high-pressure and temperature. *Science*, **266**, 1678–1680.

Goncharov, A. F., Kong, L., Mao, H.-k. (2019). High-pressure integrated synchrotron infrared spectroscopy system at the Shanghai Synchrotron Radiation Facility. *Review of Scientific Instruments*, **90**, 093905.

Goncharov, A. F., Struzhkin, V. V., Somayazulu, M. S., Hemley, R. J., Mao, H. K. (1996). Compression of ice to 210 gigapascals: infrared evidence for a symmetric hydrogen-bonded phase. *Science*, **273**, 218–220.

Gregoryanz, E., Goncharov, A. F., Matsuishi, K., Mao, H., Hemley, R. J. (2003). Raman spectroscopy of hot dense hydrogen. *Physical Review Letters*, **90**(17).

Gregoryanz, E., Goncharov, A. F., Sanloup, C., Somayazulu, M., Mao, H.-k., Hemley, R. J. (2007). High P-T transformations of nitrogen to 170 GPa. *Journal of Chemical Physics*, **126**, 184505.

Hemley, R. J., Mao, H.-k. (1988). Phase-transition in solid molecular-hydrogen at ultrahigh pressures. *Physical Review Letters*, **61**, 857–860.

Hemley, R. J., Mao, H.-k., Struzhkin, V. V. (2005). Synchrotron radiation and high pressure: new light on materials under extreme conditions. *Journal of Synchrotron Radiation*, **12**, 135–154.

Hirose, K., Fei, Y. W., Ma, Y. Z., Mao, H. K. (1999). The fate of subducted basaltic crust in the Earth's lower mantle. *Nature*, **397**, 53–56.

Hu, Q. Y., Kim, D. Y., Yang, W. G., et al. (2016). FeO2 and FeOOH under deep lower-mantle conditions and Earth's oxygen-hydrogen cycles. *Nature*, **534**, 241–244.

Jephcoat, A. P., Mao, H. K., Bell, P. M. (1986). Static compression of iron to 78-GPa with rare-gas solids as pressure-transmitting media. *Journal of Geophysical Research – Solid Earth and Planets*, **91**, 4677–4684.

Ji, C., Li, B., Liu, W., et al. (2019). Ultrahigh-pressure isostructural electronic transitions in hydrogen. *Nature*, 573, 558–562.

Li, J., Struzhkin, V. V., Mao, H. K., et al. (2004). Electronic spin state of iron in lower mantle perovskite. *Proceedings of the National Academy of Sciences of the United States of America*, **101**, 14027–14030.

Lin, J. F., Struzhkin, V. V., Jacobsen, S. D., et al. (2005). Spin transition of iron in magnesiowustite in the Earth's lower mantle. *Nature*, **436**, 377–380.

Lin, Y. H., Hu, Q. Y., Meng, Y., Walter, M., Mao, H. K. (2020). Evidence for the stability of ultrahydrous stishovite in Earth's lower mantle. *Proceedings of the National Academy of Sciences of the United States of America*, **117**, 184–189.

Liu, J., Hu, Q. Y., Bi, W. L., et al. (2019). Altered chemistry of oxygen and iron under deep Earth conditions. *Nature Communications*, **10**(1), 1–8.

Liu, J., Hu, Q. Y., Kim, et al. (2017). Hydrogen-bearing iron peroxide and the origin of ultralow-velocity zones. *Nature*, **551**, 494–497.

Liu, L. G. (1976). Orthorhombic perovskite phases observed in olivine, pyroxene and garnet at high-pressures and temperatures. *Physics of the Earth and Planetary Interiors*, **11**, 289–298.

Loubeyre, P., LeToullec, R., Hausermann, D., et al. (1996). X-ray diffraction and equation of state of hydrogen at megabar pressures. *Nature*, **383**, 702–704.

Mao, H. K., Bassett, W. A., Takahash, T. (1967). Effect of pressure on crystal structure and lattice parameters of iron up to 300 kbar. *Journal of Applied Physics*, **38**, 274–276.

Mao, H. K., Bell, P. M. (1978). High-pressure physics – sustained static generation of 1.36 to 1.72 megabars. *Science*, **200**, 1145–1147.

Mao, H. K., Bell, P. M. (1979). Equations of state of MgO and epsilon-Fe under static pressure conditions. *Journal of Geophysical Research*, **84**, 4533–4536.

Mao, H. K., Bell, P. M., Shaner, J. W., Steinberg, D. J. (1978). Specific volume measurements of Cu, Mo, Pd, and Ag and calibration of ruby r1 fluorescence pressure gauge from 0.06 to 1 Mbar. *Journal of Applied Physics*, **49**, 3276–3283.

Mao, H.-k., Chen, B., Chen, J., et al. (2016). Recent advances in high-pressure science and technology. *Matter and Radiation at Extremes*, **1**, 59–75.

Mao, H. K., Chen, L. C., Hemley, R. J., Jephcoat, A. P., Wu, Y., Bassett, W. A. (1989). Stability and equation of state of CaSio3-perovskite to 134-GPa. *Journal of Geophysical Research – Solid Earth and Planets*, **94**, 17889–17894.

Mao, H. K., Hemley, R. J. (1994). Ultrahigh-pressure transitions in solid hydrogen. *Reviews of Modern Physics*, **66**, 671–692.

Mao, H.-k., Hemley, R. J. (1996). Energy dispersive X-ray diffraction of micro-crystals at ultrahigh pressures. *International Journal of High Pressure Research*, **14**, 257–267.

Mao, H. K., Hemley, R. J., Wu, Y., et al. (1988a). High-pressure phase-diagram and equation of state of solid helium from single-crystal X-ray-diffraction to 23.3-GPa. *Physical Review Letters*, **60**, 2649–2652.

Mao, H. K., Hu, Q. Y., Yang, L. X., et al. (2017). When water meets iron at Earth's core-mantle boundary. *National Science Review*, **4**, 870–878.

Mao, H. K., Jephcoat, A. P., Hemley, R. J., et al. (1988b). Synchrotron X-ray-diffraction measurements of single-crystal hydrogen to 26.5 gigapascals. *Science*, **239**, 1131–1134.

Mao, H.-k., Kao, C., Hemley, R. J. (2001a). Inelastic X-ray scattering at ultrahigh pressures. *Journal of Physics: Condensed Matter*, **13**, 7847.

Mao, H. K., Shen, G. Y., Hemley, R. J. (1997). Multivariable dependence of Fe–Mg partitioning in the lower mantle. *Science*, **278**, 2098–2100.

Mao, H. K., Shu, J. F., Shen, G. Y., Hemley, R. J., Li, B. S., Singh, A. K. (1998). Elasticity and rheology of iron above 220 GPa and the nature of the Earth's inner core. *Nature*, **396**, 741743.

Mao, H. K., Takahashi, T., Bassett, W. A., Kinsland, G. L., Merrill, L. (1974). Isothermal compression of magnetite to 320 kbar and pressure-induced phase-transformation. *Journal of Geophysical Research*, **79**, 1165–1170.

Mao, H. K., Wu, Y., Chen, L. C., Shu, J. F., Jephcoat, A. P. (1990). Static compression of iron to 300 GPa and Fe0.8Ni0.2 alloy to 260 GPa – implications for composition of the core. *Journal of Geophysical Research-Solid Earth and Planets*, **95**, 21737–21742.

Mao, H. K., Xu, J., Bell, P. M. (1986). Calibration of the ruby pressure gauge to 800-kbar under quasi-hydrostatic conditions. *Journal of Geophysical Research-Solid Earth and Planets*, **91**, 4673–4676.

Mao, H. K., Xu, J., Struzhkin, V. V., et al. (2001b). Phonon density of states of iron up to 153 gigapascals. *Science*, **292**, 914–916.

Mao, W. L., Mao, H.-k., Meng, Y., et al. (2006a). X-ray–induced dissociation of H2O and formation of an O2–H2 alloy at high pressure. *Science*, **314**, 636–638.

Mao, W. L., Mao, H.-k., Sturhahn, W., et al. (2006b). Iron-rich post-perovskite and the origin of ultralow-velocity zones. *Science*, **312**, 564–565.

Meng, Y., Mao, H.-k., Eng, P. J., et al. (2004). The formation of sp(3) bonding in compressed BN. *Nature Materials*, **3**, 111–114.

Merkel, S., Wenk, H. R., Shu, J. F., et al. (2002). Deformation of polycrystalline MgO at pressures of the lower mantle. *Journal of Geophysical Research – Solid Earth*, **107**, (B11), ECV 3.1–EVC3.17.

Piermarini, G. J. (2001). High pressure X-ray crystallography with the diamond cell at NIST/NBS. *Journal of Research of the National Institute of Standards and Technology*, **106**, 889–920.

Saxena, S., Dubrovinsky, L., Häggkvist, P., Cerenius, Y., Shen, G., Mao, H. (1995). Synchrotron X-ray study of iron at high pressure and temperature. *Science*, **269**, 1703–1704.

Sharma, S. K., Mao, H. K., Bell, P. M. (1980). Raman measurements of hydrogen in the pressure range 0.2–630 kbar at room-temperature. *Physical Review Letters*, **44**, 886–888.

Shen, G., Mao, H.-k., Hemley, R. J. (1996). Laser-heated diamond anvil cell technique: double-sided heating with multimode Nd: YAG laser. *Computer*, **1**, L2.

Shen, G., Wang, L., Ferry, R., Mao, H.-k., Hemley, R. J. (2010). A portable laser heating microscope for high pressure research. *Journal of Physics: Conference Series*, **215**, 012191.

Shen, G. Y., Mao, H. K. (2017). High-pressure studies with X-rays using diamond anvil cells. *Reports on Progress in Physics*, **80**, 1–53.

Shen, G. Y., Mao, H. K., Hemley, R. J., Duffy, T. S., Rivers, M. L. (1998). Melting and crystal structure of iron at high pressures and temperatures. *Geophysical Research Letters*, **25**, 373–376.

Shieh, S. R., Mao, H. K., Hemley, R. J., Ming, L. C. (1998). Decomposition of phase D in the lower mantle and the fate of dense hydrous silicates in subducting slabs. *Earth and Planetary Science Letters*, **159**, 13–23.

Somayazulu, M., Dera, P., Goncharov, A. F., et al. (2010). Pressure-induced bonding and compound formation in xenon–hydrogen solids. *Nature Chemistry*, **2**, 50–53.

Stixrude, L., Hemley, R. J., Fei, Y., Mao, H. K. (1992). Thermoelasticity of silicate perovskite and magnesiowustite and stratification of the Earth's mantle. *Science*, **257**, 1099–1101.

Vos, W. L., Finger, L. W., Hemley, R. J., Mao, H. K. (1993). Novel H2-H2O clathrates at high-pressures. *Physical Review Letters*, **71**, 3150–3153.

Wang, Y., Ying, J., Zhou, Z., et al. (2018). Emergent superconductivity in an iron-based honeycomb lattice initiated by pressure-driven spin-crossover. *Nature Communications*, **9**, 1–7.

Yagi, T., Bell, P., Mao, H. (1979). Phase relations in the system MgO-FeO-SiO2 between 150 and 700 kbar at 1000 C. *Year Book Carnegie Institute Washington*, **78**, 614–618.

Yagi, T., Mao, H. K., Bell, P.M. (1978). Structure and crystal-chemistry of perovskite-type MgSiO3. *Physics and Chemistry of Minerals*, **3**, 97–110.

Yoshimura, Y., Stewart, S. T., Somayazulu, M., Mao, H.-k., Hemley, R. J. (2006). High-pressure X-ray diffraction and Raman spectroscopy of ice VIII. *Journal of Chemical Physics*, **124**, 024502.

Zeng, Z., Yang, L., Zeng, Q., et al. (2017). Synthesis of quenchable amorphous diamond. *Nature Communications*, **8**, 1–7.

Zha, C. S., Mao, H. K., Hemley, R. J. (2000). Elasticity of MgO and a primary pressure scale to 55 GPa. *Proceedings of the National Academy of Sciences of the United States of America*, **97**, 13494–13499.

Zhang, L., Yuan, H. S., Meng, Y., Mao, H. K. (2018). Discovery of a hexagonal ultradense hydrous phase in (Fe,Al)OOH. *Proceedings of the National Academy of Sciences of the United States of America*, **115**, 2908–2911.

Zhao, J., Sturhahn, W., Lin, J.-f., Shen, G., Alp, E. E., Mao, H.-k. (2004). Nuclear resonant scattering at high pressure and high temperature. *High Pressure Research*, **24**, 447–457.

Zhou, Y., Wu, J., Ning, W., et al. (2016). Pressure-induced superconductivity in a three-dimensional topological material ZrTe5. *Proceedings of the National Academy of Sciences*, **113**, 2904–2909.

Zhu, J., Zhang, J., Kong, P., et al. (2013). Superconductivity in topological insulator Sb2Te3 induced by pressure. *Scientific Reports*, **3**, 1–6.

2

Development of Static High-Pressure Techniques and the Study of the Earth's Deep Interior in the Last 50 Years and Its Future

TAKEHIKO YAGI

Development of static high-pressure techniques over the last 50 years is reviewed from the perspective of the study of the Earth's deep interior. Fifty years ago, laboratory high-pressure and -temperature experiments were limited to the conditions corresponding to that of near the surface of the Earth. In high-pressure mineral physics, extension of the pressure range directly made possible the study of deeper parts of the Earth, and many scientists spent great effort to improve various experimental techniques. As a result, currently it is possible to do precise X-ray experiments at the conditions corresponding to the center of the Earth: 6,400 km depth from the surface, about 360 GPa, and more than 5,000 K. Two quite different types of high-pressure apparatus are widely used these days. One is the large-volume high-pressure apparatus, and the other is the diamond anvil cell. Although the latter has the advantage of covering wider pressure and temperature conditions together with optical access to the sample, the former has the advantage of much larger sample volume, and, using internal resistance heaters, very stable and uniform high-temperature conditions can be achieved. Many different types of experimental techniques are combined with these high-pressure devices, and rich information about various properties of minerals and melts can now be obtained. Advancement of synchrotron radiation played a key role for such studies, and our knowledge about the Earth's deep interior has increased considerably. Further efforts are continuing to extend the pressure range beyond the limits of existing high-pressure devices.

2.1 Introduction

In the last half-century, there have been enormous developments in static high-pressure techniques. Fifty years ago, the achievable pressure and temperature conditions of the laboratory high-pressure experiments were limited to the conditions corresponding to that of only about 100 km from the surface of the Earth. Nowadays it is possible to get high-quality X-ray diffraction data on materials at the conditions corresponding to the center of the Earth, or 6,400 km from the surface. These advances have been the result of numerous efforts by many people to extend the pressure range of static high-pressure experiments. In this article, I briefly review the history of these technical developments from the viewpoint of the study of the Earth's deep interior.

There have been many good reviews on the history of static high-pressure techniques. For example, R. C. Liebermann (2011a) wrote a comprehensive review of the history of the multi-anvil apparatus. He also wrote a review article (Liebermann, 2011b) focused on the relationship of his laboratory in the US to multi-anvil laboratories in Japan. E. Ito (2009) described in detail the development and experimental techniques of the "Kawai-type" multi-anvil apparatus, which is now the most popular large-volume apparatus to study the Earth's deep interior. On another important high-pressure apparatus for Earth science, the diamond anvil cell, W. A. Bassett (2009) wrote a very good review of its 50-year history from its very beginning. I started my scientific carrier using multi-anvil apparatuses such as the tetrahedral press and the "DIA-type" cubic anvil press. After getting my Ph.D degree at the University of Tokyo, I joined the high-pressure group at the Geophysical Laboratory of the Carnegie Institution of Washington, led by H. K. Mao and P. M. Bell, as a postdoctoral fellow. There, I also had chance to work with A. V. Valkenburg, one of the pioneers of the diamond anvil cell (DAC). At the time Mao and Bell were working intensively to extend the pressure range and succeeded, for the first time, to achieve pressures above a megabar. Because of these experiences, I learned the advantages and disadvantages of both multi-anvil apparatus and diamond anvil cell, and since then I have continued my research in my laboratory at the Institute for Solid State Physics (ISSP), University of Tokyo, using both of these types of apparatus.

In high-pressure Earth science, extension of the pressure range has directly resulted in a better understanding of the processes of the deeper parts of the Earth. Therefore, many scientists spent great effort for the development of high-pressure techniques. Because of these efforts, our basic knowledge of the Earth's deep interior, such as the nature of the transition zone in the upper mantle, structure of the bottom of the lower mantle, and crystal structure of iron in the center of the Earth, was clarified by laboratory high-pressure and high-temperature experiments. The purpose of this article is to describe the history of static high-pressure experiments and how these important findings were made possible by new developments.

2.2 Early Days of the High-Pressure Experiments to Study the Earth's Deep Interior

Modern high-pressure experiments were started by P. W. Bridgman of Harvard University in the early twentieth century, and he was awarded the Nobel Prize for physics in 1946 for his pioneering contributions to high-pressure science. Most of his high-pressure works were made using the piston-cylinder apparatus. This apparatus can measure the applied pressure and the sample volume precisely. Using the pressure–volume (P–V) data, he clarified the compression behavior of many materials and discovered numerous pressure-induced phase transformations. Using the piston-cylinder apparatus, it was possible to heat the sample above 1,000°C, and studies of minerals under the conditions in the Earth were thus possible for the first time. However, the attainable pressure was limited to about 3 GPa, which corresponds to depths of about 100 km from the surface of the Earth. As a result, application to studies of the Earth's interior was limited to the crust and the shallow part of the upper mantle.

In the 1950s, efforts were made in several countries to synthesize diamonds in the laboratory. In the US, the General Electric Co. (GE) formed a team to pursue this project, and they soon realized that pressures above 5–6 GPa are required to make diamonds in the laboratory. One of the members, H. T. Hall, modified the piston-cylinder apparatus and invented the belt-type apparatus (Hall, 1960) to extend the pressure range to above 6 GPa. Using this apparatus, they succeeded in synthesizing diamonds and published a report on "man-made diamond" (Bundy et al., 1955). Hall left GE, moved to Brigham Young University, and started setting up his own high-pressure laboratory. However, the experimental technique, including the belt apparatus, was protected by the patent of GE and he was not allowed to use it. So, he invented different type of apparatus, the tetrahedral press (Figure 2.1), in which a tetrahedral-shaped solid-pressure transmitting medium was compressed by four hydraulic rams connected to each other (Hall, 1958; 1962). He continued

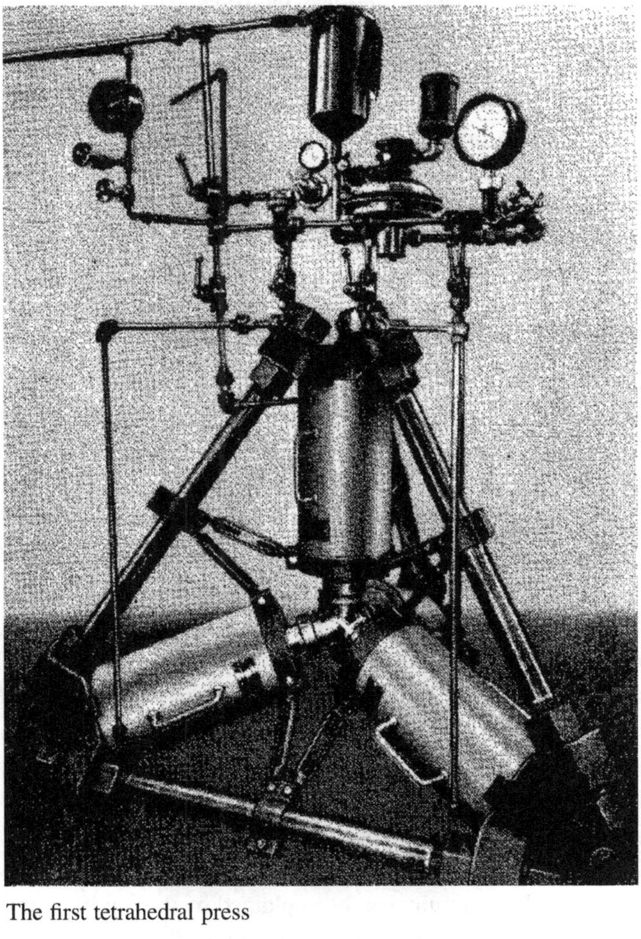

Figure 2.1 The first tetrahedral press (Hall, 1962).

developing this tetrahedral press and also invented the cubic press using six rams. Using these devices, it became possible to extend the pressure range to around 10 GPa. These devices were manufactured and used in various government institutions and big research groups in many countries. But it was so expensive that for many small research groups in US universities, it was unaffordable. Instead, in the US many research groups utilized the diamond anvil cell (DAC) apparatus, as will be described later.

For the study of the Earth's deep interior, one of the first important research targets was to clarify the origin of the seismic discontinuity observed at around 400 km depth in the upper mantle. Based on crystallographic and thermodynamic considerations, there was a prediction that this discontinuity could be explained by the phase transformation of olivine, the most abundant mineral in the upper mantle, into the spinel structure. However, there was no experimental evidence that such a phase transition occurs in olivine by increasing pressure. A. E. Ringwood of the Australian National University (ANU) was the first to make an experimental study on this problem. He succeeded to show that fayalite, Fe_2SiO_4 with the olivine structure, transforms into the spinel structure between 3–4 GPa at 400–600°C (Ringwood, 1959). The apparatus he used was the so-called "squeezer." A tiny amount of powdered sample was wrapped in a metal foil and squeezed to high pressure by two opposed anvils made of hard steel. Then the entire assembly, including the metal parts, was heated to the desired temperature. Although this apparatus could generate pressures beyond 3 GPa and heat the sample, a large uncertainty in pressure was unavoidable, and it was difficult to increase the temperature above 700°C because not only the sample but also the metal parts were heated to high temperatures. Some other high-pressure and -temperature apparatus was needed to study the olivine–spinel transition in silicates in more detail.

2.3 Developments of Multi-Anvil High Pressure Devices in Japan

The first large high-pressure apparatus constructed in Japan was the tetrahedral press, which followed the Hall's design. The Institute for Solid State Physics (ISSP) of the University of Tokyo was established in 1957 as a research center of solid-state physics in Japan. As one of the important new research initiatives, the construction of a large-volume high-pressure apparatus was begun by S. Akimoto. At that time, the only information he could get was just a single picture of the tetrahedral press in Hall's paper (Figure 2.1). With the help of engineers of Mitsubishi Heavy Industry, he first constructed a working model of 1/5 scale. Based on the various tests using this model, in 1963 he constructed a big press, which had four 1,000 ton hydraulic rams (Figure 2.2). After several minor modifications, this tetrahedral press started working routinely and was used for various research projects up to about 10 GPa and 1,200°C. Akimoto was mainly interested in the olivine–spinel transition in the Mg_2SiO_4–Fe_2SiO_4 system to clarify the origin of 400 km discontinuity. Taking advantage of precise temperature control of the multi-anvil apparatus, he made a detailed phase diagram of this system at three different temperatures, 800, 1,000, and 1,200°C (Akimoto and Fujisawa, 1968). This quantitative work clearly showed that the 400 km discontinuity can be explained by the olivine–spinel transition and made it possible to estimate the temperature at around this depth. He also synthesized

Figure 2.2 The first large-volume high-pressure apparatus constructed in Japan following Hall's design. It was installed at the Institute for Solid State Physics (ISSP) of the University of Tokyo in 1963.

various high-pressure minerals such as silicate spinel and collaborated with many research groups in the world to study their properties, including the group of William Bassett and Taro Takahashi at the University of Rochester, where Ho-kwang Mao was a graduate student (Mao et al., 1969).

From the 1960s to the 1970s, there were two other large high-pressure groups in Japanese universities. One was in Osaka University led by N. Kawai, and the other was in Nagoya University led by M. Kumazawa. Kawai's group started from a split sphere apparatus. They covered the assembled split sphere anvils by a rubber jacket and compressed the assembly in an oil bath. They made many improvements in this apparatus and invented the double-stage technique; the eight second stage cube-shaped anvils made of tungsten carbide (WC) were compressed by six outer first stage anvils made of hardened steel (Kawai and Endo, 1970). This double-stage compression technique made it possible to extend the pressures using tungsten carbide (WC) anvils, from around 10 GPa to above 20 GPa. Although the compression in oil bath had an advantage that all six first-stage anvils are compressed quite uniformly, people had to put the entire assembly in oil and take it out after each experiment, which was tedious and dirty work. Later they developed a technique to fix the first stage anvils to two guide blocks and compress them using a uniaxial press (Kawai et al., 1973). This way it became possible to do experiments without using the oil bath, and experiments became much easier. They constructed a very big 15,000 ton press and made various high-pressure experiments. E. Ito of Kawai's group moved to the Institute of Thermal Springs (at that time) of Okayama University in Misasa and

Figure 2.3 "Kawai-type apparatus" using a 5,000 ton press and E. Ito at the Institute of Thermal Springs, Okayama University, Misasa. For the color version, refer to the plate section.

constructed his laboratory. He used a 5,000-ton press (Figure 2.3) to drive this double-stage multi-anvil apparatus (called a 6-8 apparatus at that time) and studied various properties of high-pressure minerals in the mantle. In those days, the origin of the discontinuity in the mantle at around 670 km was termed the "postspinel transition" and was a research target of many groups in the world. As will be described in the next section, L. G. Liu (1976) succeeded to show that the nature of the postspinel transition is the transition of various silicates into a perovskite structure; this was later confirmed by the work of E. Ito (1977). At that time, Ito's laboratory was the only place in the world where silicate perovskite could be synthesized in milligram quantities. It was possible to make it by laser-heated diamond anvils, but the quantity was only order of micrograms, and because of the nonuniform heating, the crystallinity was lower than those synthesized by a large-volume press. As a result, many scientists who visited Japan wanted to visit Misasa (although it was a remote place and not so easy to reach), and asked Ito to provide perovskite samples to study its various properties (e.g., Ito and Weidner, 1986). Ito himself concentrated on studying the phase equilibria, and all these studies contributed much to clarify the nature of the silicate perovskite and consequently the lower mantle of the Earth. This double-stage large-volume apparatus became a very important high-pressure apparatus for the study of the Earth's deep interior, but it was not easy to make this apparatus and operate it. Ito helped people to install and operate this "Kawai-type apparatus" at Stony Brook University in the US and Bayrerishes Geoinstitut in Germany, as described in detail by Liebermann (2011a, 2011b).

Meanwhile, Kumazawa's group was developing a very unique multiple-anvil sliding system (Kumazawa, 1977), which is similar to a Kawai-type apparatus, but instead of simple compression the eight second-stage anvils slide past one another and reduce the space in the middle. This way, in principle, there was no stroke limit of the second-stage anvils, and Kumazawa expected to reach higher pressures. However, there were many technical problems, and although pressures similar to that of the Kawai-type apparatus were achieved, the system was not developed further. The Nagoya laboratory constructed several big presses, including a 17,000 ton press, and made many studies on the mineralogy of the transition zone. Many leading scientists in this field, including H. Sawamoto, E. Ohtani, T. Irifune, and T. Inoue, are from this active research group.

At that time, many Japanese companies were also interested in high-pressure investigations and invented various types of multi-anvil devices. A cubic-anvil type apparatus driven by a uniaxial press using 45° slopes in a guide block was invented and sold by Kobe Steel Co. (Osugi et al., 1964) by the commercial name "DIA-type apparatus." It was a simple and compact apparatus and easy to combine with X-ray systems. In this apparatus, the guide block is compressed in the direction parallel to the edge of the cube-shaped pressure medium, i.e., the (100) direction. Another apparatus with compression in the (111) direction was invented by the Toshiba Co. (Ichinose et al., 1975). It has the advantage that the efficiency is higher and the mechanical environment of all six anvils are equal, and precision of the advancement of anvils is higher compared to that of the DIA-type apparatus. These two types of apparatus and its modifications are widely used in many laboratories in the world.

In 1990, D. Walker et al. (1990) published a paper entitled "Some Simplifications to Multi-Anvil Devices for High Pressure Experiments." They showed that instead of constructing a special press, by putting both first- and second-stage anvils in a specially designed module and compressing it in a uniaxial press, it is possible to do high-pressure and -temperature experiments up to around 25 GPa. At that time, the laboratories where the silicate perovskite could be synthesized using a large-volume apparatus were quite limited. This "Walker module" made it possible to study silicate perovskites using existing uniaxial presses for the piston-cylinder apparatus, just by replacing the inside unit with a Walker module (Liebermann, 2011a). These double-stage multi-anvil devices originally developed in Kawai's group are now known as "Kawai-type apparatus" and used many laboratories across the world. Although attainable pressure is limited compared to that of diamond anvils, the larger sample chamber and generation of stable and uniform high temperatures make it possible to do many experiments that are difficult using diamond anvils. Intensive efforts continue in Japan and elsewhere to extend the pressure range further. Use of the recently developed very hard WC as the anvil material made it possible to generate pressures up to 65 GPa (Ishii et al., 2017), and by using sintered diamond anvils, pressures above 100 GPa (Yamazaki and Ito, 2020) have been achieved.

2.4 Invention and Development of Diamond Anvil Apparatus

Another important high-pressure apparatus for the study of the Earth's deep interior, the DAC, was invented and developed mainly in the US. Since diamond is transparent, it became possible to make optical observations of the sample under pressure, and the DAC has extended the variety of high-pressure studies enormously. In 1959, four people at the National Bureau of Standards (NBS) published a paper (Weir et al., 1959) "Infrared Studies in the 1- to 15-Micron Region to 30,000 Atmospheres," and described a newly invented diamond anvil apparatus. The basic design of the apparatus was the same as that of today (Figure 2.4), and they succeeded in making optical measurement up to 3 GPa. Although diamond is the hardest material, it is brittle and breaks easily when the two flat faces of

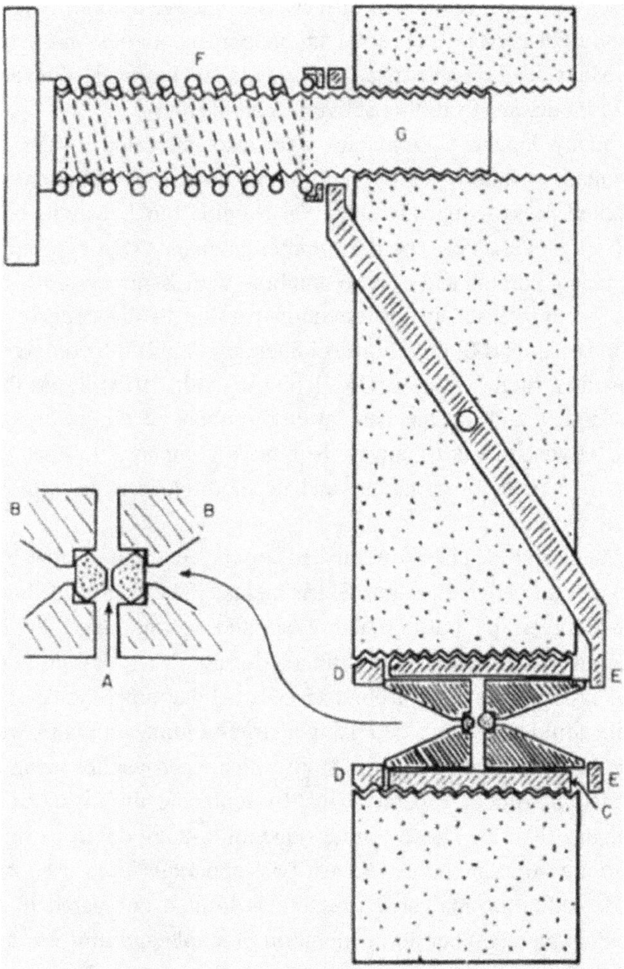

Figure 2.4 The first diamond anvil cell used at the National Burau of Standard (NBS) (Weir et al., 1959). There was no mechanism to adjust alignment of anvils, and the pressure range was limited up to about 3 GPa.

diamond anvil, which squeeze the sample and are called culets, are not parallel each other. One of the four authors of this paper, A. V. Valkenburg, told me later in the 1970s that

> No one of us have thought that this apparatus would become such an important high-pressure apparatus. For reasons that the pressure was quite limited and uncertain, the sample was so small and measurable properties were quite limited, and high-temperature experiments were difficult.

All these problems were, however, overcome by the efforts of many people, and the DAC became one of the most important tools to study the Earth's deep interior.

W. A. Bassett of the University of Rochester became interested in this apparatus and started working with it in early 1960s. The important improvement he and his colleague, T. Takahashi, made was to introduce a mechanism to adjust the angle of two diamonds and make them parallel (Figure 2.5) (Bassett et al., 1967). By this improvement, they succeeded in extending the pressure range by a factor of 10 and established a routine experimental technique to make X-ray diffraction studies up to about 30 GPa. They squeezed powder samples directly between two anvils and then irradiated a thin X-ray beam to the center of

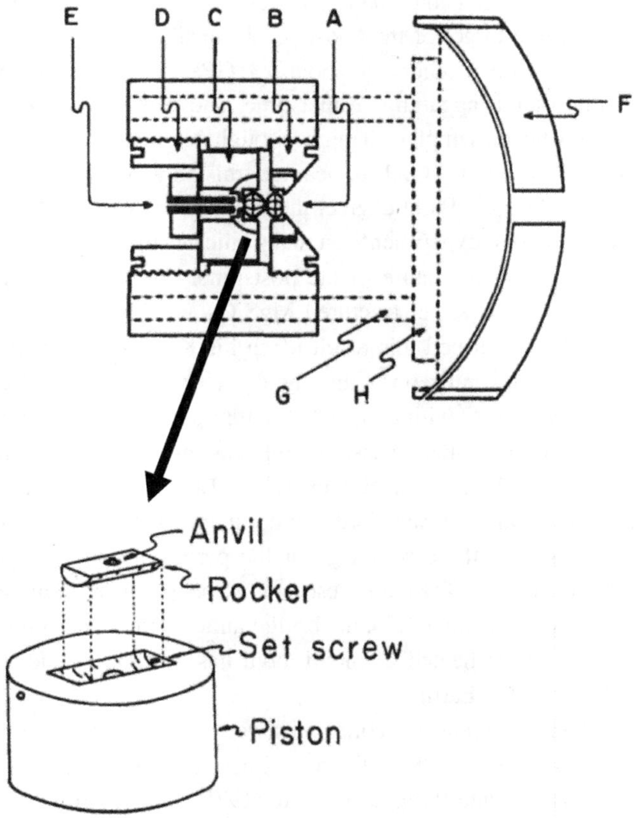

Figure 2.5 Improved DAC for X-ray diffraction study (Bassett et al., 1967). Using rocker, two diamond faces were aligned completely parallel and the pressure range was extended to 30 GPa.

the sample while under pressure. Using this technique, H. K. Mao, the first doctoral student of Bassett's group, studied the compression behavior of various silicate spinels and submitted his Ph.D thesis in 1967. At that time, the laboratory X-ray source was weak, and when a thin beam about 50 μm in diameter was used, more than 300 hours of exposure time was required to get one data point. Nevertheless, the quality of the diffraction pattern was high enough to calculate the compressibility precisely (Mao et al., 1969). Many of the samples Mao used were synthesized and provided by S. Akimoto of the University of Tokyo, because at that time available high-pressure devices in most US universities were either piston-cylinder apparatus or room temperature diamond anvil cells, and it was difficult to get such high-pressure minerals from other groups in the US.

Another important technique that Bassett's group invented was laser heating to heat the samples in the DAC up to several thousand degrees (Ming and Bassett, 1974). They irradiated a tiny graphite sample squeezed in a diamond anvil cell with a pulsed ruby laser and succeeded in transforming it into diamond. They also tried a continuous yttrium-aluminum garnet (YAG) laser and measured the sample temperature by the spectroscopic method. This technique made it possible to study the pressure-induced phase transformations in various silicates up to about 30 GPa. It was the time that many groups in the world were working hard on the target of the "postspinel transition" to clarify the origin of the 670 km discontinuity, which occurs at about 23 GPa. This pressure was beyond the pressure limit of multi-anvil apparatus at that time, and using diamond anvils there was no way to heat the silicate samples to high enough temperature. One of the graduate students of Bassett's group, L. G. Liu, learned this new laser-heated diamond anvil technique and after getting his Ph.D degree, then moved to ANU in Australia. He set up a DAC laboratory and started experiments on many silicates up to about 30 GPa and at high temperature. He found that the nature of the postspinel transition is a decomposition of spinel into a mixture of perovskite-structured $MgSiO_3$ and MgO (Liu, 1976). Soon after that, E. Ito successfully confirmed this result using his Kawai-type apparatus (Ito, 1977). By that time, there were several reports that the spinel type Mg_2SiO_4 decomposes into a mixture of simple oxides: MgO with rock salt structure plus SiO_2 with rutile structure. The calculated density of the lower mantle assuming this assemblage was in harmony with that obtained from the seismic observation. However, both Liu and Ito made convincing studies and clarified that all the major silicates in the upper mantle (olivine, pyroxene, and garnet) transform into perovskite or the assemblage of the perovskite plus simple oxides at the pressures of the lower mantle. Through these works, people were convinced that silicate perovskite, now known as the mineral name bridgmanite, is the major mineral in the lower mantle. Since then, the laser-heated diamond anvil has become an indispensable tool to study the deep interior of the Earth.

At the NBS, G. Piermarini and his colleagues were working hard to improve the DAC experimental techniques. Among these, the most important contribution was the establishment of the ruby pressure scale (Piermarini et al., 1975). At that time, the pressure of the sample was estimated from the applied force to the anvil, or by measuring the unit cell volume of the pressure standard material such as NaCl mixed with the sample using X-ray diffraction. The former has a large uncertainty because only part of the applied force is used to compress the sample and there exists large pressure gradients within the sample. The latter is more accurate, but to measure the pressure, very long X-ray exposure times were required, and a large part of the sample chamber was occupied by the pressure marker material. Piermarini and his colleagues were measuring the fluorescence of various materials and found that the ruby fluorescence is very strong and gives a clear spectrum. They mixed ruby powder with NaCl and calibrated the wavelength shift of the ruby fluorescence line against the pressures calculated from the equation of state of NaCl. They found that the wavelength shifts linearly with pressure up to about 20 GPa, within experimental uncertainty. In this technique, only very tiny ruby chips, typically 10 μm size, were required, and only a few minutes were needed for each pressure measurement. Moreover, it became possible to measure the pressure distribution within the sample by scattering multiple ruby chips. Ruby is easy to get and chemically very stable. For these reasons, the ruby fluorescence technique became a common tool to measure the pressure, and the diamond anvil cell became a popular scientific tool for high-pressure science.

When the powdered sample was squeezed directly between two diamonds, a very large pressure gradient is formed from the center to the edge of the culet. Further, the anvil culets deform into a cup shape at high pressure, and the edges of the two culets touch directly at around 30 GPa, which resulted in the fracture of the anvils. This limited the pressure range of DAC at that time. The NBS group developed the use of a metal gasket and succeeded in reducing the pressure gradient in the sample. Moreover, the gasket protects the edges of the anvils from fracture. Using metal gaskets, they measured the pressure distribution in the sample chamber by the ruby technique and showed that pressures above 50 GPa can be achieved.

After getting his Ph.D degree in Bassett's group, H. K. Mao moved to the Geophysical Laboratory of the Carnegie Institution of Washington and started setting up a new diamond anvil laboratory with P. M. Bell. He made many improvements in the design of the lever-type diamond anvil cell developed at the NBS, so that the parallelism of two culets remain unchanged even when a large force is applied. He also invented a beveled anvil by adding small facets around the culet so that the edges of the culets do not touch each other even when the culet deforms into a cup shape. After making all these improvements, Mao prepared several sets of anvils and tried to increase the pressure as much as possible by squeezing the stainless-steel gasket until the anvils broke. At low pressures, he carefully checked the pressure distribution within the culet, and if it was not concentric with the culet, he lowered the pressure and improved the alignment. He repeated these experiments and successfully extended the pressure limit from run to run. In the fifth run, he reached 170 GPa, where he observed slight change in color of one of the diamonds. He also noticed that the pressure does not go up even when applied load was increased further. He realized something had happened and started decreasing the pressure. When he examined the recovered anvils, he found that one of the anvils was plastically deformed. This was the beginning of the static "megabar experiment" (Mao and Bell, 1978). However, there were many criticisms of this work, such as "the pressure was estimated by the large extrapolation of ruby scale," or, "what they did was simply squeeze the metal gasket alone, and what kind of science can we do?" Mao and Bell had to work hard to address these points, and finally people were convinced with their results.

Since then, many people started doing megabar experiments, and efforts to extend the pressure range further were also made. When Mao and Bell first achieved 170 GPa, they used anvils with a culet diameter of 150 μm. Further extension of the pressure was made possible mainly by reducing the culet size. From around the 1980s, it became possible to use synchrotron radiation, and the brightness of the X-rays was increased by several orders of magnitude. This dramatically changed the status of high-pressure X-ray studies in both the multi-anvil and diamond anvil apparatus.

2.5 Development of Laser Heating in Diamond Anvil Cell and Melting Experiments

Since the first report of laser heating in the diamond anvil cell by Ming and Bassett (1974), the technique has advanced tremendously and it became possible to perform quantitative high-pressure and high-temperature experiments using the diamond anvil cell. In the early stage, most of the laser heating was made by scanning a thin YAG laser beam, because the sample size was much bigger than the laser beam and there was no way to study the small, heated spot of the sample alone. This technique was good enough to form thermodynamically stable silicate phases at very high pressures. Many high-pressure phases of silicates were clarified by heating the sample using this technique and examining it after recovering the sample to ambient condition, for example as used for the first synthesis of silicate perovskite (Liu, 1976). However, in this technique each portion of the sample has experienced a complicated thermal history, and accurate temperature measurement was impossible. Efforts were made to make temperature measurement of the heated spot by spectroscopic method, and it was clarified that very large temperature gradients exist even in a very tiny sample. The large gradient exists not only in the radial direction of the heated spot but also in the direction perpendicular to it, because diamond is a very good heat conductor and thin samples are cooled by the anvils from both sides. In order to overcome this problem, many improvements were made, such as heating the sample from the both sides (Shen et al., 1996, Mao et al., 1998) or modulating the beam profile of the laser (Prakapenka et al., 2008). High-pressure and -temperature in situ observations became possible when laser heating was combined with synchrotron radiation. At the beginning, the diameter of the X-ray beam was several tens of microns, and a CO_2 laser with a large heating area was used (Yagi and Susaki, 1992), but by the advance of synchrotron technique, a much smaller X-ray beam of the order of ten microns or less was available, and currently X-ray diffraction data from a uniformly heated sample can be obtained (e.g., Tateno et al., 2010; Anzellini et al., 2013).

Efforts to measure the temperature of the laser-heated spot using the spectrum of the incandescent light were made (Ming and Bassett, 1974; Heintz and Jeanloz, 1987), and quantitative measurements of the melting curves of various materials were performed (e.g., Zerr and Boehler, 1994; Shen and Lazor, 1995; Hirose et al., 1999). These measurements were made directly at high-pressure conditions corresponding to that of the lower mantle and used to estimate the temperature profile in the lower mantle. The melting temperature of pure iron is particularly important to estimate the temperature at the core–mantle boundary, but the reported results differed by more than 1,000 K among several groups at the corresponding pressure, leading to extensive debate (Boehler et al., 1990; Williams et al., 1991). When large temperature gradients exist around the heated spot, chromatic aberration of the measuring optics can bias the temperature measurement seriously, and efforts were made to solve this problem. Other than the measurements, stability and uniformity of the temperature realized by the laser heating are problems. In laser heating, the temperature of the sample is determined by the amount of the laser energy absorbed by the sample and not by the energy irradiated to the sample. Therefore, by increasing the stability of irradiating laser, sample temperature can be better stabilized, but it changes significantly when the absorption coefficient of the sample varies with time. In this regard, high-temperature experiments using a large-volume high-pressure apparatus are much more advantageous compared to those using the diamond anvil cell. Nevertheless, as the next section describes, various technical improvements are still ongoing, and by combining laser heating and synchrotron radiation, many quantitative high-pressure and high-temperature experiments are now made even at the conditions of the Earth's core.

2.6 Combination of High-Pressure Apparatus with Synchrotron Radiation

From the end of the 1970s, application of synchrotron X-ray sources to high-pressure research began and dramatically changed the situation of high-pressure X-ray studies. In the US, several groups constructed a beamline at the Stanford Synchrotron Radiation Lightsource (SSRL) at Stanford and the Cornell High Energy Synchrotron Source (CHESS) at Cornell and installed diamond anvil cells to obtain diffraction patterns at high pressures and room temperature. In Japan, construction of the first hard X-ray synchrotron radiation facility, the Photon Factory in Tsukuba, started at the end of the 1970s and people in the high-pressure community commenced discussions for the design of a new beamline. After many arguments, they decided to construct a new large-volume high-pressure apparatus that had never before been combined with synchrotron radiation. In an X-ray study, it is necessary to align the sample and X-ray beam positions with micrometer precision. For laboratory X-ray experiments using a large-volume press, this can be easily done by moving the position of an X-ray tube. At the synchrotron, however, the beam position is fixed, and so a big press weighing more than 1,000 kg had to be moved with μm precision, which was not so easy at that time. There were many other problems, but eventually we solved all of them and constructed a high-pressure and high-temperature X-ray diffraction system by combining a DIA-type cubic anvil guide block with a 500 ton hydraulic press. The system, named MAX-80, was installed at the beamline of the Photon Factory in 1982 (Figure 2.6) (Shimomura et al., 1985). At that time, it was common in high-pressure X-ray diffraction experiments with laboratory sources to use very long exposure times, from more than several hours to several days, to get one diffraction pattern. Moreover, the resolution of observed diffraction pattern was very low, and only the crystals with high symmetry, such as cubic, tetragonal, or hexagonal, could be studied. Materials with weak scattering power such as liquids, amorphous materials, and light elements were almost impossible to study. The MAX-80 changed all this "commonsense of high-pressure

Figure 2.6 MAX-80, the first large-volume high-pressure apparatus combined with synchrotron radiation, installed at the Photon Factory in Tsukuba in 1982 (Shimomura et al., 1985).

X-ray diffraction" completely, and it became possible to obtain good diffraction patterns in less than a minute. Various new experiments, such as time-resolved observation of the phase transformations, detailed study of low-symmetry materials, structure study of molten silicates, and in situ observation of the rapid graphite-diamond transition at temperatures above 1,500°C, were made (see Yagi, 1988). Very intense X-ray beams with small divergence also made it possible to get clear radiography even for very tiny materials and opened a possibility of new measurements other than X-ray diffraction. All these studies were impossible by other apparatus, including diamond anvil cells, and thus MAX-80 attracted the attention of many scientists. As a result, virtually identical devices were exported to other countries such as the US and Germany. Afterward, many large-volume devices were constructed at third-generation synchrotron facilities; SPring-8 in Japan, Advanced Photon Source (APS) in US, and the European Synchrotron Radiation Facility (ESRF) in Europe. Although the diamond anvil cell can achieve higher pressures, the advantages of large-volume apparatus, such as stable and uniform temperature conditions and a large sample chamber, made it possible to perform many new and unique studies.

One example is the viscosity measurement of molten silicates. By observing the sinking velocity of small platinum balls in a sample chamber, changes of the viscosity of molten silicates under pressure have been studied in detail (e.g., Suzuki et al., 2002). Another example is the simultaneous measurement of ultrasonic wave velocity, X-ray diffraction, and X-radiography (e.g., Kung et al., 2002). By measuring the specimen length by X-radiography, the crystalline density by X-ray diffraction, and the wave velocity by ultrasonic measurements on the same sample, the accuracy of the measurement of elasticity of minerals under pressure has increased considerably. Moreover, it opens up the possibility to determine the equations of state without relying on any existing pressure scales (see also Li and Liebermann, 2014; Wang et al., 2015).

Diamond anvil experiments combined with synchrotron radiation were limited to room temperature at first, but the pressure range was largely extended. Using a laboratory X-ray source, the diameter of the X-ray beam was 50–100 μm, and so the culet size had to be larger than 300 μm. The high brightness of the X-ray beam of the synchrotron allows the use of much smaller X-ray beams, and using the anvils with a culet size of 100 μm or less, multimegabar experiments became possible. For example, Mao et al. (1990) determined the equation of state of iron up to 300 GPa, close to the inner-core pressure. Quality X-ray diffraction data were obtained even for low-Z materials such as MgO at pressures over 200 GPa (Duffy et al., 1995).

By the end of the 1990s, people started installing laser-heated diamond anvils at the third-generation synchrotron facilities to study the deep interior of the Earth (e.g., Shen et al., 2001; Watanuki et al., 2001; Schultz et al., 2005; Meng et al., 2006; Ohishi et al., 2008; Prakapenka et al., 2008). In these facilities, developments in X-ray focusing optics made it possible to reduce the X-ray beam size down to a few μm, which contributed much to increase the pressure range and also increase the quality of the X-ray diffraction. Developments of X-ray detectors, from energy dispersive to angle dispersive, from film to image plate, and then to CCD detector, also contributed to shortening the measurement time and resolution. As a result, very high-quality in situ X-ray diffraction data were obtained even at the conditions of the deep lower mantle. The combination of laser-heated diamond anvil cell, highly focused X-ray beam, and fast 2D area detector has revolutionized the study of Earth's interior by in situ X-ray diffraction measurements. Significant progress has been made in the pressure, volume, and temperature (P–V–T) measurements of mantle and core materials (e.g., Ricolleau et al., 2009; Wolf et al., 2015), determination of melting by in situ diffraction measurements (e.g., Fiquet et al., 2010; Anzellini et al., 2013), and study of chemical reactions and phase transformations at high pressure and temperature (e.g., Ozawa et al., 2011; Andrault et al., 2012).

One noticeable achievement is the discovery of the postperovskite phase in $MgSiO_3$ by Hirose's group at SPring-8 (Murakami et al., 2004). It was really a big surprise, because for a long time people had believed that perovskite-structured silicate is stable throughout the entire range of the lower mantle. From crystallographic considerations, perovskite was known as the densest structure among various ABO_3-type compounds, and several experimental studies by that time reported that there is no phase transition in perovskite-structured silicates up to the pressure of the mantle–core boundary. Hirose's group clearly showed, however, that $MgSiO_3$ with perovskite structure transforms into a further dense phase at the conditions corresponding to the bottom of the lower mantle. It was such an unexpected discovery that there were some doubts about these results, but soon several theoretical works were made to compare the relative stability of the perovskite structure and the "postperovskite structure" (Iitaka et al., 2004; Oganov and Ono, 2004; Tsuchiya et al., 2004). Calculations in these works clearly showed that this postperovskite structure becomes energetically more favorable above about 120 GPa, the pressure corresponding to the depth just above the mantle–core boundary at 2,900 km. Seismological observations have clarified that in this region, which was called the D" layer, very odd properties exist that were difficult to explain. This new phase had the potential to solve these problems. Many people across the world started working to clarify the properties of this postperovskite phase, and now our understanding of the structure of mantle–core boundary region has advance considerably.

To study the Earth's core, efforts were continued to further extend the pressure range. The pressures required are 130–360 GPa, and by using anvils with a culet size of 20–30 μm it became possible to achieve these pressures. Experiments with laser-heating systems also made substantial progress, and in situ X-ray diffraction data were obtained for iron at the conditions of the inner core. In 2010, it was reported that hexagonal closed-pack (hcp) iron is stable at about 356 GPa and 5,500 K (Tateno et al., 2010), which corresponds to the condition at the center of the Earth. Therefore, as the pressure and temperature conditions are concerned, it became possible to reproduce the conditions anywhere in the Earth in the laboratory.

The high intensity of the X-ray beams from the third-generation synchrotron sources also made it possible to do not only diffraction studies but also spectroscopic studies under pressure. Shen and Mao (2017) provide an extensive review of the applications of various synchrotron spectroscopic techniques. Notably, X-ray emission spectroscopy (Badro et al., 2003) and synchrotron Mössbauer spectroscopy (nuclear forward scattering) (e.g., Nasu, 1996) provide new ways to investigate electronic transitions at high pressure. Studies of the spin state of iron and the site occupancies or ferrous and ferric iron in mantle minerals has significantly advanced our understanding the role of iron in the mantle (e.g., Badro et al., 2004; Jackson et al., 2005; Solomatova et al., 2016). Phonon densities of states (DOS) of iron were measured by nuclear resonant inelastic X-ray scattering (NRIXS). The development is critical for measuring wave velocity of iron at megabar pressures (Mao et al., 2001a). The technique has also been used to obtain phonon DOS for many iron compounds, including iron oxides, iron sulfides, iron carbides, and iron hydride (e.g., Struzhkin et al., 2001; Lin et al., 2004; Mao et al., 2004; Gao et al., 2011).

High-energy resolution inelastic X-ray scattering (HERIXS) provides another way to obtain sound velocity of iron (Fiquet et al., 2001). It is not possible to get full elastic constants for hcp–Fe, so the elasticity of single-crystal cobalt with hcp structure was measured to about 40 GPa to estimate the property of hcp iron in the core (Antonangeli et al., 2004). However, compressional sound velocity can be more directly measured for polycrystalline iron and iron alloys by HERIXS (e.g., Antonangeli et al., 2018). X-ray Raman measurement made it possible to examine the detailed pressure-induced electronic bonding transitions of Earth materials (Mao et al., 2001b; Lee et al., 2014). X-ray absorption spectra (XAS) also provide important information, including structural and elastic properties of material, particularly for noncrystalline materials (Rosa et al., 2020). Simultaneous measurements of the X-ray diffraction and Brillouin scattering (Murakami et al., 2009) were made to obtain the elasticity of light transparent materials in the megabar region. These various X-ray spectroscopic techniques have opened opportunities to probe various properties of materials at conditions relevant to the deep Earth.

2.7 Efforts to Extend the Pressure Range beyond the Limit of Diamond Anvils

As far as the studies of the Earth are concerned, it is not necessary to extend the pressure range any further. However, there are many interesting research targets in high-pressure science other than the Earth's interior, such as physics, materials science, and planetary science. For example, the metallization of hydrogen at room temperature is expected to occur at about 600 GPa. Inside of the giant planets such as Jupiter and Saturn, and interiors of "super Earth" extrasolar planets, pressures are above one terapascal (TPa). Thus, many investigators have tried to extend the pressure range in the DAC as much as possible. Despite extensive efforts, however, by about 2010 no successful experiments were reported above about 400 GPa and people thought that this is the upper limit of the diamond anvil.

In 2012, L. S. Dubrovinsky et al. (2012) reported that by adopting a double-stage diamond anvil cell (ds-DAC) technique, they achieved pressures as high as 650 GPa. In their experiments, they synthesize tiny spheres made of nanopolycrystalline diamond using glassy carbon spheres as starting material, and converted them into diamond at around 25 GPa and 2,000 K. Then they stacked two of these small spheres about 10~15 μm in diameter with tiny sample (rhenium foil) in between, put it in a sample chamber of a conventional diamond anvil cell together with a pressure-transmitting medium, and compressed it. In large-volume apparatus, it is well known that by adopting double-stage compression techniques the pressure range can be greatly extended because the confining pressure applied to the second-stage anvils helps to strengthen the anvil material. Inspired by this work, many groups have tried to reproduce their results of ds-DAC experiments, but unfortunately no other groups had succeeded even after eight years following the first report; however, the same group reported that pressures above 1 TPa had been achieved (N. Dubrovinskaia et al., 2016) using the same ds-DAC technique.

For the studies of the Dubrovinsky/Dubrovinskaiai laboratory in Bayreuth, many arguments were made, but the real situation remains unclear and the pressure of 1 TPa reported has not yet been confirmed by other groups. We have tried to establish a reproducible technique for the ds-DAC experiments by fabricating μm-size second-stage anvils from nanopolycrystalline diamond using the focused ion beam (FIB) (Figure 2.7). We repeated more than 20 experiments by changing designs and various parameters of the anvil shape.

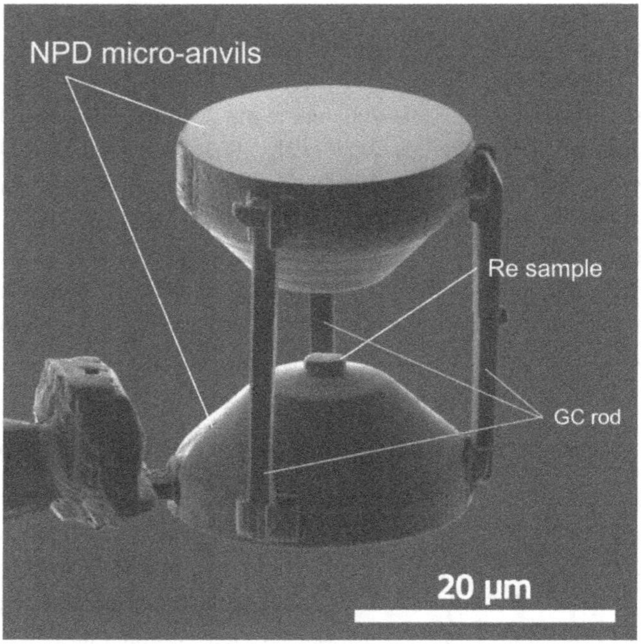

Figure 2.7 Micron-size second-stage diamond anvils for a ds-DAC experiment. It was fabricated from a large block of nanopolycrystalline diamond (NPD) using focused ion beam (FIB) technique
(Sakai et al., 2018).

In almost all the runs, the pressure reached 300 GPa or above, but the anvils always fractured below 500 GPa (Sakai et al., 2018, 2020). Several groups, including ours, have pointed out that the equations of state (EoS) of rhenium that Dubrovinsky et al. (2012) used for the pressure calculation in their first report have serious problems, and the pressure value was likely to be overestimated. In later work, they used a different pressure marker, gold, and claimed that they achieved 1.065 TPa. The EoS of gold they used (Yokoo et al., 2009) was calculated based on shock compression data and was claimed to be reliable up to about 400 GPa with the uncertainty of a few percent. Naturally, extrapolation of this EoS more than double of the pressure range has very large uncertainties. But the biggest problem the other groups encountered was the clear and pure X-ray diffraction pattern reported by the Bayreuth team. In the ds-DAC experiments, the culet size is usually less than 10 μm, and when the sample is compressed, it becomes very thin, around 100 nm or less. Sample thickness is 1–2 μm at the beginning, but a large part of the sample is squeezed out of the culet and remains around the culet. Although it is surrounded by the pressure-transmitting medium that is compressed by the first-stage anvils, those pressures are much lower than the sample compressed directly by two second-stage anvils. The diameter of the X-ray beam used in these studies is nominally 2–3 μm, but there is a "tail," and if we include the area with an intensity about a few percent of the peak intensity, the diameter becomes more than five times. When such an X-ray beam is irradiated to the samples squeezed in ds-DAC, the volume of the low-pressure samples irradiated by the tail of the X-ray beam becomes much bigger compared to the sample squeeze between the small culets, and diffractions from the lower-pressure sample always appears together with the high-pressure sample. In the work by Dubrovinsky et al. (2012) and Dubrovinskaia et al. (2016), however, they always reported beautiful peaks from the high-pressure samples alone, and they claim that it is because of the peculiar property of the anvil material they used. Their argument is that the nanocrystalline diamond they synthesized is soft enough and deforms into a concave shape at the beginning of the compression but becomes very hard and tough at high pressures, so that the sample is compressed without being squeezed out from the culet, as discussed in more detail in Yagi et al. (2020). Unfortunately, there is no proof of such deformation, and no one can judge if it is really possible or not. Some efforts were made to synthesize the nanopolycrystalline diamond using the same starting material and at the same pressure and temperature (P–T) conditions (Sakai et al., 2020). Although the result was negative, the property of materials synthesized under high P–T conditions varies a lot with many parameters, not only by P–T conditions but also by increasing and decreasing the rate and duration time of pressure and temperature, difference of the purity of starting materials, etc. Moreover, the material synthesized by the Bayreuth method is very small, only 10–20 μm, and it is not easy to measure its mechanical properties; therefore, at present there is no way to judge if the anvils the Dubrovinsky/Dubrovinskaiai laboratory used have really deformed as they claim.

Another effort to extend the pressures beyond the limit of the conventional diamond anvil has also been made by modifying the top shape of diamond anvil using FIB. In opposed anvils made of tungsten carbide, a toroidal anvil is often used, which has groove(s) around the culet (Khvostantsev et al., 2004). The groove works to reduce the outflow of gasket materials and helps to keep the sample thick. In diamond anvil cells, culets seriously deform with increasing pressure into a concave shape, and the edges of the two culets almost touch each other at around 400 GPa even when using beveled anvils. By modifying the shape of anvils and adding a groove, there is the possibility to reduce such deformation, and it may be possible to extend the pressure limit beyond 400 GPa. Actually, Dewaele et al. (2018) reported the generation of 603 GPa using this technique and the gold pressure scale. Jenei et al. (2018) also made very similar experiments and reported that pressures up to 615 GPa were achieved, although this pressure was estimated by the rhenium EoS proposed by Dubrovinsky et al. (2012) and is probably largely overestimated. In these experiments, the anvils are made of single crystal diamond, which is a big advantage because one can make not only X-ray diffraction but also optical studies such as absorption spectrum and Raman spectroscopy. However, in both the Dewaele and Jenei studies over many trials, only one run could reach above 500 GPa and the success rate is very low. Further studies are required to make this technique routine.

2.8 Future Perspectives

As we have seen, experimental techniques for high-pressure mineralogy have made enormous progress, and now it is possible to reproduce any pressure and temperature conditions within the Earth in the laboratory. So far, however, the experiments have been mostly limited to X-ray diffraction, and stable structures of materials were clarified. In order to understand the Earth's deep interior, much more information is required, and measurements by other means such as Raman or Brillouin scattering, X-ray spectroscopy, and other techniques are required. Furthermore, not only the static behavior but also dynamic properties such as viscosity and deformation will be required to understand dynamics in the Earth. Information we can get from the tiny sample in the diamond anvil cell has previously been very limited, but with the advancement of electron microscopy and other technique such as FIB to handle or fabricate very small materials, we can now obtain rich information from the samples squeezed in diamond anvil cells. The third-generation synchrotron facilities made it possible to do not only the diffraction but also various spectroscopic studies. However, for these studies, the intensities and coherence of X-ray beams in existing facilities are not sufficient, and further developments are required. On the other hand, even though the pressure is limited, the much larger sample volumes of multi-anvil apparatus are attractive for many studies. Efforts are continuing to extend pressures using the Kawi-type apparatus. This apparatus, however, is not easy to use in this pressure range, and thus other efforts are being made to achieve megabar pressures using the Paris–Edinburgh type large-volume press combined with diamond anvils (Kono et al., 2016). These new high-pressure techniques cover almost the entire P–T conditions of the lower mantle, and they will provide new possibilities for the studies of this region of Earth's interior.

For the experiments beyond the pressure range of conventional diamond anvils, optimizing the culet shape using FIB looks promising to extend the pressure range up to 600–700 GPa. However, the ideal shape of the culet remains unclear, and it is a time- and cost-consuming process to find it by repeating many experiments. Thus far, theoretical simulations using numerical calculations are difficult because it is not a simple problem of anvil deformation alone, but flow of the sample/gasket materials must be also considered. The boundary conditions between the elastically deformed diamond and plastically deformed sample/gasket remain unknown. But as reported in Li et al. (2018), recent progress in synchrotron radiation made it possible to make nanosize-thin X-ray beams and make 2D absorption mapping. This technique provides us important information about the deformation of anvils and flow of sample/gasket materials, as shown in Figure 2.8. Combining these observational data with the theoretical calculations to clarify the stress and strain distribution in the anvil, it may become possible to find the optimum shape of the anvils.

In principle, the ds-DAC looks promising to achieve much higher pressures. At present, however, the behavior of micro-anvils remains unclear, and it is necessary to clarify how the second-stage anvils deform under pressure. The 2D X-ray absorption technique used by Li et al. (2018) also looks very promising. It is also necessary to study the property of nanopolycrystalline diamond in more detail. If we can find the methods to make it stiffer than the existing material, it will contribute a great deal to extending the pressure range of ds-DAC experiments.

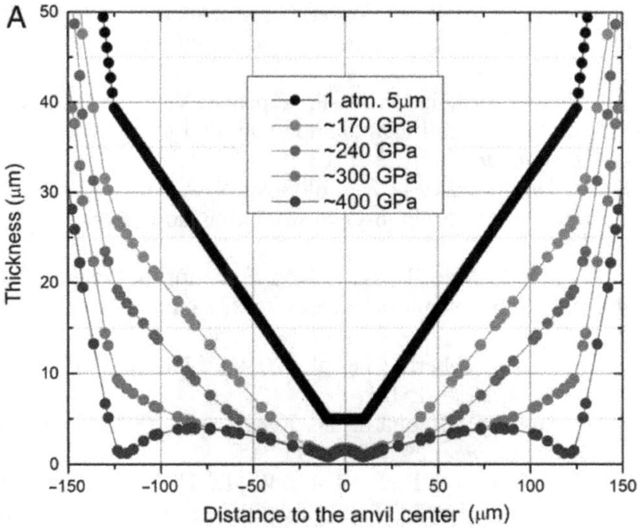

Figure 2.8 Deformation of the diamond anvil with increasing pressure measured by the 2D X-ray absorption mapping using a nanometer-thin X-ray beam (Li et al., 2018).

References

Akimoto, S., Fujisawa, H. (1968). Olivine-spinel solid solution equilibria in the system Mg_2SiO_4-Fe_2SiO_4, *Journal of Geophysical Research*, **73**, 1467–1479.

Andrault, D., Petitgirard, S., Lo Nigro, G., et al. (2012). Solid-liquid iron partitioning in Earth's deep mantle. *Nature*, **487**, 354–357.

Antonangeli, D., Krisch, M., Fiquet, G., et al. (2004). Elasticity of cobalt at high pressure studied by inelastic X-ray scattering, *Physical Review Letters*, **93**, 215505.

Antonangeli, D., Morard, G., Paolasini, L., et al. (2018). Sound velocities and density measurements of solid hcp-Fe and hcp-Fe-Si (9 wt.%) alloy at high pressure: constraints on the Si abundance in the Earth's inner core. *Earth and Planetary Science Letters*, **482**, 446–453.

Anzellini, S., Dewaele, A., Mezouar, M., Loubeyre, P. (2013). Melting of iron at Earth's inner core boundary based on fast X-ray diffraction. *Science*, **340**, 464–466.

Badro, J., Fiquet, G., Guyot, F., et al. (2003). Iron partitioning in Earth's mantle: toward a deep lower mantle discontinuity. *Science*, **300**, 789–791.

Badro, J., Rueff, J.-P., Vanko, G., Monaco, G., Fiquet, G.,Guyot, F. (2004). Electronic transitions in perovskite: possible nonconvecting layers in the lower mantle. *Science*, **305**, 383–386.

Bassett, W. A. (2009). Diamond anvil cell, 50th birthday. *High Pressure Research*, **29**, 163–186.

Bassett, W. A., Takahashi, T., Stock, P. W. (1967). X-ray diffraction and optical observations on crystalline solids up to 300 kbar. *Review of Scientific Instruments*, **38**, 37–42.

Boehler, R., Bargen, N. V., Chopelas, A. (1990). Melting, thermal expansion, and phase transitions of iron at high pressures. *Journal of Geophysical Research*, **95**, 21731–21736.

Bundy, F. P., Hall, H. T., Strong, H. M., Wentrof, R. H. (1955). Man-made diamonds. *Nature*, **176**, 51–55.

Dewaele, A., Loubeyre, P., Occelli, F., Marie, O., Mezouar, M. (2018). Toroidal diamond anvil cell for detailed measurements under extreme static pressures. *Nature Communications*, **9**, 2913.

Dubrovinsky, L. S., Dubrovinskaia, N. A., Prakapenka, V. B., Abakumov, A. M. (2012). Implementation of micro-ball nanodiamond anvils for high-pressure studies above 6 Mbar. *Nature Communications*, **3**, 1163.

Dubrovinskaia N. A., Dubrovinsky L. S., Solopova N. A., et al. (2016). Terapascal static pressure generation with ultrahigh yield strength nanodiamond. *Science Advances,* **2**, e1600341, 1–12.

Duffy, T. S., Hemley, R. J., Mao, H. K. (1995). Equation of state and shear strength at multimegabar pressures: magnesium oxide to 227 GPa. *Physical Review Letters*, 74, 1371–1374.

Fiquet, G., Auzende, A. L., Siebert, J., et al. (2010). Melting of peridotite to 140 gigapascals. *Science*, **329**, 1516–1518.

Fiquet, G., Badro, J., Guyot, F., Requardt, H., Krisch, M. (2001). Sound velocities in iron to 110 gigapascals. *Science*, **291**, 468–471.

Gao, L., Chen, B., Zhao, J., Alp, E. E., Sturhahn, W., Li, J. (2011). Effect of temperature on sound velocities of compressed Fe_3C, a candidate component of the Earth's inner core. *Earth and Planetary Science Letters*, **309**, 213–220.

Hall, H. T. (1958). Some high-pressure, high-temperature apparatus design considerations: equipment for use at 100 000 atmospheres and 3000 C. *Review of Scientific Instruments*, **29**, 267–275.

Hall, H. T. (1960). Ultra-high-pressure, high-temperature apparatus: the "Belt." *Review of Scientific Instruments*, **31**, 125–131.

Hall, H. T. (1962). Anvil guide device for multiple-anvil high pressure apparatus. *Review of Scientific Instruments*. **33**, 1278–1280.

Heinz, D. L., Jeanloz, R. (1987). Temperature measurements in the laser heated diamond cell, in M. H. Manghnani, Y. Syono, eds., *High Pressure Researches in Mineral Physics*. AGU, pp. 113–127.

Hirose, K., Fei, Y., Ma, Y., Mao, H. K. (1999). The fate of subducted basaltic crust in the Earth's lower mantle. *Nature*, **397**, 53–56.

Ichinose, K., Wakatsuki, M., Aoki, T. (1975). A new sliding type cubic anvil high pressure apparatus. *Journal of the High Pressure Institute of Japan*, **13**, 244–253.

Iitaka, T., Hirose, K., Kawamura, K., Murakami, M. (2004). The elasticity of the $MgSiO_3$ post-perovskite phase in the Earth's lowermost mantle. *Nature*, **430**, 442–445.

Ishii, T., Yamazaki, D., Tsujino, N., et al. (2017). Pressure generation to 65 GPa in Kawai-type multi-anvil apparatus with tungsten carbide anvils. *High Pressure Research*, **37**, 505–515.

Ito, E. (1977). The absence of oxide mixture in high-pressure phases of Mg-silicates. *Geophysical Research Letters*, **4**, 72–74.

Ito, E. (2009) Multianvil cells and high pressure experimental methods, in G. D. Price ed., *Treaties on Geophysics*, Vol.2: Mineral Physics, Elsevier, pp. 197–230.

Ito, E., Weidner, D. J. (1986). Crystal growth of $MgSiO_3$ perovskite. *Geophysical Research Letters*, **13**, 464–466.

Jackson, J. M., Sturhahn, W., Shen, G. Y., et al. (2005). A synchrotron Mössbauer spectroscopy study of $(Mg,Fe)SiO_3$ perovskite up to 120 GPa. *American Mineralogist*, 90, 199–205.

Jenei, Z., O'Bannon, E. F., Weir, S. T., Cynn, H., Lipp, M. J., Evans, W. J. (2018). Single crystal toroidal diamond anvils for high pressure experiments beyond 5 megabar. *Nature Communications*, **9**, 3563.

Kawai, N., Endo, S. (1970). The generation of ultrahigh hydrostatic pressures by a split sphere apparatus. *Review of Scientific Instruments*, **41**, 1178–1181.

Kawai, N., Togaya, M., Onodera, A. (1973). A new device for high pressure vessels. *Proceedings of the Japan Academy*, **49**, 623–626.

Khvostantsev, T. G., Slesarev, V. N., Brazhkin, V. V. (2004). Toroid type high-pressure device: history and prospect. *High Pressure Research*, **24**, 371–383.

Kono, Y., Kenney-Benson, C., Ikuta, D., Shibazaki, Y., Wang, Y., Shen, G. (2016). Ultrahigh-pressure polyamorphism in GeO2 glass with coordination number> 6. *Proceedings of the National Academy of Sciences,* **113**, 3436–3441.

Kumazawa, M. (1977). A novel device to reach higher pressure in larger volume, in M. H. Manghnani, S. Akimoto, eds., *High-Pressure Research: Applications in Geophysics*, Academic Press, pp. 563–572.

Kung, J., Li, B., Weidner, D. J., Zhang, J., Liebermann, R. C. (2002). Elasticity of ($Mg_{0.83}$, $Fe_{0.17}$)O ferropericlase at high pressure: ultrasonic measurements in conjunction with X-radiation techniques. *Earth and Planetary Science Letters*, **203**, 557–566.

Lee, S. K., Eng, P. J., Mao, H. K. (2014). Probing of pressure-induced bonding transitions in crystalline and amorphous earth materials: insights from X-ray Raman scattering at high pressure. *Review in Mineralogy & Geochemistry*, **78**, 139–174.

Li, B., Liebermann R. C. (2014). Study of the Earth's interior using measurements of sound velocities in minerals by ultrasonic interferometry. *Physics of the Earth and Planetary Interiors*, **233**, 135–153.

Li, B., Ji, C., Yang, W., et al. (2018). Diamond anvil cell behavior up to 4 Mbar. *Proceeding of National Academy of Science,* **115**, 1713–1717.

Liebermann, R. C. (2011a). Multi-anvil, high pressure apparatus: a half-century of development and progress. *High Pressure Research*, **31**, 493–532.

Liebermann, R. C. (2011b). Bob-san and high pressure science and technology in Japan: a 40-year history. *Review of High Pressure Science and Technology*, **21**, 115–126.

Lin, J. F., Fei, Y., Sturhahn, W., Zhao, J., Mao, H. K., Hemley, R. J. (2004). Magnetic transition and sound velocities of Fe_3S at high pressure: implications for Earth and planetary cores. *Earth and Planetary Science Letters*, **226**, 33–40.

Liu, L. G. (1976). The post-spinel phase of forsterite. *Nature*, **262**, 770–772.

Mao, H. K., Bell, P. M. (1978). High pressure physics: sustained static generation of 1.36 to 1.72 megabars. *Science*, **200**, 1145–1147.

Mao, H. K., Kao, C., Hemley, R. J. (2001b). Inelastic X-ray scattering at ultrahigh pressures. *Journal of Physics: Condensed Matter*, **13**, 7847.

Mao, H. K., Shen, G., Hemley, R. J., Duffy, T. S. (1998). X-ray diffraction with a double hot plate laser heated diamond cell, in M. H. Manghnani, T. Yagi, eds., Properties of Earth and Planetary Materials, AGU, pp. 27–34.

Mao, H. K., Takahashi, T., Bassett, W. A., Weaver, J. S., Akimoto, S. (1969). Effect of pressure and temperature on the molar volumes of wustite and of three $(Fe,Mg)_2SiO_4$ spinel solid solutions. *Journal of Geophysical Research*, **74,** 1061–1069.

Mao, H. K., Wu, Y., Chen, L. C., Shu, J. F. Jephcoat, A. P. (1990). Static compression of iron to 300 GPa and $Fe_{0.8}Ni_{0.2}$ alloy to 260 GPa: implications for composition of the core. *Journal of Geophysical Research*, **95**, 21737–21742.

Mao, H. K., Xu, J., Struzhkin, et al. (2001a). Phonon density of states of iron up to 153 gigapascals. *Science*, **292**, 914–916.

Mao, W. L., Sturhahn, W., Heinz, D. L., Mao, H. K., Shu, J., Hemley, R. J. (2004). Nuclear resonant X-ray scattering of iron hydride at high pressure. *Geophysical Research Letters*, **31**, L15618.

Meng, Y., Shen, G., Mao, H. K. (2006). Double-sided laser heating system at HPCAT for in situ X-ray diffraction at high pressures and high temperatures. *Journal of Physics: Condensed Matter*, **18,** S1097–S1103.

Ming, L.C., Bassett, W. A. (1974). Laser heating in the diamond anvil press up to 2000°C sustained and 3000°C pulsed at pressures up to 260 kilobars. *Review of Scientific Instruments*, **45**, 1115–1118.

Murakami, M., Asahara, Y., Ohishi, Y., Hirao, N., Hirose, K. (2009). Development of in situ Brillouin spectroscopy at high pressure and high temperature with synchrotron radiation and infrared laser heating system: application to the Earth's deep interior. *Physics of the Earth and Planetary Interiors*, **174**, 282–291.

Murakami, M., Hirose, K., Kawamura, K., Sata, N., Ohishi, Y. (2004). Post-perovskite transition in $MgSiO_3$. *Science*, **304**, 855–858.

Nasu, S. (1996). High-pressure Mossbauer spectroscopy with nuclear forward scattering of synchrotron radiation. *High Pressure Research*, **14**, 405–412.

Oganov, A. R., Ono, S. (2004). Theoretical and experimental evidence for a post-perovskite phase of $MgSiO_3$ in Earth's D" layer. *Nature*, **430**, 445–448.

Ohishi, Y., Hirao, N., Sata, N., Hirose, K., Takata, M. (2008). Highly intense monochromatic X-ray diffraction facility for high-pressure research at SPring-8. *High Pressure Research*, **28**, 163–173.

Osugi, J., Shimizu, K., Inoue, K., Yasunami, K. (1964). A compact cubic anvil high pressure apparatus. *Review of Physics and Chemistry Japan*, **34**, 1–6.

Ozawa, H., Takahashi, F., Hirose, K., Ohishi, Y., Hirao, N. (2011). Phase transition of FeO and stratification in Earth's outer core. *Science*, **334**, 792–794.

Piermarini, G. J., Block, S., Barnett, J. D., Forman, R. A. (1975). Calibration of the pressure dependence of the R_1 ruby fluorescence line to 195 kbar. *Journal of Applied Physics*, **46**, 2774–2780.

Prakapenka, V. B., Kubo, A., Kuznetsov, A., et al. (2008). Advanced flat top laser heating system for high pressure research at GSECARS: application to the melting behavior of germanium. *High Pressure Research*, **28**, 225–235.

Ricolleau, A., Fei, Y., Cottrell, E., et al. (2009). Density profile of pyrolite under the lower mantle conditions. *Geophysical Research Letters*, **36**, L06302.

Ringwood, A. E. (1959). The olivine-spinel inversion in fayalite. *American Mineralogist*, **44**, 659–661.

Rosa, A. D., Mathon, O., Torchio, R., et al. (2020). Nano-polycrystalline diamond anvils: key device for XAS at extreme conditions: their use, scientific impact, present status and future needs. *High Pressure Research*, **40**, 65–81.

Sakai, T., Yagi, T., Irifune, T., et al. (2018). High pressure generation using double-stage diamond anvil technique: problems and equations of state of rhenium. *High Pressure Research*, **38**, 107–119.

Sakai, T, Yagi, T., Takeda, R., et al. (2020). Conical support for double-stage diamond anvil apparatus. *High Pressure Research*, **40**, 12–21.

Schultz, E., Mezouar, M., Crichton, W., et al. (2005). Double-sided laser heating system for in situ high pressure and high temperature monochromatic x-ray diffraction at the ESRF. *High Pressure Research*, **25,** 71–83.

Shen, G., Lazor, P. (1995). Measurement of melting temperatures of some minerals under lower mantle pressures, *Journal of Geophysical Research* **B, 100,** 17699–17713.

Shen, G., Mao, H. K. (2017). High-pressure studies with X-rays using diamond anvil cells. *Reports on Progress in Physics,* **80**, 1–53.

Shen, G, Mao, H. K., Hemley, R. J. (1996). Laser-heating diamond- anvil cell technique: double-sided heating with multimode Nd:YAG laser, *Advanced Materials'96 – New Trends in High Pressure Research*, NIRIM, pp. 149–152.

Shen, G., Rivers, M. L., Wang, Y., Sutton, S. R. (2001). A laser heated diamond cell system at the Advanced Photon Source for in situ x-ray measurements at high pressure and temperature. *Review of Scientific Instruments*, **72**, 1273–1282.

Shimomura, O., Yamaoka, S., Yagi, T., et al. (1985). Multi-anvil type X-ray system for synchrotron radiation, in S. Minomura, ed., *Solid State Physics under Pressure-Recent Advance with Anvil Devices*, Terrapub, pp. 351–356.

Solomatova, N. V., Jackson, J. M., Sturhahn, W., et al. (2016). Equation of state and spin crossover of (Mg, Fe)O at high pressure, with implications for explaining topographic relief at the core–mantle boundary. *American Mineralogist*, **101**, 1084–1093.

Struzhkin, V. V., Mao, H. K., Hu, J., et al. (2001). Nuclear inelastic X-ray scattering of FeO to 48 GPa. *Physics Review Letters*, **87**, 255501.

Suzuki, A., Ohtani, E., Funakoshi, K., Terasaki, H., Kubo, T. (2002). Viscosity of albite melt at high pressure and high temperature. *Physics and Chemistry of Minerals*, **29**, 159–165.

Tateno, S., Hirose, K., Ohishi, Y., Tatsumi, Y. (2010). The structure of iron in Earth's inner core. *Science*, **330,** 359–361.

Tsuchiya, T., Tsuchiya, J., Umemoto, K., Wentzcovitch, R. M. (2004). Phase transition in $MgSiO_3$ perovskite in the earth's lower mantle. *Earth and Planetary Science Letters*, **224**, 241–248.

Walker, D., Carpenter, M. A., Hitch, C. M. (1990). Some simplifications to multianvil devices for high pressure experiments. *American Mineralogist*, **75**, 1020–1028.

Watanuki, T., Shimomura, O., Yagi, T., Kondo, T., Isshiki, M. (2001). Construction of laser-heated diamond anvil cell system for in situ X-ray diffraction study at SPring-8. *Review of Scientific Instruments*, **72**, 1289–1292.

Wang, X., Chen, T., Qi, X. et al. (2015). Acoustic travel time gauges for in-situ determination of pressure and temperature in multi-anvil apparatus. *Journal of Applied Physics*, **118**, 065901.

Weir, C. E., Lippincott, E. R., Valkenburg, A. V., Bunting, E. N. (1959). Infrared studies in the 1-to 15-micron region to 30,000 atmospheres. *Journal of Research NBS***, 63A**, 55–62.

Williams, Q., Nittle, E., Jeanloz, R. (1991). The high-pressure melting curve of iron: a technical discussion. *Journal of Geophysical Research*, **96**, 2171–2184.

Wolf, A. S., Jackson, J. M., Dera, P., Prakapenka, V. B. (2015). The thermal equation of state of (Mg, Fe)SiO_3 bridgmanite (perovskite) and implications for lower mantle structures. *Journal of Geophysical Research*, **120**, 7460–7489.

Yagi, T. (1988). MAX80: large-volume high-pressure apparatus combined with synchrotron radiation. *EoS*, **69**, 18–19,27.

Yagi, T., Sakai, T., Kadobayashi, H., Irifune, T. (2020). Review: high pressure generation technique beyond the limit of conventional diamond anvils. *High Pressure Research*, **40**, 148–161.

Yagi, T., Susaki, J, (1992). A laser heating system for diamond anvil using CO_2 laser, in Y. Syono, M. H. Manghnani, eds., *High-Pressure Research: Application to Earth and Planetary Sciences*, Terrapub/AGU, pp. 51–54.

Yamazaki, D., Ito, E. (2020). High pressure generation in the Kawai-type multianvil apparatus equipped with sintered diamond anvils. *High Pressure Research*, **40**, 3–11.

Yokoo, M., Kawai, N., Nakamura, K. G., Kondo, K., Tange, Y., Tsuchiya, T. (2009). Ultrahigh-pressure scales for gold and platinum at pressures up to 550 GPa. *Physical Review B*. **80**, 104114.

Zerr, A., Boehler, R. (1994) Constraints on the melting temperature of the lower mantle from high-pressure experiments on MgO and magnesioüstite. *Nature*, **371**, 506–508.

3

Applications of Synchrotron and FEL X-Rays in High-Pressure Research

GUOYIN SHEN AND WENDY L. MAO

Accelerator-based hard X-ray sources (storage-ring synchrotron radiation, and X-ray free electron laser, or FEL) provide X-ray beams with high energy, high brilliance, short tens-of-picosecond-to-femtosecond pulses, and high coherence that are well suited for high-pressure studies. Developments in high-pressure technology, advanced X-ray optics and detectors, and synergies with theoretical computations have helped drive the rapid growth of high-pressure research using synchrotron and FEL X-rays. In this chapter, we present a brief review of the research field from a historical perspective, illustrated by selected aspects on research using the diamond anvil cell. We then highlight a few of the active areas in high-pressure X-ray research, including ultrahigh-pressure generation, amorphous materials at high pressure, phase transition kinetics, and materials metastability. Finally, an outlook on future directions and opportunities with the upgrades in both synchrotron and FEL facilities worldwide is presented.

3.1 Introduction

Applying pressure increases the energy density of a material and induces changes such as modifications in atomic arrangement, broadening of electronic bandwidths, and/or changes in the magnetic exchange interactions, profoundly altering a material's physical and chemical properties. Since the early twentieth century (Bridgman, 1909), pressure has been an effective means, both theoretically and experimentally, to study material behavior at the conditions of planetary interiors and to discover functional materials with exceptional properties. In computational studies, pressure is now widely used as an effective tuning parameter for establishing predictive models and for providing guidance in the search for novel materials. In experimental work, the development of high-pressure technology has enabled the generation of a vast pressure range, far exceeding that at the center of the Earth (364 GPa), via various pressure devices and platforms that can then be combined with a suite of probing tools for material characterization.

The diamond anvil cell (DAC) is one of the most widely used high-pressure devices with which extremely high pressures can be generated at the expense of reduced sample volume. The arrival of accelerator-based X-ray sources (synchrotron radiation and free electron laser, or FEL) provides small, yet powerful, probes to reach the minute samples by penetrating through the diamond (or in some cases an X-ray transparent gasket) windows. Since the 1970s, great efforts have been made to take advantage of accelerator-based X-ray sources for high-pressure research, including the following: (1) high intensity over a broad energy range; (2) extremely low emittance (small source size and angular divergence); (3) tunable energy and bandwidth control; (4) various timing structures (from nanosecond to femtosecond pulses); (5) polarization of the radiation; and (6) coherence of the X-ray beam.

In this chapter, we present a brief review of the developments in high-pressure research using synchrotron and FEL X-rays from a historical perspective (Section 3.2). This is such a dynamic, wide-ranging, and rapidly evolving field that it is impossible to touch all of the exciting research being conducted, so instead we chose to highlight just a few of the active areas in high-pressure X-ray research (Section 3.3) and provide an outlook on future developments in high-pressure X-ray studies in view of the upgrades to both synchrotron and FEL facilities worldwide (Section 3.4).

3.2 A Brief History of High-Pressure X-Ray Studies

3.2.1 High-Pressure X-Ray Diffraction

The DAC was first invented in 1959 by two independent groups at the National Bureau of Standards (Weir et al., 1959) and the University of Chicago (Jamieson et al., 1959), based on the Bridgeman-type opposed anvil design. X-ray diffraction (XRD) was among the first probes using the newly invented DAC (Jamieson et al., 1959; Piermarini and Weir, 1962). In the early years, the XRD measurement using X-ray tubes was tedious and time consuming for DAC samples, often requiring more than 15 hours to make one measurement. Using an intense and broad energy spectrum of synchrotron radiation at the Deutsches Elektronen-Synchrotron (DESY), Buras et al. (1977) first applied the energy dispersive XRD technique to study phase transitions in TeO_2 at high pressures using a DAC, with dramatically reduced collection time of less than 15 minutes for each measurement. Despite the successful demonstration of the application of synchrotron radiation in high-pressure research, it took until the early 1980s for several groups worldwide to develop high-pressure synchrotron research programs. In Europe, early research activities were mainly at DESY using XRD coupled with DACs (Buras et al., 1977; Olsen et al., 1984; Benedict et al., 1986; Staun Olsen et al., 1986). In Japan, the large-volume press (LVP) was first coupled with synchrotron radiation in the mid-1980s at the Photon Factory for high-pressure diffraction studies (Shimomura et al., 1985). The development enabled the studies of phase equilibria for important mantle minerals and phase transition kinetics of weak X-ray scatterers (such as carbon) at in situ high-pressure, high-temperature conditions (Susaki et al., 1985; Yamaoka et al., 1986; Kikegawa et al., 1987; Yagi et al., 1987). In the United States, high-pressure energy dispersive XRD using DAC was established for studying phase transitions and structure determinations of materials at the Cornell High Energy Synchrotron Source (CHESS) (Bassett, 1980; Baublitz et al., 1981) and Stanford Synchrotron Radiation Lightsource (SSRL) (Skelton et al., 1982; Ming et al.,

1983). At the National Synchrotron Light Source (NSLS-I), high-pressure XRD measurements were performed with the established high-pressure facilities at a superconducting Wiggler beamline using both DAC (Mao et al., 1988; Mao et al., 1989) and LVP (Wang et al., 1991; Vaughan, 1993).

The finely collimated synchrotron radiation provides small X-ray beams to match the small samples in DACs. An important development in the mid-1980s was precise XRD measurements using the small X-ray beam for samples at ultrahigh pressures in the multi-megabar range. Samples of ~10 μm in diameter subjected to megabar pressures were used to detect crystallographic phase transitions in alkali halides and germanium (Vohra et al., 1986a, 1986b; Mao et al., 1989). Under extreme compression above 200 GPa, CsI was found to have a hexagonal close-packed crystal structure with an ideal c/a ratio of 1.63 ± 0.01 (Mao et al., 1989) and to display the structure and compressional behavior that converged with those of solid xenon, which is isoelectronic with CsI. Under pressures above 300 GPa, direct pressure-volume measurements on pure iron and an iron-nickel alloy provided a direct comparison with solid inner core densities, placing constraints on the thermal models of the Earth's interior (Mao et al., 1990).

A combination of single-crystal XRD and energy dispersive techniques was employed for high-pressure single-crystal studies using a DAC (Mao et al., 1988). The diffraction geometry in energy dispersive mode provides high spatial resolution at the sample and effectively reduces the background signal arising from scattered radiation from the diamond anvils. Thus, the developed single-crystal XRD technique was particularly useful for the measurements of the high-pressure crystal structure and equation of state of low scattering materials, such as solid hydrogen (Loubeyre et al., 1996; Mao et al., 1988; Zha et al., 1993) and van der Waals compounds in solid nitrogen-helium mixtures (Vos et al., 1992) and methane-hydrogen mixtures (Somayazulu et al., 1996).

In 1992, the imaging plate was first applied as a detector for high-pressure angle dispersive XRD experiments using synchrotron radiation (Shimomura et al., 1992). The imaging plate coupled with angle dispersive XRD allows for retrieval of more reliable diffraction intensity and provides an order of magnitude improvement in the relative precision ($\Delta d/d$) in determining diffraction peak positions compared to those with the energy dispersive technique.

These highly efficient XRD measurements using synchrotron radiation enabled the integration of other experimental variables, such as temperature, with high-pressure devices. High-pressure low-temperature XRD experiments were performed using a DAC in a cryostat (Olsen et al., 1993). Resistively heated DACs were developed for high-pressure high-temperature XRD studies (Mao et al., 1991; Fei and Mao, 1993, 1994; Fei et al., 1992). A double-sided laser-heated DAC system was developed for in situ X-ray measurements using synchrotron radiation (Shen et al., 1996; Mao et al., 1998). The system was applied to study element partitioning of the lower-mantle minerals (Mao et al., 1997) and the melting of iron at high pressures (Shen et al., 1998).

By the mid-1990s, the power of synchrotron radiation was increasingly recognized for high-pressure research. Yet, dedicated high-pressure beamlines were still scarce, with only a handful of expert groups having access to the limited synchrotron facilities. The arrival of

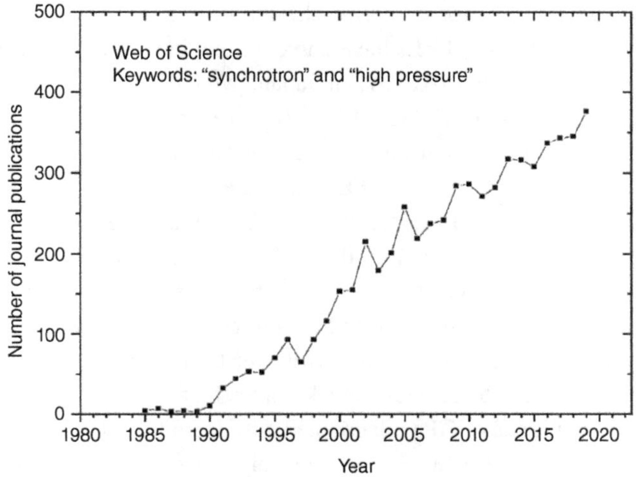

Figure 3.1 Number of papers in high-pressure research using synchrotron radiation published in scientific journals per year. The number was based on the search via the Web of Science using keywords of "high pressure" and "synchrotron." The actual number of publications may be even higher.

third-generation synchrotron facilities gave a boost to the establishment of dedicated high-pressure beamlines, providing the access to the advanced high-pressure synchrotron facilities for general research groups worldwide (Figure 3.1). At the European Synchrotron Radiation Facility (ESRF) in France, two high-pressure beamlines were among the first few beamlines established for general user operation (Hausermann and Hanfland, 1996). At the Advanced Photon Source (APS) outside Chicago, high-pressure beamlines soon became operational as soon as the APS facility first delivered X-ray beams (Rivers et al., 1998). At the SPring-8 facility in Japan, several beamlines have been equipped with high-pressure devices since its early operation. At these established high-pressure beamlines, a number of high-pressure synchrotron techniques have been developed, including those at ESRF (Mezouar, 2010; Mezouar et al., 2017), APS (Shen et al., 2001; Shen and Wang, 2014; Shen and Sinogeikin, 2015), and SPring-8 (Ohishi et al., 2008; Hirao et al., 2020). More recently, high-pressure beamlines have been established in many newer synchrotron facilities, such as Petra-III in Germany (Liermann et al., 2015), Diamond in UK, Soleil in France (Itié et al., 2015), Elettra in Italy (Lausi et al., 2015), Beijing Synchrotron Radiation Facility (BSRF) in Beijing (Liu, 2016), the Advanced Light Source (ALS) in California (Kunz et al., 2005a, 2005b; Stan et al., 2018), and the Brazilian Synchrotron Light Laboratory (LNLS) in Brazil (Lima et al., 2016), among others.

In the last decade, while high-pressure research has flourished at synchrotron facilities, the recent arrival of X-ray FELs has enabled novel studies of structural dynamics at the angstrom-femtosecond spatial and temporal scales. X-ray FELs generate coherent X-ray pulses of a few to 100 fs for a wavelength range extending from about 100 nm to less than 1 Å. The unprecedented peak brightness is approximately a billion times higher than synchrotron sources based on storage rings. These combined characteristics make the X-ray FEL a unique tool for exploring the ultrafast and ultrasmall (Pellegrini et al., 2016). A number of hard X-ray FELs have come online, including the SPring-8 Angstrom Compact Free Electron Laser (SACLA) in Japan, which began operations in 2012; the Pohang Accelerator Laboratory (PAL) FEL in South Korea, which began lasing in 2016; the Swiss FEL, which also started lasing in 2016; and the European XFEL (EuXFEL), which began operations in 2017. X-ray FELs provide a number of opportunities for high-pressure studies, which will be discussed further in subsequent sections. This includes both dynamic compression (using optical laser drivers) and static compression (using DACs). At the Linac Coherent Light Source (LCLS), the world's first hard X-ray FEL that has been operational at SLAC National Accelerator Laboratory since 2009, one of the end stations houses the Matter in Extreme Conditions (MEC) instrument (Nagler et al., 2015). High-power optical-laser systems to generate shock waves that result in dynamic high-pressure conditions of up to hundreds of GPa are coupled with the ultrafast X-ray source, which has enabled ultrafast XRD experiments on a large range of material systems (Bostedt et al., 2016).

The development in high-pressure synchrotron and FEL techniques and their impact in advancing high-pressure science have been summarized in several review articles (Duffy, 2005; Mao and Mao, 2007; Shen and Mao, 2016; McMahon, 2018; Mao et al., 2019). Here, we highlight a few of them.

Small X-ray beams and ultrahigh-pressure generation: For megabar experiments, DACs with small culets in beveled or toroidal-shaped design are generally used (Dewaele et al., 2018; Li et al., 2018). The micrometer and submicrometer synchrotron X-ray beams provide the necessary small probes for characterizing ultrahigh pressures and the pressure distribution in a DAC. The latter is particularly important because achieving peak extreme pressures inevitably involves steep pressure gradients. Understanding and controlling the pressure gradients and how they are distributed in the corresponding regions relative to the sampling volume are critical for ultrahigh pressure generation and reliable data interpretation. Figure 3.2 shows the pressure distribution as a function of the radial distance from the center of the culet (Li et al., 2018). The submicron X-ray probe provides the pressure distribution that is concentrated on the culet area (<20 μm in diameter) with the bevel design. Such data provide critical information on the relationships between the pressure distribution and anvil shapes and configurations. Optimizing anvil designs for reliably generating pressures over 400 GPa (Dubrovinsky et al., 2012; Sakai et al., 2015; Dewaele et al., 2018; Jenei et al., 2018; Li et al., 2018; Yagi et al., 2020) has been a subject of extensive studies in recent years.

Double-sided laser-heated DAC for in situ X-ray probes at high-pressure and high-temperature conditions: To study material behavior at simultaneously high-pressure, high-temperature conditions, a uniformly heated volume much larger than the probing X-ray beam size is a prerequisite for reliable X-ray measurements. Significant progress in shaping and defining the temperature distribution in the laser-heated DAC has been made in a double-sided heating configuration

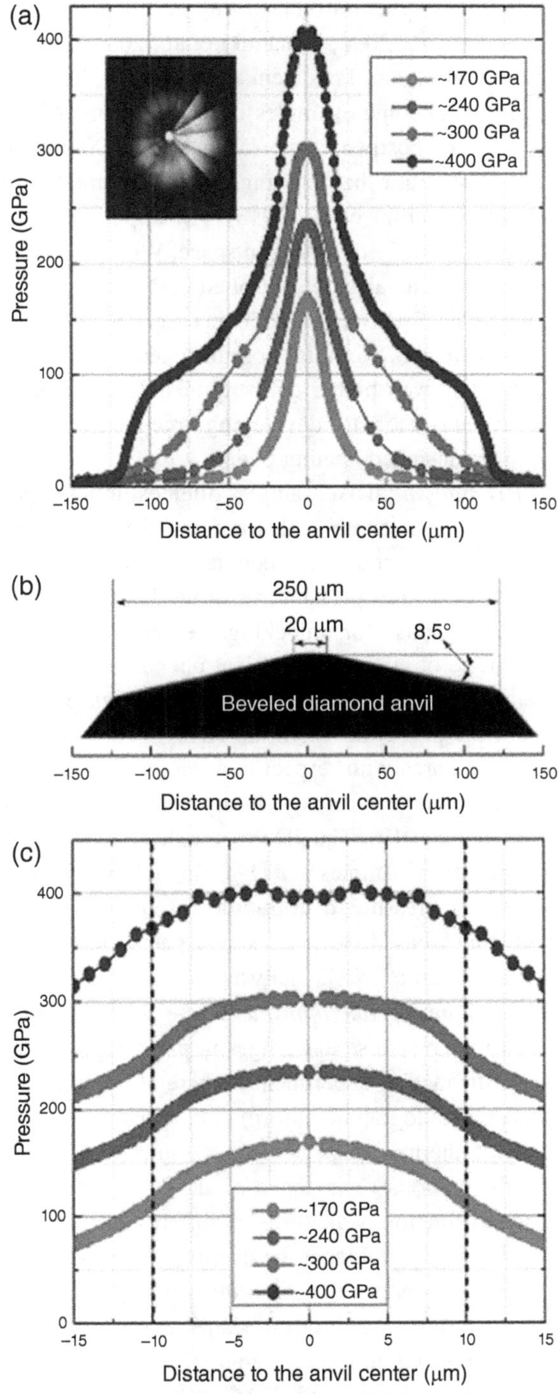

Figure 3.2 Pressure distribution as a function of the radial distance from the center of the culet in a DAC for (a) a large region covering entire anvil and (c) a smaller range covering culet region. (b) is a schematic side view of the diamond anvil with a 20 μm culet beveled to 250 μm at 8.5° as a reference.

From Li et al., *Proceedings of the National Academy of Science*, 115, 1713, 2018, reprinted with permission from the American Association for the Advancement of Science (AAAS).

(Shen et al., 2001). Many synchrotron facilities are equipped with double-sided laser-heated systems (Shen et al., 2001; Watanuki et al., 2001; Zhao et al., 2004; Schultz et al., 2005; Meng et al., 2006; Prakapenka et al., 2008; Liermann et al., 2015; Liu, 2016; Stan et al., 2018). Some examples using the double-sided laser-heated DAC include the discovery of postperovskite (Mao et al., 2004; Murakami et al., 2004; Oganov and Ono, 2004) and the determination of the melting curve of iron (Shen et al., 1998, 2004a; Anzellini et al., 2013). An important advantage of the laser-heating technique is its localized heating in space. With pulsed or modulated heating, the local heating by laser can also be controlled in the time domain (Goncharov et al., 2010; Montgomery et al., 2018). Fast-modulated laser heating in a DAC can suppress thermally activated diffusion, minimize possible chemical reactions of the sample and environment, and reach even higher temperatures. The timing structure of synchrotron and FEL X-rays is suitable for developing time-resolved X-ray measurements of samples following predesigned heating paths.

Two-dimensional XRD mapping: DAC samples often exhibit a large degree of complexity and inhomogeneity, on length scales ranging from nanometers to hundreds of microns. The ability to perform fast and detailed spatially resolved characterization of the inhomogeneity under high pressure, or in the high-pressure temperature-treated samples, holds great promise for unlocking the complexity and for gaining understanding of a multitude of emerging physical phenomena in high-pressure sciences. With developments in efficient XRD data collection, high-pressure XRD can now be performed as a full 2D probe, in which the XRD is collected on a two-dimensional grid by fly-scanning an area with respect to a microfocused X-ray beam using a fast (millisecond) X-ray area detector in synchrony with the sample positioners (Hrubiak et al., 2019; Smith et al., 2019). The 2D mapping capability can provide information about the inhomogeneity of samples in a DAC, including (1) phase mapping based on structural features, (2) pressure distribution based on material compression, (3) microstructure mapping based on peak shape parameters, and (4) grain size and preferred orientation based on XRD analysis.

Single-crystal XRD and multigrain approach: The most reliable information about crystal structure and the response to changes in pressure and temperature is obtained from single-crystal diffraction experiments. There are a number of developments in obtaining detailed crystallographic information, such as structure models (atomic position, occupancy, thermal displacement parameter) (Lavina et al., 2014) and electron density distributions (Yamanaka et al., 2007; Shen et al., 2012; Li et al., 2019). Because the diffraction is measured at different orientations, the single-crystal XRD technique also has advantages in dealing with phases with low symmetry. However, growing and/or preserving high-quality single crystals at very high pressures is difficult. With small synchrotron beams, traditional powder samples may be treated as multiple grains. Grain growth can be further promoted by laser heating and other heating methods integrated with DAC, often resulting in "spotty" patterns in diffraction images. An approach using "multigrain crystallography" has been developed with the diffraction signals of hundreds of single crystals simultaneously

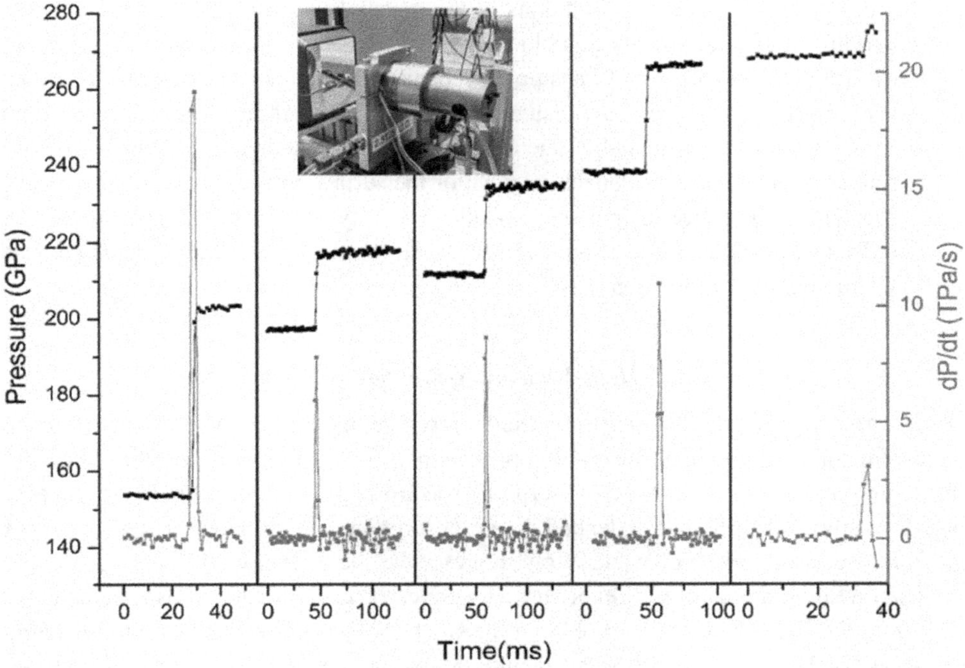

Figure 3.3 Sharp pressure jumps on a Mo sample between 150 and 280 GPa with a compression rate as high as 35 TPa s^{-1} using a piezo-driven DAC (inset photo) at HPCAT, APS.

collected. The multigrain approach is an effective way to increase redundancy and completeness in data collection, thus significantly improving data quality, as demonstrated in recent structural studies (Dubrovinsky et al., 2010; Langrand et al., 2017; Bykov et al., 2018; Li et al., 2019; Zhang et al., 2019).

In situ XRD studies under rapid compression and decompression: Traditionally, DAC is a static device, with pressure controlled mechanically by turning screws. The development of the dynamic DAC (dDAC) has enabled studies of material behavior under rate-controlled pressurization (Evans et al., 2007; Sinogeikin et al., 2015). Electromechanical piezoelectric actuators are used to replace screws and enable rapid and precise control of time-dependent load. A compression rate as high as >35 TPa s^{-1} has been reached for a Mo sample in the megabar pressure region (Figure 3.3) (Shen and Mao, 2016). Under rapid compression and decompression conditions, materials often display interesting properties and metastability. The integration of fast XRD capabilities with the fast loading and unloading high-pressure devices (Sinogeikin et al., 2015; Smith et al., 2015) provides a structural probe that can monitor dynamic structural changes for studying transition kinetics, reaction rates, and metastable phases (Jacobsen et al., 2015; Lin et al., 2018; Shen et al., 2019b).

Ultrafast XRD coupled with dynamic compression: In dynamic compression experiments at X-ray FELs, by varying the laser pulse and peak power one can tune the pressure and temperature conditions in the material and collect X-ray snapshots at varying delay times to allow us to capture the dynamic response to provide phase transformation kinetics and transition mechanism information – including nucleation and growth of new phases under extreme conditions. Additionally, since changes from an insulating to metallic state are accompanied by a dramatic change in optical properties, one can simultaneously monitor the surface reflectivity of samples using standard velocimetry techniques, so-called velocity interferometer system for any reflector (VISAR). See Sections 3.3 and 3.4 for examples of ultrafast XRD coupled with dynamic compression.

3.2.2 High-Pressure X-Ray Spectroscopy

X-ray spectroscopy provides element-specific information of electronic properties, such as band structure and bonding. Several X-ray spectroscopy techniques have been developed for high-pressure studies: X-ray absorption spectroscopy (XAS), including near-edge spectroscopy (XANES) and extended X-ray absorption fine structure (EXAFS); X-ray emission spectroscopy (XES); and X-ray fluorescence spectroscopy (XFS).

The first high-pressure XAS experiments were carried out in late 1970s (Ingalls et al., 1978) on the EXAFS of the Fe K-edge in FeS_2 and FeF_2 using a channel-cut monochromator at SSRL. Also using the SSRL facility, Shimomura et al. (1978) measured both the XANES and the EXAFS of GaAs in a DAC as the sample cross the covalent-to-metallic transition under high pressures. Early high-pressure XAS studies include oxidation states and coordination numbers of elements in glasses (Fleet et al., 1984; Itie et al., 1989) and local structural distortions in various perovskites (Andrault et al., 1988).

One limitation in high-pressure XAS studies arises from the absorption and diffraction glitches coming from the diamond anvils. To greatly reduce the amount of anvil material in the X-ray beam path, counterbores (perforations) can be drilled into the diamond anvils, as used in Bassett-type hydrothermal DAC for studying ions in aqueous solutions at moderately high-pressure high-temperature conditions (Bassett et al., 2000; Mayanovic et al., 2007) with applicable X-ray energies down to 5 keV. When the incident X-ray energies are over 20 keV, using nanocrystalline diamond anvils can avoid diffraction glitches from the single-crystal diamond anvils (Baldini et al., 2011; Ishimatsu et al., 2012). An alternative beam path through high-strength gasket (beryllium, cBN, or boron) has been also used for high-pressure XAS studies (Hong et al., 2009; Park et al., 2015).

With a focused polychromatic beam (~5 μm) of extremely high flux (Pascarelli and Mathon, 2010), the parallel detection of the whole XAS spectrum can be quickly collected. The fast XAS measurement has enabled tracking rapid changes in the local and electronic structure of selective atoms. The fast measurement and small beam size also allow 2D mapping of heterogeneous samples with micron resolution, where each pixel contains full XAS information (Aquilanti et al., 2009). Laser-heated DAC has been integrated with the fast XAS measurements. For example, the loss of prepeak and of the oscillations in the 7130–7140 eV energy range are used for signifying melting of iron at high pressures (Aquilanti et al., 2015).

The first high-pressure XES study revealed a pressure-induced high-to-low-spin transition in FeS (Rueff et al., 1999), evidenced by the disappearance of the low-energy satellite in the Fe K_β emission spectrum of FeS. Soon after, this technique was widely applied for studying pressure-induced spin transitions (Figure 3.4). Examples of XES studies include the predicted high-spin–low-spin transitions in iron oxides and silicates, such as (Mg,Fe)O and Fe_2O_3 (Badro et al., 1999, 2002, 2003), bridgmanites (Li et al., 2004, 2012), and silicate postperovskite (Yamanaka et al., 2012) to above 100 GPa.

XES data are typically collected by a Rowland circle spectrometer with synchronized $\theta-2\theta$ scans of the analyzer and detector. Multiple analyzers have been used to increase solid angle coverage in data collection (Xiao et al., 2016). A short-distance spectrometer of large, solid angle coverage improves the efficiency in XES data, reducing collection time from hours to only several minutes (Pacold et al., 2012). The XES technique has been integrated with laser-heated DAC techniques (Lin et al., 2005b).

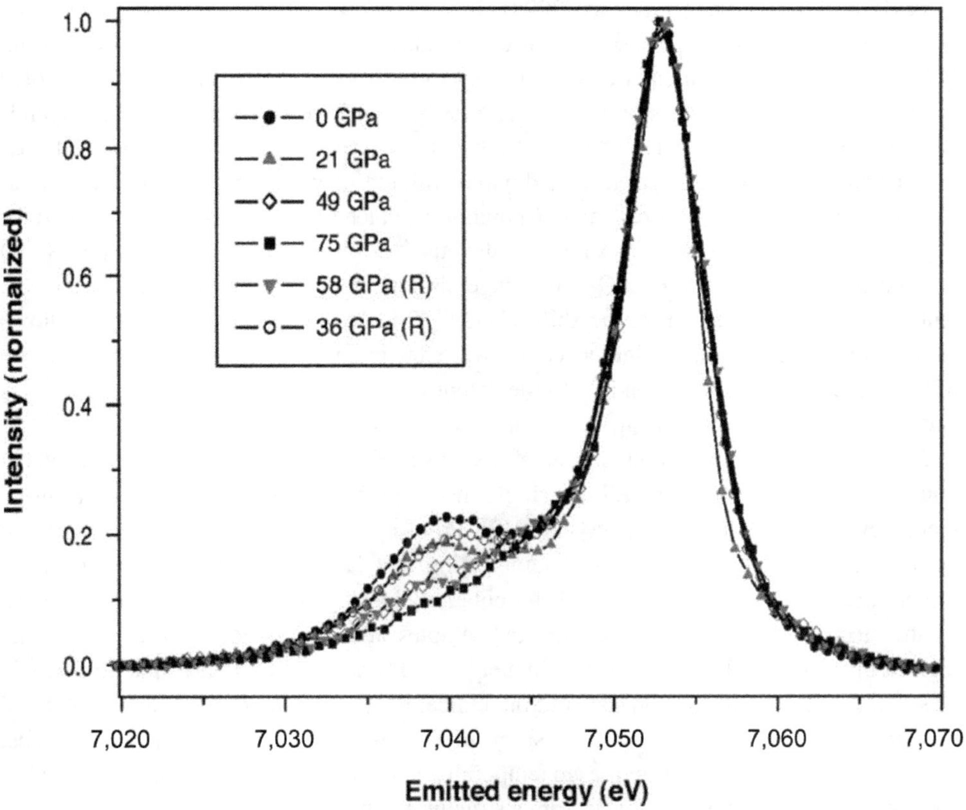

Figure 3.4 Fe K_β X-ray emission spectra collected on a ferropericlase sample $(Mg_{0.83}Fe_{0.17})O$ at high pressures. The satellite peak on the low-energy side is characteristic of iron in the high-spin state. The absence of the satellite, which occurs between 58–75 GPa, demonstrates a high-spin to a low-spin transition.

From Badro et al., *Science,* 300, 789, 2003, reprinted with permission from AAAS.

The emission signal can be collected as the incident beam energy is changed incrementally or selectively across an absorption edge, resulting in a 2D data set of the incident energy versus emission energy. The projection to the incident and emission planes gives the partial fluorescence yield (PFY) and the resonant XES (RXES), respectively. Because the lifetime broadening of the final state is considerably smaller than that of the core excited state, this resonant method significantly enhances footprints of electron states and has a remarkable sharpening effect in projected spectra in PFY (Rueff, 2010). The sharpening effect makes the PFY technique suitable for studying mixed-valence properties or the multiconfigurational nature of *f*-electron systems, especially for actinides where conventional XAS cannot resolve the fine distribution of the unoccupied electronic states due to the large lifetime broadening of the $2p_{3/2}$ level (Booth et al., 2012; Kvashnina et al., 2014; Nasreen et al., 2016). High-pressure RXES has been used for quantitatively measuring the development of multiple electronic configurations with differing $4f$ occupation numbers, revealing information on the delocalization of the strongly correlated $4f$ electrons (Bradley et al., 2012; Lipp et al., 2012) and for obtaining subtle signals from intermediate spin state of Fe_3O_4 at 15–16 GPa (Ding et al., 2008).

XES measurements have also been used to study valence electrons (Struzhkin et al., 2006; Li et al., 2015), which is complementary to those provided by XAS on unoccupied states. Valence-to-core emission features in XES are sensitive to the type of metal ligands and can be used for characterization of the valence electronic levels (Gallo and Glatzel, 2014). Contrary to standard valence band photoemission spectroscopy that provides similar information but usually requires vacuum conditions, valence-to-core XES is well suited for high-pressure studies. A recent example is the study of the Ge coordination number (CN) and mean Ge-O distances in GeO_2 from the emission energy and the intensity of the K_β" emission line. Measurement of the shift of the K_β" line in compressed GeO_2 glass allows for tracking the CN of Ge at high pressures, while the intensity of the K_β" line is a measure of the Ge-O distance corresponding to the extent of hybridization between O 2s electrons and Ge 4p valence electrons (Spiekermann et al., 2019).

Recent developments in femtosecond XES at X-ray FELs such as LCLS have enabled the potential to conduct ultrafast XES in a single shot to probe electronic structure in transition metal compounds (Alonso-Mori et al., 2012). Proof-of-principle measurements have been performed at the MEC instrument demonstrating that XES can be combined with laser drivers and simultaneous ultrafast XRD for obtaining information on the lattice structure and on the electronic structure of laser-shocked samples in a single shot (Figure 3.5). At the highest pressure of 80 GPa and approximately 2,000 K and a 10 ns delay time, the sample was found to display a low spin state in iron, indicating an upper bound in both pressure and time for when the electronic spin transition in Fe^{2+} in the material took place. The fast measurements provide opportunities to temporally study the onset of an electronic transition and resolve whether it is decoupled from a structural transition.

The first high-pressure XFS experiments measured the composition and solubility of AgCl in water at high-pressure high-temperature conditions (Schmidt and Rickers, 2003). XFS can be used as multi-element probes for studies of dissolution kinetics (Schmidt et al., 2007; Tanis et al., 2016). A typical detection limit for high-pressure XFS is at a

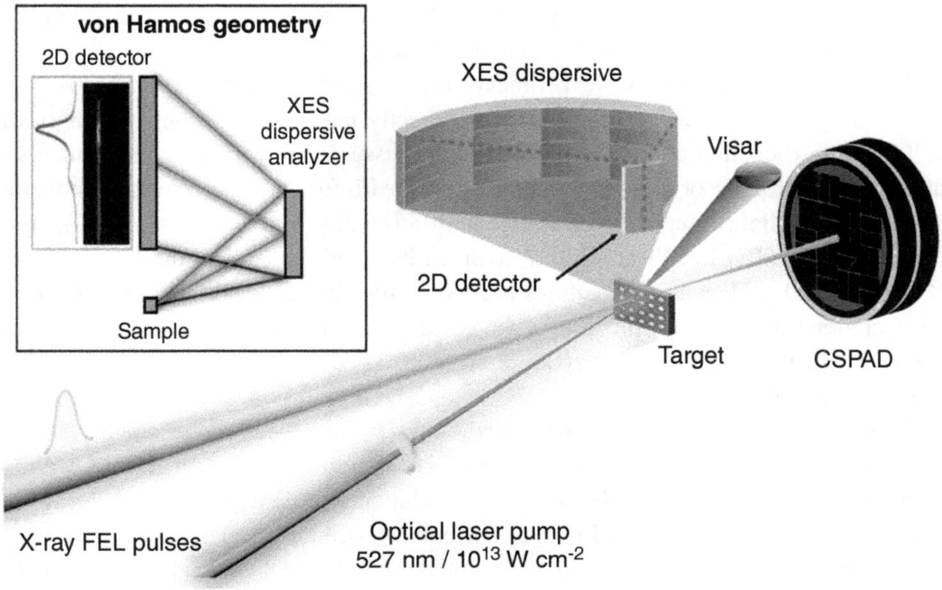

Figure 3.5 Schematic of an ultrafast XES setup at MEC, LCLS. An optical laser pump generates a shock wave in a target, and an X-ray FEL pulse probes the sample at extreme conditions using an XES spectrometer combined with XRD detectors (CSPAD). For the color version, refer to the plate section.

few-parts-per-million level of concentration for elements down to atomic number of 22 (Ti) and to a pressure of 10 GPa at high temperatures to at least 1,273 K (Petitgirard et al., 2009; Tanis et al., 2012; Tanis et al., 2016). The use of a confocal geometry reduces unwanted background signals (Wilke et al., 2010), further improving efficiency and consequently improving the detection limit.

3.2.3 High-Pressure Inelastic X-Ray Scattering

Inelastic X-ray scattering (IXS) measures the dynamic structure factor $S(E,Q)$, which is a function of momentum transfer Q and energy transfer E. The dynamic structure factor contains information of all electronic excitations, including phonons, magnons, core-electron excitations, plasmons, the collective fluctuations of valence electrons, and Compton scattering. Compared to other inelastic scattering techniques (inelastic neutron scattering, inelastic electron scattering, light Brillouin, and Raman scattering), IXS is well suited for high-pressure sample environments because of the penetrating power of hard X-rays; small, focused beam size from highly collimated X-rays; and large Q coverage. Over the past two decades, several methods have been developed for high-pressure studies: nonresonant IXS or X-ray Raman scattering, resonant IXS (RIXS), nuclear resonant IXS (NRIXS), and nuclear forward scattering (NFS). Depending on the energy resolution required, different X-ray spectrometers have been constructed: low-energy resolution IXS

(LERIX) with a typical energy resolution of 1 eV, medium-energy resolution IXS (MERIX) with 30–300 meV, and high-energy resolution IXS (HERIX) with 0.3–3 meV.

The first high-pressure IXS experiments measured the dynamic structure factor of electrons in solid ^{4}He and provided the first directly measured electronic excitations in solid helium (Schell et al., 1995). In later high-pressure applications, measurements of near core-electron absorption edge features revealed information on chemical bonding, particularly for light elements such as carbon (Mao et al., 2003; Zeng et al., 2017), boron (Meng et al., 2004; Lee et al., 2008), hydrocarbons (Fister et al., 2009), and oxygen (Cai et al., 2005). Because the features in IXS are sensitive to local "short-range" order, the technique has been useful in revealing structural changes in glasses (SiO_2, B_2O_3, GeO_2, $MgSiO_3$) at high pressure (Lin et al., 2007; Lee et al., 2018; Kim et al., 2019; Lee et al., 2019).

Although it has been long recognized that X-rays can be used to probe lattice dynamics, it was not until the arrival of third-generation synchrotron facilities that HERIX with the required mega-electronvolt (MeV)-energy resolution was established for measuring phonon dispersion, first at HASYLAB in Germany (Burkel et al., 1987), and then at ESRF (Sette et al., 1995), APS (Alp et al., 2002), and SPring-8 (Baron et al., 2001). The first high-pressure HERIX experiments measured the dispersion of longitudinal acoustic phonons in iron up to 110 GPa (Fiquet et al., 2001), with the determined longitudinal wave velocity (V_p) following the Birch law. Since then, HERIX has been applied to studying the elasticity anisotropy of *f*-electron metals (Manley et al., 2003; Wong et al., 2003), sound velocities of materials at pressure conditions over 1 Mbar (Antonangeli et al., 2010, 2011; Mao et al., 2012) and at high-pressure high-temperature conditions using a laser-heating system (Fukui et al., 2013; Ohtani et al., 2013; Kamada et al., 2014; Sakamaki et al., 2016; Takahashi et al., 2019). Using single-crystal samples, complete phonon dispersions in the Brillouin zone have been mapped by HERIX at high pressures (Bosak et al., 2008; Farber et al., 2006; Krisch et al., 2011).

Momentum-resolved HERIX can also be applied to liquids for extracting information of sound velocity and viscosity (Sinn et al., 2003; Scopigno et al., 2005; Alatas et al., 2008). The potential of studying liquids at high-pressure high-temperature conditions using HERIX was illustrated using a Paris–Edinburgh large-volume press (Falconi et al., 2004) and a resistively heated DAC (Alatas et al., 2008).

Using NRIXS, a variety of thermodynamic parameters can be derived from the measured density of phonon states for selected isotopes, such as the Debye temperature, Grüneisen parameter, mean sound velocity, and the lattice contribution to entropy and specific heat (Sturhahn et al., 1995). The first high-pressure NRIXS experiments measured the lattice dynamics of hcp–iron at pressures from 22 to 42 GPa in a DAC (Lübbers et al., 2000). Using a modified magabar DAC, NRIXS experiments on hcp-iron were conducted at extended pressures to 153 GPa (Mao et al., 2001), and at in situ high-pressure high-temperature conditions (Shen et al., 2004b; Lin et al., 2005c). The measurement of Debye sound velocity distinguishes compression and shear wave velocities as well as their temperature and pressure dependence, addressing anisotropy of sound velocities and mode Grüneisen constants (Murphy et al., 2011; Jackson et al., 2013b).

High-pressure NFS was first performed using a DAC (Nasu, 1996), in which a pressure-induced magnetic hyperfine interaction at ^{57}Fe in $SrFeO_{2.97}$ was detected at 44 GPa. The NFS hyperfine signals are very sensitive to internal magnetic fields, electric field gradients, and isomer shifts (Takano et al., 1991; Nasu, 1996; Chefki et al., 1998) and has been used to study high-pressure behavior of materials at megabar pressures, such as magnetic collapse (Li et al., 2004; Lin et al., 2005a), site occupancy (Catalli et al., 2011; Lin et al., 2012), valence and spin state (Jackson et al., 2005; Speziale et al., 2005; Shim et al., 2009; Chen et al., 2012), and as a probe for superconductivity (Troyan et al., 2016). Fast NFS experiments have been performed to determine the high-pressure melting of iron (Jackson et al., 2013a), using the Lamb–Mössbauer factor, which describes the probability of recoilless absorption. In NFS, the timing circuit removes prompt events, but records the time elapsed between excitation and reemission. This delayed forward scattering measures the Mössbauer effect in the time domain. At SPring-8, synchrotron Mössbauer spectrometers in the energy domain have been developed for high-pressure research, suitable for megabar pressures using a DAC (Mitsui et al., 2009; Seto et al., 2010; Masuda et al., 2016).

Recently an experimental platform to perform IXS experiments at FELs was developed at MEC, LCLS. The instrument resolution was determined to be $\sim$50 meV for single-shot measurements on a plastic, polymethyl methacrylate (PMMA), and longitudinal acoustic modes in polycrystalline diamond were also collected. Future work to couple this IXS setup with high power lasers can potentially enable IXS measurements during dynamic compression at MEC and the new high energy density (HED) instrument at the European XFEL (McBride et al., 2018). Future planned instruments at LCLS-II high-energy upgrade (see Section 4.2), such as the Dynamic X-Ray Spectroscopy (DXS) instrument, is designed for high-resolution IXS. Current high-resolution IXS measurements in a DAC can be severely flux limited. LCLS-II HE will provide two to three orders higher average brightness delivered to the sample than that at synchrotron sources, making the DXS instrument a potential game-changer for high-pressure IXS studies at extended pressures.

3.2.4 High-Pressure X-Ray Imaging

X-ray imaging bridges information at various scales from atomistic scale, mesoscale, to millimeter scale. Both the full-field imaging approach and the position-mapping approach with a small, focused beam are widely used. A variety of high-pressure X-ray imaging techniques have been developed for various purposes, such as fast radiography, tomography, and phase contrast imaging. More recently, high-pressure coherent X-ray diffraction imaging has been developed for nanosized particles.

Since the beginning of high-pressure X-ray experiments, X-ray radiography has been used for locating high-pressure sample position and defining sample shape. By selecting proper monochromatic beams, radiography from the absorption contrast can be used for density determination of amorphous and liquid materials (Shen et al., 2002; Sato and Funamori, 2010; Petitgirard et al., 2019b). While X-ray radiography measures the absorption (or density) contrast, phase contrast imaging (PCI) records variations in the absorption and the X-ray refractive index of a sample, making it useful in imaging weakly scattering materials, where the density contrast may be weak. For example, PCI was used to study the immiscibility of two or more liquids in a system (Kono et al., 2015); fast PCI measurements were used for studying the dynamic response of materials at extreme conditions (Jensen et al., 2012; Luo et al., 2012; Schropp et al., 2012; Parab et al., 2016; Feng et al., 2018; Olbinado et al., 2018).

X-ray computed tomography (CT) has become a powerful tool for the real-time characterization of microstructural and morphological evolution under high pressure. With the third-generation synchrotron light sources, it is possible to obtain X-ray radiographs at the millisecond level and collect tomographic data by rotating the sample in seconds. Microtomography has been conducted using an opposing anvil device with X-ray transparent containment ring or using Paris–Edinburgh cells loaded with a hydraulic press to reach high-pressure conditions of 10 GPa and higher and temperatures up to ~2,000 K (Lesher et al., 2009; Yu et al., 2016). Developments in coupling transmission X-ray microscopy to DACs requires even higher spatial resolution for direct imaging of submicron features. Using micron resolution tomography, a direct density measurement method from reconstructing the linear absorption coefficient at various pressures was established and applied to a study of amorphous Se under high pressure (Liu et al., 2008). For a typical sample size with dimensions of ~100 μm in a DAC at moderate pressure, 1 μm spatial resolution only provides 10^{-2} resolution in volume, which is an order of magnitude lower compared to XRD for crystalline materials. At higher pressures (>10 GPa), the sample thickness is dramatically reduced to tens of microns or less. In this case, micron resolution tomography will not provide volume measurements with sufficient accuracy for EoS measurements. 3D tomography using nanoscale transmission X-ray microscopes (nano-TxM) with tens of nanometers spatial resolution enables volume measurements with accuracy rivaling X-ray diffraction of crystalline solids, and additional applications in visualization of multiphase assemblages and spectroscopic mapping in 3D, opening many exciting opportunities for high-pressure research.

As a benchmark for establishing nano-TxM for EoS determination, the volume change with increasing pressure for a crystalline Sn sample was determined in a DAC using nano-TxM. Results were compared to XRD and found to have comparable error bars (Wang et al., 2012; Shi et al., 2013). Subsequent application of nano-TxM for determination of the EoS for a wide variety of amorphous and poorly crystalline phases has been demonstrated for germania (Lin et al., 2013), bulk metallic glasses (Zeng et al., 2014, 2016; Chen et al., 2015), and also weak scatterers such as glassy carbon and SiO_2 (Figure 3.6). The interconnectivity of iron melt networks in a silicate matrix were visualized at lower mantle pressures for implications into planetary core formation (Shi et al., 2013). The energy tunability of incident X-rays from a synchrotron source enables the quantitative visualization of chemical species and valence states of materials in situ. By performing XANES-based 3D tomography as a function of pressure, real-time chemical and elemental mapping information with up to 30 nm spatial resolution can be quantitatively constructed. This allows a rare opportunity for understanding the mechanisms of pressure-induced phase growth, phase boundary transformations, formation of domains, or charge transfer dynamics in situ and in operando. The first such application was the visualization of a $BiNiO_3$

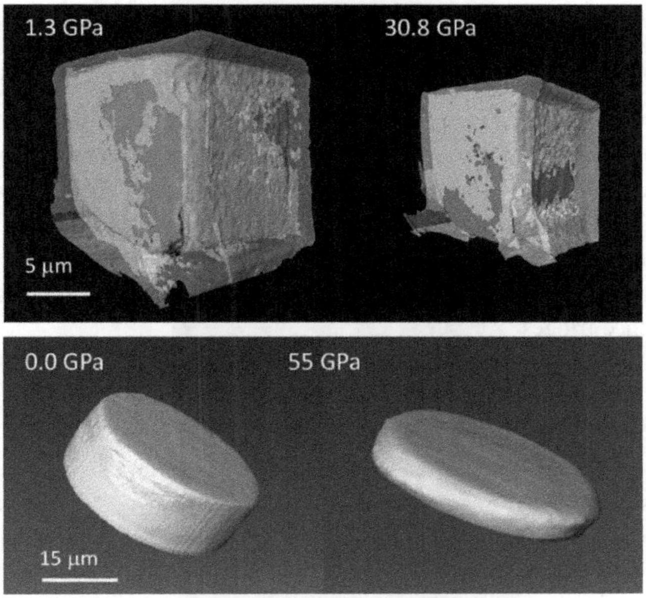

Figure 3.6 Top panel: representative 3D renderings from a high-pressure nano- transmission X-ray microscopy (TxM) experiment on a Pt-coated SiO_2 cube. The Pt coating is shown in gray and the "hollow" volume representing the SiO_2 cube is shown in red. The volume has decreased by a factor of two from 1.3 to 30.8 GPa. Bottom panel: representative 3D renderings from high-pressure nano-TxM experiment on an uncoated SiO_2 cylinder. For the color version, refer to the plate section.
Adapted from Mao et al. (2019).

powder sample at high pressures as it crossed a structural transition that is accompanied by an intermetallic charge transfer between Bi and Ni ions (Liu et al., 2014). The use of in situ 3D spectroscopic tomography at the Ni absorption edge at varying pressures allows one to visualize grain boundary disappearance and growth as the low-pressure phase transforms into the high-pressure phase.

By utilizing the coherent fraction from third-generation synchrotron sources and high coherent flux of X-ray FELs to illuminate samples, X-ray based coherent diffractive imaging (CDI) can further improve the spatial resolution limited by X-ray optics, and provide an imaging tool for viewing even smaller nanoparticles (Miao et al., 2015). In Bragg CDI (BCDI), the diffraction patterns from a submicron-sized crystal are collected by rotating the sample around a Bragg peak (Harder and Robinson, 2013). In the resulting 3D diffraction intensity distribution, called a fringe pattern, the spacing of fringes contains information on the crystal shape, and iterative algorithms can retrieve the phase information from the distribution of diffraction intensity to map lattice distortions. BCDI enables imaging of the shape and lattice strain in crystals with nanometer-scale spatial resolution, picometer-scale lattice distortion sensitivity, and strain resolution of ~10^{-4} (Hruszkewycz et al., 2017). While the feasibility of high pressure DAC studies have been demonstrated for over a decade (Le Bollach et al., 2009), there have been only a limited number of reports to fairly low pressure – notably on the 3D strain distribution in a 400 nm gold nanocrystal to 6.4 GPa (Yang et al., 2013) and deformation twinning in a silver nanocube to below 3 GPa (Huang et al., 2015). DAC experiments are difficult due to the complicated sample environment (having to pass through either the diamond anvils or the beryllium gasket) that leads to absorption of the incident X-rays and can introduce distortions that affect the wavefront of the beam. Finding the same submicron grain after pressure changes is also a technical challenge, as the nanocrystals drift or rotate even when compressed in a nominally solid pressure-transmitting medium. Moreover, significant deformation may occur under pressure, which can destroy the facets on the nanocrystals, resulting in a loss of the interference fringes.

3.3 Highlights from High-Pressure Research Using Synchrotron and FEL X-Rays

High-pressure research has become an important field for the application of large accelerator-based facilities. Integrating various X-ray probes with hydrostatic or uniaxial stress conditions, controlled shear stresses, rate-controlled compression and decompression, laser-driven dynamic compression, laser and resistive heating, and cryogenic cooling has enabled investigations of the structural, vibrational, electronic, and magnetic properties of materials and their dynamics over a wide range of extreme conditions. These studies address longstanding problems whose answers are now within our grasp, promise the discovery of surprising and novel phenomena that challenge our conventional wisdom and have potential for significant societal impact in various fields. Several recent review articles (Shen and Mao, 2016; McMahon, 2018; Sanloup and de Grouchy, 2018; Mao et al., 2019) summarize progress in several fields. In this section, we select a few examples to highlight the active research areas in recent years.

3.3.1 Ultrahigh-Pressure Generation

The maximum pressure range of ~400 GPa was reached and optimized 30 years ago in DAC with the "beveled anvil" geometry (Mao et al., 1989, 1990) of 30–50 μm flat culets (e.g., beveled at 8.5° to 300 μm diameter). Such geometry has been used successfully and extensively ever since and continues to be responsible for many of the recent discoveries in high-pressure research. The 300–400 GPa ceiling is not an intrinsic limit of the diamond strength; rather, it is a geometric limit of the "optimized anvil" when the anvils are elastically deformed to a point that completely negates the bevels. As a result, further increasing of the load does not increase the pressure on the sample (Li et al., 2018). Indeed, toroid configurations have been used as modified diamond tips, the so-called "toroidal DAC," with which the maximum attainable pressure is over 500 GPa (Dewaele et al., 2018; Jenei et al., 2018). With the use of double-stage DAC, it is reported that pressure over 1 TPa can be reached (Dubrovinsky et al., 2012; Dubrovinskaia et al., 2016). However, the result from double-stage DAC is still under debate. Only one group has claimed the extended pressure range, with no other group successfully reproducing the results despite numerous attempts (Vohra et al., 2015; Sakai et al., 2020; Yagi et al., 2020).

Usually, a small beam size of ∼1/10 of the tip size is required to obtain clear XRD peaks from the high-pressure sample (Sakai et al., 2018). Whether it is a toroidal or double-stage configuration, the anvil's tip size is typically less than 10 μm in diameter. Therefore, it is essential to use a submicron X-ray beam for characterizing the sample with small tip size. In most beamlines currently dedicated for high-pressure experiments, the X-ray beam size is about 1–3 μm. It is anticipated that high-pressure beamlines with a submicron beam optimized for the high-pressure experiments using DAC will be available with the storage ring upgrade such as at the ESRF, APS, and SPring-8, thus making it possible to monitor anvil deformation and distribution of sample materials under extreme pressure. Such studies are essential, together with computational simulations, to design optimal anvil shapes for routinely generating ultrahigh pressures over 500 GPa.

The extended pressure range will enlarge the scope of high-pressure research. For example, the quest of metallic hydrogen can be directly addressed by fully determining the electronic structure and bonding of solid and fluid hydrogen at multimegabar pressures (McMahon et al., 2012). Fundamental thermodynamic properties, such as phase relations and equations of state, can be determined at a greatly extended pressure–temperature range with higher precision and accuracy using the enabling submicron X-ray probes. The extended pressure range will cover the conditions in the deep interiors of Jovian planets, super-Earths, and other exoplanets, enabling the studies of relevant materials at planets larger than Earth.

3.3.2 Amorphous Materials at High Pressure

The deepest and most interesting unsolved problem in solid state theory is probably the theory of the nature of glass and the glass transition.

—*P. W. Anderson*

Applying pressure is an effective means to test theoretical models. The effect of pressure on the structure of amorphous materials is often manifested in three ways: the modification of network structure (or intermediate-range order), the evolution of local environment (or short-range order), and the compression of interatomic distances. The observation of "halos" in XRD from liquids and glasses mean that noncrystalline systems possess distinct network structures. The pair distribution function, extracted from XRD or EXAFS, is often used for structural description and directly compared to theory and simulation results. A particular case for the effect of pressure is the occurrence of polyamorphism (i.e., multiple structures in amorphous materials with the same composition separated by a first-order-like transition) even in simple systems, such as in Si (Deb et al., 2001), H_2O (Mishima et al., 1984; Shen et al., 2019a), and SiO_2 (Hemley et al., 1988). Recently, experiments on sulfur (Henry et al., 2020) show a first-order liquid–liquid transition and an associated liquid–liquid critical point, manifested by a sharp density jump between the low- and high-density liquids and by distinct features in the pair distribution function. In bismuth, an intermediate metastable liquid phase was observed in a solid–solid transition in Bi under decompression (Lin et al., 2017), which provides an interesting process in understanding the mechanism and kinetics of solid–solid transformations. Akin to a supercooled liquid, such pressure-driven metastable liquids may be ubiquitous in systems with *V*-shaped melting lines, particularly under rapid compression or decompression routes.

One challenge in experiments on amorphous material relates to their relatively weak X-ray scattering. A recently developed opposed-type double-stage large-volume cell, with second-stage diamond anvils, allows the generation of pressures over 100 GPa with a sample volume more than ~100 times larger than that of a conventional Mbar DAC (Kono et al., 2020a). The double-stage large-volume cell has enabled studies of the structure of silicate glasses at ultrahigh pressures of >100 GPa (Kono et al., 2018, 2020b). For example, in SiO_2 glass, it is found that the Si coordination number, CN(Si), increases from 4 to 6 with increasing pressure in a range of ~15–50 GPa (Benmore et al., 2010; Sato and Funamori, 2010). However, further structural changes in CN(Si) to coordination >6 at ultrahigh-pressure conditions are still a subject of extensive debate (Sato and Funamori, 2010; Prescher et al., 2017; Lee et al., 2019; Murakami et al., 2019; Petitgirard et al., 2019a), largely because of experimental challenges in obtaining the accurate real-space functions of silicate glasses at very high pressures. With the relatively large sample volume in a double-stage cell, more accurate pair distribution functions can be obtained, revealing direct experimental evidence of CN(Si) > 6 in SiO_2 glass at pressures above 106 GPa. The obtained real-space functions also allow for calculating packing fractions of Si and O atoms, which provide robust indicators in evaluating the local structural evolution from CN(Si) = 6 to >6. Considering similarities in pressure-induced structural changes between SiO_2 glass and silicate melts, the structural results of SiO_2 glass imply that silicate melts with CN(Si) > 6 structures may have higher densities than silicate minerals that have CN (Si) = 6 throughout the Earth's mantle (see Figure 3.7). Hence, from the local structure point of view, silicate melts could be gravitationally stable in the deep mantle.

Dynamic compression experiments along the principal Hugoniot access simultaneous high-pressure high-temperature conditions that cross the melt boundary, allowing direct measurements on melts rather than analog glasses. A recent study that compared static high-pressure XRD measurements on silicate glasses in a DAC (up to 157 GPa) at room temperature with dynamic compression experiments at an X-ray FEL on silicate melts and glasses covering conditions down to the Earth's core–mantle boundary (~130 GPa, up to 6,000 K) found that they were quite similar over a large pressure–temperature range (Morard et al., 2020).

3.3.3 Transition Kinetics and Materials Metastability

Different from the traditional thermal path, which is mostly entropy driven, rate-controlled pressurization provides density-driven pathways, often involving atomic rearrangement and bonding changes, for new materials discovery and synthesis and for investigating materials metastability, phase transition mechanisms, and transformation kinetics. The development of the dDAC, using piezoelectric actuators, has enabled studies of materials behavior under rate-controlled pressurization (Evans et al., 2007; Sinogeikin et al., 2015), with controllable compression rates up to >35 TPa s^{-1}. A membrane-controlled

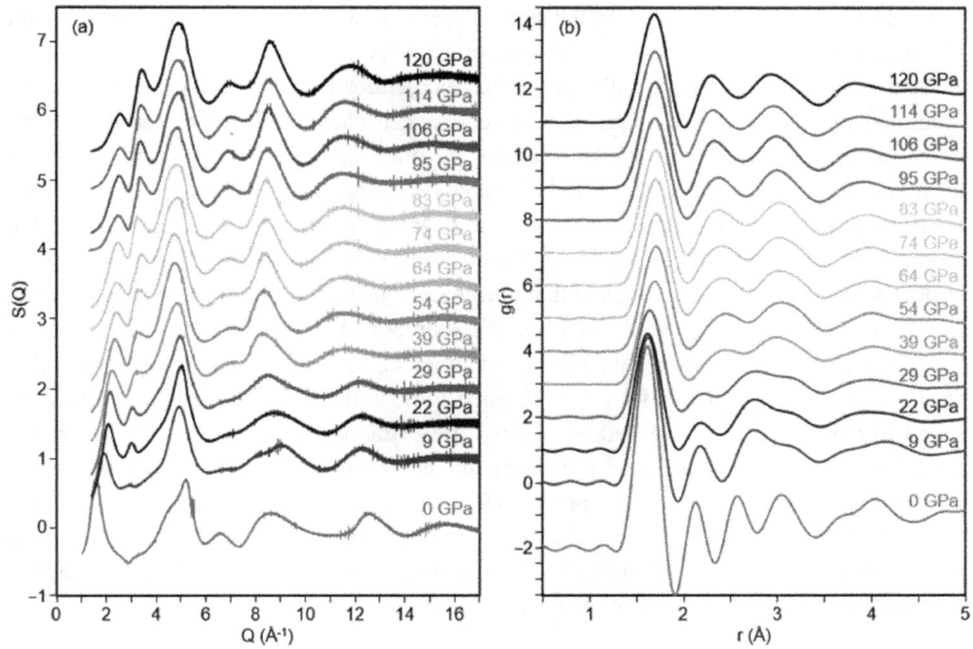

Figure 3.7 (Left) Structure factors of SiO_2 glass at high pressures up to 120 GPa, determined by EDXD using a double-stage large-volume press. (Right) The deduced pair distribution functions from the structure factor data.
After Kono et al. (2020b).

mechanism has also been used for rapid compression and decompression coupled with fast X-ray detectors, with a controllable rate from nearly 0 to ~100 GPa s^{-1} (Sinogeikin et al., 2015). The advantages of using membrane control include the compactness of the devices, the option of double-sided controls for compression and decompression, and a large range of pressure coverage.

The rate-controlled techniques have been used for studies of the nucleation of phase transitions, phase growth, and metastable phases (Haberl et al., 2015; Jacobsen et al., 2015; Konôpková et al., 2015; Lin et al., 2016; Shen et al., 2019b). With femtosecond XRD at X-ray FEL facilities, phase transition and melting kinetics have been studied under laser-driven compression for a wide range of materials systems. An investigation of SiO_2 phase transition kinetics demonstrated for the first time the crystallization from an amorphous-starting material under shock compression using in situ XRD. On compression, all starting materials of silica transformed to crystalline stishovite; upon release of the shock wave, however, diffraction showed amorphization to an intermediate- and high-density amorphous phase. Time-resolved XRD observed the nucleation and growth of nanocrystalline stishovite grains and determined the grain size as a function of time after shock loading (Gleason et al., 2015, 2017b). Additional examples of dynamic studies at X-ray FELs include the freezing kinetics of H_2O (Gleason et al., 2017a), time-resolved observations of melting and phase transitions in shock-compressed bismuth (Gorman et al., 2015; Gorman et al., 2018), and a recent report on the structural evolution of shock-compressed iron with ~50 picosecond temporal resolution (Hwang et al., 2020).

Many functional materials with open crystal structures (e.g., tetrahedral coordination) often undergo phase transformations to densely packed phases under compression. Under decompression, however, the transformations are often irreversible and the material displays rich metastability depending on the applied decompression rate (Haberl et al., 2016). Pressure-induced amorphization occurs in a material when it is brought outside the thermodynamic stability field below the glass transition temperature (T_g) by compression or decompression. It has been argued that vitrification generally occurs when an isothermal (de)compression path below T_g crosses the extrapolated melting curve in pressure–temperature space (Mishima et al., 1984; Hemley et al., 1988). It is interesting to note that at temperatures above T_g under rate-controlled (de)compressions, metastable liquids are found to exist when pressure pathways cross the extrapolated melting curve (Lin et al., 2017, 2018). The rate-controlled pressurization has become a key technique to access pressure–temperature regions that previously were unreachable for studying metastable materials (Figure 3.8) (Shen et al., 2019b), providing a potentially advantageous pathway for investigating phase transition mechanisms, kinetics, and materials metastability, and for discovery and synthesis of new materials.

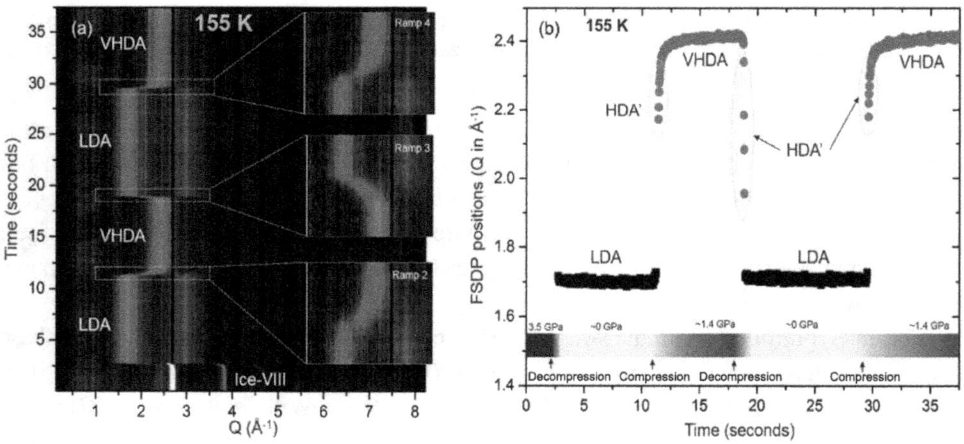

Figure 3.8 (a) X-ray diffraction data of H_2O under multistage ramps of rapid compression and decompression processes at 155.2 ± 0.1 K as a function of time, with the diffraction intensity represented in "temperature" scale. The polyamorphic transformations from low-density amorphous (LDA) to very-high-density amorphous (VHDA) and the reverse transformations from VHDA to LDA are repeatable and manifested by large shifts in the first sharp diffraction peak (FSDP). The inset figures are the enlarged views near the loading or unloading ramps, displaying an intermediate phase of high-density amorph (HDA′). (b) The FSDP positions of amorphous H_2O. Pressure conditions are marked in the bar below the FSDP data. The highlighted areas represent HDA′ with rapidly, but continuously, changing FSDP positions. Noticeable gaps in FSDP can be seen between LDA and HDA′ with larger gaps under compression condition.
After Shen et al. (2019b).

3.4 Outlook on Future Developments

We are in the midst of numerous exciting opportunities for advancing high-pressure X-ray studies, including vast improvements in accelerator-based X-ray sources, novel high-pressure technologies, advanced X-ray optics and detectors, and new lessons from machine learning, including effective data mining and synergetic integrations with theoretical computations. Particularly, a new avenue for creating high-brightness X-ray beams has been adopted in several storage ring-based synchrotron sources worldwide, based on the use of multibend achromat (MBA) magnet lattices. Further, the new developments in FEL facilities offer increased X-ray energies well suited for high-pressure studies. Indeed, we are looking at a new era in high-pressure X-ray science, one that promises an in-depth understanding of matter under even more extreme conditions across a much broader hierarchy of length and time scales.

3.4.1 High-Pressure Research at MBA Storage Ring Facilities

One important characteristic of MBA-based storage rings is the significantly reduced horizontal emittance, resulting in an ideal "round" X-ray beam for developing powerful new capabilities for high-pressure experiments. The MBA sources will provide an increase of two to three orders of magnitude in brightness and coherence of high-energy X-rays, enabling advanced X-ray probes with orders of magnitude improvements in both spatial and temporal resolution.

The low emittance and high brightness of X-ray beams will dramatically improve the ability to study smaller samples and faster processes with greater precision and accuracy. Ultrahigh pressure frontiers will be extensively explored using ever-smaller X-ray beams for peak pressures with precisely mapped stress gradients at multimegabar pressures. Coupled with the advancement in rapid (de)compression and modulated laser heating (and associated quenching techniques), the enhanced temporal resolution will broadly address the dynamic processes and properties of materials, including phase transformation processes (nucleation, growth, metastable, or intermediate phases), new forms of metastable phases via nonequilibrium transformations (e.g., superheating/cooling, over-/underpressurization), and underlying details revealing transition hysteresis and incomplete transformation.

The high coherence fraction will allow new coherence-based X-ray techniques for high-pressure studies: CDI and X-ray ptychography for quantitative 3D images of the real and imaginary contributions, yielding information about magnetization, composition, bonding configuration, and strain; and X-ray photon correlation spectroscopy (XPCS) techniques for studying the dynamics of various equilibrium and nonequilibrium processes at high pressure.

Perhaps the biggest changes will come from the advanced capabilities of nanometer-spatial resolution to examine in situ the hierarchical structures from the electron level to the macroscopic level, and to study how they change in time and in response to external pressure–temperature conditions. Property-specific 3D images, such as composition, phase, valence, spin, strain, and shape, will be measured by mapping high-pressure samples using the small X-ray beam with the corresponding highly efficient X-ray probes or by applying coherent imaging techniques. Total crystallography on complex, bulk samples will reveal crystallographic data for individual grains, regarding their positions, crystal orientations, lattice distortions, thermal parameters, and even detailed atomic positions.

3.4.2 High-Pressure Research at X-Ray FELs

High-pressure research at X-ray FELs need not be limited to laser-shock experiments. The main requirement for enabling DAC studies to be compatible with X-ray FELs is higher photon energies to penetrate the DAC. In 2019, the first DAC experiments using an X-ray FEL were demonstrated at the high-energy density (HED) instrument at the EuXFEL which can generate photon energies >20 keV. The LCLS-II upgrade of LCLS is poised to provide ultrafast X-rays at a much higher repetition rate of 1 MHz (compared to 120 Hz for LCLS) but only up to 5 keV. A further extension of LCLS-II, i.e., the high-energy upgrade (LCLS-II HE) will push the X-ray energies to ~13.5 keV (with higher energies attainable by using higher harmonics – although at the expense of flux). LCLS-II HE will not only provide hard X-rays with high peak brightness, but also with unprecedented average brightness far exceeding existing storage rings and planned diffraction limited storage rings. This offers exciting opportunities for ultrafast DAC studies. LCLS-II HE is planned for 2027–2028. An initial high energy upgrade to >20 keV using the Cu-Linac (120 Hz) and the new LCLS-II undulators will be available in early 2021, providing initial opportunities for DAC experiments to be coupled to LCLS instruments.

Acknowledgments

GS acknowledges the support of HPCAT operations by Department of Energy (DOE)-NNSA's Office of Experimental Sciences. WM acknowledges support by the Department of Energy, Office of Science, Basic Energy Sciences, Materials Sciences and Engineering Division, under Contract DE-AC02–76SF00515.

References

Alatas, A., Sinn, H., Zhao, J. Y., et al. (2008). Experimental aspects of inelastic X-ray scattering studies on liquids under extreme conditions (P–T). *High Pressure Research*, **28**, 175–183.

Alonso-Mori, R., Kern, J., Gildea, R. J., et al. (2012). Energy-dispersive X-ray emission spectroscopy using an X-ray free-electron laser in a shot-by-shot mode. *Proceedings of the National Academy of Science.*, **109**, 19103–19107.

Alp, E. E., Sturhahn, W., Toellner, T. S., Zhao, J., Hu, M., Brown, D. E. (2002). Vibrational dynamics studies by nuclear resonant inelastic x-ray scattering. *Hyperfine Interactions*, **144**, 3-20.

Andrault, D., Itie, J. P., Poirier, J. P. (1988). High-pressure synchrotron radiation EXAFS study of perovskites. *Chemical Geology*, **70**, 60–60.

Antonangeli, D., Siebert, J., Aracne, C. M., et al. (2011). Spin crossover in ferropericlase at high pressure: a seismologically transparent transition? *Science*, **331**, 64–67.

Antonangeli, D., Siebert, J., Badro, J., Farber, D. L., Fiquet, G., Morard, G., Ryerson, F. J. (2010). Composition of the Earth's inner core from high-pressure sound velocity measurements in Fe–Ni–Si alloys. *Earth and Planet. Science Letters*, **295**, 292–296.

Anzellini, S., Dewaele, A., Mezouar, M., Loubeyre, P., Morard, G. (2013). Melting of iron at Earth's inner core boundary based on fast X-ray diffraction. *Science*, **340**, 464–466.

Aquilanti, G., Pascarelli, S., Mathon, O., Munoz, M., Narygina, O., Dubrovinsky, L. (2009). Development of micro-XANES mapping in the diamond anvil cell. *Journal of Synchrotron Radiation*, **16**, 376–379.

Aquilanti, G., Trapananti, A., Karandikar, A., et al. (2015). Melting of iron determined by X-ray absorption spectroscopy to 100 GPa. *Proceedings of the National Academy of Science.*, **112**, 12042–12045.

Badro, J., Fiquet, G., Guyot, F., et al. (2003). Iron partitioning in Earth's mantle: toward a deep lower mantle discontinuity. *Science*, **300**, 789–791.

Badro, J., Fiquet, G., Struzhkin, V. V., et al. (2002). Nature of the high-pressure transition in Fe_2O_3 hematite. *Physical Review Letters*, **89**, 205504.

Badro, J., Struzhkin, V. V., Shu, J., et al. (1999). Magnetism in FeO at megabar pressures from X-ray emission spectroscopy. *Physical Review Letters*, **83**, 4101–4104.

Baldini, M., Yang, W., Aquilanti, G., et al. (2011). High-pressure EXAFS measurements of crystalline Ge using nanocrystalline diamond anvils. *Physical Review B*, **84**, 014111.

Baron, A. Q. R., Tanaka, Y., Miwa, D., et al. (2001). Early commissioning of the SPring-8 beamline for high resolution inelastic X-ray scattering. *Nuclear Instruments and Methods in Physics Research A*, **467–468**, *Part 1*, 627–630.

Bassett, W. A. (1980). Synchrotron radiation, an intense X-ray source for high pressure diffraction studies. *Physics of the Earth and Planetary Interiors*, **23**, 337–340.

Bassett, W. A., Anderson, A. J., Mayanovic, R. A., Chou, I. M. (2000). Modified hydrothermal diamond anvil cells for XAFS analyses of elements with low energy absorption edges in aqueous solutions at sub- and supercritical conditions. *Z. Kristall.*, **215**, 711–717.

Baublitz, J. M. A., Arnold, V., Ruoff, A. L. (1981). Energy dispersive X-ray diffraction from high pressure polycrystalline specimens using synchrotron radiation. *Review of Scientific Instruments*, **52**, 1616–1624.

Benedict, U., Grosshans, W. A., Holzapfel, W. B. (1986). Systematics of electron delocalization in lanthanide and actinide elements under pressure. *Physica B+C*, **144**, 14–18.

Benmore, C. J., Soignard, E., Amin, S. A., et al. (2010). Structural and topological changes in silica glass at pressure. *Physical Review B*, **81**, 054105.

Booth, C. H., Jiang, Y., Wang, D. L., et al. (2012). Multiconfigurational nature of 5f orbitals in uranium and plutonium intermetallics. *Proceedings of the National Academy of Science.*, **109**, 10205–10209.

Bosak, A., Hoesch, M., Antonangeli, D., Farber, D. L., Fischer, I., Krisch, M. (2008). Lattice dynamics of vanadium: Inelastic X-ray scattering measurements. *Physical Review B*, **78**, 020301.

Bostedt, C., Boutet, S., Fritz, D. M., et al. (2016). Linac coherent light source: the first five years. *Review of Modern Physics*, **88**, 015007.

Bradley, J. A., Moore, K. T., Lipp, M. J., et al. (2012). 4f electron delocalization and volume collapse in praseodymium metal. *Physical Review B*, **85**, 100102.

Bridgman, P. W. (1909). The measurement of high hydrostatic pressure. I. A simple primary gauge. *Proceedings of the American Academy of Arts and Sciences*, **44**, 201–217.

Buras, B., Olsen, J. S., Gerward, L., Will, G., Hinze, E. (1977). X-ray energy-dispersive diffractometry using synchrotron radiation. *Journal of Applied Crystallography*, **10**, 431–438.

Burkel, E., Peisl, J., Dorner, B. (1987). Observation of inelastic X-ray scattering from phonons. *Europhysics Letters*, **3**, 957.

Bykov, M., Bykova, E., Aprilis, G., et al. (2018). Fe-N system at high pressure reveals a compound featuring polymeric nitrogen chains. *Nature Communications*, **9**, 2756.

Cai, Y. Q., Mao, H. K., Chow, P. C., et al. (2005). Ordering of hydrogen bonds in high-pressure low-temperature H_2O. *Physics Review Letters*, **94**, 025502.

Catalli, K., Shim, S.-H., Dera, P., et al. (2011). Effects of the Fe^{3+} spin transition on the properties of aluminous perovskite – new insights for lower-mantle seismic heterogeneities. *Earth and Planetary Science Letters*, *310*, 293–302.

Chefki, M., Abd-Elmeguid, M. M., Micklitz, H., et al. (1998). Pressure-induced transition of the sublattice magnetization in $EuCo_2P_2$: change from local moment Eu(4f) to itinerant Co(3d) magnetism. *Physical Review Letters*, **80**, 802–805.

Chen, B., Jackson, J. M., Sturhahn, W., et al. (2012). Spin crossover equation of state and sound velocities of $(Mg_{0.65}Fe_{0.35})O$ ferropericlase to 140 GPa. *Journal of Geophysical Research*, **117**, B08208.

Chen, D. Z., Shi, C. Y., An, Q., et al. (2015). Fractal atomic-level percolation in metallic glasses. *Science*, **349**, 1306–1310.

Deb, S. K., Wilding, M., Somayazulu, M., McMillan, P. F. (2001). Pressure-induced amorphization and an amorphous-amorphous transition in densified porous silicon. *Nature*, **414**, 528–530.

Dewaele, A., Loubeyre, P., Occelli, F., Marie, O., Mezouar, M. (2018). Toroidal diamond anvil cell for detailed measurements under extreme static pressures. *Nature Communications*, **9**, 2913.

Ding, Y., Haskel, D., Ovchinnikov, S. G., et al. (2008). Novel pressure-induced magnetic transition in magnetite (Fe_3O_4). *Physical Review Letters.*, **100**, 045508.

Dubrovinskaia, N., Dubrovinsky, L., Solopova, N. A., et al. (2016). Terapascal static pressure generation with ultrahigh yield strength nanodiamond. *Science Advances*, **2**, e1600341.

Dubrovinsky, L., Boffa-Ballaran, T., Glazyrin, K., et al. (2010). Single-crystal X-ray diffraction at megabar pressures and temperatures of thousands of degrees. *High Pressure Research*, **30**, 620–633.

Dubrovinsky, L., Dubrovinskaia, N., Prakapenka, V. B., Abakumov, A. M. (2012). Implementation of micro-ball nanodiamond anvils for high-pressure studies above 6 Mbar. *Nature Communications*, **3**, 1163.

Duffy, T. S. (2005). Synchrotron facilities and the study of the Earth's deep interior. *Reports on Progress in Physics*, **68**, 1811–1859.

Evans, W. J., Yoo, C.-S., Lee, G. W., Cynn, H., Lipp, M. J., Visbeck, K. (2007). Dynamic diamond anvil cell (dDAC): a novel device for studying the dynamic-pressure properties of materials. *Review of Scientific Instruments*, **78**, 073904.

Falconi, S., Monaco, G., Crichton, W. A., Mezouar, M., Verbeni, R. (2004). Collimator for inelastic X-ray scattering experiments at high temperature and pressure conditions. *High Pressure Research*, **24**, 463–469.

Farber, D. L., Krisch, M., Antonangeli, D., et al. (2006). Lattice dynamics of molybdenum at high pressure. *Physical Review Letters*, **96**, 115502.

Fei, Y. W., Mao, H. K. (1993). Static compression of $Mg(OH)_2$ to 78-GPa at high-temperature and constraints on the equation of state of fluid H_2O. *Journal of Geophysical Research–Solid Earth*, **98**, 11875–11884.

Fei, Y. W., Mao, H. K. (1994). In-situ determination of the NiAs phase of FeO at high-pressure and temperature. *Science*, **266**, 1678–1680.

Fei, Y. W., Mao, H. K., Shu, J. F., Parthasarathy, G., Bassett, W. A., Ko, J. D. (1992). Simultaneous high-P, high-T X-ray-diffraction study of beta-$(Mg,Fe)_2SiO_4$ to 26-GPa and 900-K. *Journal of Geophysical Research–Solid Earth*, **97**, 4489–4495.

Feng, Z. D., Zhou, Y. H., Tan, R., et al. (2018). Dynamic damage and fracture of a conductive glass under high-rate compression: a synchrotron based study. *Journal of Non-crystalline Solids*, **494**, 40–49.

Fiquet, G., Badro, J., Guyot, F., Requardt, H., Krisch, M. (2001). Sound velocities in iron to 110 Gigapascals. *Science*, **291**, 468–471.

Fister, T. T., Nagle, K. P., Vila, F. D., et al. (2009). Intermediate-range order in water ices: nonresonant inelastic X-ray scattering measurements and real-space full multiple scattering calculations. *Physical Review B*, **79**, 174117.

Fleet, M. E., Herzberg, C. T., Henderson, G. S., Crozier, E. D., Osborne, M. D., Scarfe, C. M. (1984). Coordination of Fe, Ga and Ge in high pressure glasses by Mössbauer, Raman and X-ray absorption spectroscopy, and geological implications. *Geochim. Cosmochim. Acta*, **48**, 1455–1466.

Fukui, H., Sakai, T., Sakamaki, T., et al. (2013). A compact system for generating extreme pressures and temperatures: an application of laser-heated diamond anvil cell to inelastic X-ray scattering. *Review of Scientific Instruments*, **84**(11), 10.1063/1.4826497.

Gallo, E., Glatzel, P. (2014). Valence to core X-ray emission spectroscopy. *Advanced Materials*, **26**, 7730–7746.

Gleason, A. E., Bolme, C. A., Galtier, E., et al. (2017a). Compression freezing kinetics of water to ice VII. *Physical Review Letters*, **119**, 025701.

Gleason, A. E., Bolme, C. A., Lee, H. J., et al. (2017b). Time-resolved diffraction of shock-released SiO2 and diaplectic glass formation. *Nature Communications*, **8**, 1481.

Gleason, A. E., Bolme, C. A., Lee, H. J., et al. (2015). Ultrafast visualization of crystallization and grain growth in shock-compressed SiO_2. *Nature Communications*, **6**(1), 8191.

Goncharov, A. F., Prakapenka, V. B., Struzhkin, V. V., Kantor, I., Rivers, M. L., Dalton, D. A. (2010). X-ray diffraction in the pulsed laser heated diamond anvil cell. *Review of Scientific Instruments*, **81**, 113902.

Gorman, M. G., Briggs, R., McBride, E. E., et al. (2015). Direct observation of melting in shock-compressed bismuth with femtosecond X-ray diffraction. *Physical Review Letters*, **115**, 095701.

Gorman, M. G., Coleman, A. L., Briggs, R., et al. (2018). Femtosecond diffraction studies of solid and liquid phase changes in shock-compressed bismuth. *Scientific Reports*, **8**, 16927.

Haberl, B., Guthrie, M., Sinogeikin, S. V., Shen, G., Williams, J. S., Bradby, J. E. (2015). Thermal evolution of the metastable r8 and bc8 polymorphs of silicon. *High Pressure Research.*, **35**, 99–116.

Haberl, B., Strobel, T. A., Bradby, J. E. (2016). Pathways to exotic metastable silicon allotropes. *Applied Physics Reviews*, **3**, 040808.

Harder, R., Robinson, I. K. (2013). Coherent X-ray diffraction imaging of morphology and strain in nanomaterials. *JOM*, **65**, 1202–1207.

Hausermann, D., Hanfland, M. (1996). Optics and beamlines for high-pressure research at the European synchrotron radiation facility. *High Press. Research*, **14**, 223–234.

Hemley, R. J., Jephcoat, A. P., Mao, H. K., Ming, L. C., Manghnani, M. H. (1988). Pressure-induced amorphization of crystalline silica. *Nature*, **334**, 52–54.

Henry, L., Mezouar, M., Garbarino, G., Sifré, D., Weck, G., Datchi, F. (2020). Liquid–liquid transition and critical point in sulfur. *Nature*, **584**, 382–386.

Hirao, N., Kawaguchi, S. I., Hirose, K., Shimizu, K., Ohtani, E., Ohishi, Y. (2020). New developments in high-pressure X-ray diffraction beamline for diamond anvil cell at SPring-8. *Matter and Radiation at Extremes*, **5**, 018403.

Hong, X. G., Newville, M., Prakapenka, V. B., Rivers, M. L., Sutton, S. R. (2009). High quality X-ray absorption spectroscopy measurements with long energy range at high pressure using diamond anvil cell. *Review of Scientific Instruments*, **80**(7), 073908.

Hrubiak, R., Smith, J. S., Shen, G. (2019). Multimode scanning X-ray diffraction microscopy for diamond anvil cell experiments. *Review of Scientific Instruments*, **90**, 025109.

Hruszkewycz, S. O., Cha, W., Andrich, P., et al. (2017). In situ study of annealing-induced strain relaxation in diamond nanoparticles using Bragg coherent diffraction imaging. *APL Materials*, **5**, 26105.

Huang, X., Yang, W., Harder, R., et al. (2015). Deformation twinning of a silver nanocrystal under high pressure. *Nano Letters*, **15**, 7644–7649.

Hwang, H., Galtier, E., Cynn, H., et al. (2020). Subnanosecond phase transition dynamics in laser-shocked iron. *Science Advances*, **6**, eaaz5132.

Ingalls, R., Garcia, G. A., Stern, E. A. (1978). X-ray absorption at high pressure. *Physical Review Letters*, **40**, 334.

Ishimatsu, N., Matsumoto, K., Maruyama, H., et al. (2012). Glitch-free X-ray absorption spectrum under high pressure obtained using nano-polycrystalline diamond anvils. *Journal of Synchrotron Radiation*, **19**, 768-772.

Itié, J. P., Girard, E., Guignot, N., Le Godec, Y., Mezouar, M. (2015). Crystallography under high pressure using synchrotron radiation. *Journal of Physics D: Applied Physics*, **48**, 504007.

Itié, J. P., Polian, A., Calas, G., Petiau, J., Fontaine, A., Tolentino, H. (1989). Pressure-induced coordination changes in crystalline and vitreous GeO_2. *Physical Review Letters*, **63**, 398.

Jackson, J. M., Sturhahn, W., Lerche, M., et al. (2013a). Melting of compressed iron by monitoring atomic dynamics. *Earth and Planetary Science Letters*, **362**, 143–150.

Jackson, J. M., Sturhahn, W., Lerche, M., et al. (2013b). Melting of compressed iron by monitoring atomic dynamics. *Earth and Planetary Science Letters*, **362**, 143–150.

Jackson, J. M., Sturhahn, W., Shen, et al. (2005). A synchrotron Mossbauer spectroscopy study of (Mg,Fe)SiO_3 perovskite up to 120 GPa. *American Mineralogist*, **90**, 199–205.

Jacobsen, M. K., Velisavljevic, N., Sinogeikin, S. V. (2015). Pressure-induced kinetics of the α to ω transition in zirconium. *Journal of Applied Physics*, **118**, 025902.

Jamieson, J. C., Lawson, A. W., Nachtrieb, N. D. (1959). New device for obtaining X-ray diffraction patterns from substances exposed to high pressure. *Review of Scientific Instruments*, **30**, 1016–1019.

Jenei, Z., O'Bannon, E. F., Weir, S. T., Cynn, H., Lipp, M. J., Evans, W. J. (2018). Single crystal toroidal diamond anvils for high pressure experiments beyond 5 megabar. *Nature Communications*, **9**, 3563.

Jensen, B. J., Luo, S. N., Hooks, D. E., et al. (2012). Ultrafast, high resolution, phase contrast imaging of impact response with synchrotron radiation. *AIP Advances*, **2**, 012170.

Kamada, S., Ohtani, E., Fukui, H., et al. (2014). The sound velocity measurements of Fe_3S. *American Mineralogist.*, **99**, 98–101.

Kikegawa, T., Iwasaki, H., Fujimura, T., et al. (1987). Synchrotron-radiation study of phase-transitions in phosphorus at high-pressures and temperatures. *Journal of Applied Crystallography*, **20**, 406–410.

Kim, Y. H., Yi, Y. S., Kim, H. I., et al. (2019). Structural transitions in MgSiO3 glasses and melts at the core–mantle boundary observed via inelastic X-ray scattering. *Geophysical Research Letters.*, **46**, 13756–13764.

Kono, Y., Kenney-Benson, C., Shen, G. (2020a). Opposed type double stage cell for Mbar pressure experiment with large sample volume. *High Pressure Research*, **40**, 175–183.

Kono, Y., Kenney-Benson, C., Shibazaki, Y., Park, C., Wang, Y. B., Shen, G. Y. (2015). X-ray imaging for studying behavior of liquids at high pressures and high temperatures using Paris–Edinburgh press. *Review of Scientific Instruments*, **86**(7), 072207.

Kono, Y., Shibazaki, Y., Kenney-Benson, C., Wang, Y., Shen, G. (2018). Pressure-induced structural change in MgSiO₃ glass at pressures near the Earth's core–mantle boundary. *Proceedings of the National Academy of Science.*, **115**, 1742–1747.

Kono, Y., Shu, Y., Kenney-Benson, C., Wang, Y., Shen, G. (2020b). Structural evolution of SiO_2 glass with Si coordination number greater than 6. *Physical Review Letters*, **125**, 205701.

Konôpková, Z., Rothkirch, A., Singh, A. K., Speziale, S., Liermann, H.-P. (2015). In situ X-ray diffraction of fast compressed iron: analysis of strains and stress under non-hydrostatic pressure. *Physics Review B*, **91**, 144101.

Krisch, M., Farber, D. L., Xu, R., et al. (2011). Phonons of the anomalous element cerium. *Proceedings of the National Academy of Science*, **108**, 9342–9345.

Kunz, M., Caldwell, W. A., Clark, S. M., et al. (2005a). ALS beamline 12.2.2, a high-pressure X-ray user facility at the US-West-Coast. *Acta Crystallographica A*, **61**, C374–C374.

Kunz, M., MacDowell, A. A., Caldwell, W. A., et al. (2005b). A beamline for high-pressure studies at the Advanced Light Source with a superconducting bending magnet as the source. *Journal of Synchrotron Radiation*, **12**, 650–658.

Kvashnina, K. O., Kvashnin, Y. O., Butorin, S. M. (2014). Role of resonant inelastic X-ray scattering in high-resolution core-level spectroscopy of actinide materials. *Journal of Electron Spectroscopy and Related Phenomena*, **194**, 27–36.

Langrand, C., Hilairet, N., Nisr, C., et al. (2017). Reliability of multigrain indexing for orthorhombic polycrystals above 1 Mbar: application to MgSiO3 post-perovskite. *Journal of Applied Crystallography*, **50**, 120–130.

Lausi, A., Polentarutti, M., Onesti, S., et al. (2015). Status of the crystallography beamlines at Elettra. *European Physical Journal Plus*, **130**, 43.

Lavina, B., Dera, P., Downs, R. T. (2014). Modern X-ray diffraction methods in mineralogy and geosciences. *Reviews in Mineralogy and Geochemistry.*, **78**, 1–31.

Le Bolloch, D., Itie, J. P., Polian, A., Ravy, S. (2009). Combining high pressure and coherent diffraction: a first feasibility test. *High Pressure Research*, **29,** 635--638. https://doi.org/10.1080/08957950903421693.

Lee, S. K., Eng, P. J., Mao, H. K., Shu, J. F. (2008). Probing and modeling of pressure-induced coordination transformation in borate glasses: Inelastic X-ray scattering study at high pressure. *Physical Review B*, **78**, 214203.

Lee, S. K., Kim, Y. H., Chow, P., Xiao, Y. M., Ji, C., Shen, G. Y. (2018). Amorphous boron oxide at megabar pressures via inelastic X-ray scattering. *Proceedings of the National Academy of Sciences of the United States of America*, **115**, 5855–5860.

Lee, S. K., Kim, Y. H., Yi, Y. S., et al. (2019). Oxygen quadclusters in SiO_2 glass above megabar pressures up to 160 GPa revealed by X-ray Raman scattering. *Physical Review Letters*, **123**(23), 235701.

Lesher, C. E., Wang, Y. B., Gaudio, S., Clark, A., Nishiyama, N., Rivers, M. (2009). Volumetric properties of magnesium silicate glasses and supercooled liquid at high pressure by X-ray microtomography. *Physics of the Earth and Planetary Interiors*, **174**, 292–301.

Li, B., Ji, C., Yang, W., et al. (2018). Diamond anvil cell behavior up to 4 Mbar. *Proceedings of the National Academy of Science*, **115**, 1713–1717.

Li, J., Struzhkin, V. V., Mao, H.-k., et al. (2004). Electronic spin state of iron in lower mantle perovskite. *Proceedings of the National Academy Science.*, **101**, 14027–14030.

Li, R., Liu, J., Bai, L., Tse, J. S., Shen, G. (2015). Pressure-induced changes in the electron density distribution in α-Ge near the α-β transition. *Applied Physics Letters*, **107**, 072109.

Li, R., Liu, J., Popov, D., Park, C., Meng, Y., Shen, G. (2019). Experimental observations of large changes in electron density distributions in β-Ge. *Physical Review B*, **100**, 224106.

Liermann, H. P., Konopkova, Z., Morgenroth, W., et al. (2015). The extreme conditions beamline P02.2 and the extreme conditions science infrastructure at PETRA III. *Journal of Synchrotron Radiation*, **22**, 908–924.

Lima, F. A., Saleta, M. E., Pagliuca, R. J. S., et al. (2016). XDS: a flexible beamline for X-ray diffraction and spectroscopy at the Brazilian synchrotron. *Journal of Synchrotron Radiation*, **23**, 1538–1549.

Lin, C., Smith, J. S., Sinogeikin, S. V., et al. (2016). Kinetics of the B1-B2 phase transition in KCl under rapid compression. *Journal of Applied Physics*, **119**, 045902.

Lin, C., Smith, J. S., Sinogeikin, S. V., et al. (2017). A metastable liquid melted from a crystalline solid under decompression. *Nature Communications*, **8**, 14260.

Lin, C., Smith, J. S., Sinogeikin, S. V., Shen, G. (2018). Experimental evidence of low-density liquid water upon rapid decompression. *Proceedings of National Academy Science.*, **115**, 2010–2015.

Lin, J.-F., Alp, E. E., Mao, Z., et al. (2012). Electronic spin states of ferric and ferrous iron in the lower-mantle silicate perovskite. *American Mineralogist*, **97**, 592–597.

Lin, J.-F., Fukui, H., Prendergast, D., et al. (2007). Electronic bonding transition in compressed SiO2 glass. *Physical Review B*, **75**, 012201.

Lin, J.-F., Struzhkin, V. V., Jacobsen, S. D., et al. (2005a). Spin transition of iron in magnesiowüstite in the Earth's lower mantle. *Nature*, **436**, 377–380.

Lin, J.-F., Struzhkin, V. V., Jacobsen, S. D., et al. (2005b). X-ray emission spectroscopy with a laser-heated diamond anvil cell: a new experimental probe of the spin state of iron in the Earth's interior. *Journal of Synchrotron Radiation*, **12**, 637–641.

Lin, J. F., Sturhahn, W., Zhao, J., Shen, G., Mao, H. K., Hemley, R. J. (2005c). Sound velocities of hot dense iron: Birch's law revisited. *Science*, **308**, 1892–1894.

Lin, Y., Zeng, Q., Yang, W., Mao, W. L. (2013). Pressure-induced densification in GeO2 glass: a transmission X-ray microscopy study. *Applied Physics Letters*, **103**, 261909.

Lipp, M. J., Sorini, A. P., Bradley, J., et al. (2012). X-ray emission spectroscopy of cerium across the γ-α volume collapse transition. *Physical Review Letters*, **109**, 195705.

Liu, H., Wang, L., Xiao, X., et al. (2008). Anomalous high-pressure behavior of amorphous selenium from synchrotron X-ray diffraction and microtomography. *Proceedings of the National Academy of Science.*, **105**, 13229–13234.

Liu, J. (2016). High pressure X-ray diffraction techniques with synchrotron radiation. *Chinese Physics B*, **25**, 076106.

Liu, Y., Wang, J., Azuma, M., Mao, W. L., Yang, W. (2014). Five-dimensional visualization of phase transition in BiNiO3 under high pressure. *Applied Physics Letters*, **104**, 043108.

Loubeyre, P., LeToullec, R., Hausermann, D., et al. (1996). X-ray diffraction and equation of state of hydrogen at megabar pressures. *Nature*, **383**, 702–704.

Lübbers, R., Grünsteudel, H. F., Chumakov, A. I., Wortmann, G. (2000). Density of states in iron at high pressure. *Science*, **287**, 1250–1253.

Luo, S. N., Jensen, B. J., Hooks, D. E., et al. (2012). Gas gun shock experiments with single-pulse X-ray phase contrast imaging and diffraction at the Advanced Photon Source. *Review of Scientific Instruments*, **83**, 073903.

Manley, M. E., Lander, G. H., Sinn, H., et al. (2003). Phonon dispersion in uranium measured using inelastic X-ray scattering. *Physical Review B*, **67**, 052302.

Mao, H. K., Hemley, R. J., Chen, L. C., Shu, J. F., Finger, L. W., Wu, Y. (1989). X-ray diffraction to 302 gigapascals: high-pressure crystal structure of cesium iodide. *Science*, **246**, 649–651.

Mao, H. K., Hemley, R. J., Fei, Y., et al. (1991). Effect of pressure, temperature, and composition on lattice-parameters and density of (Fe,Mg)SiO_3-perovskites to 30 GPa. *Journal of Geophysical Research–Solid Earth and Planets*, **96**, 8069–8079.

Mao, H. K., Jephcoat, A. P., Hemley, R. J., et al. (1988). Synchrotron X-ray diffraction measurements of single-crystal hydrogen to 26.5 gigapascals. *Science*, **239**, 1131–1134.

Mao, H. K., Mao, W. L. (2007). Diamond-anvil cells and probes for high P–T mineral physics, in G. D. Price, ed., *Treatise on Geophysics*, Elsevier, pp. 231–268.

Mao, H. K., Shen, G. Y., Hemley, R. J. (1997). Multivariable dependence of Fe-Mg partitioning in the lower mantle. *Science*, **278**, 2098–2100.

Mao, H. K., Shen, G., Hemley, R. J., Duffy, T. S. (1998). X-ray diffraction with a double hot plate laser heated diamond cell, in M. H. Manghnani, T. Yagi, eds., *Properties of Earth and Planetary Materials*, AGU, pp. 27–34.

Mao, H. K., Wu, Y., Chen, L. C., Shu, J. F., Jephcoat, A. P. (1990). Static compression of iron to 300 GPa and $Fe_{0.8}Ni_{0.2}$ alloy to 260 GPa: implications for composition of the core. *Journal of Geophysical Research*, **95**, 21737–21742.

Mao, H. K., Xu, J., Struzhkin, V., et al. (2001). Phonon density of states of iron up to 153 gigapascals. *Science*, **292**, 914–916.

Mao, W. L., Lin, Y., Liu, Y., Liu, J. (2019). Applications for nanoscale X-ray imaging at high pressure. *Engineering*, **5**, 479–489.

Mao, W. L., Mao, H. K., Eng, P. J., et al. (2003). Bonding changes in compressed superhard graphite. *Science*, **302**, 425–427.

Mao, W. L., Shen, G., Prakapenka, V. B., et al. (2004). Ferromagnesian postperovskite silicates in the D" layer of the Earth. *Proceedings of the National Academy of Science*, **101**, 15867–15869.

Mao, Z., Lin, J.-F., Liu, J., et al. (2012). Sound velocities of Fe and Fe-Si alloy in the Earth's core. *Proceedings of the National Academy of Science*., **109**, 10239–10244.

Masuda, R., Kobayashi, Y., Kitao, S., Kurokuzu, et al. (2016). Synchrotron radiation based Mössbauer absorption spectroscopy of various nuclides. *Hyperfine Interactions*, **237**, 43.

Mayanovic, R. A., Anderson, A. J., Bassett, W. A., Chou, I. M. (2007). Synchrotron X-ray spectroscopy of Eu/HNO3 aqueous solutions at high temperatures and pressures and Nb-bearing silicate melt phases coexisting with hydrothermal fluids using a modified hydrothermal diamond anvil cell and rail assembly. *Review of Scientific Instruments*, **78**, 053108.

McBride, E. E., White, T. G., Descamps, A., et al. (2018). Setup for meV-resolution inelastic X-ray scattering measurements and X-ray diffraction at the Matter in Extreme Conditions endstation at the Linac Coherent Light Source. *Review of Scientific Instruments*, **89**, 10F104.

McMahon, J. M., Morales, M. A., Pierleoni, C., Ceperley, D. M. (2012). The properties of hydrogen and helium under extreme conditions. *Review of Modern Physics*, **84**, 1607–1653.

McMahon, M. I. (2018). Synchrotron and FEL studies of matter at high pressures, in E. Jaeschke, S. Khan, J. R. Schneider, J. B. Hastings, eds., *Synchrotron Light Sources and Free-Electron Lasers: Accelerator Physics, Instrumentation and Science Applications*, Springer International Publishing, pp. 1–40.

Meng, Y., Mao, H. K., Eng, P., et al. (2004). The formation of sp^3 bonding in compressed BN. *Nature Materials.*, **3**, 111–114.

Meng, Y., Shen, G., Mao, H.-k. (2006). Double-sided laser heating system at HPCAT for in situ X-ray diffraction at high pressures and high temperatures. *Journal of Physics: Condensed Matter*, **18**, s1097–s1103.

Mezouar, M. (2010). Synchrotron high-pressure high-temperature techniques. Paper presented at the High-Pressure Crystallography conference, Dordrecht.

Mezouar, M., Giampaoli, R., Garbarino, G., et al. (2017). Methodology for in situ synchrotron X-ray studies in the laser-heated diamond anvil cell. *High Pressure Research*, **37**, 170–180.

Miao, J., Ishikawa, T., Robinson, I. K., Murnane, M. M. (2015). Beyond crystallography: diffractive imaging using coherent X-ray light sources. *Science*, **348**, 530.

Ming, L. C., Manghnani, M. H., Balogh, J., Qadri, S. B., Skelton, E. F., Jamieson, J. C. (1983). Gold as a reliable internal pressure calibrant at high temperatures. *Journal of Applied Physics*, **54**, 4390–4397.

Mishima, O., Calvert, L. D., Whalley, E. (1984). "Melting ice" I at 77 K and 10 kbar: a new method of making amorphous solids. *Nature*, **310**, 393.

Mitsui, T., Hirao, N., Ohishi, Y., et al. (2009). Development of an energy-domain Fe-57-Mossbauer spectrometer using synchrotron radiation and its application to ultrahigh-pressure studies with a diamond anvil cell. *Journal of Synchrotron Radiation*, **16**, 723–729.

Montgomery, J. M., Lipp, M. J., Jenei, Z., Meng, Y., Evans, W. J. (2018). A simple and portable multi-channel pyrometer allowing temperature measurements down to 800 K on the microsecond scale. *Review of Scientific Instruments*, **89**, 125117.

Morard, G., Hernandez, J.-A., Guarguaglini, M., et al. (2020). In situ X-ray diffraction of silicate liquids and glasses under dynamic and static compression to megabar pressures. *Proceedings of the National Academy of Science*, **117**, 11981.

Murakami, M., Hirose, K., Kawamura, K., Sata, N., Ohishi, Y. (2004). Post-perovskite phase transition in $MgSiO_3$. *Science*, **304**, 855–858.

Murakami, M., Kohara, S., Kitamura, N., et al. (2019). Ultrahigh-pressure form of SiO_2 glass with dense pyrite-type crystalline homology. *Physical Review B*, **99**, 045153.

Murphy, C. A., Jackson, J. M., Sturhahn, W., Chen, B. (2011). Melting and thermal pressure of hcp-Fe from the phonon density of states. *Physics of Earth and Planetary Interiors*, **188**, 114–120.

Nagler, B., Arnold, B., Bouchard, G., et al. (2015). The matter in extreme conditions instrument at the Linac Coherent Light Source. *Journal of Synchrotron Radiation*, **22**, 520–525.

Nasreen, F., Antonio, D., VanGennep, D., et al. (2016). High pressure effects on U L3 X-ray absorption in partial fluorescence yield mode and single crystal X-ray diffraction in the heavy fermion compound UCd_{11}. *Journal of Physics: Condensed Matter*, **28**, 105601.

Nasu, S. (1996). High pressure Mossbauer spectroscopy with nuclear forward scattering of synchrotron radiation. *High Pressure Research*, **14**, 405–412.

Oganov, A. R., Ono, S. (2004). Theoretical and experimental evidence for a post-perovskite phase of MgSiO3 in Earth's "D'' layer. *Nature*, **430**, 445–448.

Ohishi, Y., Hirao, N., Sata, N., Hirose, K., Takata, M. (2008). Highly intense monochromatic X-ray diffraction facility for high-pressure research at SPring-8. *High Pressure Research*, **28**, 163–173.

Ohtani, E., Shibazaki, Y., Sakai, T., et al. (2013). Sound velocity of hexagonal close-packed iron up to core pressures. *Geophysical Research Letters*, **40**, 2013GL057667.

Olbinado, M. P., Cantelli, V., Mathon, O., et al. (2018). Ultra high-speed X-ray imaging of laser-driven shock compression using synchrotron light. *Journal of Physics D: Applied Physics*, *51*.

Olsen, J. S., Buras, B., Gerward, L., et al. (1984). A new high-pressure phase and the equation of state of YbH2. *Physica Scripta*, *29*, 503–507.

Olsen, J. S., Gerward, L., Dancausse, J. P., Gering, E. (1993). Developments and new possibilities in high-pressure powder diffraction with synchrotron-radiation – results for cerium metal and U_6Fe. *Physica B*, *190*, 92–97.

Pacold, J. I., Bradley, J. A., Mattern, B. A., et al. (2012). A miniature X-ray emission spectrometer (miniXES) for high pressure studies in a diamond anvil cell. *Journal of Synchrotron Radiation*, *19*, 245–251.

Parab, N. D., Roberts, Z. A., Harr, M. H., et al. (2016). High speed X-ray phase contrast imaging of energetic composites under dynamic compression. *Appl.ied Physics Letters*, **109**, 131903.

Park, C., Popov, D., Ikuta, D., et al. (2015). New developments in micro-X-ray diffraction and X-ray absorption spectroscopy for high-pressure research at 16-BM-D at the Advanced Photon Source. *Review of Scientific Instruments*, *86*, 072205.

Pascarelli, S., Mathon, O. (2010). Advances in high brilliance energy dispersive X-ray absorption spectroscopy. *Physical Chemistry Chemical Physics*, *12*, 5535–5546.

Pellegrini, C., Marinelli, A., Reiche, S. (2016). The physics of X-ray free-electron lasers. *Review of Modern Physics*, *88*, 015006.

Petitgirard, S., Daniel, I., Dabin, Y., Cardon, H., Tucoulou, R., Susini, J. (2009). A diamond anvil cell for X-ray fluorescence measurements of trace elements in fluids at high pressure and high temperature. *Review of Scientific Instruments*, **80**(3), 033906.

Petitgirard, S., Sahle, C. J., Weis, C., Gilmore, K., Spiekermann, G., Tse, J. S., et al. (2019a). Magma properties at deep Earth's conditions from electronic structure of silica. *Geochemical Perspectives Letters*, **9**, 32–37.

Petitgirard, S., Spiekermann, G., Glazyrin, K., Garrevoet, J., Murakami, M. (2019b). Density of amorphous GeO2 to 133 GPa with possible pyritelike structure and stiffness at high pressure. *Physical Review B*, **100**, 214104.

Piermarini, G. J., Weir, C. (1962). A diamond cell for X-ray diffraction studies at high pressures. *Journal of Research of the National Bureau of Standards A: Physics and Chemistry*, **66A**, 325–331.

Prakapenka, V. B., Kubo, A., Kuznetsov, A., et al. (2008). Advanced flat top laser heating system for high pressure research at GSECARS: application to the melting behavior of germanium. *High Pressure Research*, **28**, 225–235.

Prescher, C., Prakapenka, V. B., Stefanski, J., Jahn, S., Skinner, L. B., Wang, Y. (2017). Beyond sixfold coordinated Si in SiO2 glass at ultrahigh pressures. *Proceedings of the National Academy of Science*, 114, 10041–10046.

Rivers, M. L., Duffy, T. S., Wang, Y., Eng, P. J., Sutton, S. R., Shen, G. (1998). A new facility for high-pressure research at the Advanced Photon Source, in M. H. Manghnanai, T. Yagi, eds., *Properties of Earth and Planetary Materials at High Pressure and Temperature*, American Geophysical Union, pp. 79–87.

Rueff, J. P. (2010). An Introduction to inelastic X-ray scattering, in E. Beaurepaire, J. P. Kappler, G. Krill, F. Scheurer, eds.*Magnetism and Synchrotron Radiation*, Springer-Verlag, pp. 263.

Rueff, J. P., Kao, C. C., Struzhkin, V. V., et al. (1999). Pressure-induced high-spin to low-spin transition in FeS evidenced by X-ray emission spectroscopy. *Physical Review Letters.*, **82**, 3284–3287.

Sakai, T., Yagi, T., Irifune, T., et al. (2018). High pressure generation using double-stage diamond anvil technique: problems and equations of state of rhenium. *High Pressure Research*, **38**, 107–119.

Sakai, T., Yagi, T., Ohfuji, H., et al. (2015). High-pressure generation using double stage micro-paired diamond anvils shaped by focused ion beam. *Review of Scientific Instruments.*, **86**, 033905.

Sakai, T., Yagi, T., Takeda, R., et al. (2020). Conical support for double-stage diamond anvil apparatus. *High Pressure Research*, **40**, 12–21.

Sakamaki, T., Ohtani, E., Fukui, H., et al. (2016). Constraints on Earth's inner core composition inferred from measurements of the sound velocity of hcp-iron in extreme conditions. *Science Advances*, **2**(2), e1500802.

Sanloup, C., de Grouchy, C. J. L. (2018). X-ray diffraction structure measurements, in Y. Kono, C. Sanloup, eds., *Magmas under Pressure*, Elsevier, pp. 137–153.

Sato, T., Funamori, N. (2010). High-pressure structural transformation of SiO_2 glass up to 100 GPa. *Physical Review B*, **82**, 184102.

Schell, N., Simmons, R. O., Kaprolat, A., Schulke, W., Burkel, E. (1995). Electronic excitations in hcp He-4 at 61.5 MPa and 4.3 k studied by inelastic X-ray-scattering spectroscopy. *Physical Review Letters*, **74**, 2535–2538.

Schmidt, C., Rickers, K. (2003). In-situ determination of mineral solubilities in fluids using a hydrothermal diamond-anvil cell and SR-XRF: solubility of AgCl in water. *American Mineralogist*, **88**, 288–292.

Schmidt, C., Rickers, K., Bilderback, D. H., Huang, R. (2007). In situ synchrotron-radiation XRF study of REE phosphate dissolution in aqueous fluids to 800 degrees C. *Lithos*, **95**, 87–102.

Schropp, A., Patommel, J., Seiboth, F., et al. (2012). Developing a platform for high-resolution phase contrast imaging of high pressure shock waves in matter, in S. P. Moeller, M. Yabashi, S. P. HauRiege, eds., *X-Ray Free-Electron Lasers: Beam Diagnostics, Beamline Instrumentation, and Applications*, vol. 8504, SPIE, https://doi.org/10.1117/12.929882.

Schultz, E., Mezouar, M., Crichton, W., et al. (2005). Double-sided laser heating system for in situ high pressure & high temperature monochromatic X-ray diffraction at the ESRF. *High Pressure Research*, **25**, 71–83.

Scopigno, T., Ruocco, G., Sette, F. (2005). Microscopic dynamics in liquid metals: the experimental point of view. *Review of Modern Physics.*, **77**, 881–933.

Seto, M., Masuda, R., Higashitaniguchi, S., et al. (2010). Mössbauer spectroscopy in the energy domain using synchrotron radiation. *Journal of Physics: Conference Series*, **217**, 012002.

Sette, F., Ruocco, G., Krisch, M., et al. (1995). Collective dynamics in water by high energy resolution inelastic X-ray scattering. *Physical Review Letters*, **75**, 850–853.

Shen, G., Ikuta, D., Sinogeikin, S., Li, Q., Zhang, Y., Chen, C. (2012). Direct observation of a pressure-induced precursor lattice in silicon. *Physical Review Letters*, **109**, 205503.

Shen, G., Mao, H. K. (2016). High-pressure studies with X-rays using diamond anvil cells. *Reports on Progress in Physics*, **80**, 016101.

Shen, G., Mao, H. K., Hemley, R. J. (1996). Laser-heating diamond-anvil cell technique: Double-sided heating with multimode Nd:YAG laser. Paper presented at the Advanced Materials'96 -New Trends in High Pressure Research, NIRIM, Tsukuba.

Shen, G., Prakapenka, V. B., Rivers, M. L., Sutton, S. R. (2004a). Structure of liquid iron at pressures up to 58 GPa. *Physical Review Letters*, **92**, 185701.

Shen, G., Rivers, M. L., Wang, Y., Sutton, S. J. (2001). A laser heated diamond cell system at the Advanced Photon Source for in situ X-ray measurements at high pressure and temperature. *Review of Scientific Instruments*, **72**, 1273–1282.

Shen, G., Sata, N., Newville, M., Rivers, M. L., Sutton, S. R. (2002). Molar volumes of liquids measured in a diamond anvil cell. *Applied Physics Letters*, **81**, 1411–1413.

Shen, G., Sinogeikin, S. (2015). Preface: high-pressure studies with X-rays. *Review of Scientific Instruments*, **86**, 071901.

Shen, G., Smith, J. S., Kenney-Benson, C. (2019a). Nature of polyamorphic transformations in H2O under isothermal compression and decompression. *Physical Review Materials*, **3**, 073404.

Shen, G., Smith, J. S., Kenney-Benson, C., Ferry, R. A. (2019b). In situ x-ray diffraction study of polyamorphism in H2O under isothermal compression and decompression. *Journal of Chemical Physics*, **150**, 244201.

Shen, G., Sturhahn, W., Alp, E. E., et al. (2004b). Phonon density of states in iron at high pressures and high temperatures. *Physics and Chemistry of Minerals.*, **31**, 353–359.

Shen, G. Y., Mao, H. K., Hemley, R. J., Duffy, T. S., Rivers, M. L. (1998). Melting and crystal structure of iron at high pressures and temperatures. *Geophysical Research Letters.*, **25**, 373–376.

Shen, G. Y., Wang, Y. B. (2014). High-pressure apparatus integrated with synchrotron radiation, in G. S. Henderson, D. R. Neuville, R. T. Downs, eds., *Spectroscopic Methods in Mineralogy and Materials Sciences*, vol. 78, Cambridge University Press, pp. 745–777.

Shi, C. Y., Zhang, L., Yang, W., et al. (2013). Formation of an interconnected network of iron melt at Earth's lower mantle conditions. *Nature Geoscience*, **6**, 971–975.

Shim, S.-H., Bengtson, A., Morgan, D., et al. (2009). Electronic and magnetic structures of the postperovskite-type Fe_2O_3 and implications for planetary magnetic records and deep interiors. *Proceedings of the National Academy of Science.*, **106**, 5508–5512.

Shimomura, O., Fukamachi, T., Kawamura, T., Hosoya, S., Hunter, S., Bienenstock, A. (1978). EXAFS measurement of high-pressure metallic phase of GaAs by use of a diamond anvil cell. *Japan Journal of Applied Physics*, **17**, 221.

Shimomura, O., Takemura, K., Fujihisa, H., et al. (1992). Application of an imaging plate to high-pressure X-ray study with a diamond anvil cell (invited). *Review of Scientific Instruments*, **63**, 967–973.

Shimomura, O., Yamaoka, S., Yagi, T., et al. (1985). Multi-anvil type X-ray system for synchrotron radiation, in S. Minomura, ed., *Solid State Physics under Pressure: Recent Advance with Anvil Devices*, KTK Scientific Publisher, pp. 351–356.

Sinn, H., Glorieux, B., Hennet, L., et al. (2003). Microscopic dynamics of liquid aluminum oxide. *Science*, **299**, 2047–2049.

Sinogeikin, S. V., Smith, J. S., Rod, E., Lin, C., Kenney-Benson, C., Shen, G. (2015). Online remote control systems for static and dynamic compression and decompression using diamond anvil cells. *Review of Scientific Instruments*, **86**, 072209.

Skelton, E. F., Kirkland, J., Qadri, S. B. (1982). Energy-dispersive measurements of diffracted synchrotron radiation as a function of pressure: applications to phase transitions in KCl and KI. *Journal of Applied Crystallography*, **15**, 82–88.

Smith, J. S., Rod, E. A., Shen, G. (2019). Fly scan apparatus for high pressure research using diamond anvil cells. *Review of Scientific Instruments*, **90**, 015116.

Smith, J. S., Sinogeikin, S. V., Lin, C., Rod, E., Bai, L., Shen, G. (2015). Developments in time-resolved high pressure X-ray diffraction using rapid compression and decompression. *Review of Scientific Instruments*, **86**, 072208.

Somayazulu, M. S., Finger, L. W., Hemley, R. J., Mao, H. K. (1996). High-pressure compounds in methane-hydrogen mixtures. *Science*, **271**, 1400–1402.

Speziale, S., Milner, A., Lee, V. E., Clark, S. M., Pasternak, M. P., Jeanloz, R. (2005). Iron spin transition in Earth's mantle. *Proceedings of the National Academy of Science*, **102**, 17918–17922.

Spiekermann, G., Harder, M., Gilmore, K., et al. (2019). Persistent octahedral coordination in amorphous GeO_2 up to 100 GPa by Kβ" X-ray emission spectroscopy. *Physical Review X*, **9**, 011025.

Stan, C. V., Beavers, C. M., Kunz, M., Tamura, N. (2018). X-ray diffraction under extreme conditions at the Advanced Light Source. *Quantum Beam Science*, **2**, 4.

Staun Olsen, J., Steenstrup, S., Gerward, L., Benedict, U., Itié, J. P. (1986). High-pressure structural studies of uranium and thorium compounds with the rocksalt structure. *Physica B+C*, **139–140**, 308–310.

Struzhkin, V. V., Mao, H. K., Lin, J.-F., Hemley, R. J., Tse, J. S., Ma, Y., et al. (2006). Valence band X-ray emission spectra of compressed germanium. *Physical Review Letters.*, **96**, 137402.

Sturhahn, W., Toellner, T. S., Alp, E. E et al. (1995). Phonon density of states measured by inelastic nuclear resonant scattering. *Physical Revew Letters*, **74**, 3832–3835.

Susaki, J., Akaogi, M., Akimoto, S., Shimomura, O. (1985). Garnet-perovskite transformation in $CaGeO_3$ – in situ X-ray measurements using synchrotron radiation. *Geophysical Research Letters*, **12**, 729–732.

Takahashi, S., Ohtani, E., Sakamaki, T., et al. (2019). Sound velocity of Fe3C at high pressure and high temperature determined by inelastic X-ray scattering. *Comptes Rendus Geoscience*, **351**, 190–196.

Takano, M., Nasu, S., Abe, T., et al. (1991). Pressure-induced high-spin to low-spin transition in $CaFeO_3$. *Physical Review Letters*, **67**, 3267–3270.

Tanis, E. A., Simon, A., Tschauner, O., et al. (2012). Solubility of xenotime in a 2 M HCl aqueous fluid from 1.2 to 2.6 GPa and 300 to 500 °C. *American Mineralogist*, **97**, 1708–1713.

Tanis, E. A., Simon, A., Zhang, Y., et al. (2016). Rutile solubility in NaF–NaCl–KCl-bearing aqueous fluids at 0.5–2.79 GPa and 250–650 °C. *Geochimica et Cosmochimica Acta*, **177**, 170–181.

Troyan, I., Gavriliuk, A., Rüffer, R., et al. (2016). Observation of superconductivity in hydrogen sulfide from nuclear resonant scattering. *Science*, **351**, 1303–1306.

Vaughan, M. T. (1993). In situ X-ray diffraction using synchrotron radiation at high P and T in a multi-anvil device, in R. W. Luth, ed., *Experiments at High Pressure and Applications to the Earth's Mantle, Short Course Handbook*, Mineralogical Association of Canada, pp. 95–130.

Vohra, Y. K., Brister, K. E., Desgreniers, S., Ruoff, A. L., Chang, K. J., Cohen, M. L. (1986a). Phase-transition studies of germanium to 1.25 mbar. *Physical Review Letters*, **56**, 1944–1947.

Vohra, Y. K., Brister, K. E., Weir, S. T., Duclos, S. J., Ruoff, A. L. (1986b). Crystal structures at megabar pressures determined by use of the Cornell synchrotron source. *Science*, **231**, 1136–1138.

Vohra, Y. K., Samudrala, G. K., Moore, S. L., Montgomery, J. M., Tsoi, G. M., Velisavljevic, N. (2015). High pressure studies using two-stage diamond micro-anvils grown by chemical vapor deposition. *High Pressure Research*, **35**, 282–288.

Vos, W. L., Finger, L. W., Hemley, R. J., Hu, J. Z., Mao, H. K., Schouten, J. A. (1992). A high-pressure Van-der-Waals compound in solid nitrogen helium mixtures. *Nature*, **358**, 46–48.

Wang, J., Yang, W., Wang, S., et al. (2012). High pressure nano-tomography using an iterative method. *Journal of Applied Physics*, **111**, 112626–112625.

Wang, Y. B., Weidner, D. J., Liebermann, R. C., et al. (1991). Phase-transition and thermal-expansion of $MgSiO_3$ perovskite. *Science*, **251**, 410–413.

Watanuki, T., Shimomura, O., Yagi, T., Kondo, T., Isshiki, M. (2001). Construction of laser-heated diamond anvil cell system for in situ X-ray diffraction study at SPring-8. *Review of Scientific Instruments*, **72**, 1289–1292.

Weir, C. E., Lippincott, E. R., Van Valkenburg, A., Bunting, E. N. (1959). Infrared studies in the 1- to 15-micron region to 30,000 atmospheres. *Journal of Research of the National Bureau of Standards A: Physics and Chemistry*, **63A**, 55–62.

Wilke, M., Appel, K., Vincze, L., Schmidt, C., Borchert, M., Pascarelli, S. (2010). A confocal set-up for micro-XRF and XAFS experiments using diamond-anvil cells. *Journal of Synchrotron Radiation*, **17**, 669–675.

Wong, J., Krisch, M., Farber, D. L., et al. (2003). Phonon dispersions of fcc δ-plutonium-gallium by inelastic X-ray scattering. *Science*, **301**, 1078–1080.

Xiao, Y., Chow, P., Shen, G. (2016). High pressure X-ray emission spectroscopy at the advanced photon source. *High Pressure Research*, **36**, 315–331. https://doi.org/10.1080/08957959.2016.1209498.

Yagi, T., Akaogi, M., Shimomura, O., Suzuki, T., Akimoto, S. (1987). In situ observation of the olivine-spinel phase-transformation in Fe_2SiO_4 using synchrotron radiation. *Journal of Geophysical Research–Solid Earth and Planets*, **92**, 6207–6213.

Yagi, T., Sakai, T., Kadobayashi, H., Irifune, T. (2020). Review: high pressure generation techniques beyond the limit of conventional diamond anvils. *High Pressure Research*, 40, 148–161.

Yamanaka, T., Hirose, K., Mao, W. L., et al. (2012). Crystal structures of $(Mg_{1-x},Fe_x)SiO_3$ postperovskite at high pressures. *Proceedings of the National Academy Science*, **109**, 1035–1040.

Yamanaka, T., Komatsu, Y., Nomori, H. (2007). Electron density distribution of $FeTiO_3$ ilmenite under high pressure analyzed by MEM using single crystal diffraction intensities. *Physics and Chemistry of Minerals*, **34**, 307–318.

Yamaoka, S., Shimomura, O., Akaishi, M., et al. (1986). X-ray-observation of the formation of diamond and cubic boron-nitride at high-pressure and temperature. *Physica B & C*, **139**, 668–670.

Yang, W., Huang, X., Harder, R., Clark, J. N., Robinson, I. K., Mao, H.-k. (2013). Coherent diffraction imaging of nanoscale strain evolution in a single crystal under high pressure. *Nature Communications*, *4*, 1680.

Yu, T., Wang, Y., Rivers, M. L. (2016). Imaging in 3D under pressure: a decade of high-pressure X-ray microtomography development at GSECARS. *Progress in Earth and Planetary Science*, **3**, 17.

Zeng, Q., Kono, Y., Lin, Y., et al. (2014). Universal fractional noncubic power law for density of metallic glasses. *Physical Review Letters*, **112**, 185502.

Zeng, Q., Lin, Y., Liu, Y., et al. (2016). General 2.5 power law of metallic glasses. *Proceedings of the National Academy of Science.*, **113**, 1714–1718.

Zeng, Z. D., Yang, L. X., Zeng, Q. S., et al. (2017). Synthesis of quenchable amorphous diamond. *Nature Communications*, **8**, 322.

Zha, C. S., Duffy, T. S., Mao, H. K., Hemley, R. J. (1993). Elasticity of hydrogen to 24-GPa from single-crystal brillouin-scattering and synchrotron X-ray-diffraction. *Physical Review B*, **48**, 9246–9255.

Zhang, L., Yuan, H., Meng, Y., Mao, H.-K. (2019). Development of high-pressure multi-grain X-ray diffraction for exploring the Earth's interior. *Engineering*, **5**, 441–447.

Zhao, J., Sturhahn, W., Lin, J. F., Shen, G., Alp, E. E., Mao, H. K. (2004). Nuclear resonant scattering at high pressure and high temperature. *High Pressure Research*, **24**, 447–457.

4

Development of Large-Volume Diamond Anvil Cell for Neutron Diffraction: The Neutron Diamond Anvil Cell Project at ORNL

REINHARD BOEHLER, BIANCA HABERL, JAMIE J. MOLAISON, AND MALCOM GUTHRIE

Ten years ago, Dave Mao, director of Energy Frontier Research in Extreme Environments (EFree), a Department of Energy (DOE) energy frontier, recognized the importance of neutron science for energy research. The subsequent establishment of a neutron group within EFree lead to the formation of an Instrument Development Team for SNAP, the dedicated high-pressure beamline at the Spallation Neutron Source at Oak Ridge National Laboratory in Tennessee. The core concept was to develop novel high-pressure techniques to expand the pressure range for neutron diffraction. A quite ambitious goal was set to reach half megabar levels (50 GPa), which at the time was considered extremely challenging. Here we will give a brief overview of the developments during the last decade in this novel area of research. An important factor was that during this period multicarat diamond anvils have become available grown by chemical vapor deposition (CVD), making research in this pressure range and beyond rather routine. This chapter shows the latest developments in large anvil and anvil support designs, compact multiple ton diamond cells, and new neutron methodologies. Achievements are illustrated with some examples of high-quality neutron diffraction patterns collected on sample sizes much small than conventional sizes.

4.1 Introduction

Neutron scattering is the technique of choice for structural studies of materials that contain light elements such as hydrogen, deuterium, carbon, or oxygen for which X-ray or other optical techniques may be insufficient. Neutrons cannot only distinguish between elements of similar Z-contrast but also detect light elements next to heavy ones, such as in oxides. It is also possible to obtain unique insight into local bonding environments via isotope substitutions where elemental isotopes with different scattering cross-sections are measured under the same conditions. Furthermore, neutrons carry a magnetic moment, giving rise to magnetic scattering in addition to the nuclear Bragg scattering.

Especially in combination with in situ experiments at extreme conditions, neutron diffraction has emerged as a tool for a wide range of science areas such as materials science, and specifically quantum materials, chemistry, physics, and geosciences. A particular focus lies thereby on high-pressure neutron diffraction since pressure is a good tuning parameter for quantum materials, is a useful tool for synthesis, and is critical for understanding the behaviors inside icy planetary matter where high pressure and temperatures persist.

In contrast to X-ray scattering, neutron scattering suffers from relatively low flux neutron sources. Nevertheless, there has been high-pressure activity since the 1960s, and today a wide array of pressure cells are in use at neutron sources, each optimized for a certain application (Klotz, 2013). For example, cylindrical gas pressure cells allow for very large sample volumes (~1–2 cc) under hydrostatic conditions or in reactive environments (H_2, D_2, CO_2, etc.) up to ~1 GPa (see (Dos Santos et al., 2018, and the references. therein). Similarly, so-called clamp cells enable volumes around 0.1 cc while allowing for a maximum pressure of 4 GPa (Fujiwara et al., 2007). Both devices can be easily combined with other extreme conditions such as high magnetic fields or ultralow temperature (Podlesnyak et al., 2018). Several neutron cell designs are based on multi-anvil cells, such as the palm cubic cell that has enabled ~7 GPa, or include large presses to enable pressure/temperature conditions of 10 GPa/2,000 K or even 16 GPa/1,273 K (Abe et al., 2010; Sano-Furukawa et al., 2014). A key breakthrough has been the development of the Paris–Edinburgh cell that in standard toroidal-anvil configuration typically allows for diffraction experiments close to 30 GPa (with sample volumes of around 0.02 cc) while remaining relatively portable (Klotz, 2013). Particularly noteworthy are several recent experiments at J-PARC's PLANET beamline that have achieved up to 40 GPa (Hattori et al., 2019; Klotz et al., 2019).

Higher pressures and higher temperatures, however, are best facilitated in a diamond anvil cell (DAC). These cells have reached several 100 GPa and up to ~5,000 K for X-ray and other optical measurements (Boehler, 2000). While initial developments focused on gem cells, for example using sapphire, the use of diamond cells for neutron diffraction was pioneered in Russia (Glazkov et al., 1988) and later developed further by Goncharenko et al. (2005). These DACs were used for powder as well as single-crystal studies. They achieved up to ~40 GPa and, most noteworthy, successfully determined the lattice parameters of D_2 at these pressures (Goncharenko, 2005). More recently, a number of studies have focused on the use of neutron DACs for single-crystal diffraction. For example, work at the Australian Nuclear Science and Technology Organisation's (ANSTO) KOALA diffractometer demonstrated sufficient data quality on a single-crystal sample loaded into a Merrill–Basset-type diamond cell for full refinement (Binns et al., 2016). In this study, however, no pressure was applied. Another study at the Heinz

Maier-Leibnitz Zentrum's HEiDi diffractometer achieved high-quality in situ single-crystal data in a panoramic-style DAC up to 1 GPa (Grzechnik et al., 2018). Yet another set of work conducted at IBR-2's DN-6 diffractometer demonstrated the use of a typical Boehler Almax DAC for single-crystal diffraction (Kozlenko et al., 2018). Furthermore, powder diffraction studies in neutron DACs have also been attempted at ISIS's PEARL diffractometer (Kozlenko et al., 2018), and pressure records up to ~70 GPa are being pushed using nanodiamonds at the Japan Proton Accelerator Research Complex's (J-PARC) PLANET diffractometer (Komatsu et al., 2020).

However, none of these efforts have achieved the same high pressures and temperatures as available in optical/spectroscopic and X-ray studies that employ standard DACs. Over the last decade, a significant effort to bridge this pressure gap has been undertaken at the Oak Ridge National Laboratory (ORNL) and specifically at its dedicated high-pressure diffractometer, Spallation Neutrons at Pressure (SNAP), located at the Spallation Neutron Source (SNS). In this contribution, this past development is reviewed, including the pressure record for neutron diffraction, 94 GPa (Boehler et al., 2013), and the pressure record for neutron Rietveld refinem ent, 62 GPa (Guthrie et al., 2019). Additionally, recent developments are described, particularly the surprising usefulness of rhenium as neutron DAC gasket. This has the unique advantage that gasket technology has now reached a level that enables laser heating in the neutron DAC, and with this the high pressure and high temperature conditions necessary for planetary sciences and mineral physics.

4.2 Neutron Diamond Cells at Oak Ridge National Laboratory

In 2006, the initial construction of the Spallation Neutron Source at ORNL was completed. One of the early instruments commissioned was the SNAP diffractometer. Due to the unprecedented flux available at SNS, SNAP was constructed in a manner more similar to a synchrotron beamline rather than a typical neutron beamline, with a particular focus on eventually enabling work in neutron DACs.

Key partners of this development effort were John Parise as well as a center funded by the Department of Energy, EFree, located at the Geophysical Laboratory and led by Dave Mao. This collaboration brought together EFree scientists Malcolm Guthrie and Reinhard Boehler with SNAP scientists Chris Tulk, Antonio dos Santos, Jamie Molaison, and many more (see Figure 4.1). This collaboration was clearly the incubator for much of the ORNL DAC successes.

The first design of a SNAP neutron DAC was based on the panoramic DAC (Mao et al., 2001) as originally developed for X-ray scattering (see Figure 4.2). This required unusually large diamonds because first tests showed that at the time, the minimum sample size for neutron scattering required culets of at least 1.5 mm in diameter. The required high forces of several tons could only be sustained with conical anvil support (Boehler and De Hantsetters, 2004). These supports consisted of an elaborate binding-ring setup made of an inner polycrystalline diamond (PCD) ring, a supporting tungsten carbide (WC) ring, and an outer steel ring (Boehler et al., 2013). The first anvils used were cut from natural diamonds and had a diameter of 4 mm. Pressure was applied through a gas-driven

Figure 4.1 In 2010, EFree Partners focused on SNAP are attending a meeting at the SNS hosted by Dean Myles (ORNL). The meeting was attended by (in order from left to) Yang Ding (EFree/APS), Jamie Molaison (ORNL), Dave Mao (EFree Director), Maria Baldini (EFree), Jeff Yarger (ASU), Maddury Somayazulu (EFree), Malcolm Guthrie (EFree Chief Scientist), John Parise (Stonybrook, SNAP PI), Chris Tulk (ORNL), Reini Boehler (EFree Chief Scientist), Xiaojia Chen (EFree), Neelam Pradhan (ORNL), and Antonio dos Santos (ORNL). Note that affiliations given are those held in 2010. For the color version, refer to the plate section.

membrane up to a load of ~10 metric tons. The neutron beam entered the cell along the pressure direction, and diffraction patterns were collected at a scattering angle centered on 90°. The gaskets were made from steel as the rather limited diffraction aperture required steel's relatively good neutron transmission properties (e.g., the absorption cross-section of Fe is 2.6 barn and of Cr 3.1 barn for 2200 m/s neutrons (Sears, 1992). With this design, a record pressure of 94 GPa was achieved (Boehler et al., 2013). This was the first and only time to date that pressures that close to 100 GPa were reached with neutron diffraction. This first generation of ORNL neutron DACs suffered, however, from the fact that the diffraction aperture was limited (e.g., the largest d-spacing measurable was only ~2.4 Å) and that anvil/seat setups were expensive, difficult to manufacture, and typically broke on each run.

A significant breakthrough was made possible by the emergence of very large synthetic single-crystal diamonds grown by chemical vapor deposition (CVD). A range of such CVD anvils laser-cut to shape and polished for culets/facets is shown in Figure 4.3. These anvils cut into a conical design now allow for a new cell design based on precision-machined pistons and new seats made from steel that allow for large diffraction apertures with d-spacings up to at least 5 Å accessible on SNAP (also see Figure 4.3) (Boehler et al., 2017). Furthermore, these large anvils enable larger culets and thus larger sample sizes.

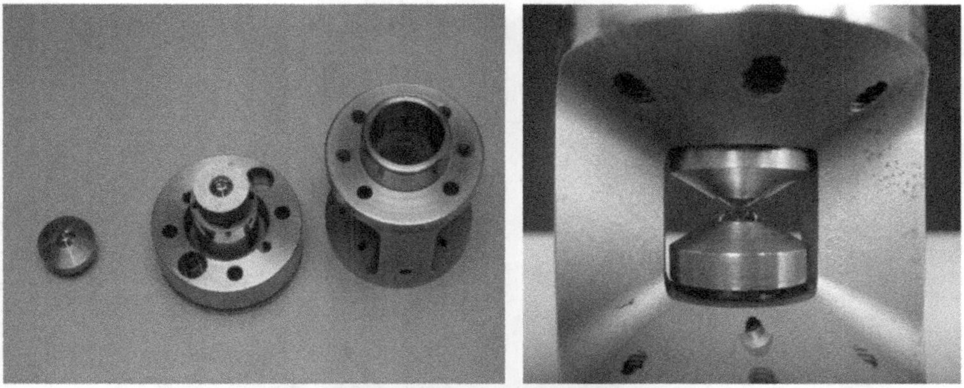

Photo credit: Carlos Jones

Photo credit: Mark Loguillo

Figure 4.2 Pictures of the various generations of ORNL DACs. Top: the panoramic DAC shown disassembled as well as with a close-up view of anvils in steel seats. Middle: the new SNAP DAC with CVD anvils mounted into seats. Bottom: an assembly of two clamped CuBe DAC with Versimax^{TM} anvils and gasket components. For the color version, refer to the plate section.

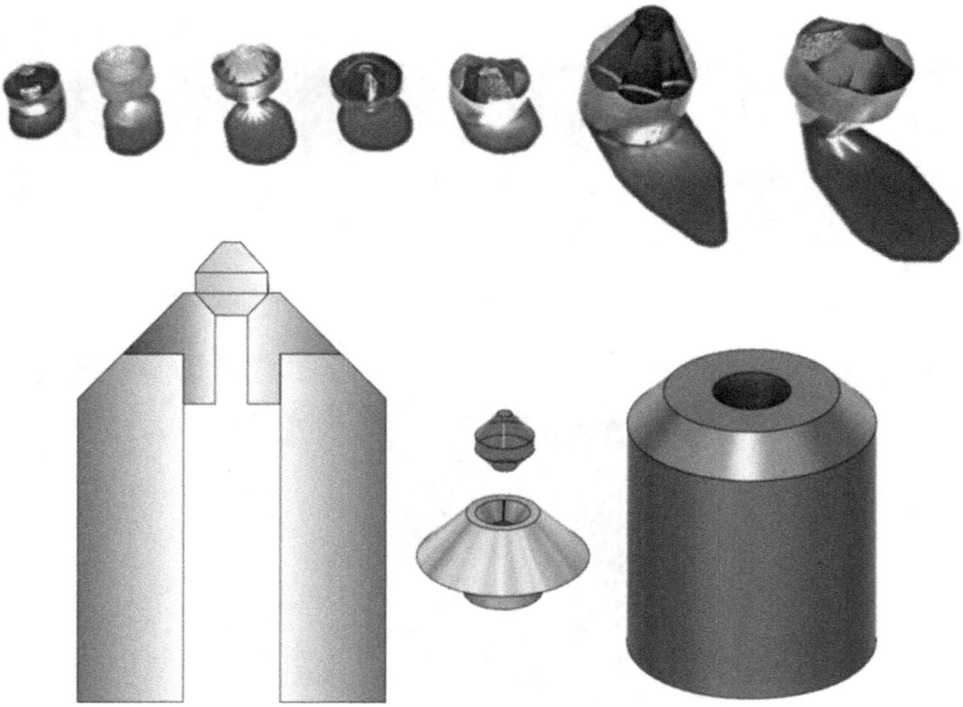

Figure 4.3 Top: a photo of various synthetic diamonds grown by chemical vapor deposition lined up together. For reference of scale, the anvil on the right is a 10 mm diameter CVD anvil. Bottom: schematic images of precision-machined pistons, steel seats, and conical CVD anvils used on the current generation of SNAP DACs with a typical anvil diameter of 6 mm. For the color version, refer to the plate section.

A typical anvil setup uses a 2 mm diameter culet with a 1 mm diameter sample chamber. This allows for a sample volume of 0.15 mm^3 up to ~40 GPa pressure. Pressure continues to be applied through a membrane press, and gaskets used were made from steel. Such a cell using a 1.5 mm diameter culet was used to pressurize D_2O-ice to 62 GPa. The data were of unprecedented quality, and the large d-range allowed a full Rietveld refinement, the highest pressure neutron refinement performed on any material to date (Guthrie et al., 2019). This cell remains the standard SNAP cell for many experiments in the user program to date.

In addition to experiments on SNAP, the new ultralarge CVD anvils also enabled neutron scattering experiments at neutron instruments at ORNL other than SNAP. In fact, these ultralarge anvils were first developed and purchased through an ORNL-funded project targeted at developing high-pressure neutron spectroscopy. This led to the development of a clamped DAC that allows for pressures of ~10 GPa on a 3 mm diameter culet single-crystal diamond. The resulting sample volume of about 1 mm^3 is sufficient for vibrational neutron spectroscopy at SNS's VISION beamline. As this beamline has a different diffraction geometry, the single-crystal diamonds could be replaced with inexpensive PCD anvils, in particular a Co-free version, Versimax™. These cells do not allow for in situ pressure increase, and pressures are applied in a press and clamped in offline.

While the first generation of these cells was made from steel, a later generation was made from copper-beryllium for improved cooling and potentially future magnetic measurements. Interestingly, the setup of the VISION beamline required a very large scattering aperture of 120°. This made the cell design also very interesting for single-crystal diffraction, and it has now been adapted for a range of beamlines at SNS, e.g., for the CORELLI beamline, but also at the High Flux Isotope Reactor (HFIR), e.g., with the four-circle diffractometer and the Imagine beamline (Haberl et al., 2017, 2018).

Thus, high-pressure neutron scattering in a DAC at ORNL's two facilities, SNS and HFIR, is now widespread beyond the initial incubator SNAP. However, pressures in a DAC above ~10 GPa will always require the high flux, tightly collimated beam, and ease of access that are available on the SNAP beamline. Furthermore, previous designs relied on the use of steel gaskets, which are not ideal for reactive gases and also do not provide the stability needed for laser heating or extremely high pressures. Thus, development continues on the SNAP beamline to tackle these issues.

4.3 Advances in Neutron Diamond Cells

While the new precision-machined pistons, cells, steel seats, and ultralarge CVD diamonds have helped to address several issues in neutron DAC experiments, namely sample size and diffraction aperture, other problems remained unsolved. These concerned particularly the pressure application in the membrane press and the optimization of the gasket. Over the last two years, these issues have been addressed to further improve reliability of our neutron DACs as well as open new avenues of research.

Firstly, a new compact hydraulic press was developed to apply load online to the neutron SNAP DAC (see Figure 4.4). The main reason for this development was the need to reliably apply loads well above 10 tons. Above these loads, the gas membranes previously used tended to fail, resulting in expensive diamond failure. The new press increases the load capacity from 12 tons to at least 20 tons. The new compact hydraulic press is operated using a Teledyne Isco pump whereby each 15 bars applied equates to 1 metric ton. This press is now routinely used for all room temperature DAC experiments, while the membrane press is still used for low-temperature experiments conducted with a closed cycle refrigerator.

Another key improvement was the change in the gasket material from steel to rhenium for significantly higher mechanical stability, lower reactivity, and thermal stability during laser heating. For X-ray scattering, the gasket materials of choice are tungsten and rhenium. However, neither is ideal in terms of neutron properties (e.g., the neutron absorption of W is 18.3 barn, and of Re it is a rather high 89.7 barn). W is barely acceptable, but it tends to flake and is somewhat brittle for the required thickness. Re is mechanically superior but neutron absorption is very high. However, the increased diffraction aperture enabled by the ultralarge CVD anvils and the new cell/seat design surprisingly enabled the use of Re gaskets.

Figure 4.5 shows a schematic of the anvil assembly with the incoming neutron beam, a (shielding) Re gasket, and the scattered beam as per the typical SNAP detector setting for DAC experiments. Clearly, the Re gasket clips the edge of the low-angle West detector

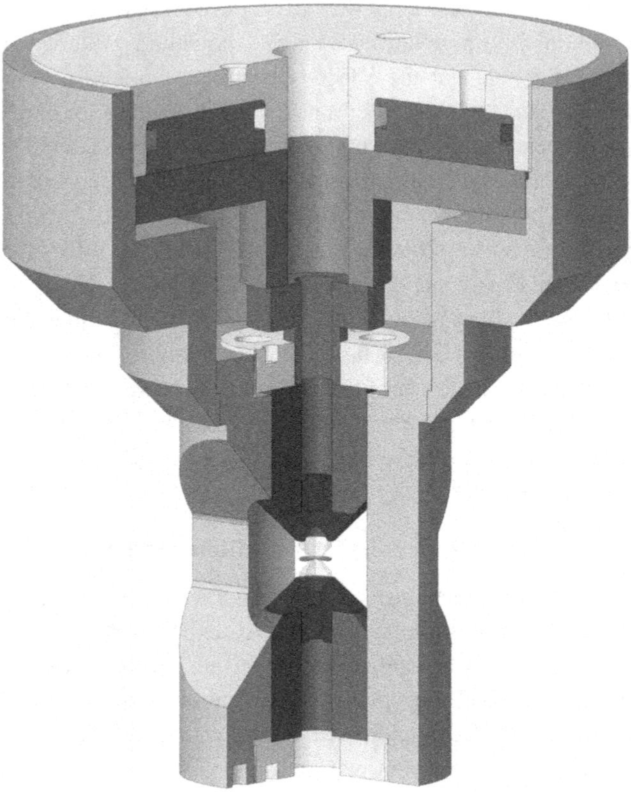

Figure 4.4 Schematic of the compact hydraulic press fitted to the recent neutron DACs. It employs a double O-ring seal to achieve loads up to 20 tons. For reference of size, the outer diameter of the top can, i.e., the largest horizontal extent of the press, is 115 mm. For the color version, refer to the plate section.

(on the right in the schematic) and obscures some of the high-angle East detector (on the left in the schematic). Furthermore, in addition to neutron absorption considerations, it was also important to prevent the gasket from tilting or cupping over one anvil. Thus, additional aluminum wedges are inserted to stabilize the gasket. The entire assembly is then slipped into an Al ring for stability. Note that Al does not readily absorb neutrons (it has an absorption cross-section of 0.2 barn) and is thus largely "invisible" compared to the Re gasket. Clearly, it appears that the majority of the detector is free of interference from the absorbing Re gasket. Nonetheless, having a strip around 90° fully blocked would be a major issue for a monochromatic beamline since a part of the d-range would be excluded. SNAP, however, is a white Laue time-of-flight beamline. Thus, dependent on neutron energy, diffraction from a given d-spacing is spread across a large range of angular space of the detector. In the particular case of an obscured region around 90° due to a Re gasket, we found the entire d-range was observed.

To illustrate these points, an intensity map of the resulting 2D diffraction as seen on the SNAP detectors is shown in Figure 4.5. The SNAP detectors consist of two banks

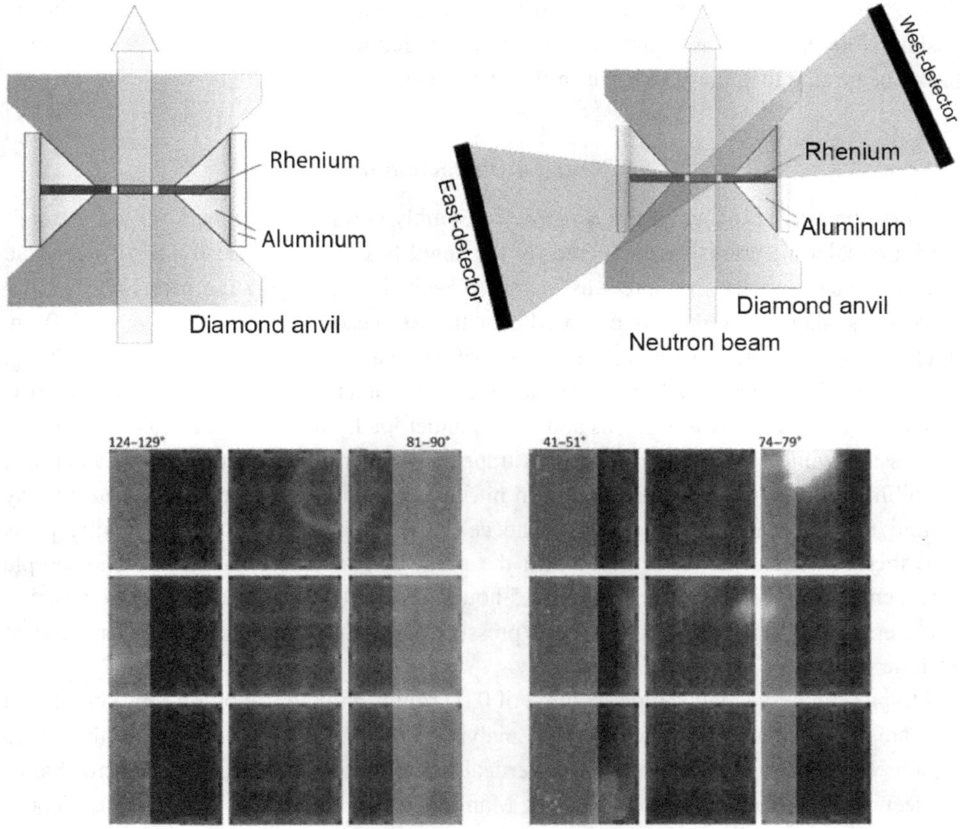

Figure 4.5 Top-left: close-up of the Re gasket, Al wedges, and ring assembled around the diamond anvils. Top-right: schematics of the diffraction geometry: incoming neutron beam (yellow), illuminated sample (red), and scattered beam (orange) as captured by the typical detector setting on SNAP. Bottom: 2D detector intensity map of the two SNAP detector banks West (left) and East (right) as seen in Mantid. Each detector is made up of 3 × 3 modules, and the black lines correspond to their boundaries. The data shown are obtained from ice at ~45 GPa contained inside a Re gasket and the diffraction signal is constricted to a d-range of 1.98–2.10 Å. Blue corresponds to a low diffraction intensity and yellow to a high intensity. See text for further detail. For the color version, refer to the plate section.

(named West and East) that are made up of 3 × 3 modules each. An example of an intensity map of the diffracted signal as seen in the detector banks in Mantid, the neutron data analysis platform used, is shown in the figure. The data used are taken from ice at 45 GPa contained inside a Re gasket (details of the experiment are provided later). The d-range shown in the detector banks is reduced to 1.98–2.10 Å, yet clearly a diffraction signal (yellow region) is detected in the majority of both detectors.

However, in the angular range from 41–51°, no sample signal is detected in this d-range due to the low energies (and low d-values) corresponding to these angles. Similarly, no sample signal is detected in the highest angular range (124–129°) due to absorption of cadmium placed on seats and anvils for shielding. In contrast, in the angular range from

74–79° on the West and 81–90° on the East, the diffraction signal is blocked by the Re gasket. Note that these angular ranges differ somewhat from an ideal gasket perpendicular to the beam due to some gasket tilt in the experiment.

4.4 Neutron Diffraction on Ice

The first experiment using such a Re gasket assembly was performed on D_2O-ice. A set of 1.15 mm culet diamonds was prepared for the new SNAP DAC (Vascomax 350 steel seat, 6 mm diameter anvil). Pressure was increased with the new hydraulic press, and a four-piece Re gasket assembly was prepared. For the Re gasket, a 6 mm Re disk of 250 μm thickness was indented to 140 μm. A hole of 450 μm was drilled using an electric discharge machining (EDM) microdriller. Deuterated water was loaded with a syringe. The cell was then pressurized in a press to 3 tons and closed under load. This served to seal in the sample at pressure while ensuring no water was trapped between the gasket and anvil interface. A collimator made from hexagonal boron nitride with a 450 μm bore was placed directly behind the upstream diamond anvil. The cell was then placed in the hydraulic press described in Figure 4.4, and the load in the press was increased to 3 tons. The sample was then measured at this pressure for 11.5 hours. Thereafter, the pressure was increased in two steps to a load of 6.5 tons. At this pressure, the sample was measured for close to 15 hours.

Despite the very small sample volume of 0.022 mm^3 and the use of a Re gasket, the ice data are of sufficient quality for detailed analysis. The time-of-flight diffraction data were separately reduced for each of the six vertical columns spanning the two detector banks (as seen at the bottom of Figure 4.5) using Mantid (Arnold et al., 2014), taking into account corrections for diamond absorption (Guthrie et al., 2017) and other instrument-dependent corrections. These data sets were simultaneously Rietveld refined, using GSAS-II (Toby and Von Dreele, 2013), against a structural model in the $Pn\underline{3}m$ space group, with the D atom on 8*e* Wyckoff sites at (*x*,*x*,*x*) and using a simple single-site model for the O atom, which was located on the 2*a* Wyckoff site at $\left(\frac{1}{4}, \frac{1}{4}, \frac{1}{4}\right)$ (Kuhs et al., 1984). In initial refinements, the lattice parameter and D-atom position were varied (isotropic models were used for the thermal motion of both atoms, but these parameters were kept fixed at reasonable values). In addition, separate Gaussian peak-width parameters were refined for each of the six detector banks. The resultant fits had a weighted R-factor $R_w = 1.77\%$ (averaged across all banks) and yielded a best-fitting lattice parameter of $a = 2.7711$ (4) Å. This corresponds to a pressure of 61.1 GPa on the basis of a previously published Birch–Murnaghan equation of state for ice VII (Hemley et al., 1987). This is a significantly higher pressure than previously obtained with steel gaskets while still giving sufficient data quality for refinements.

A remarkable aspect of the data is that the combination of the time-of-flight method together with large angular detectors and the wide scattering angle afforded by the cell provides direct information on angular-dependent effects such as preferred orientation and uniaxial strain. This is currently unique on any high-pressure neutron diffractometer. Indeed, inspection of the separate Rietveld fits to individual detector bank data revealed some significant angular-dependent misfits of peak positions, suggesting the presence of uniaxial strain. This could be well fitted by varying the D11 elastic strain parameters in GSAS-II, reducing the weighted R-factor to $R_w = 1.59\%$. The relative intensities of sample diffraction peaks also showed a clear angular dependence, and this could be further fitted using GSAS-II cylindrical texture model, which is based on a spherical-harmonics decomposition. It was necessary to use an eighth-order harmonic model to obtain a good description of the relative intensities finally yielding $R_w = 1.43\%$ with a texture index of 2.06. The resultant fits are shown separately along with the data for each bank in Figure 4.6.

Alternatively, all angular banks can be combined to give good statistics for final refinement as shown in Figure 4.7.

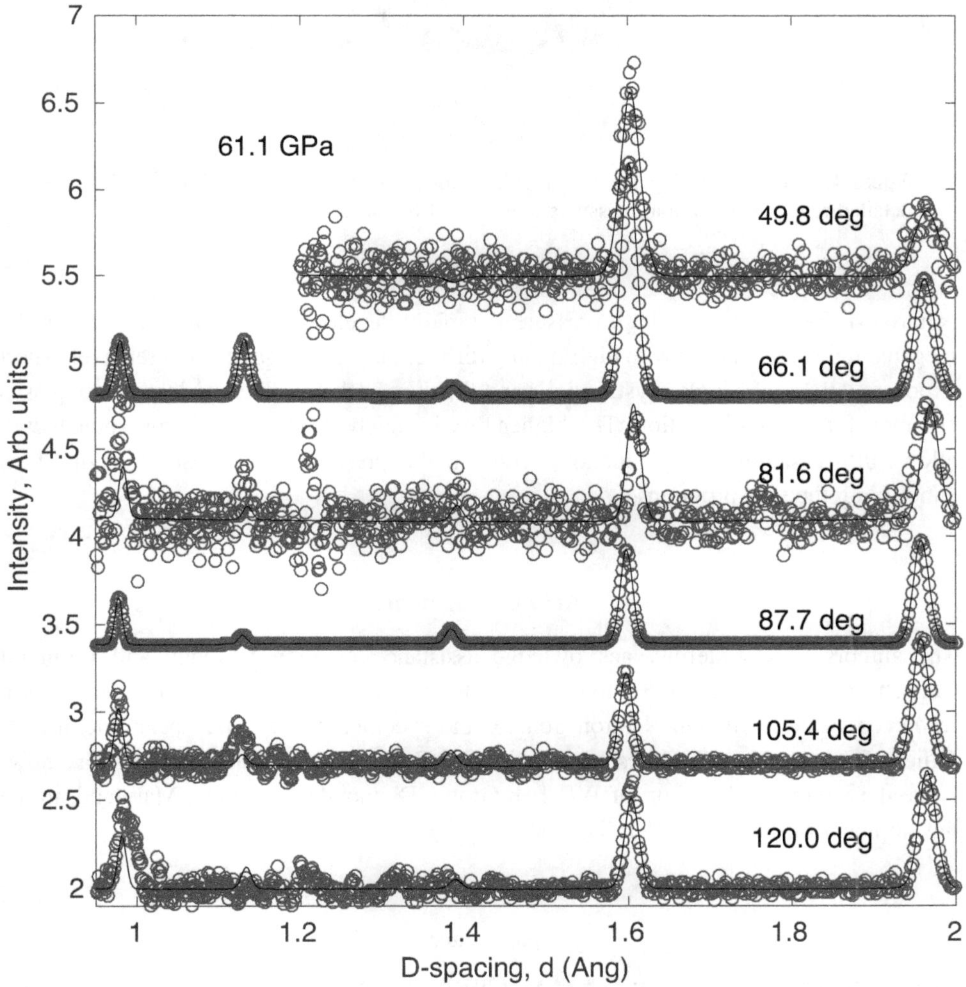

Figure 4.6 A full diffraction pattern is measured in each of the angular banks in a neutron DAC equipped with a Re gasket. A clear variation in the relative intensities of the 110 and 111 peaks (at 1.96 Å and 1.60 Å respectively) is observed and can be used to determine preferred orientation in the sample.

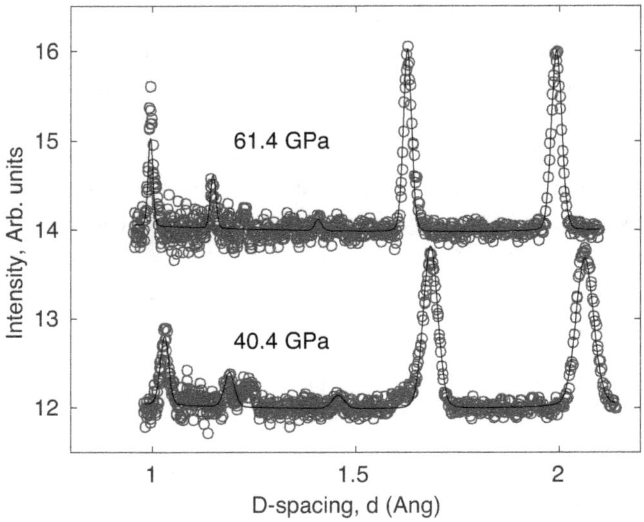

Figure 4.7 By integrating the full angular range, statistics can be optimized, allowing detailed refinements even at pressures in excess of 60 GPa.

4.5 Conclusions

Neutron diffraction at very high pressure is rapidly developing. Routine pressures in the megabar range are feasible with high-quality diffraction patterns obtained in relatively short exposure durations. Single-crystal anvils and rhenium gaskets will allow laser heating as is common for X-ray diffraction. The higher flux in newly developed neutron sources will allow further sample size reduction, reducing the presently high cost of anvils, thus simplifying the user programs.

Acknowledgments

The authors very gratefully acknowledge assistance of John Loveday and Bernhard Massani (both University of Edinburgh, UK) during the ice experiment. This research used resources at the Spallation Neutron Source, and specifically the SNAP beamline, a DoE Office of Science User Facilities operated by Oak Ridge National Laboratory. We acknowledge the interest and support of WD Lab Grown Diamonds, Beltsville, Maryland, in the development of the large CVD diamond anvils.

References

Abe, J., Arakawa, M., Hattori, T., et al. (2010). A cubic-anvil high-pressure device for pulsed neutron powder diffraction. *Review of Scientific Instruments*, **81**(4), 043910. https://doi.org/10.1063/1.3384238.

Arnold, O., Bilheux, J. C., Borreguero, J. M., et al. (2014). Mantid – data analysis and visualization package for neutron scattering and μSR experiments. *Nuclear*

Instruments and Methods in Physics Research, Section A: Accelerators, Spectrometers, Detectors and Associated Equipment, **764**, 156–166. https://doi.org/10.1016/j.nima.2014.07.029

Binns, J., Kamenev, K. V., McIntyre, G. J., Moggach, S. A., Parsons, S. (2016). Use of a miniature diamond-anvil cell in high-pressure single-crystal neutron Laue diffraction. *IUCrJ*, **3**(3), 168–179. https://doi.org/10.1107/S2052252516000725.

Boehler, R. (2000). High-pressure experiments and the phase diagram of lower mantle and core materials. *Reviews of Geophysics*, **38**(2), 221–245. https://doi.org/10.1029/1998RG000053.

Boehler, R., De Hantsetters, K. (2004). New anvil designs in diamond-cells. *High Pressure Research*, **24**(3), 391–396. https://doi.org/10.1080/08957950412331323924.

Boehler, R., Guthrie, M., Molaison, J. J., S., et al. (2013). Large-volume diamond cells for neutron diffraction above 90 GPa. *High Pressure Research*, **33**(3), 546–554. https://doi.org/10.1080/08957959.2013.823197.

Boehler, R., Molaison, J. J., Haberl, B. (2017). Novel diamond cells for neutron diffraction using multi-carat CVD anvils. *Review of Scientific Instruments*, **88**(8), 083905. https://doi.org/10.1063/1.4997265.

Dos Santos, A. M., Molaison, J. J., Haberl, B., et al. (2018). The high pressure gas capabilities at Oak Ridge National Laboratory's neutron facilities. *Review of Scientific Instruments*, **89**, 092907. https://doi.org/10.1063/1.5032096.

Fujiwara, N., Matsumoto, T., Nakazawab, K., Hisada, A., Uwatoko, Y. (2007). Fabrication and efficiency evaluation of a hybrid NiCrAl pressure cell up to 4 GPa. *Review of Scientific Instruments*, **78**(7), 073905. https://doi.org/10.1063/1.2757129.

Glazkov, V. P., Besedin, S. P., Goncharenko, I. N., Irodova, A. V. (1988). Neutron-diffraction study of the equation of state of molecular deuterium at high pressures. *JETP Letters*, **47**(12), 661–664.

Goncharenko, I., Gukasov, A., Loubeyre, P., IUCr. (2005). Single crystal neutron experiments under pressures up to 38 GPa. *Acta Crystallographica Section A: Foundations of Crystallography*, **61**(a1), 134.

Goncharenko, I., Loubeyre, P. (2005). Neutron and X-ray diffraction study of the broken symmetry phase transition in solid deuterium. *Nature*, **435**(7046), 1206–1209. https://doi.org/10.1038/nature03699.

Grzechnik, A., Meven, M., Friese, K. (2018). Single-crystal neutron diffraction in diamond anvil cells with hot neutrons. *Journal of Applied Crystallography*, **51**(2), 351–356. https://doi.org/10.1107/S1600576718000997.

Guthrie, M., Boehler, R., Molaison, J. J., Haberl, B., Dos Santos, A. M., Tulk, C. (2019). Structure and disorder in ice VII on the approach to hydrogen-bond symmetrization. *Physical Review B*, **99**(18), 184112. https://doi.org/10.1103/PhysRevB.99.184112.

Guthrie, M., Pruteanu, C. G., Donnelly, M. E., et al. (2017). Radiation attenuation by single-crystal diamond windows. *Journal of Applied Crystallography*, **50**(1), 76–86. https://doi.org/10.1107/S1600576716018185.

Haberl, B., Dissanayake, S., Wu, Y., et al. (2018). Next-generation diamond cell and applications to single-crystal neutron diffraction. *Review of Scientific Instruments*, **89**(9), 092902. https://doi.org/10.1063/1.5031454.

Haberl, B., Dissanayake, S., Ye, F., et al. (2017). Wide-angle diamond cell for neutron scattering. *High Pressure Research*, **37**, 495–506.

Hattori, T., Sano-Furukawa, A., Machida, S., et al. (2019). Development of a technique for high pressure neutron diffraction at 40 GPa with a Paris–Edinburgh press. *High Pressure Research*, **39**(3), 417–425.

Hemley, R. J., Jephcoat, A. P., Mao, H. K., Zha, C. S., Finger, L. W., Cox, D. E. (1987). Static compression of H_2O-ice to 128 GPa (1.28 Mbar). *Nature*, **330**(6150), 737–740. https://doi.org/10.1038/330737a0.

Klotz, S. (2013). *Techniques in High Pressure Neutron Scattering*. CRC Press, Taylor & Francis Group.

Klotz, S., Casula, M., Komatsu, K., Machida, S., Hattori, T. (2019). High-pressure structure and electronic properties of YbD_2 to 34 GPa. *Physical Review B*, **100**, 020101(R).

Komatsu, K., Klotz, S., Nakano, S., et al. (2020). Developments of nano-polycrystalline diamond anvil cells for neutron diffraction experiments. *High Pressure Research*, **40**, 184–193. https://doi.org/10.1080/08957959.2020.1727465.

Kozlenko, D., Kichanov, S., Lukin, E., Savenko, B. (2018). The DN-6 neutron diffractometer for high-pressure research at half a megabar scale. *Crystals*, **8**(8), 331. https://doi.org/10.3390/cryst8080331.

Kuhs, W. F., Finney, J. L., Vettier, C., Bliss, D. V. (1984). Structure and hydrogen ordering in ices VI, VII and VIII by neutron powder diffraction. *Journal of Chemical Physics,* **81**, 3612.

Mao, H. K., Xu, J., Struzhkin, V. V., et al. (2001). Phonon density of states of iron up to 153 gigapascals. *Science,* **292**(5518), 914–916.

Podlesnyak, A., Loguillo, M., Rucker, G. M., et al. (2018). Clamp cell with in situ pressure monitoring for low-temperature neutron scattering measurements. *High Pressure Research*, 38(4), 482–492.

Sano-Furukawa, A., Hattori, T., Arima, H., et al. (2014). Six-axis multi-anvil press for high-pressure, high-temperature neutron diffraction experiments. *Review of Scientific Instruments,* **85**(11), 113905. https://doi.org/10.1063/1.4901095.

Sears, V. F. (1992). Neutron scattering lengths and cross sections. *Neutron News*, **3**(3), 26–37. https://doi.org/10.1080/10448639208218770.

Toby, B. H., Von Dreele, R. B. (2013). GSAS-II: The genesis of a modern open-source all purpose crystallography software package. *Journal of Applied Crystallography*, **46**(2), 544–549. https://doi.org/10.1107/S0021889813003531.

5

Light-Source Diffraction Studies of Planetary Materials under Dynamic Loading

SALLY J. TRACY

Fundamental data on planetary materials under extreme conditions are required to establish physics-based models of planetary interiors and impact events. Dynamic compression experiments provide a means of studying material properties under the conditions of planetary interiors. Experimental shock wave studies also present a unique capability to study impact phenomena in real time, providing insight into hypervelocity collisions relevant to planetary formation and evolution. Recent experimental developments have extended the types of measurements that are possible during the nanosecond to microsecond duration of shock experiments – opening entirely new lines of inquiry. New facilities that couple dynamic compression platforms with high-flux X-ray sources have allowed for in situ X-ray diffraction under dynamic loading. Such experiments can address a range of longstanding questions, including the following: What crystallographic phases are stable under what conditions? What is their thermoelastic behavior? When do they melt or vaporize? And what phases will form on release? Answers to these questions and others will provide input for next-generation models of the structure, dynamics, and evolution of planetary interiors as well and natural impact processes.

5.1 Introduction

Laboratory shock wave experiments have long played an important role in characterizing the properties of geophysical materials under high-pressure, high-temperature conditions (Asimow, 2015). Dynamic compression experiments generate pressure–temperature states comparable to planetary adiabats, permitting studies of deep planetary interiors. In addition, shock wave experiments present a unique capability to study natural impact events. Shock experiments present advantages over static studies, often providing distinct and complementary information. Most noteworthy, dynamic experiments provide access to a broad region of pressure–temperature space and can readily achieve pressures well beyond what is accessible with static methods. Shock compression experiments permit measurements of equations of state (McQueen 1962; Akins and Ahrens, 2002; Zhang et al., 2014), phase transitions (Brown and McQueen, 1986), melting

(Williams et al., 1987), sound velocities (Brown and McQueen, 1986), the Grüneisen parameter (Thomas and Asimow, 2013), and transport properties (Kondo et al., 1981). However, an historical limitation of shock wave studies has been that the in situ crystal structure of the high-pressure phase formed under dynamic compression is generally not known. Conclusive determination of the high-pressure phases that form under shock loading has required the development of new diagnostics including ultrafast X-ray diffraction (XRD).

In recent years, capabilities for dynamic compression have grown rapidly worldwide. The development of a range of new facilities that couple pulsed XRD with different dynamic compression platforms enables the direct determination of the atomic-level structure of materials under dynamic loading. The first efforts to combine XRD with shock loading were relatively low-pressure single-crystal investigations, typically focused on tracking a single Bragg peak (Johnson et al., 1970). The development of light-source-based dynamic compression facilities at synchrotrons (Gupta et al., 2012) and free electron lasers (Milathianaki et al., 2013) has expanded this capability substantially. New facilities extend studies of planetary materials into the multimegabar regime and have allowed for the study of a wider range of samples, including polycrystals (Newman et al., 2018a), low-Z materials (Kraus et al., 2017), and melts (Morard et al., 2020). At these facilities, diffraction is combined with simultaneous velocimetry measurements to provide a complete picture of the material response from the atomic length scale to the continuum level. Such experiments can, for the first time, resolve longstanding ambiguities concerning the structure of planetary materials under dynamic loading and release for nanosecond to microsecond durations.

While this chapter focuses on light-source diffraction studies, parallel developments using plasma X-ray sources (Wark et al., 1987) at high-powered laser facilities, including the Omega Laser (University of Rochester) and the National Ignition Facility (Lawrence Livermore National Laboratory), now permit the collection of in situ XRD at conditions into the terapascal regime (Wicks et al., 2018; Lazicki et al., 2021). Moreover, other X-ray diagnostics such as absorption spectroscopy and imaging have now been reported at synchrotrons (Luo et al., 2012; Torchio et al., 2016), X-ray free electron lasers (Harmand et al., 2015; Schropp et al., 2015), and large-scale laser facilities (Denoeud et al., 2014). These advances can, in many cases, provide unique and complementary data to the studies highlighted in this chapter.

5.2 Shock Wave Experiments

A shock wave is a supersonic disturbance that propagates through a medium as a discontinuous jump in pressure, density, and particle velocity (Figure 5.1a). A material's Hugoniot defines the locus of pressure-density states achieved via shock loading. Conservation of mass, energy, and momentum across the shock front gives a set of equations known as the Rankine–Hugoniot equations, which relate the velocity of the

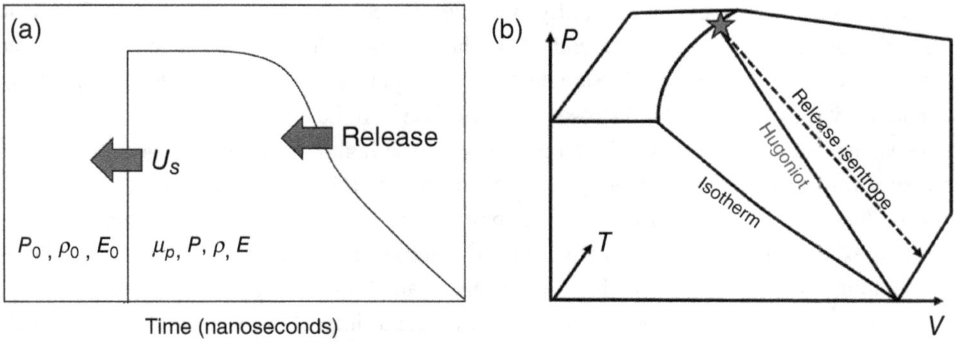

Figure 5.1 (a) Wave profile showing a shock traveling at velocity U_S and rapidly bringing a material from an ambient state (P_O, ρ_O, E_O) to the high-pressure state on the Hugoniot (u_p, P, ρ, E). (b) Schematic illustration of the Hugoniot compared to the isothermal EoS. Also shown is a release isentrope, illustrating the high-temperature release path.

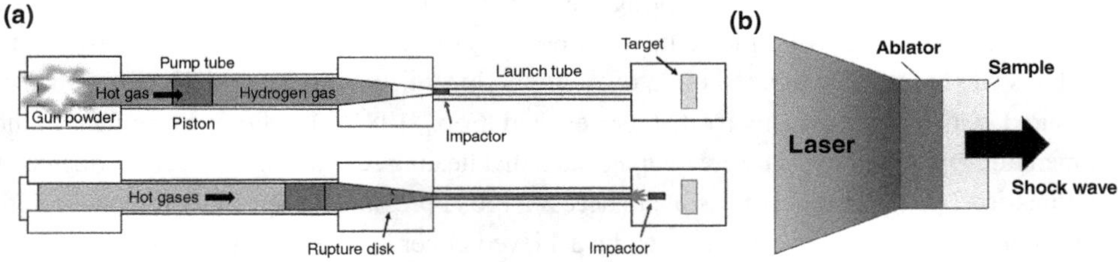

Figure 5.2 (a) Schematic of a two-stage gas gun experiment (after Fredriksson, CCBY-SA3.0). (b) Schematic of a laser-ablation shock experiment.

shock wave, U_S, and the particle velocity behind it, u_p, to the pressure, P, density, ρ, and specific internal energy, E:

$$\rho_0 U_S = \rho\left(U_S - u_p\right),$$

$$P - P_0 = \rho_0 U_S u_p,$$

$$E - E_0 = \frac{1}{2}\left(\frac{1}{\rho_0} - \frac{1}{\rho}\right)(P + P_0).$$

Here, the subscript 0 refers to the initial state ahead of the shock front. The Rankine–Hugoniot equations are central to shock physics because they relate the thermodynamic state variables of interest to velocities that can be measured. In this way, shock studies provide a means of directly assessing the equation of state (EoS) of a material without relying on an independent pressure standard.

Impact launchers provide one means of generating shock waves in the laboratory. In a two-stage light-gas-gun experiment (Figure 5.2a), an explosive charge drives a piston into a gas-filled pump tube compressing a gas (hydrogen or helium). Above some threshold pressure, a rupture disk bursts, separating the gas-filled pump tube from an evacuated launch tube. The gas expands rapidly into the launch tube, accelerating a projectile to velocities up to 7–8 km/s. At the end of the launch tube, the projectile impacts the target, sending a shock wave through the sample. Gas-gun experiments have typical load durations of hundreds of nanoseconds to microseconds and use ~mm-thick samples with ~cm lateral dimensions. Alternatively, laser-driven experiments use high-power lasers to generate a compression wave (Figure 5.2b). A pulsed laser is focused on a low-Z ablation layer, generating a plasma that expands backward sending a compression wave forward into the sample. Laser ablation experiments typically have durations of nanoseconds and provide a means of achieving multimegabar pressures with 10–100 μm thick samples of ~mm lateral dimensions.

Whether generated by a gas gun, a laser, or a natural impact event, a shock wave rapidly brings a material from ambient conditions to a transient high-pressure high-temperature state. As compared to the isothermal EoS measured in static compression experiments, the Hugoniot defines both a steeper pressure-density trend and involves significant heating (Figure 5.1b). Shock wave experiments generally have a duration of nanoseconds to microseconds, while natural shock events can extend to seconds, depending on the size of the impactor. In either case, a release wave will eventually bring the material back to ambient pressure.

While the Rankine–Hugoniot equations do not constrain the temperature, shock temperatures can be calculated based on known thermodynamic parameters (McQueen, 1962) or measured with optical pyrometry (McQueen and Fritz, 1982). In shock experiments, the temperature and shock pressure are coupled such that heating scales with the relative degree of compression (ρ/ρ_0). Thus, it is possible to access a broad pressure–temperature region through control of the initial conditions. This can be achieved either by altering the initial temperature (Stewart and Ahrens, 2005; Fat'yanov et al., 2018) or the initial density. High-temperature states can be explored with porous samples (Jeanloz, 1979), while lower temperatures can be accessed either via precompression methods (Jeanloz et al., 2007; Seagle and Lopez, 2015) or by using dense, high-pressure polymorphs (Luo et al., 2002; Millot et al., 2020).

Release from the Hugoniot follows an approximately isentropic path, where pressure quenches rapidly, but the temperature can remain high for some time (Figure 5.1b). As a result, understanding high-temperature alterations that may occur on unloading becomes critical for the interpretation of recovered material and modeling impact events. High-pressure phases formed on compression may not be retained or may be substantially modified as a result of the high temperatures that occur on pressure release. In some instances, the high-pressure phase is quenchable, but more commonly, the material reverts to its ambient structure or another phase (crystalline or amorphous). For sufficiently high shock pressures, the release path may cross the melting or vaporization curve. Certain materials may also be prone to thermally induced decomposition upon release. Despite these challenges, the observation of high-pressure phases and the presence of certain microstructural features in recovered samples are often used to constrain natural impact conditions (Stöffler and Langenhorst 1994; Langenhorst and Deutsch, 2012; Gillet and El Goresy, 2013). With this type of postmortem analysis, there is often ambiguity as to which features and phases formed during compression and which represent modifications that occurred during the unloading process. Consequently, the study of recovered samples cannot be used to conclusively identify the structure of the high-pressure phase or capture the structural evolution on unloading.

5.3 Continuum Diagnostics

Wave profile measurements (u_p versus time) utilize velocimetry techniques to collect continuous records of particle velocity during loading and release by tracking the Doppler shift of laser light reflected off the sample. In certain cases, the sample is backed by a transparent window material (often lithium fluoride) and the sample–window interface velocity is measured. Interferometry-based velocimetry diagnostics include the velocity interferometer system for any reflector (VISAR) (Barker and Hollenbach, 1972) and photon Doppler velocimetry (PDV) (Strand et al., 2006). The shock velocity (U_s) can be determined by measuring the shock wave transit time through a sample of known thickness. With known shock velocity and initial density, the particle velocity profile can be expressed as the evolution of the pressure-density state during the loading and release process via the Rankine–Hugoniot equations. With this type of U_s–u_p measurement, the Hugoniot curve can be mapped out through a series of experiments (i.e., shots) to determine how the continuum density of the material varies with increasing shock pressure. Here, the continuum density describes the average macroscopic state of the material, as opposed to the atomic-level crystal density.

It is known empirically that most materials exhibit a linear U_s–u_p relationship. Phase transitions under shock loading give rise to changes in the compressibility with increasing pressure, appearing as kinks in the pressure-density Hugoniot or first-order discontinuities in the U_s–u_p trend. Figure 5.3a shows Hugoniot data for forsterite (Mg_2SiO_4) (Jackson and Ahrens, 1979; Syono et al., 1981; Watt and Ahrens, 1983; Mosenfelder et al., 2009). Based on the breaks in the linear U_s–u_p curve (Figure 5.3b), these data have been interpreted as indicating a phase transition initiating at ~50 GPa with a mixed-phase region extending to ~100 GPa (Kalashnikov et al., 1973; Van Thiel, 1977), although the structure of the high-pressure phase is unknown.

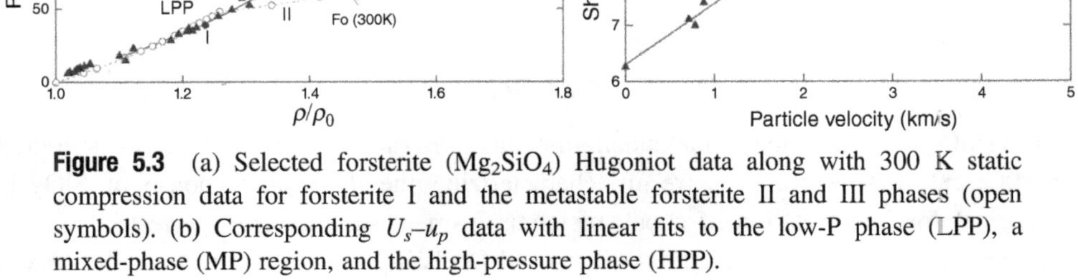

Figure 5.3 (a) Selected forsterite (Mg_2SiO_4) Hugoniot data along with 300 K static compression data for forsterite I and the metastable forsterite II and III phases (open symbols). (b) Corresponding U_s–u_p data with linear fits to the low-P phase (LPP), a mixed-phase (MP) region, and the high-pressure phase (HPP).

While it is generally straightforward to identify phase transitions based on this type of continuum Hugoniot data, there is often considerable ambiguity in the atomic-level structure of the high-pressure phase(s). It is often assumed that the high-pressure phase formed under shock compression is the stable structure observed under static compression or the structure predicted using density functional theory, although in most cases, there is no direct evidence demonstrating this. Due to the fast time scales of shock wave experiments, kinetic effects often result in shifts in phase boundaries compared to what is observed in static studies, metastable phases, or even in different sequences of phase transitions. Moreover, for phase changes with small volume changes, it can be challenging to discern transitions by relying solely on this type of continuum data.

5.4 In Situ X-Ray Diffraction under Plate Impact Shock Loading

The Dynamic Compression Sector (DCS) is a beamline at the Advanced Photon Source (APS), a third-generation synchrotron facility at Argonne National Laboratory (Gupta et al., 2012). DCS makes use of the tunable, high-energy X-ray capabilities at APS for single-pulse diffraction and imaging studies of materials during shock loading and release. DCS has a range of experimental stations coupling different dynamic loading platforms with time-resolved diagnostics, including an impact launcher station and a recently opened laser-shock station (Broege et al., 2019).

The impact station at DCS contains a two-stage light gas gun providing a range of projectile velocities up to 6.5 km/s. Targets are typically impacted with a 10 mm diameter lithium fluoride (LiF) single crystal mounted in a polycarbonate projectile and diffraction is carried out in transmission geometry (Figure 5.4a), collecting four XRD frames over the ~600 ns loading and release process. The incoming projectile interrupts a fast photodiode directly in front of the impact surface, triggering the gated collection of a series of XRD images as the shock traverses the sample and during the subsequent release. A central point VISAR or PDV probe collects wave profile data. Multiple radially distributed PDV probes register the wave arrival at the rear surface to determine the timing and impact tilt.

To achieve sufficient flux for single-pulse experiments, DCS makes use of an ~1 keV bandwidth "pink" X-ray beam, delivering $\sim 10^9$ photons/bunch ($\sim$100 ps duration). The detector system utilizes a fast phosphor scintillator that converts scattered X-rays into visible light, which is then directed to one of four intensified CCD cameras. This allows for the collection of four XRD patterns using consecutive synchrotron pulses (153.4 ns frame spacing). The ability to collect multiple frames during each shot enables the study of the temporal evolution of structural changes during shock compression and release.

5.4.1 Silica

Silica (SiO_2) is one of the most abundant minerals of Earth's crust and is widely distributed in different rock types. As a result, characterizing the dynamic response of SiO_2 is important for interpretation of shock metamorphism in samples from terrestrial impact sites (Stöffler and Langenhorst 1994; Sharp and DeCarli 2006). Furthermore, silica serves

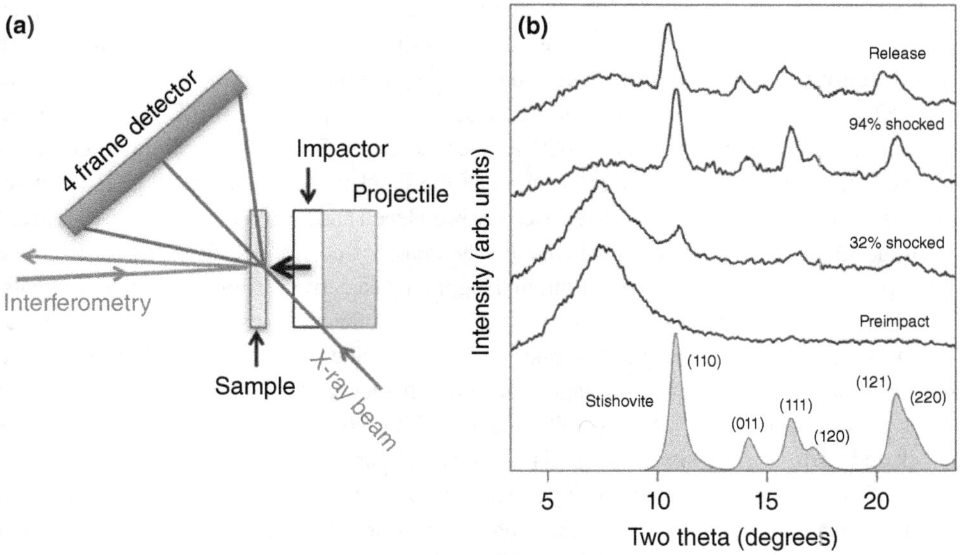

Figure 5.4 (a) Experimental setup for plate impact XRD experiments at DCS. A two-stage gas gun is used to impact a mineral sample with a projectile at high velocity. Synchrotron X-rays are directed through the sample, allowing its structure to be determined by XRD. (b) Series of four XRD frames collected for fused silica shock compressed to ~40 GPa, showing shock crystallization of stishovite. A simulated XRD pattern for the stishovite structure is shown below the experimental data.

as an archetype for silicate crystals and melts of planetary interiors. Hence, characterizing the high-pressure high-temperature behavior of SiO_2 is important for understanding potential silica-rich regions of the deep Earth and other terrestrial planets.

Silica has been one of the most extensively examined materials under dynamic compression. Early continuum gas-gun studies established that under shock loading, quartz transforms through an assumed mixed-phase region between 15–40 GPa to a dense high-pressure phase (Wackerle, 1962; Trunin et al., 1971; Grady et al., 1974). Based on these results, it has often been assumed that this high-pressure phase corresponds to the octahedrally coordinated stishovite structure observed in static experiments (McQueen et al., 1963; Akins and Ahrens, 2002). However, until recently, there was no direct evidence demonstrating this. Recovery experiments and samples collected at natural impact sites are found to consist primarily of amorphous material with trace quantities of stishovite (Sharp and DeCarli, 2006). Because of the high-temperature release path, it was an open question whether amorphization occurs during compression or release. On more general grounds, it has been suggested that reconstructive transformations involving tetrahedral to octahedral coordination changes are kinetically limited on shock wave time scales, and hence there is insufficient time for the formation of stishovite during rapid compression (Langenhorst and Deutsch, 2012). Accordingly, it has been contended that the high-pressure phase of quartz instead corresponds to a metastable intermediate or a dense amorphous structure (Chaplot and Sikka, 2000; Panero et al., 2003).

Recent experiments at DCS demonstrated shock crystallization of stishovite from silica glass (Tracy et al., 2018). This work was the first in situ determination of the crystal structure of the high-pressure phase of a shocked silicate in a gas-gun experiment. Figure 5.4b shows four XRD frames captured during shock loading of silica glass to ~40 GPa where the calculated shock temperature is ~3,600 K. Here, the first frame shows the ambient glass prior to impact. Frames 2 and 3 were captured as the shock wave traveled through the sample, and the final frame was captured on release. The broadened and asymmetric peak profiles result from the structure of the pink beam source. The densities determined from fits to stishovite XRD patterns correspond well to continuum Hugoniot data derived from U_s–u_p measurements. Stishovite grain sizes inferred from diffraction peak widths are consistent with the extrapolation of results from molecular dynamics simulations (Shen et al., 2016), which identify a homogenous nucleation and two-stage grain growth model for the formation of stishovite under shock compression. These results established a unified understanding of the previous shock and static compression data of this fundamental material.

In a follow-up gas-gun study using crystalline starting material, Tracy et al. employed a similar experimental design to reveal a phase transformation to a disordered metastable phase as opposed to crystalline stishovite along the α-quartz Hugoniot (Tracy et al., 2020). This result challenges longstanding assumptions about the dynamic response of quartz and suggests kinetics associated with reconstructive phase transformations can affect the high-pressure phase stability of silicate minerals under dynamic loading.

In another recent laser compression study, Gleason et al. reported evidence for the formation of stishovite from fused silica on nanosecond time scales and a reversion to amorphous silica on release (Gleason et al., 2015, 2017b). Time series taken during the compression process were used to estimate a nucleation time for stishovite of 1.4 ns, in good agreement with molecular dynamics simulations (Shen et al., 2016). While the highest-pressure datum in Gleason's study (35 GPa) overlaps the gas-gun results, stishovite peaks were also observed at 19 GPa and below, where gas-gun results showed only amorphous material despite two orders of magnitude longer compression times. Differences in these studies may result from strain-rate-dependent effects or systematic offsets in pressure determination.

5.4.2 Forsterite

Magnesium silicates are believed to comprise the bulk of Earth's mantle and are expected to be major phases in terrestrial planets. Forsterite (Mg_2SiO_4) is the Mg-rich endmember of the olivine series and represents a key planetary building block (Mustard et al., 2005). Forsterite (Fo) is also found in meteorites (Mason, 1963), comets (Hanner, 1999), and in accretion disks around young stars (van Boekel et al., 2004). The dynamic behavior of olivine is also of interest for understanding shock metamorphism generated by hypervelocity impacts on planetary bodies (Langenhorst, 2002) and for interpreting the dynamic history of meteorites (Gillet et al., 2007).

Experimental shock wave data on forsterite has been applied to understanding the interiors of Earth and other planets. However, valid interpretation of these data hinges on assuming a correct phase assemblage under dynamic compression. In heated static compression experiments, forsterite breaks down above ~20 GPa to form an assemblage of periclase (MgO) and perovskite ($MgSiO_3$) (Ito and Takahashi 1989; Fei et al., 2004). Figure 5.3 shows a collection of forsterite Hugoniot data (Jackson and Ahrens, 1979; Syono et al., 1981; Watt and Ahrens, 1983; Mosenfelder et al., 2007). These data have been interpreted as a transformation to a high-pressure phase through a mixed-phase region that begins around 50 GPa and is completed near 100 GPa.

Until recently, the phase transition under shock loading was believed to be the same disproportionation reaction observed in static experiments. The significant overpressure required to complete the phase transition was attributed to sluggish kinetics. This interpretation was revisited in a recent plate-impact study carried out at DCS (Newman et al., 2018b). In contrast to the previous interpretation, this study reported that forsterite does not break down into periclase and perovskite but instead undergoes a transformation to a metastable forsterite III phase for shock pressures between 44–73 GPa. The Fo III phase is a six-coordinated structure that has been reported from 300 K single-crystal XRD (Finkelstein et al., 2014). Figure 5.3a shows the 300 K static compression curve for Forsterite, illustrating a series of phase transitions through the metastable Fo II and Fo III phases.

The recent DCS study used sintered polycrystalline starting material. The Fo III structure was verified and lattice parameters were refined using fits to the diffraction profiles. This result demonstrates a breakthrough capability to refine a high-pressure crystal structure of a low-symmetry phase using a polycrystalline sample. Moreover, this study indicates that kinetics can dominate material properties on shock wave time scales, affecting the interpretation of the observed EoS. This result has important implications for the interpretation of dynamic loading data for systems that are expected to undergo subsolidus phase segregation.

5.4.3 Diamond

Diamond is another important planetary material. Understanding the phase transformation behavior of carbon under shock loading is critical to interpretations of shock metamorphism at meteorite impact sites. The observation of nanometer-sized grains of the high-pressure hexagonal diamond polymorph (lonsdaleite) is often used as a marker of meteorite impact sites. Theoretical investigations indicate a higher activation barrier for lonsdaleite formation from graphitic carbon compared to cubic diamond (Tateyama et al., 1996; Khaliullin et al., 2011), and there has been continued debate over the (meta)stability and formation conditions of the hexagonal diamond phase.

In a recent study at DCS, Turneaure et al. collected in situ XRD for graphite under plate-impact shock loading and reported a full transformation to the hexagonal lonsdaleite phase at 50 GPa (Turneaure et al., 2017). Highly oriented pyrolytic graphite was shocked along the c-axes, and in situ XRD data were interpreted as a strained hexagonal diamond structure with the (100) plane parallel to the graphite basal plane. This finding conflicted with a recent free electron laser study reporting cubic diamond formation for graphite laser-shocked above 55 GPa and hexagonal diamond for shock stresses only above ~170 GPa (Kraus et al., 2016). In another DCS study, Volz et al. (2020) reconciled these observations through a comparison of the phase transformation behavior of highly oriented pyrolytic graphite and turbostratic carbon with a high degree of mosaic spread. This study established that highly oriented pyrolytic graphite transforms to textured hexagonal diamond at ~50 GPa, while turbostratic graphite transforms to nanograined cubic diamond at ~60 GPa. These results suggest that the observed high-pressure structure of shocked graphite depends strongly on the initial crystal structure and demonstrates the capability of DCS to collect high-quality diffraction data for low-Z materials.

5.5 Laser-Shock Studies at X-Ray Free Electron Laser Sources

In recent years, X-ray free electron laser (XFEL) facilities have redefined what is possible in ultrafast science (McNeil and Thompson, 2010; Bostedt et al., 2016). XFEL sources generate femtosecond X-ray bursts with peak X-ray brightness 10 orders of magnitude higher than that of the best synchrotron sources. In self-amplified spontaneous emission mode (SASE), the XFEL delivers quasimonochromatic (1% $\Delta E/E$) 50 fs pulses containing $\sim 10^{12}$ photons/pulse. These intense, ultrafast pulses make XFELs an excellent platform for characterizing the time-dependent structural response of materials undergoing shock-induced phase transformations. A number of XFEL facilities have now established dedicated dynamic compression beamlines to explore the dynamic response of materials under ultrafast loading. Such facilities can be used to collect high-quality XRD data on the nanosecond timescales of laser-shock experiments. The first such facility was the Matter in Extreme Conditions (MEC) beamline located at Linac Coherent Light Source (LCLS) at SLAC National Accelerator Laboratory (Nagler et al., 2015). Since LCLS came online, additional hard-XFEL sources have built dedicated laser-shock stations, including the SPring-8 Angstrom Compact Free Electron Laser (SACLA) in Japan, and more recently, the HED beamline of the European XFEL (EuXFEL) in Hamburg, Germany.

New beamlines at LCLS, SACLA, and EuXFEL combine the unparalleled XFEL brightness with a high-powered optical drive laser. These facilities present a unique opportunity to study weakly scattering materials with complex diffraction patterns. A variety of studies (Gleason et al., 2015, 2017a; Kraus et al., 2016; Albertazzi et al., 2017; Schoelmerich et al., 2020) have now demonstrated the capabilities to collect diffraction data at high pressures for various types of materials, including those with low crystalline symmetry.

The long-pulse laser at MEC can deliver up to 60 J over 5–20 ns, achieving an irradiance of ~10^{12} W/cm^2. At MEC, the frequency-doubled Nd:glass drive laser ($\lambda = 527$ nm) can be run at a repetition rate of one shot every seven minutes. For shock experiments, the laser is focused onto the sample with a spot size of 100–600 μm using phase plates for beam smoothing and conditioning. The pressure is controlled by both changing the laser energy and tuning the laser spot size. At MEC, it is possible to routinely achieve shock pressures above 200 GPa with ~nanosecond load durations. LCLS recently completed a major upgrade and now supplies higher-energy X-rays up to 25 keV. Higher energies allow for the use of thicker samples and provide considerably more two theta coverage. This can be important for the unique identification and structural solution of complex diffraction patterns from low-symmetry crystal structures.

Figure 5.5a shows a schematic of the experimental configuration at MEC. The drive laser impinges on the target as two overlapping beams at 20° to the target normal. The

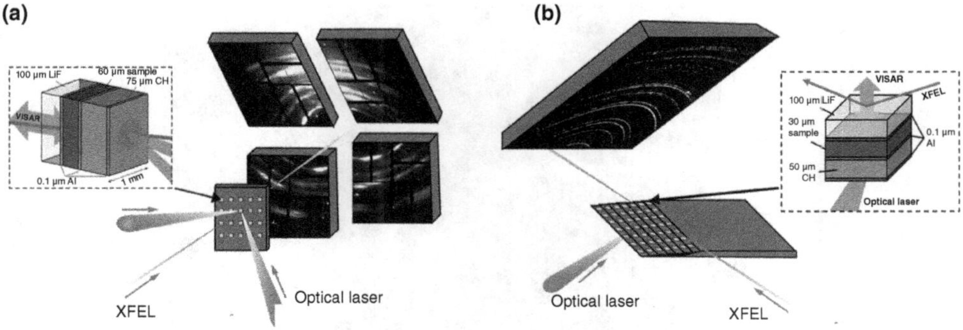

Figure 5.5 (a) Experimental configuration for laser-driven shock experiments at MEC showing the set of detectors positioned about direct beam and the translatable target cassette. Two arms of the drive laser impinge on the target at 20° to the target normal. (b) Experimental configuration at SACLA's BL3-EH5 showing the detector positioned at an angle to intercept diffracted X-rays in a grazing incidence refection geometry. Insets show schematics of typical target packages with a polycarbonate (CH) ablator, a sample, and LiF window. For the color version, refer to the plate section.
After Schoelmerich et al. (2020).

XFEL is focused with Be compound refractive lenses to a ~20 μm diameter spot size at the center of the laser spot. Angular-dispersive XRD is carried out in transmission geometry using multiple pixel-array detectors positioned about the incident beam for wide angular coverage in both two theta and the azimuthal direction. Detectors were developed at SLAC specifically for XFEL applications and provide low noise and an ultrahigh dynamic range (Philipp et al., 2011; Blaj et al., 2019). Figure 5.5b shows the experimental configuration at SACLA's BL3-EH5 beamline. Similar to MEC, BL3 houses a nanosecond Nd:Yag laser (532 nm) with a maximum energy around 60 J for a 10 ns pulse. In comparison to MEC, the geometry at SACLA enables experiments in a reflection geometry with a grazing-incidence angle of between the sample surface and XFEL beam (Figure 5.5b).

At both MEC and BL3-EH5 targets are affixed to a translatable cassette mounted on a motorized target stage with six degrees of freedom. The target stage and detectors are housed within a large vacuum chamber such that several tens of targets can be aligned to the target center and shot without venting the chamber. The line-imaging VISAR provides continuum information, allowing for pressure determination and wave profile data collection. The VISAR can also provide information on X-ray timing and positioning as well as for identifying any nonplanarity of the load. The line VISAR at MEC relays a spatially resolved image through the interferometer and onto the slit of a streak camera, providing an up to 300 μm field of view with sweep durations up to 50 ns. Shifts in the recorded fringe pattern correspond to phase shifts, which can be used to calculate the particle velocity after the shock reaches the rear sample surface. Two VISAR systems are used with different velocity sensitivities. This provides redundant data and resolves ambiguities associated with sharp velocity jumps that exceed the time response of the system. Figure 5.6 shows an example interferogram collected at MEC along with the associated analyzed wave profile.

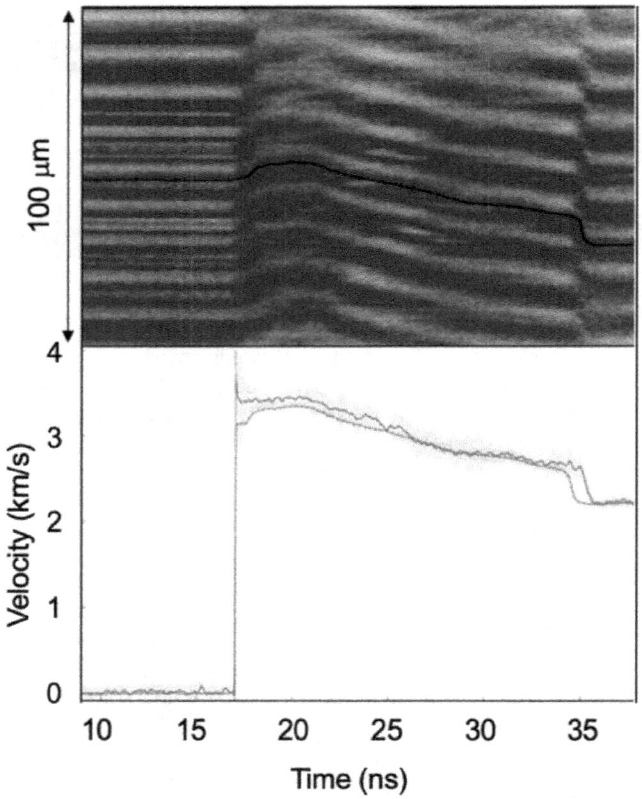

Figure 5.6 Line VISAR data showing the 2D interferogram from the streak camera with the wave profiles below. The two traces are from independent VISARs with different sensitivities. For the color version, refer to the plate section.

5.5.1 Silicate Liquids and Glasses

The phase relations and EoS of silicates at planetary interior conditions are necessary to constrain the formation, evolution, and internal structure of Earth and other rocky planets. Measuring structural properties of silicate melts over the pressure–temperature range of planetary interiors presents a significant challenge but is critical to modeling the solidification dynamics of the Earth (Stixrude et al., 2009; Elkins-Tanton, 2012). In addition, super-Earth exoplanets may possess long-lived magma oceans that can affect their magnetic field and potential dynamo formation (Harada et al., 2014; Soubiran and Militzer, 2018). Thus, understanding the structure and properties of molten silicates is important for both understanding Earth's history and for interpreting the structure and potential habitability of terrestrial exoplanets.

Silicate glasses are often adopted as analogs for understanding the high-pressure behavior of silicate melts at the conditions of planetary interiors. A recent laser-shock study carried out at MEC used in situ XRD to show that $MgSiO_3$ glass does not crystallize along the Hugoniot and remains amorphous up until the expected melt transition between 90–110 GPa (Morard et al., 2020). Beyond an increase in silicon coordination at ~20 GPa, there was no further indication of any major structural changes along the Hugoniot up to

130 GPa and 6,000 K. A comparison of the diffraction data collected under shock loading (both below and above melting) to 300 K static compression data shows comparable overall features. Although the limited Q range ($Q = 4\pi\sin\theta/\lambda$) at MEC precludes a full structural analysis, the data show no evidence for major differences between the liquid and cold-compressed silicate glass. These results suggest glasses provide a suitable proxy for silicate liquids under pressure and demonstrate a capability to collect amorphous and liquid diffraction for silicate materials at the conditions of planetary interiors.

New developments at the Eu-XFEL and LCLS-II provide higher-energy photons (up to 25 keV) and improved detector coverage, substantially increasing both azimuthal and Q-coverage. With these new capabilities, in the near future it may be possible to revisit the high-pressure structure of silicate liquids and glasses and derive quantitative structures via pair distribution function analysis.

5.5.2 Hydrocarbons

The carbon–hydrogen system is important for planetary science. It is believed that along with water and ammonia, methane comprises a major component of the ice giants. This includes Uranus and Neptune and their satellites as well as icy-giant exoplanets. Understanding the phase relations and chemical stability of planetary ices is critical for modeling the structure and thermal evolution of these bodies. At high pressure and temperature, hydrocarbons such as methane are expected to decompose and precipitate diamond (Ross, 1981). Carbon–hydrogen separation may result in the formation of diamond-containing layers in the deep planetary interiors of ice giants (Benedetti, 1999). Additionally, this type of separation could influence convective heat transport (Nettelmann et al., 2016). Recent laser-shock experiments carried out at MEC compress polystyrene samples to 150 GPa and 5,000 K and show experimental evidence for phase separation and diamond formation using in situ XRD (Kraus et al., 2017). These results provide constraints on the internal structure of ice giants and potentially provide data that can improve our understanding of the thermal evolution of planets such as Uranus and Neptune.

5.5.3 Carbides

Stars with sufficiently high carbon-to-oxygen ratios may host exoplanets composed largely of carbides rather than the silicates and oxides of terrestrial planets. The interior of this class of exoplanets likely includes a mantle rich in silicon carbide (SiC). The behavior of SiC has been explored under both static and shock compression. However, extrapolations of existing studies diverge strongly when extended to higher pressures. Early experiments identified a phase change under 300 K static compression to a rocksalt-type (B1) structure near 100 GPa (Yoshida et al., 1993). This structural transformation involves a change from fourfold to sixfold coordination and is accompanied by a ~20% volume collapse (Daviau and Lee, 2017; Kidokoro et al., 2017; Miozzi et al., 2018). Gas-gun shock wave experiments showed evidence of a similar phase transition near 100 GPa, with a mixed-phase region extending up to ~140 GPa (Sekine and Kobayashi, 1997; Vogler et al., 2006).

A phase transformation involving a significant increase in density should play a major role in the interior structure, dynamics, and heat loss rate of a carbon planet.

A recent laser-shock study at LCLS demonstrates the transformation to the high-pressure B1 phase under shock loading (Tracy et al., 2019). This study represents a significant extension of previous gas-gun and diamond cell results, reaching pressures exceeding 200 GPa. A coexistence of the low- and high-pressure phases was observed in a mixed-phase region, and complete transformation to the B1 phase was observed above 150 GPa. The densities measured by XRD are in agreement with both continuum gas-gun studies and a theoretical B1 Hugoniot derived from static-compression data. Time-resolved measurements during shock loading and release reveal a large hysteresis upon unloading, with the B1 phase retained for several nanoseconds to pressures as low as 5 GPa before reverting to the ambient structure (Figure 5.7).

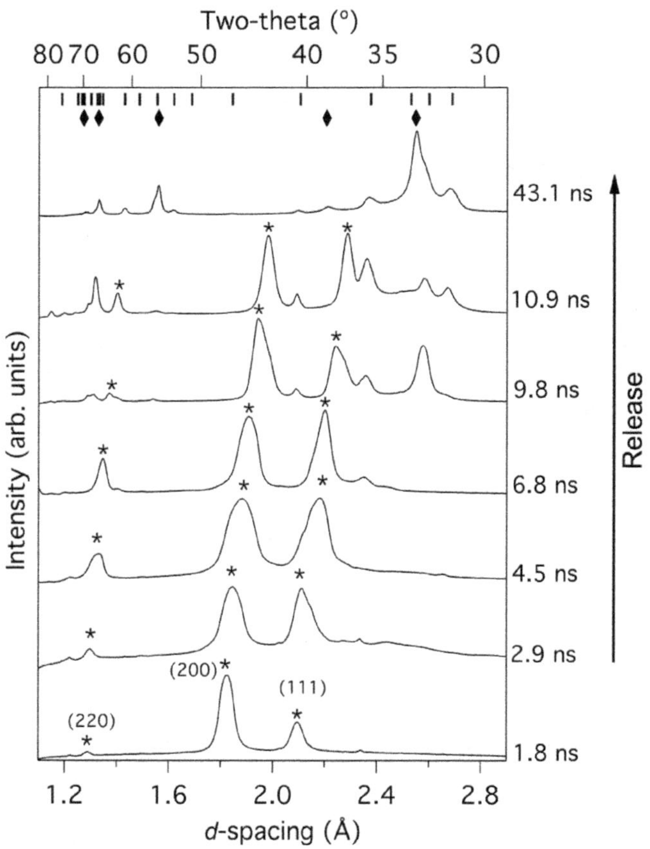

Figure 5.7 XRD patterns for a time series collected at a peak stress of ~175 GPa for single-crystal α-SiC starting material (4H phase). X-ray probe times after the shock enters SiC are listed at the right. The shock transit time through the sample is ~ 2 ns. The bottom pattern was collected during a 175 GPa uniform stress state. The rest of the patterns were collected on release. High-pressure B1 peaks are marked with asterisks. For the latest probe time (43.1 ns), the pattern can be indexed with a combination of peaks from the low-pressure 3C (diamonds) and 4H (lines) phases, indicated by symbols above the pattern.

5.6 Conclusions and Outlook

Capabilities for dynamic compression are growing rapidly worldwide, creating new opportunities to study planetary materials over a range of pressure, temperature, and strain rates. New large-scale user facilities include hard-XFEL sources (LCLS-II, EuXFEL, SACLA) and beamlines at synchrotron facilities (ESRF upgrade with dynamic compression capabilities, DCS). The European XFEL promises the capability of high-power laser experiments with shot repetition rates up to 10 Hz, considerably enhancing the magnitude of data collected and potentially altering the design and operation of experiments. The higher photon energies available at both the EuXFEL and LCLS-II coupled with improved detector coverage should increase the Q coverage substantially. This opens new capabilities to perform quantitative studies of liquids. This is also critical for the unique identification and structural solution of complex diffraction patterns from low-symmetry crystal structures. This is particularly important at higher pressures, where all reflections shift to higher Q. Furthermore, iron-bearing minerals suffer from fluorescence at lower energies. By increasing the X-ray energy, the absorption length increases, allowing for the use of thicker samples and significantly improving the diffraction signal.

The development of light-source-based facilities for ultrafast diffraction under dynamic loading is a major advancement in dynamic compression science, providing unprecedented data on the crystal structures and phase stability of planetary materials under extreme conditions. The lack of in situ crystal structure information under dynamic loading has long presented a major limitation, resulting in an ambiguous interpretation of the continuum shock wave data for a range of key planetary materials. New facilities that couple single-pulse XRD with different dynamic loading platforms permit unambiguous structure determination and identification of phase transitions under conditions of planetary interiors and natural impact events. Together with theoretical calculations and astronomical observations, diagnostic advances for dynamic compression will continue to forward our understanding of planets in our solar system and beyond.

References

Akins, J. A., Ahrens, T. J. (2002). Dynamic compression of SiO_2: a new interpretation. *Geophysical Research Letters*, **29**(10), 31-1–31-4.

Albertazzi, B., Ozaki, N., Zhakhovsky, V., et al. (2017). Dynamic fracture of tantalum under extreme tensile stress. *Science Advances*, **3**(6), e1602705.

Asimow, P. D. (2015). Dynamic compression, in G. Schubert, ed., *Treatise on Geophysics*, Elsevier, pp. 393–416.

Barker, L. M., Hollenbach, R. E. (1972). Laser interferometer for measuring high velocities of any reflecting surface. *Journal of Applied Physics*, **43**(11), 4669–4675.

Benedetti, L. R. (1999). Dissociation of CH4 at high pressures and temperatures: diamond formation in giant planet interiors? *Science*, **286**(5437), 100–102.

Blaj, G., Dragone, A., Kenney, C. J., et al. (2019). Performance of ePix10K, a high dynamic range, gain auto-ranging pixel detector for FELs. *AIP Conference Proceedings*, **2054**(1), 060062.

Bostedt, C., Boutet, S., Fritz, D. M., et al. (2016). Linac coherent light source: the first five years. *Reviews of Modern Physics*, **88**(1), 015007.

Broege, D., Fochs, S., Brent, G., et al. (2019). The Dynamic Compression Sector laser: a 100-J UV laser for dynamic compression research. *Review of Scientific Instruments*, **90**(5), 053001.

Brown, J. M., McQueen, R. G. (1986). Phase transitions, Grüneisen parameter, and elasticity for shocked iron between 77 GPa and 400 GPa. *Journal of Geophysical Research: Solid Earth*, **91**(B7), 7485–7494.

Chaplot, S. L., Sikka, S. K. (2000). Molecular-dynamics simulation of shock-stress-induced amorphization of α-quartz. *Physical Review B*, **61**(17), 11205–11208.

Daviau, K., Lee, K. K. M. (2017). Zinc-blende to rocksalt transition in SiC in a laser-heated diamond-anvil cell. *Physical Review B*, **95**(13), 134108.

Denoeud, A., Benuzzi-Mounaix, A., Ravasio, A., et al. (2014). Metallization of warm dense SiO 2 studied by XANES spectroscopy. *Physical Review Letters*, **113**(11), 116404.

Elkins-Tanton, L. T. (2012). Magma oceans in the inner solar system. *Annual Review of Earth and Planetary Sciences*, **40**(1), 113–139.

Fat'yanov, O. V., Asimow, P. D., Ahrens, T. J. (2018). Thermodynamically complete equation of state of MgO from true radiative shock temperature measurements on samples preheated to 1850 K. *Physical Review B*, **97**(2), 024106.

Fei, Y., Van Orman, J., Li, J., et al. (2004). Experimentally determined postspinel transformation boundary in Mg_2SiO_4 using MgO as an internal pressure standard and its geophysical implications. *Journal of Geophysical Research: Solid Earth*, **109**(B2). doi:10.1029/2003JB002562

Finkelstein, G. J., Dera, P. K., Jahn, S., et al. (2014). Phase transitions and equation of state of forsterite to 90 GPa from single-crystal X-ray diffraction and molecular modeling. *American Mineralogist*, **99**(1), 35–43.

Gillet, P., El Goresy, A. (2013). Shock events in the solar system: the message from minerals in terrestrial planets and asteroids. *Annual Review of Earth and Planetary Sciences*, **41**(1), 257–285.

Gillet, P., El Goresy, A., Beck, P., Chen, M. (2007). High-pressure mineral assemblages in shocked meteorites and shocked terrestrial rocks: mechanisms of phase transformations and constraints to pressure and temperature histories, *Geological Society of America Special Papers*, 421, 57–82.

Gleason, A. E., Bolme, C. A., Galtier, E., et al. (2017a). Compression freezing kinetics of water to ice VII. *Physical Review Letters*, **119**(2), 025701.

Gleason, A. E., Bolme, C. A., Lee, H. J., et al. (2015). Ultrafast visualization of crystallization and grain growth in shock-compressed SiO_2. *Nature Communications*, **6**(1), 8191.

Gleason, A. E., Bolme, C. A., Lee, H. J., et al. (2017b). Time-resolved diffraction of shock-released SiO_2 and diaplectic glass formation. *Nature Communications*, **8**(1), 1481.

Grady, D. E., Murri, W. J., Fowles, G. R. (1974). Quartz to stishovite: wave propagation in the mixed phase region. *Journal of Geophysical Research*, **79**(2), 332–338.

Gupta, Y. M., Turneaure, S. J., Perkins, K., et al. (2012). Real-time, high-resolution X-ray diffraction measurements on shocked crystals at a synchrotron facility. *Review of Scientific Instruments*, **83**(12), 123905.

Hanner, M. S. (1999). The silicate material in comets, in K. Altwegg, P. Ehrenfreund, J. Geiss, W. F. Huebner, eds., *Composition and Origin of Cometary Materials*. Springer Netherlands, pp. 99–108.

Harada, Y., Goossens, S., Matsumoto, K., et al. (2014). Strong tidal heating in an ultralow-viscosity zone at the core–mantle boundary of the Moon. *Nature Geoscience*, **7**(8), 569–572.

Harmand, M., Ravasio, A., Mazevet, S., et al. (2015). X-ray absorption spectroscopy of iron at multimegabar pressures in laser shock experiments. *Physical Review B*, **92**(2), 024108.

Ito, E., Takahashi, E. (1989). Postspinel transformations in the system Mg_2SiO_4-Fe_2SiO_4 and some geophysical implications. *Journal of Geophysical Research: Solid Earth*, **94**(B8), 10637–10646.

Jackson, I., Ahrens, T. J. (1979). Shock wave compression of single-crystal forsterite. *Journal of Geophysical Research*, **84**(B6), 3039.

Jeanloz, R. (1979). Properties of iron at high pressures and the state of the core. *Journal of Geophysical Research*, **84**(B11), 6059.

Jeanloz, R., Celliers, P. M., Collins, G. W., et al. (2007). Achieving high-density states through shock-wave loading of precompressed samples. *Proceedings of the National Academy of Sciences*, **104**(22), 9172–9177.

Johnson, Q., Mitchell, A., Keeler, R. N., Evans, L. (1970). X-ray diffraction during shock-wave compression. *Physical Review Letters*, **25**(16), 1099–1101.

Kalashnikov, N. G., Pavlovskii, M. N., Simakov, G. V., Trunin, R. F. (1973). Dynamic compressibility of the minerals of calcite group. *Izv. Akad. Nauk SSSR. Fiz. Zemli*, **2**, 23–29.

Khaliullin, R. Z., Eshet, H., Kühne, T. D., Behler, J., Parrinello, M. (2011). Nucleation mechanism for the direct graphite-to-diamond phase transition. *Nature Materials*, **10** (9), 693–697.

Kidokoro, Y., Umemoto, K., Hirose, K., Ohishi, Y. (2017). Phase transition in SiC from zinc-blende to rock-salt structure and implications for carbon-rich extrasolar planets. *American Mineralogist*, **102**(11), 2230–2234.

Kondo, K., Sawaoka, A., Ahrens, T. J. (1981). Electrical measurements on fused quartz under shock compression. *Journal of Applied Physics*, **52**(8), 5084–5089.

Kraus, D., Ravasio, A., Gauthier, M., et al. (2016). Nanosecond formation of diamond and lonsdaleite by shock compression of graphite. *Nature Communications*, **7**(1), 10970.

Kraus, D., Vorberger, J., Pak, A., et al. (2017). Formation of diamonds in laser-compressed hydrocarbons at planetary interior conditions. *Nature Astronomy*, **1**(9), 606–611.

Langenhorst, F. (2002). Shock metamorphism of some minerals: basic introduction and microstructural observations. *Bulletin of the Czech Geological Survey*, **77**, 265–282.

Langenhorst, F., Deutsch, A. (2012). Shock metamorphism of minerals. *Elements*, **8**(1), 31–36.

Lazicki, A., McGonegle, D., Rygg, J. R., et al. (2021). Metastability of diamond ramp-compressed to 2 terapascals. *Nature*, **589**(7843), 532–535.

Livermore: Lawrence Livermore Laboratory. (1977). Compendium of shock wave data (No. Report UCRL-50108).

Luo, S. N., Jensen, B. J., Hooks, D. E., et al. (2012). Gas gun shock experiments with single-pulse X-ray phase contrast imaging and diffraction at the Advanced Photon Source. *Review of Scientific Instruments*, **83**(7), 073903.

Luo, S.-N., Mosenfelder, J. L., Asimow, P. D., Ahrens, T. J. (2002). Direct shock wave loading of stishovite to 235 GPa: implications for perovskite stability relative to an oxide assemblage at lower mantle conditions. *Geophysical Research Letters*, **29**(14), 36-1–36-4.

Mason, B. (1963). Olivine composition in chondrites. *Geochimica et Cosmochimica Acta*, **27**, 1011–1023.

McNeil, B. W. J., Thompson, N. R. (2010). X-ray free-electron lasers. *Nature Photonics*, **4** (12), 814–821.

McQueen, R. G. (1962). Laboratory techniques for very high pressures and the behavior of metals under dynamic loading (No. LADC-5561, CONF-331-4, 4013771).

McQueen, R. G., Fritz, J. N. (1982). Some techniques and results from high-pressure shock-wave experiments utilizing the radiation from shocked transparent materials. *AIP Conference Proceedings*, **78**(1), 193–207.

McQueen, R. G., Fritz, J. N., Marsh, S. P. (1963). On the equation of state of stishovite. *Journal of Geophysical Research*, **68**(8), 2319–2322.

Milathianaki, D., Boutet, S., Williams, G. J., et al. (2013). Femtosecond visualization of lattice dynamics in shock-compressed matter. *Science*, **342**(6155), 220–223.

Millot, M., Zhang, S., Fratanduono, D. E., et al. (2020). Recreating giants impacts in the laboratory: shock compression of Bridgmanite to 14 Mbar. *Geophysical Research Letters*, **47**(4), e2019GL085476.

Miozzi, F., Morard, G., Antonangeli, D., et al. (2018). Equation of state of SiC at extreme conditions: new insight into the interior of carbon-rich exoplanets. *Journal of Geophysical Research: Planets*, **123**(9), 2295–2309.

Morard, G., Hernandez, J.-A., Guarguaglini, M., et al. (2020). In situ X-ray diffraction of silicate liquids and glasses under dynamic and static compression to megabar pressures. *Proceedings of the National Academy of Sciences*, **117**(22), 11981–11986.

Mosenfelder, J. L., Asimow, P. D., Ahrens, T. J. (2007). Thermodynamic properties of Mg_2SiO_4 liquid at ultra-high pressures from shock measurements to 200 GPa on forsterite and wadsleyite. *Journal of Geophysical Research*, **112**(B6), B06208.

Mosenfelder, J. L., Asimow, P. D., Frost, D. J., Rubie, D. C., Ahrens, T. J. (2009). The $MgSiO_3$ system at high pressure: thermodynamic properties of perovskite, postperovskite, and melt from global inversion of shock and static compression data. *Journal of Geophysical Research: Solid Earth*, **114**(B1). doi:10.1029/2008JB005900

Mustard, J. F., Poulet, F., Gendrin, A., et al. (2005). Olivine and pyroxene diversity in the crust of Mars. *Science*, **307**(5715), 1594–1597.

Nagler, B., Arnold, B., Bouchard, G., et al. (2015). The matter in extreme conditions instrument at the Linac Coherent Light Source. *Journal of Synchrotron Radiation*, **22** (3), 520–525.

Nettelmann, N., Wang, K., Fortney, J. J., et al. (2016). Uranus evolution models with simple thermal boundary layers. *Icarus*, **275**, 107–116.

Newman, M. G., Kraus, R. G., Akin, M. C., et al. (2018a). In situ observations of phase changes in shock compressed forsterite. *Geophysical Research Letters*, **45**(16), 8129–8135.

Newman, M. G., Kraus, R. G., Akin, M. C., et al. (2018b). In situ observations of phase changes in shock compressed forsterite. *Geophysical Research Letters*, **45**(16), 8129–8135.

Panero, W. R., Benedetti, L. R., Jeanloz, R. (2003). Equation of state of stishovite and interpretation of SiO_2 shock-compression data. *Journal of Geophysical Research: Solid Earth*, **108**(B1). doi:10.1029/2001JB001663.

Philipp, H. T., Hromalik, M., Tate, M., Koerner, L., Gruner, S. M. (2011). Pixel array detector for X-ray free electron laser experiments. *Nuclear Instruments and Methods in Physics Research Section A: Accelerators, Spectrometers, Detectors and Associated Equipment*, **649**(1), 67–69.

Ross, M. (1981). The ice layer in Uranus and Neptune – diamonds in the sky? *Nature*, **292** (5822), 435–436.

Schoelmerich, M. O., Tschentscher, T., Bhat, S., et al. (2020). Evidence of shock-compressed stishovite above 300 GPa. *Scientific Reports*, **10**(1), 10197.

Schropp, A., Hoppe, R., Meier, V., et al. (2015). Imaging shock waves in diamond with both high temporal and spatial resolution at an XFEL. *Scientific Reports*, **5**(1), 11089.

Seagle, C. T., Lopez, A. J. (2015). Pre-compressed targets for gas gun studies (No. SAND-2015-8118). Retrieved from www.osti.gov/servlets/purl/1504849/

Sekine, T., Kobayashi, T. (1997). Shock compression of 6H polytype SiC to 160 GPa. *Physical Review B*, **55**(13), 8034–8037.

Sharp, T. G., DeCarli, P. S. (2006). Shock effects in meteorites, in D. S. Lauretta, H. Y. McSween. Jr., eds. *Meteorites and the Early Solar System II*. University of Arizona Press, pp. 653–677.

Shen, Y., Jester, S. B., Qi, T., et al. (2016). Nanosecond homogeneous nucleation and crystal growth in shock-compressed SiO_2. *Nature Materials*, **15**(1), 60–65.

Soubiran, F., Militzer, B. (2018). Electrical conductivity and magnetic dynamos in magma oceans of Super-Earths. *Nature Communications*, **9**(1), 3883.

Stewart, S. T., Ahrens, T. J. (2005). Shock properties of H_2O ice. *Journal of Geophysical Research: Planets*, **110**(E3). doi:10.1029/2004JE002305

Stixrude, L., de Koker, N., Sun, N., Mookherjee, M., Karki, B. B. (2009). Thermodynamics of silicate liquids in the deep Earth. *Earth and Planetary Science Letters*, **278**(3–4), 226–232.

Stöffler, D., Langenhorst, F. (1994). Shock metamorphism of quartz in nature and experiment: I. Basic observation and theory. *Meteoritics*, **29**(2), 155–181.

Strand, O. T., Goosman, D. R., Martinez, C., Whitworth, T. L., Kuhlow, W. W. (2006). Compact system for high-speed velocimetry using heterodyne techniques. *Review of Scientific Instruments*, **77**(8), 083108.

Syono, Y., Goto, T., Sato, J., Takei, H. (1981). Shock compression measurements of single-crystal forsterite in the pressure range 15–93 GPa. *Journal of Geophysical Research: Solid Earth*, **86**(B7), 6181–6186.

Tateyama, Y., Ogitsu, T., Kusakabe, K., Tsuneyuki, S. (1996). Constant-pressure first-principles studies on the transition states of the graphite-diamond transformation. *Physical Review B*, **54**(21), 14994–15001.

Thomas, C. W., Asimow, P. D. (2013). Direct shock compression experiments on premolten forsterite and progress toward a consistent high-pressure equation of state for CaO-MgO-Al_2 O_3 -SiO_2 -FeO. *Journal of Geophysical Research: Solid Earth*, **118** (11), 5738–5752.

Torchio, R., Occelli, F., Mathon, O., et al. (2016). Probing local and electronic structure in warm dense matter: single pulse synchrotron X-ray absorption spectroscopy on shocked Fe. *Scientific Reports*, **6**(1), 26402.

Tracy, S. J., Smith, R. F., Wicks, J. K., et al. (2019). *In situ* observation of a phase transition in silicon carbide under shock compression using pulsed X-ray diffraction. *Physical Review B*, **99**(21), 214106.

Tracy, S. J., Turneaure, S. J., Duffy, T. S. (2018). *In situ* X-ray diffraction of shock-compressed fused silica. *Physical Review Letters*, **120**(13), 135702.

Tracy, S. J., Turneaure, S. J., Duffy, T. S. (2020). Structural response of α-quartz under plate-impact shock compression. *Science Advances*, **6**(35), eabb3913.

Trunin, R. F., Simakov, G. V., Podurets, M. A., Moiseyev, B. N., Popov, L. V. (1971). Dynamic compressibility of quartz and quartzite at high pressure. *Izvestiya, Academy of Science, USSR, Physics of the Solid Earth*, **1**, 8–11.

Turneaure, S. J., Sharma, S. M., Volz, T. J., Winey, J. M., Gupta, Y. M. (2017). Transformation of shock-compressed graphite to hexagonal diamond in nanoseconds. *Science Advances*, **3**(10), eaao3561.

van Boekel, R., Min, M., Leinert, Ch., et al. (2004). The building blocks of planets within the 'terrestrial' region of protoplanetary disks. *Nature*, **432**(7016), 479–482.

van Thiel, M. (1977). Compendium of shock wave data. Lawrence Livermore Laboratory Report, UCRL-50108, pp. 408–409.

Vogler, T. J., Reinhart, W. D., Chhabildas, L. C., Dandekar, D. P. (2006). Hugoniot and strength behavior of silicon carbide. *Journal of Applied Physics*, **99**(2), 023512.

Volz, T. J., Turneaure, S. J., Sharma, S. M., Gupta, Y. M. (2020). Role of graphite crystal structure on the shock-induced formation of cubic and hexagonal diamond. *Physical Review B*, **101**(22), 224109.

Wackerle, J. (1962). Shock-wave compression of quartz. *Journal of Applied Physics*, **33**(3), 922–937.

Wark, J. S., Whitlock, R. R., Hauer, A., Swain, J. E., Solone, P. J. (1987). Shock launching in silicon studied with use of pulsed x-ray diffraction. *Physical Review B*, **35**(17), 9391–9394.

Watt, J. P., Ahrens, T. J. (1983). Shock compression of single-crystal forsterite. *Journal of Geophysical Research: Solid Earth*, **88**(B11), 9500–9512.

Wicks, J. K., Smith, R. F., Fratanduono, D. E., et al. (2018). Crystal structure and equation of state of Fe-Si alloys at super-Earth core conditions. *Science Advances*, **4**(4), eaao5864.

Williams, Q., Jeanloz, R., Bass, J., Svendsen, B., Ahrens, T. J. (1987). The melting curve of iron to 250 gigapascals: a constraint on the temperature at Earth's center. *Science*, **236**(4798), 181–182.

Yoshida, M., Onodera, A., Ueno, M., Takemura, K., Shimomura, O. (1993). Pressure-induced phase transition in SiC. *Physical Review B*, **48**(14), 10587–10590.

Zhang, Y., Sekine, T., He, H., Yu, Y., Liu, F., Zhang, M. (2014). Shock compression of Fe-Ni-Si system to 280 GPa: implications for the composition of the Earth's outer core. *Geophysical Research Letters*, **41**(13), 4554–4559.

6

New Analysis of Shock-Compression Data for Selected Silicates

THOMAS DUFFY

The study of minerals under shock compression provides fundamental constraints on their response to conditions of extreme pressure, temperature, and strain rate and has applications to understanding meteorite impacts and the deep Earth. The recent development of facilities for real-time in situ X-ray diffraction studies under gun- or laser-based dynamic compression provides new capability for understanding the atomic-level structure of shocked solids. Here traditional shock pressure-density data for selected silicate minerals (garnets, tourmaline, nepheline, topaz, and spodumene) are examined through comparison of their Hugoniots with recent static compression and theoretical studies. The results provide insights into the stability of silicate structures and the possible nature of high-pressure phases under shock loading. This type of examination highlights the potential for in situ atomic-level measurements to address questions about phase transitions, transition kinetics, and structures formed under shock compression for silicate minerals.

6.1 Introduction

Laboratory shock-compression experiments have wide-ranging applications in geophysics, planetary science, and materials science (Langenhorst and Hornemann, 2005; Asimow, 2015). Shock wave studies provide data relevant to understanding planetary formation, shock metamorphism, and high-pressure phases of the deep Earth. Large-impact events played a major role in the early solar system and have implications for planetary accretion and the evolution of planetary surfaces, interiors, and atmospheres (Gillet and El Goresy, 2013). The rapid loading of rocks and minerals during impact events produces unique deformation features and characteristic phases that constrain impact conditions (Sharp and DeCarli, 2006; Gillet and El Goresy, 2013). Additionally, laboratory shock experiments provide a means to create high-pressure phases and investigate geophysically important properties, such as the equation of state, over a very wide range of pressure–temperature conditions, including those expected in deep planetary interiors (Asimow, 2015).

Silicate minerals have been a particular focus of shock-wave studies over the years. More than two dozen silicate minerals and silicate-rich rocks have been examined using in situ gas-gun shock-compression experiments, often up to pressures of 100 GPa or more

(Marsh, 1980; Ahrens and Johnson, 1995a, 1995b; Trunin, 2005). Due to their geological importance, shock data for a few key minerals (e.g., quartz, olivine, orthopyroxene, and feldspars) have been extensively analyzed to interpret their shock-induced phase transitions and constrain the physical properties of their high-pressure phases (Sekine and Ahrens, 1992; Akins et al., 2004; Luo et al., 2004; Mosenfelder et al., 2007, 2009; Asimow, 2015). But shock data for many other silicate minerals have received limited examination or none at all, and the possible nature of shock polymorphism or metamorphism in these materials is poorly explored.

Standard Hugoniot measurements provide information on pressure-density states reached under shock compression. Density is determined at the continuum level, and the underlying atomic-level structure of the material is not directly constrained in traditional experimental approaches. Interpretation of shock data has therefore long relied on comparison of measured densities under shock conditions with static compression data or theoretical calculations, taking into account temperature differences between the shock and static data (Ahrens et al., 1969; Davies and Gaffney, 1973; Sekine and Ahrens, 1992). In the last few decades, the development of static high-pressure techniques, especially in combination with synchrotron X-ray diffraction, has greatly expanded our understanding of both stable and metastable high-pressure phases of silicates over a wide range of pressures (Duffy, 2005; Shen and Mao, 2017). Additionally, advances in theory now allow for ab initio prediction of high-pressure phase behavior in complex silicate mineral systems (Wentzcovitch and Stixrude, 2010). However, in many cases, the new data have not been compared in any detail to the preexisting shock results, most of which were collected in the second half of the last century.

The recent development of capabilities that allow for in situ examination of the atomic-level structure has reinvigorated shock-compression studies of geological materials (Duffy and Smith, 2019). X-ray diffraction capabilities are currently available or under development at laser-shock facilities including the Omega Laser (University of Rochester), the National Ignition Facility (Lawrence Livermore National Laboratory), the Linac Coherent Light Source (SLAC National Accelerator Laboratory), the Dynamic Compression Sector (DCS, Argonne National Laboratory), the Pohang Accelerator Laboratory, and the European X-Ray Free Electron Laser, among others. In addition, X-ray experiments can be performed on samples compressed via plate impact at DCS. The compression time scales for these experiments range from ~10 nanoseconds (lasers) to hundreds of nanoseconds (gas guns), allowing for studies of dynamic behavior at variable strain rates. Plate-impact experiments using gun-based loading at DCS have already provided insights into the nature of high-pressure phases formed under shock loading in fused silica (Tracy et al., 2018), quartz (Tracy et al., 2020), and forsterite (Newman et al., 2018). Atomic-level structural measurements under laser shock have been reported on various geologically relevant materials, including fused silica (Gleason et al., 2015), moissanite (Tracy et al., 2019), enstatite (Hernandez et al., 2020), forsterite (Kim et al., 2021), and periclase (Coppari et al., 2013). Results of these studies, which have covered pressures from tens to hundreds of GPa, show a diverse range of behaviors, including transformation to expected equilibrium phases, often kinetically delayed to higher pressures, and formation of metastable crystalline, amorphous, or partially disordered phases.

These recent experimental developments motivate a review of selected traditional continuum-level Hugoniot pressure-density results for silicate minerals to evaluate these historical data in light of recent static experimental and theoretical studies. Such an analysis can provide insights into the possible types of high-pressure phases or assemblages that may form under shock compression and provides motivation for further atomic-level structural studies to better understand the properties and behavior of silicates under shock loading.

6.2 Shock Compression

A shock wave is a large-amplitude mechanical wave that propagates supersonically through a material. Shock waves are typically produced by high-velocity impact or rapid energy deposition such as by detonation of explosives or by high-intensity laser irradiation. When a large load is applied to a surface over a short time period, a stress wave is transmitted into the material that rapidly steepens into a shock wave. Across the propagating shock front, there is a nearly step-like increase in pressure, density, and other properties (Figure 6.1).

Application of conservation of mass, momentum, and energy across a shock discontinuity leads to the Rankine–Hugoniot equations that relate the velocity of the shock wave, U_S, and the particle velocity, u_p, imparted to the material to the thermodynamic variables of pressure, P, density, ρ, and specific internal energy, E:

$$\rho_0 U_S = \rho\left(U_S - u_p\right), \tag{6.1}$$

$$P - P_0 = \rho_0 U_S u_p, \tag{6.2}$$

$$E - E_0 = \frac{1}{2}(P + P_0)\left(\frac{1}{\rho_0} - \frac{1}{\rho}\right). \tag{6.3}$$

The subscript 0 refers to the initial conditions ahead of the shock wave, and the material is assumed to be initially at rest ($u_p = 0$). In the hydrodynamic approximation used here, the longitudinal stress is taken as equivalent to the hydrostatic pressure under the assumption that deviatoric stresses are small. The locus of P–ρ states attained upon shock compression is known as the Hugoniot. Shock loading is an irreversible process that generates considerable entropy and corresponding heating, especially at higher pressures. The temperature, T, along the Hugoniot for a single shock is given by the following differential equation:

$$\frac{dT}{dV} = -T\frac{\gamma}{V} + \left[\frac{dP}{dV}(V_0 - V) + (P - P_0)\right]\frac{1}{2C_V}, \tag{6.4}$$

where V is the specific volume ($1/\rho$), C_V is the constant-volume specific heat, and γ is the Grüneisen parameter.

Minerals typically undergo a progressive sequence of changes with increasing load under shock compression. At low stresses, the material is compressed elastically to a stress level known as the Hugoniot elastic limit (HEL), above which it undergoes plastic or brittle

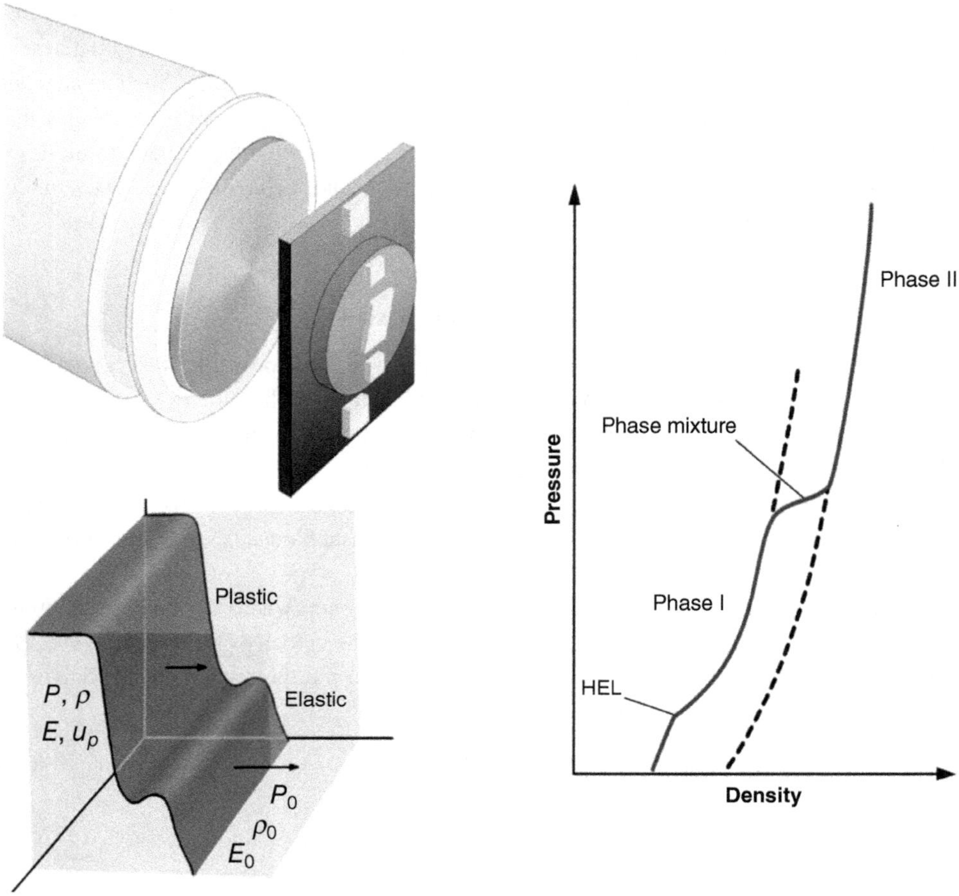

Figure 6.1 Upper left: Schematic illustration of traditional plate-impact shock wave experiment. The projectile carrying the impactor is shown on the left. The target assembly on the right schematically illustrates one common measurement approach used for silicates in which transit times and velocities are determined using flat and inclined mirrors. For details, see Ahrens (1987). Lower left: Illustration of a two-wave structure of a shock wave. Right: Schematic illustration of the major regions of a Hugoniot curve for a material undergoing a phase transformation. HEL = Hugoniot elastic limit. For the color version, refer to the plate section.

deformation to a high-pressure state (Figure 6.1). Eventually, materials will usually undergo structural phase transitions during shock loading. The shock state is maintained inertially until it is released isentropically through the propagation of unloading waves from the rear or edges of the compressed sample.

The response of silicates to shock loading has been studied since the 1960s (Wackerle, 1962; Ahrens and Gregson, 1964; McQueen et al., 1967) and detailed reviews of the field are available (Langenhorst and Hornemann, 2005; Asimow, 2015). Most previous shock wave studies were performed using gas guns to produce plate impacts at velocities up to ~6–7 km/s (Figure 6.1). This drives a shock wave into the target reaching pressures up to 100 GPa or more for ~microsecond durations with typical strain rates of ~10^6 s^{-1}. Various techniques are used to measure the velocity of the shock wave and the particle velocity behind the shock front, including shock transit-time measurements and velocity interferometry. Pressure, density, and specific internal energy are then determined from (6.1–6.3). Detailed reviews of experimental techniques under shock loading are provided elsewhere (Ahrens, 1987; Langenhorst and Hornemann, 2005 Asimow, 2015).

Laboratory shock experiments provide a means to study a range of high-pressure properties, including the equation of state, sound velocities, solid-state phase transitions, melting, and other thermodynamic properties (Asimow, 2015). A notable feature of laboratory shock experiments is that the pressure and temperature states achieved can be comparable to those expected in the deep interiors of planets. Shock wave studies also provide the underlying basis for widely used pressure calibrations in static diamond anvil cell experiments based on the ruby fluorescence scale (Mao et al., 1986) or in situ X-ray diffraction standards (Nellis, 2007).

The Hugoniot equation of state is an empirical linear relationship between shock and particle velocity that is observed for many materials:

$$U_S = c_0 + s u_p, \tag{6.5}$$

where c_O and s are constants. In the absence of phase transitions and neglecting material strength, c_O corresponds to the ambient-pressure bulk sound velocity, and s is related to the pressure derivative of the bulk modulus.

For a material whose shock compression behavior can described by (6.5), the corresponding pressure-compression ($\eta = \rho/\rho_O$) curve is given by the following:

$$P_H = \rho_0 \eta \frac{c_0^2(\eta - 1)}{(\eta - s(\eta - 1))^2}. \tag{6.6}$$

Phase transitions frequently occur under shock loading. Those that are accompanied by appreciable density changes manifest themselves as kinks or discontinuities in the U_S–u_p and P–ρ relationships (Figure 6.1). At the transition, the shock wave bifurcates into two waves moving at different velocities, often marking the existence of a mixed-phase region.

Due to the short time scales of shock experiments, transformation to thermodynamically equilibrium assemblages may be precluded or delayed, especially for reconstructive transformations that require long-range atomic diffusion. As a result, phase transitions often need to be overdriven in pressure or may result in the formation of metastable or amorphous phases on the Hugoniot.

Identification of the high-pressure phases formed under shock loading is often not straightforward. One approach is to compare the shock densities with corresponding high-pressure static results from diamond anvil cell experiments or theoretical calculations, taking into account differences in temperature between the shock states and the conditions of static or theoretical calculations. In general, such comparisons are useful for identifying candidate high-pressure structures but cannot identify the high-pressure phase unambiguously.

Shock Hugoniot data can be related to other thermodynamic compression paths through the Mie–Grüneisen equation (Asimow, 2015):

$$P_H(V) - P_S(V) = \frac{\gamma(V)}{V}[E_H(V) - E_S(V)], \tag{6.7}$$

where the subscript H refers to the Hugoniot and the subscript S refers to a reference curve, such as the principal isentrope, but could equally well be a 300 K isotherm or a 0 K cold curve. Using (6.7), the pressure along the Hugoniot as a function of specific volume for a material undergoing a phase transition can be related to that along the corresponding isentrope of the high-pressure phase from the following (McQueen et al., 1967):

$$P_H = \frac{\frac{\gamma}{V}\left[\int_{V_{02}}^{V} P_S dV - E_{tr}\right] + P_S}{1 - \frac{\gamma}{V}\frac{(V_0 - V)}{2}}, \tag{6.8}$$

where V_0 is the specific volume of the low-pressure phase and V_{02} is the volume of the high-pressure phase. E_{tr} is the specific internal energy difference between the low- and high-pressure phases. The Grüneisen parameter is taken to be a function of volume alone according to the following:

$$\gamma = \gamma_0\left(\frac{V}{V_0}\right)^q, \tag{6.9}$$

where $q = 1$ implies that γ/V is constant. Use of (6.8) allows for the construction of a calculated Hugoniot for a candidate phase based on knowledge of its equation of state along an isentrope or isotherm, or conversely the prediction of an isentrope or isotherm from the shock Hugoniot data. This approach has been widely used to interpret shock compression data in terms of possible high-pressure phases or assemblages (Sekine and Ahrens, 1992).

Another approach to studying shock-compressed materials is postmortem analysis of recovered samples. This method allows for detailed microscale examination of the shocked material using diagnostics such as X-ray diffraction and electron microscopy. However, there is often ambiguity as to which observed features and phases were formed during compression and which represent modifications during the unloading process. High-pressure phases formed along the Hugoniot may not be preserved or may be substantially modified along the quasi-isentropic unloading path, which results in high temperatures being retained as pressure is released.

In this chapter, previously reported shock pressure-density data for selected silicate minerals are examined by comparing their Hugoniot compression behavior with recent static compression and theoretical studies of relevant compositions. Such analysis can provide insights into the persistence of low-pressure four-coordinated silicate structures under shock loading. It can also provide constraints on the amount of overpressure required to drive transitions under shock loading, the extent of intermediate or mixed-phase regions, and the volume change associated with shock phase transitions. Possible high-pressure phases and assemblages are evaluated through calculation of theoretical Hugoniots based on static data.

While extensive shock data exist for several mineral groups (oxides, carbonates, fluorides, etc.), silicates are examined here due to their geological relevance and the ability to select a range of chemistries and structure types for examination. Minerals exhibiting silica polymerization ranging from isolated tetrahedra to three-dimensional framework structures are considered. Both dense and open structures are evaluated as well as a volatile-bearing composition. The choice of minerals is admittedly somewhat arbitrary and based partly on the availability of recent static or theoretical data for comparison as well as an emphasis on minerals whose shock behavior has escaped detailed consideration in previous work. The results provide insights into the stability of silicate structures and the possible nature of high-pressure phases under shock loading and serve to motivate further studies using in situ atomic-level analysis techniques for better understanding of the high-pressure phases of shocked silicates.

6.3 Selected Silicates under Shock Compression

6.3.1 Garnets

Silicate garnets are stable over an extensive range of pressures, temperatures, and geochemical environments (Geiger 2013). They are major constituents of the Earth's upper mantle and transition zone extending to 660 km depth. Ca-rich grossular garnets are found in chondritic meteorites and their calcium-aluminum inclusions (Rubin and Ma, 2017). High-pressure majorite garnets are found to occur in association with melt veins in a number of shocked meteorites (Sharp and DeCarli, 2006).

Hugoniot data for almandine-rich garnet, $(Fe_{0.79},Mg_{0.14},Ca_{0.04},Mn_{0.03})_3Al_2Si_3O_{12}$, shocked along the [100] direction, were reported to 66 GPa (Graham and Ahrens, 1973). The mean value of the Hugoniot elastic limit was found to be 9 GPa. Figure 6.2 shows the Hugoniot data in pressure-compression space, together with recent 300 K equations of state of almandine determined from X-ray diffraction and elasticity measurements. The shock results are consistent with static data for almandine to 28 GPa (Fan et al., 2009) and with the range of densities extrapolated to higher pressures from diffraction and elasticity data (Jiang et al., 2004). There is no evidence in the U_S–u_p or P–ρ/ρ_0 data for a phase transition with an appreciable volume change up to the maximum pressure reached.

In the previous shock study (Graham and Ahrens, 1973), the data were interpreted to indicate a phase transition, proposed to be to the ilmenite structure, occurring above 20 GPa. This was based on a comparison with a 300 K equation of state constructed from early ultrasonic elasticity data (Soga, 1967) which overestimated the pressure derivative of the bulk modulus, yielding a stiff 300 K compression curve. More recent static compression and elasticity data (Jiang et al., 2004; Fan et al., 2009) yield an equation of state that is in agreement with the shock data indicating no detectable phase transition up to 66 GPa (Figure 6.2).

High P–T experiments have shown that almandine decomposes to oxides above ~20 GPa (Akaogi et al., 1998). At higher pressures, laser-heated diamond cell experiments

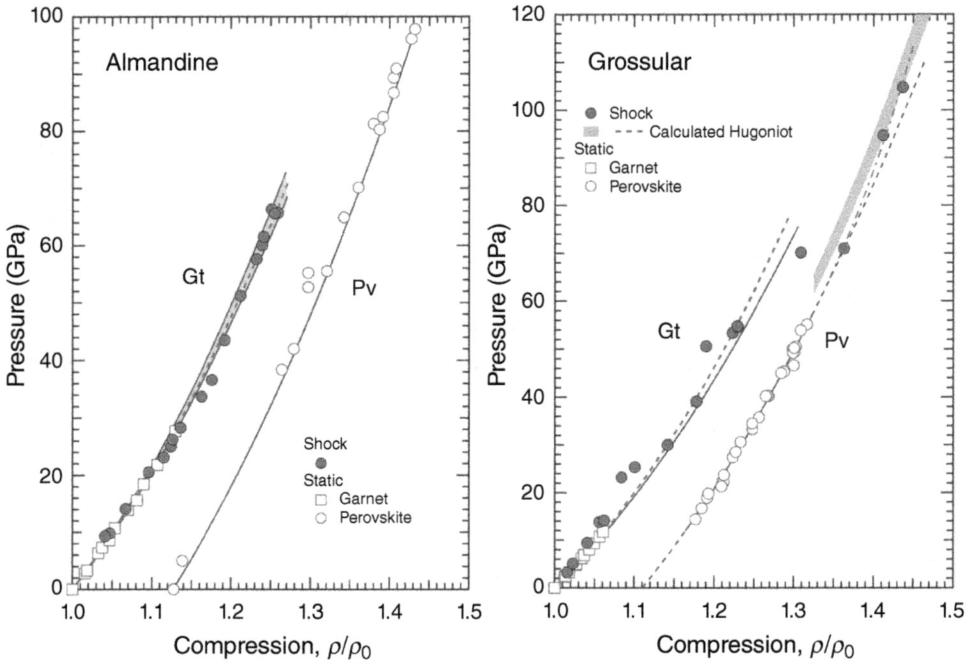

Figure 6.2 Pressure versus compression (ρ/ρ_0) data for almandine-rich (left) and grossular (right) garnet. Hugoniot data points are shown as filled red symbols. Dash-dot red line for the grossular high-pressure phase is obtained from a fit to the U_S–u_p data. Calculated Hugoniots for the low-pressure phases are shown as red dashed lines and for the high-pressure phase of grossular as a shaded pink band (assuming ambient-pressure Grüneisen parameters ranging between 1 and 2). Open symbols show 300 K static compression data for the garnet and perovskite-type phases. Blue solid lines and the blue shaded region are equation of state fits to 300 K compression data with extrapolations. The Hugoniot elastic compression data for almandine garnet are not shown. Abbreviations are Gt = garnet, Pv = perovskite.

References are, for shock data Graham and Ahrens (1973); Marsh (1980); for 300 K static data, Zhang et al. (1999); Jiang et al. (2004); Fan et al. (2009); Dorfman et al. (2012); Greaux et al. (2014).

have revealed that almandine adopts an orthorhombic perovskite structure with aluminum distributed between the octahedral and dodecahedral sites: $(Fe_{0.75}Al_{0.25})(Si_{0.75}Al_{0.25})O_3$ (Dorfman et al., 2012). The observed density of the Fe-rich perovskite phase is greater than found along the Hugoniot (Figure 6.2), ruling out the formation of a phase with octahedrally coordinated silicon upon shock loading over the pressure range examined. The garnet structure thus appears to be one that is resistant to transformation under shock. While a subtle phase transition cannot be ruled out, any structural change would not be expected to lead to an increase in cation coordination up to at least 66 GPa.

Shock-compression data have been reported for grossular ($Ca_3Al_2Si_3O_{12}$) to 105 GPa (Marsh, 1980). The detailed composition of the starting sample was not specified, and the reported density was 4% less than pure grossular. While the shock data exhibit some scatter, they are generally consistent with the static equation of state of grossular determined by extrapolation of single-crystal X-ray diffraction (Zhang et al., 1999) and elasticity data (Jiang et al., 2004) to 54 GPa (Figure 6.2).

Under static compression, grossular transforms to the orthorhombic perovskite structure ($(Ca_{0.75}Al_{0.25})(Si_{0.75}Al_{0.25})O_3$) above 21 GPa at high temperatures (Greaux et al., 2011; Yusa et al., 1995). The 300 K equation of state of Ca-Al-perovskite was determined in static experiments by synthesis at 50 GPa followed by decompression (Greaux et al., 2014). Under shock loading, a phase transition occurs above 54 GPa, and the transition appears to be complete by 70 GPa, indicating a relatively narrow mixed-phase region (Figure 6.2). The calculated shock temperature at 54 GPa is ~500 K, and the shock transition involves an ~6% density increase. The density of shock-compressed grossular is comparable to the extrapolated equation of state for Ca-Al perovskite, suggesting the shocked material has transformed to a structure with six-coordinated silicon (Figure 6.2). Theoretical Hugoniots calculated using the Mie–Grüneisen approach (6.8) and Grüneisen parameters ranging from 1–2 are shown in Figure 6.2, suggesting a high-pressure phase consistent with perovskite or a slightly denser form. In the pyrope-almandine system, it has been observed that an additional transformation to a postperovskite phase with a small density increase occurs at higher pressures (Shieh et al., 2011), but it is unknown if grossular can form such a phase. The high-pressure phase on the Hugoniot could potentially also be a dense amorphous phase or oxide decomposition products.

Synthetic rare-earth oxide garnets are a technologically important group. Shock wave studies on these materials have focused primarily on gadolinium gallium garnet ($Gd_5Ga_3O_{12}$, GGG). Experimental work has shown that GGG has a large Hugoniot elastic limit (~30 GPa) and undergoes a broad phase transition from 65–120 GPa to a so-called "quasi-incompressible phase" (Mashimo et al., 2006). Dynamic compression measurements on GGG have been extended up to 2.6 TPa (Ozaki et al., 2016). Under static compression at 300 K, GGG is stable to 70 GPa and then gradually amorphizes up to 86 GPa (Hua et al., 1996). It transforms to the perovskite structure upon heating above 1,500 K (Mao et al., 2011). However, the shock wave data for GGG indicate it is denser than expected for the perovskite phase. This suggests a further crystalline phase transition, decomposition, or formation of a dense amorphous phase on the Hugoniot (Mao et al., 2011).

Static compression studies of rare-earth oxide garnets demonstrate that the garnet structure persists to high pressure at 300 K. A number of these rare-earth garnets have been shown to eventually undergo pressure-induced amorphization at pressures ranging from 51 GPa to more than 100 GPa (Hua et al., 1996; Stan et al., 2015). In a single-crystal X-ray diffraction experiment, the silicate garnet pyrope was found to persist to 84 GPa at 300 K (Finkelstein et al., 2012). As discussed previously, under shock compression, garnets do not show evidence for a phase transition with an appreciable volume change to at least 55–65 GPa. Thus, both the static and dynamic data indicate that the garnet structure can persist to pressures well beyond its expected thermodynamic stability. On laser heating at high pressures, static experiments show that both silicate and rare-earth oxide garnets adopt single-phase perovskite structures (Dorfman et al., 2012; Greaux et al., 2014; Lin et al., 2013). Under shock compression, grossular and GGG transform to high-pressure forms consistent with the presence of octahedral silicon. While the perovskite structure is one candidate for such a phase, direct atomic-level X-ray diffraction measurements are needed to determine more conclusively the identity of the high-pressure Hugoniot phase of silicate garnets.

6.3.2 Tourmaline

Tourmalines are a large group of structurally and chemically complex borosilicate minerals found throughout the Earth's crust (Henry and Dutrow, 2018). Historically, tourmalines were used as piezoelectric pressure sensors in early laboratory shock experiments (Hearst et al., 1965). Hugoniot data on tourmaline to 113 GPa were reported in the compendium of Los Alamos shock wave data (Marsh, 1980). Although the specific composition of the samples was not specified, the reported density (3.179 g/cm^3) falls within the (Na(Mg, Fe)$_3$Al$_6$(BO$_3$)$_3$Si$_6$O$_{18}$(OH)$_4$) system between the Mg-endmember dravite (3.013 g/cm^3) and Fe-endmember schorl (3.276 g/cm^3). This system is by far the most common among natural tourmalines. The Hugoniot data are shown in Figure 6.3.

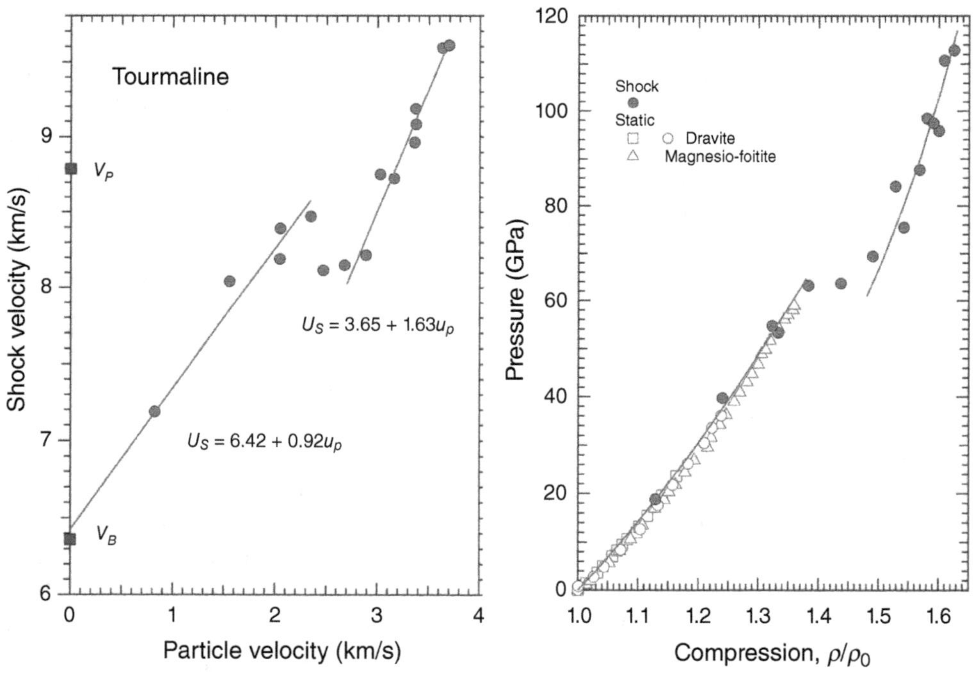

Figure 6.3 Left: Shock velocity–particle velocity relationship for tourmaline with linear fits in low- and high-pressure regions (red solid lines). Blue solid squares indicate ambient compressional and bulk sound velocities of tourmaline. Right: Pressure versus compression for tourmaline under shock (red solid) and single-crystal static (open) compression. Solid red lines are obtained from the linear U_S–u_p relationships.

References are, for shock data, Marsh (1980), and for static data, O'Bannon et al. (2018); Berryman et al. 2019).

Recently, tourmaline was investigated in several static studies at high pressures. Dravite was examined to 24 GPa by single-crystal X-ray diffraction and up to 65 GPa by luminescence spectroscopy in one study (O'Bannon et al., 2018). In another single-crystal study, the compressibility of five synthetic tourmalines, including dravite, were investigated by X-ray diffraction to as high as 60 GPa (Berryman et al., 2019). Aside from a potential subtle change in the structure of dravite around 9 GPa (O'Bannon et al., 2018), the basic tourmaline structure persists for all compositions investigated up to 60–65 GPa (Berryman et al., 2019). In simultaneous high *P–T* experiments in the diamond anvil cell, the tourmaline structure was found to persist to at least 18 GPa at 723 K (Xu et al., 2016).

A comparison of 300 K static and shock data for tourmalines shows good agreement with the shock data slightly offset to lower compression plausibly due to thermal effects (Figure 6.3). Up to 63 GPa, corresponding to 43% compression, a linear U_S–u_p relationship describes the shock data well and the intercept at $u_p = 0$ is close to the expected bulk sound velocity for dravite. This indicates that tourmaline undergoes no major phase transition and likely persists under shock loading to this pressure. In combination, the static and shock results show that the complex tourmaline structure is surprisingly persistent under compression. At higher pressures, the shock data are indicative of phase transitions occurring through a narrow mixed-phase region from 63–69 GPa. The calculated shock temperature at 63 GPa is ~750 K. A second linear U_S–u_p relationship provides a good fit the data from 69–113 GPa defining a high-pressure phase region. The density change associated with the phase transition is ~7–9%.

The nature of the high-pressure tourmaline structure is unknown as there is no static data extending above 65 GPa at room or high temperature. Further high-pressure investigations of the tourmaline structure are warranted to ascertain the nature of the structural changes occurring above 63 GPa. Almandine garnets have high shock impedance ($Z = \rho_0 U_S$) whereas tourmaline exhibits a more moderate value. Taken together, the results for tourmaline and garnets show that complex silicate structures can persist to very high pressures both statically at 300 K and dynamically under shock compression. These two minerals demonstrate a remarkable persistence of tetrahedrally coordinated silicon under both low-temperature static loading and high-strain-rate dynamic loading.

6.3.3 Nepheline

Nepheline, $NaAlSiO_4$, is a relatively low-density tectosilicate that occurs in alkali-rich, silica-poor igneous rocks. It is also found in chondritic meteorites (Rubin and Ma, 2017). Due to its role as a possible decomposition product of jadeite and plagioclase, its shock compression behavior has attracted interest (Sekine and Ahrens, 1992). Hugoniot data for $NaAlSiO_4$ nepheline were reported to 91 GPa (Simakov and Trunin, 1980). In the U_S–u_p plane, the data show a clear discontinuity indicative of a phase transition (Figure 6.4). Comparison with a 300 K equation of state from static compression (Akaogi et al., 2002; Gatta and Angel, 2007) shows that the shock data are consistent with the compression behavior of nepheline to 19 GPa but deviate to higher density by 25 GPa (Figure 6.4).

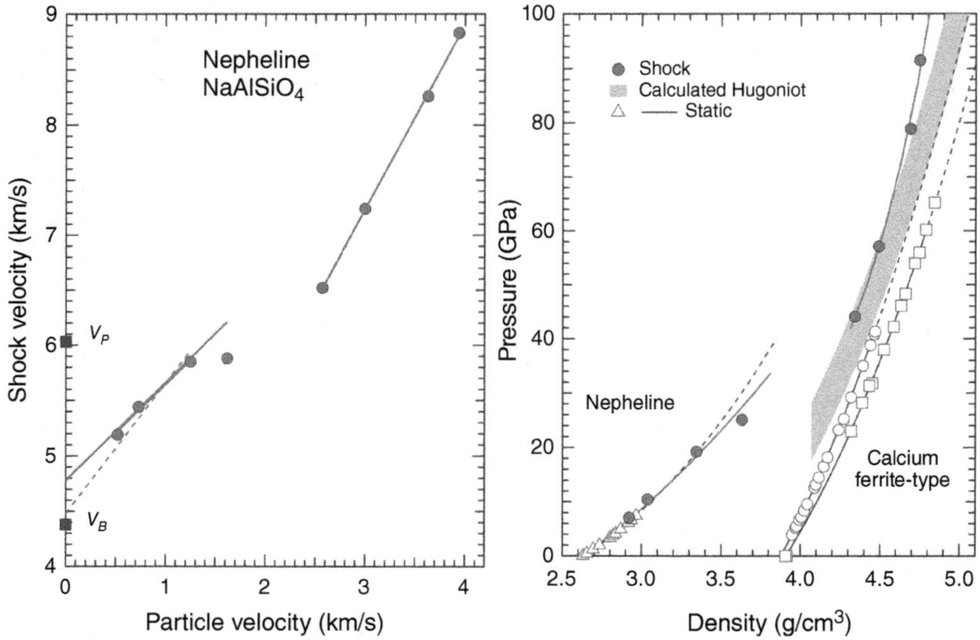

Figure 6.4 Left: Shock velocity–particle velocity relationship for $NaAlSiO_4$ showing linear fits to low-pressure and high-pressure regions (red solid lines). An alternative fit to the low-pressure data that includes the ambient bulk sound velocity (solid square) is shown as the red dashed line. Right: Pressure versus density for $NaAlSiO_4$ under shock (red solid symbols and solid lines) and static (open symbols and blue solid lines) compression. For the low-pressure phase, open triangles are single-crystal X-ray diffraction data, and the dashed line is an extrapolated equation of state. Calcium ferrite-type data are shown as open blue symbols with equations of state indicated by blue solid lines (dashed where extrapolated). A calculated Hugoniot (pink-shaded region) is based on equation of state from Guignot and Andrault (2004) and assuming zero-pressure Grüneisen parameters between 1–2.
References: Shock data: Simakov and Trunin (1980), nepheline static data: Akaogi et al. (2002);Gatta and Angel (2007), and calcium-ferrite-type static data: Dubrovinsky et al. (2002); Guignot and Andrault (2004).

The extent of a mixed-phase region is unclear due to paucity of data, but a well-defined high-pressure phase can be identified from both U_S–u_p and P–ρ plots from 44–91 GPa (Figure 6.4).

It was proposed early on that the high-pressure phase of nepheline along the Hugoniot is consistent with the calcium ferrite-type (CF) structure observed in laser-heated diamond cell experiments (Liu, 1978; Sekine and Ahrens, 1992). Subsequently, the calcium ferrite-type phase of $NaAlSiO_4$ has been examined in more detail. The stability of the CF phase has been demonstrated up to 75 GPa and 2,450 K (Tutti et al., 2000). The 300 K equation of state was reported in two separate experiments up to 40 GPa (Dubrovinsky et al., 2002) and 65 GPa (Guignot and Andrault, 2004), although the reported high-pressure volumes differ by ~2% in these studies. Calculated Hugoniots based on the isothermal data of Guignot and Andrault (2004) are consistent with the Hugoniot data at lower pressures but deviate at higher pressures (Figure 6.4). The stiffer equation of state of Dubrovinsky et al. (2002) provides better agreement with shock data, but this static data set may be affected by nonhydrostatic stress, as no pressure medium or laser annealing was used in this work.

Nepheline is an example of a relatively low-density (ρ_0 = 2.63 g/cm^3) silicate that transforms to a dense high-pressure structure at modest pressures (19–44 GPa) under shock loading. The density difference between the low- and high-pressure phases is large (~50%), and thus the material is expected to undergo substantial shock heating. $NaAlSiO_4$ is an endmember composition of the chemically complex aluminum-rich calcium ferrite-type phase expected to form in midocean ridge basalt compositions in the lower mantle (Wicks and Duffy, 2016). It is generally important for understanding high-pressure behavior of the Na_2O–Al_2O_3–SiO_2 system. The more recent static compression data confirm that the calcium ferrite phase remains a plausible candidate high-pressure phase along the Hugoniot. However, there is a need to resolve the discrepancy with existing static equations of state and to better constrain the Hugoniot curve for this material.

6.3.4 Topaz

Topaz, $Al_2SiO_4(F,OH)_2$, is a volatile-bearing, orthosilicate mineral expected to be stable under subduction zone conditions. Single-crystal diamond anvil cell experiments on natural topaz at 300 K have shown that the structure persists to at least 45 GPa (Gatta et al., 2014) (Figure 6.5). Spectroscopic measurements on synthetic topaz-OH indicate that while there may be a change in the compression mechanism near 45 GPa, the basic structure persists to at least 60 GPa (O'Bannon and Williams, 2019).

Shock compression experiments have been carried out on topaz from 15–146 GPa (Simakov et al., 1974). The reported density of the starting sample was close to the expected crystal density of topaz, but the OH/F composition was not specified. Natural topaz is fluorine dominant with <30% OH. Up to 23 GPa, the density of shocked topaz is consistent with experimental 300 K single-crystal data (Gatta et al., 2014) as well as the theoretical calculations for the OH endmember (Mookherjee et al., 2016). By 37 GPa, the density of the shocked material exceeds the 300 K data, indicating transformation to a denser phase or assemblage (Figure 6.5). The shock data from 37–146 GPa can be fit with a single linear shock velocity–particle velocity relationship: $U_S = 5.3(0.4) + 1.7(0.2)u_p$, which yields the pressure–compression relation shown in Figure 6.5. At 60 GPa, the density of shocked topaz is about 4% greater than that of the extrapolated static equation of state for the low-pressure phase (Gatta et al., 2014; Mookherjee et al., 2016).

Possible high-pressure phases or assemblages for topaz-OH can be examined assuming the Hugoniot is not strongly affected by anionic substitution. Recent static-compression experiments in the Al_2O_3–SiO_2–H_2O system indicate there are multiple possible high-pressure phases or assemblages that can be derived from an endmember topaz-OH composition. Topaz-OH II, synthesized at 14 GPa and 1,670 K (Kanzaki, 2010), bears structural similarities to diaspore and is characterized by a high degree of disorder with about 1/3 of the Si cations in octahedral coordination. An Al-rich form of the dense hydrous

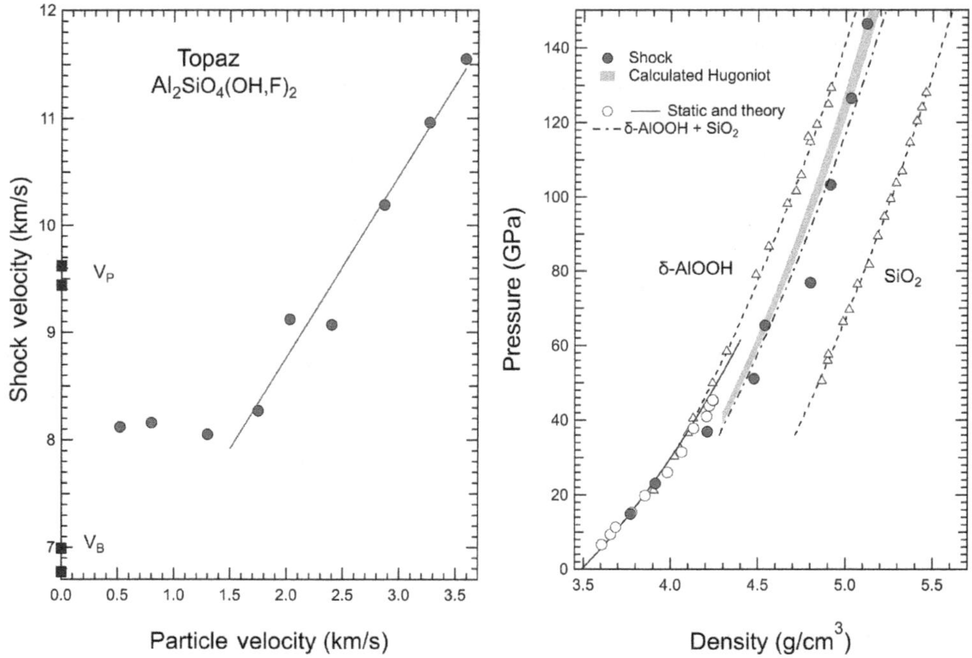

Figure 6.5 Left: Shock velocity–particle velocity relationship for topaz with linear fit in the high-pressure phase region (red solid line). Blue solid squares indicate ambient compressional and bulk sound velocities for OH and F endmembers. Right: Pressure versus density for topaz under shock (solid red symbols) and single-crystal static (open circles) compression. The solid blue line is the theoretical equation of state for topaz-OH at 300 K. Black open triangles and dashed black lines show 300 K equation of state data for δ-AlOOH and SiO_2 (stishovite, $CaCl_2$-type). The dash-dot black line is the 300 K equation of state for the δ-AlOOH and SiO_2 mixture, and the pink-shaded band shows the calculated Hugoniot for the mixture assuming ambient-pressure Grüneisen parameters between 1–2.

References: Shock data: Simakov et al. (1974), topaz static: Gatta et al. (2014), theory: Mookherjee et al. (2016), δ-AlOOH: Duan et al. (2018), and SiO_2: Andrault et al. (2015).

magnesian silicate phase D was synthesized at 26 GPa (Pamato et al., 2015), although the composition of this phase was Al-deficient relative to topaz. Theoretical calculations, however, suggest aluminum phase D ($Al_2SiO_6H_2$) may be stabilized at higher temperatures (Panero and Caracas, 2020). There are various possible decomposition products of topaz, including the formation of δ-AlOOH + SiO_2, corundum + SiO_2 + fluid, or phase egg ($AlSiO_3(OH)$) + corundum + fluid.

The high-pressure aluminous phase δ-AlOOH has a structure similar to the $CaCl_2$-type structure adopted by SiO_2 and has been shown to be stable to at least 134 GPa at moderately high temperature (Sano et al., 2008). This phase may be an important water carrier in the Earth's deep lower mantle. The 300 K equations of state for δ-AlOOH and SiO_2 are shown in Figure 6.5. The density of a mixture corresponding to a topaz stoichiometry (2 δ-AlOOH + SiO_2) was determined using the additive oxide mixture method (Al'tshuler and Sharipdzhanov, 1971; Telegin et al., 1980). Figure 6.5 shows that the calculated Hugoniot of the mixture is generally consistent with the shock data within its scatter. Further shock compression studies of topaz are warranted to understand better the nature of shock-compressed high-pressure phases in this volatile-bearing mineral, especially in view of the growing recognition of the potential importance of hydrous phases in the deep Earth (Ohtani, 2020).

6.3.5 Spodumene

Pyroxenes are chain silicates that are major rock-forming minerals in the Earth's crust and upper mantle. While Mg-rich orthopyroxene has received extensive attention under shock loading (Luo et al., 2004; Hernandez et al., 2020), Hugoniot measurements of a number of clinopyroxenes have also been reported, including diopside (Svendsen and Ahrens 1983, 1990), hedenbergite (Simakov et al., 1974), jadeite (Marsh, 1980; Takazawa et al., 1998), augite (Marsh, 1980), and spodumene (Simakov and Trunin, 1980). Here spodumene is presented as an example illustrating the relationship between the shock data and current understanding of the high-pressure behavior of clinopyroxenes.

Spodumene ($LiAlSi_2O_6$) is a lithium-bearing clinopyroxene crystallizing in space group *C2/c*. Lithium clinopyroxenes have been extensively studied to understand their crystal chemical behavior and low-pressure phase transition (Arlt and Angel, 2000; Nestola et al., 2009). In addition to having geological relevance, lithium aluminosilicates are important technologically in the glass and ceramic industry because of their high chemical resistivity and resistance to thermal shock.

High-pressure studies of spodumene have been carried out to 15 GPa using single-crystal X-ray diffraction and Raman and fluorescence spectroscopy (Pommier et al., 2003; Ullrich et al., 2009; O'Bannon and Williams, 2019). A first-order displacive phase transition from the *C2/c* to $P2_1/c$ space group occurs at 3.2 GPa (Arlt and Angel, 2000; Ullrich et al., 2009). Static data for spodumene to ~10 GPa (Ullrich et al., 2009) are plotted in Figure 6.6 together with shock Hugoniot data covering the range from 10–116 GPa (Simakov and Trunin, 1980). At 24 GPa, the density of spodumene on the Hugoniot exceeds that of the extrapolated static equations of state, indicating a transformation to a denser structure. The U_S–u_p data suggest a likely intermediate or mixed-phase region up to 33 GPa, while the data from 57–116 GPa are consistent with compression of a dense high-pressure phase.

Static compression data and theoretical calculations for $LiAlSi_2O_6$ are limited to 15 GPa, so the pressure extent of persistence of the $LiAlSi_2O_6$ structure at 300 K is unknown. In addition to a high-pressure crystalline or amorphous form of $LiAlSi_2O_6$, there are several possible decomposition products that may occur, including $LiAlSiO_4$ + SiO_2; $LiAlO_2$ + $2SiO_2$; or Li_2O + Al_2O_3 + $4SiO_2$. These can be used as model systems for examining the Hugoniot data.

The lithium-aluminum compound, $LiAlO_2$, has applications in nuclear physics and as a battery material. It is known to have a number of different polymorphs up to 5 GPa (α, β, γ, ε, β') (Lei et al., 2008). Low-pressure polymorphs have Li and Al in fourfold coordination. A dense high-pressure phase (δ-phase) with both Al and Li in sixfold coordination was

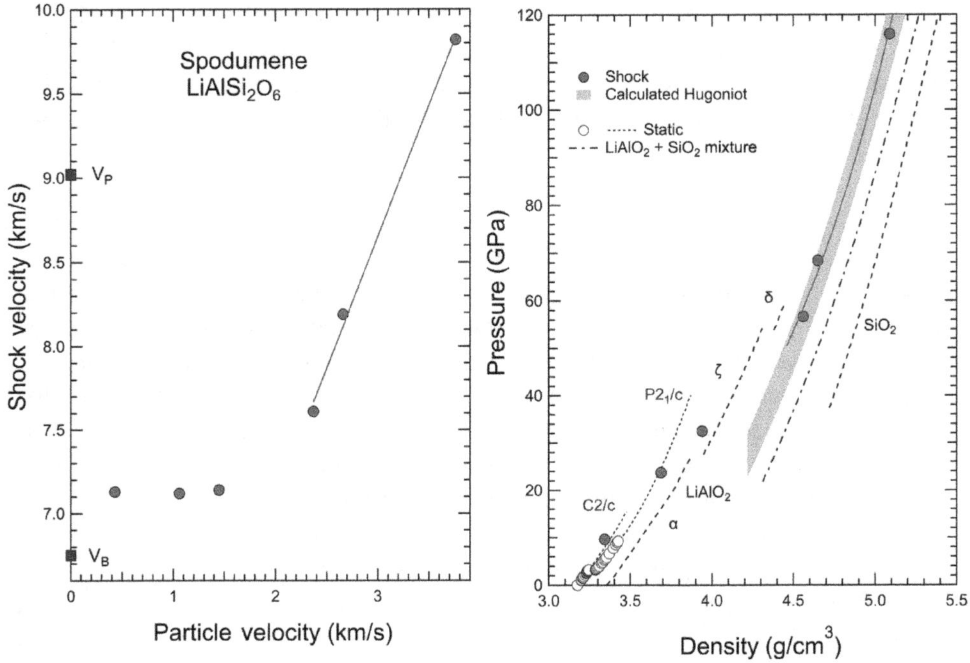

Figure 6.6 Left: Shock velocity–particle velocity relationship for $LiAlSi_2O_6$ spodumene showing linear fit to high-pressure region (red solid line). Blue solid squares indicate ambient compressional and bulk sound velocities. Right: Pressure versus density for $LiAlSi_2O_6$ under shock (solid red symbols and the solid line) and static single-crystal compression data (open blue symbols). The dotted blue lines show extrapolation of equation of state data for the C2/c and $P2_1/c$ phases. The dashed black lines show low-temperature equation of state data for $LiAlO_2$ (α, ζ, δ phases) and SiO_2, and the dash-dot black line shows the result of the mixing model. The pink shaded region is the calculated Hugoniot based on the mixing model and assuming ambient-pressure Grüneisen parameters between 1–2.

References: Shock data: Simakov and Trunin (1980), spodumene static data: Arlt and Angel (2000); Ullrich et al. (2009), $LiAlO_2$: Liu and Liu (2018), SiO_2: Andrault et al. (2015).

synthesized and recovered in shock wave experiments reaching up to as high as 48 GPa (Li et al., 2004). More recently, a theoretical study reported evidence for the stability of an additional monoclinic polymorph (ζ) predicted to be stable above 27 GPa (Liu and Liu, 2018). The high-pressure structures of $LiAlO_2$ have similar ambient densities (~3.5 g/cm^3) and bulk moduli (~158 GPa). Using 300 K static compression data for SiO_2 (stishovite/$CaCl_2$-type) (Andrault et al. 2015) and the theoretical equation of state for the high-pressure $LiAlO_2$ phase (Liu and Liu, 2018), the pressure–density relation for the $LiAlO_2$ + $2SiO_2$ mixture was calculated assuming ideal mixing. A theoretical Hugoniot calculated from the mixture equation of state fits the existing Hugoniot data well (Figure 6.6). Thus, the Hugoniot data are consistent with transformation to a fully six-coordinated structure, consistent with other silicates. In situ diffraction measurements are needed to discriminate among the possible Hugoniot assemblages.

6.4 Concluding Remarks

Reexamination of shock compression data for selected silicates in the light of more recent static experimental and theoretical studies illustrates the current state of understanding as well as open questions about the behavior of silicates under shock loading. In general, shock-compression data for low-pressure phases agree well with recent 300 K single-crystal X-ray diffraction data for these materials. A remarkable persistence of four-coordinated silicate structures to pressures well above their expected stability limit is observed in some cases under both low-temperature static loading and high-strain-rate dynamic loading. Low-pressure structures are observed to persist under shock loading to more than 60 GPa in tourmalines and garnets before transformations to high-pressure phases occur. The extent of mixed-phase regions on the Hugoniot is often not well defined, but in some cases (e.g., tourmaline), it appears to be narrow (<10 GPa).

Recent static compression studies to very high pressures allow for more direct comparison to shock wave data for many silicates. High-pressure phases of shocked silicates are consistent with close-packed, six-coordinated structures, but structural identification based on pressure-density data alone is inherently nonunique. Furthermore, shock wave measurements for many silicates are limited to a restricted number of data points, and the precision of the data is uncertain. Constraints on the elastic precursor and sample characterization may also be limited. Modern wave profile measurement techniques have not been applied to many silicates. Hence, there is a need for improved Hugoniot data for silicate minerals.

The recent development of capabilities for in situ atomic-level studies of mineral under shock compression has the potential to address many of the preceding questions. A number of such studies have already been carried out, demonstrating the capability to identify the structure of high-pressure phases, either stable or metastable, their degree of crystallinity, and possible decomposition. In recently reported X-ray diffraction studies on shocked $MgSiO_3$ and Mg_2SiO_4, transformation to metastable crystalline phases have been identified at lower pressures with amorphization occurring at higher shock pressures (Newman et al., 2018; Hernandez et al., 2020; Kim et al., 2021). In the case of SiO_2, metastable phases have been observed to form or persist under shock compression from crystalline starting material (Schoelmerich et al., 2020; Tracy et al., 2020), while the expected stishovite phase was synthesized in the case of amorphous starting material (Gleason et al., 2015; Tracy et al., 2018). These recent results may indicate that shock wave experiments provide a better route to the formation of metastable and amorphous phases than the crystalline phases of silicates found in high-temperature static compression. However, studies to date have been limited to only a few silicate structure types, and more work is needed before general conclusions can be drawn. Shock wave data remain highly relevant to understanding high-pressure behavior of silicates and other geological materials. The analysis presented here illustrates the range of silicate behavior and open questions that can be addressed with newly developed experimental capabilities.

Acknowledgments

Ian Ocampo provided assistance in the calculations of shock temperatures, and Kalliopi Monoyios assisted in figure preparation. The manuscript was improved by comments from A. Gleason and T. Sekine. This research was supported by the National Science Foundation under contract EAR-1644614 and by the DOE/NNSA under Cooperative Agreement DE-NA0003957.

References

Ahrens, T. J. (1987). Shock wave techniques for geophysics and planetary physics, in C. G. Sammis, T. L. Henyey, eds., *Methods of Experimental Physics*, Academic Press, pp. 185–235.

Ahrens, T. J., Anderson, D. L., Ringwood, A. E. (1969). Equations of state and crystal structures of high-pressure phases of shocked silicates and oxides. *Reviews of Geophysics*, **7**(4), 667–707.

Ahrens, T. J., Gregson, V. G. (1964). Shock compression of crustal rocks – data for quartz calcite + plagioclase rocks. *Journal of Geophysical Research*, **69**(22), 4839–4874.

Ahrens, T. J., Johnson, M. L. (1995a). Shock wave data for minerals, in T. J. Ahrens, ed., *Mineral Physics and Crystallography: A Handbook of Physical Constants*, Vol. 2. American Geophysical Union, pp. 143–184.

Ahrens, T. J., Johnson, M. L. (1995b). Shock wave data for rocks, in T. J. Ahrens, ed., *Mineral Physics and Crystallography: A Handbook of Physical Constants*, Vol. 3. American Geophysical Union, pp. 35–44.

Akaogi, M., Ohmura, N., Suzuki, T. (1998). High-pressure dissociation of $Fe_3Al_2Si_3O_{12}$ garnet: phase boundary determined by phase equilibrium experiments and calorimetry. *Physics of the Earth and Planetary Interiors*, **106**(1), 103–113.

Akaogi, M., Tanaka, A., Kobayashi, M., Fukushima, N., Suzuki, T. (2002). High-pressure transformations in $NaAlSiO_4$ and thermodynamic properties of jadeite, nepheline, and calcium ferrite-type phase. *Physics of the Earth and Planetary Interiors*, **130**(1), 49–58.

Akins, J. A., Luo, S. N., Asimow, P. D., Ahrens, T. J. (2004). Shock-induced melting of $MgSiO_3$ perovskite and implications for melts in Earth's lowermost mantle. *Geophysical Research Letters*, **31**(14), L14612.

Al'tshuler, L. V., Sharipdzhanov, L. L. (1971). Additive equations of state of silicates at high pressures. *Izvestiya: Physics of the Solid Earth, English Translation*, **3**, 167–177.

Andrault, D., Angel, R. J., Mosenfelder, J. L., Bihan, T. L. (2015). Equation of state of stishovite to lower mantle pressures. *American Mineralogist*, **88**(2–3), 301–307.

Arlt, T., Angel, R. J. (2000). Displacive phase transitions in C-centered clinopyroxenes: spodumene, $LiScSi_2O_6$ and $ZnSiO_3$. *Physics and Chemistry of Minerals*, **27**(10), 719–731.

Asimow, P. D. (2015). Dynamic compression, in G. Schubert, ed., *Treatise on Geophysics,* 2nd ed. Elsevier, pp. 393–416.

Berryman, E. J., Zhang, D., Wunder, B., Duffy, T. S. (2019). Compressibility of synthetic Mg-Al tourmalines to 60 GPa. *American Mineralogist*, **104**(7), 1005–1015.

Coppari, F., Smith, R. F., Eggert, J. H., et al. (2013). Experimental evidence for a phase transition in magnesium oxide at exoplanet pressures. *Nature Geoscience*, **6**(11), 926–929.

Davies, G., Gaffney, E. (1973). Identification of high-pressure phases of rocks and minerals from Hugoniot data. *Geophysical Journal of the Royal Astronomical Society*, **33**(2), 165–183.

Dorfman, S. M., Shieh, S. R., Meng, Y., Prakapenka, V. B., Duffy, T. S. (2012). Synthesis and equation of state of perovskites in the (Mg, Fe)$_3$Al$_2$Si$_3$O$_{12}$ system to 177 GPa. *Earth and Planetary Science Letters*, **357**, 194–202.

Duan, Y., Sun, N., Wang, S., et al. (2018). Phase stability and thermal equation of state of δ-AlOOH: implication for water transportation to the deep lower mantle. *Earth and Planetary Science Letters*, **494**, 92–98.

Dubrovinsky, L. S., Dubrovinskaia, N. A., Prokopenko, V. B., Bihan, T. L. (2002). Equation of state and crystal structure of NaAlSiO$_4$ with calcium-ferrite type structure in the conditions of the lower mantle. *High Pressure Research*, **22**(2), 495–499.

Duffy, T. S. (2005). Synchrotron facilities and the study of the Earth's deep interior. *Reports on Progress in Physics*, **68**(8), 1811–1859.

Duffy, T. S., Smith, R. F. (2019). Ultra-high pressure dynamic compression of geological materials. *Frontiers in Earth Science*, **7**, 23.

Fan, D. W., Zhou, W. G., Liu, C. Q., et al. (2009). The thermal equation of state of (Fe$_{0.86}$Mg$_{0.07}$Mn$_{0.07}$)$_3$Al$_2$Si$_3$O$_{12}$ almandine. *Mineral Magazine*, **73**(1), 95–102.

Finkelstein, G. J., Dera, P. K., Duffy, T. S. (2012). Single-crystal X-ray diffraction of pyrope garnet to 84 GPa. *AGU Fall Meeting Abstracts*, **43**, MR43C-2332.

Gatta, G. D., Angel, R. J. (2007). Elastic behavior and pressure-induced structural evolution of nepheline: Implications for the nature of the modulated superstructure. *American Mineralogist*, **92**(8–9), 1446–1455.

Gatta, G. D., Morgenroth, W., Dera, P., Petitgirard, S., Liermann, H.-P. (2014). Elastic behavior and pressure-induced structure evolution of topaz up to 45 GPa. *Physics and Chemistry of Minerals*, **41**(8), 569–577.

Geiger, C. A. (2013). Garnet: a key phase in nature, the laboratory, and technology. *Elements*, **9**(6), 447–452.

Gillet, P., El Goresy, A. (2013). Shock events in the solar system: the message from minerals in terrestrial planets and asteroids. *Annual Review of Earth and Planetary Sciences,* **41**, 257–285.

Gleason, A. E., Bolme, C. A., Lee, H. J., et al. (2015). Ultrafast visualization of crystallization and grain growth in shock-compressed SiO$_2$. *Nature Communications*, **6**, 8191.

Graham, E. K., Ahrens, T. J. (1973). Shock wave compression of iron-silicate garnet. *Journal of Geophysical Research*, **78**(2), 375–392.

Greaux, S., Andrault, D., Gautron, L., Bolfan-Casanova, N., Mezouar, M. (2014). Compressibility of Ca$_3$Al$_2$Si$_3$O$_{12}$ perovskite up to 55 GPa. *Physics and Chemistry of Minerals*, **41**(6), 419–429.

Greaux, S., Nishiyama, N., Kono, Y., et al. (2011). Phase transformations of Ca$_3$Al$_2$Si$_3$O$_{12}$ grossular garnet to the depths of the Earth's mantle transition zone. *Physics of the Earth and Planetary Interiors*, **185**(3), 89–99.

Guignot, N., Andrault, D. (2004). Equations of state of Na–K–Al host phases and implications for MORB density in the lower mantle. *Physics of the Earth and Planetary Interiors*, **143–144**, 107–128.

Hearst, J. R., Irani, G. B., Geesaman, L. B. (1965). Piezoelectric response of Z-cut tourmaline to shocks of up to 21 kilobars. *Journal of Applied Physics*, **36**(11), 3440–3444.

Henry, D. J., Dutrow, B. L. (2018). Tourmaline studies through time: contributions to scientific advancements. *Journal of Geosciences*, **63**(2), 77–98.

Hernandez, J.-A., Morard, G., Guarguaglini, M., et al. (2020). Direct observation of shock-induced disordering of enstatite below the melting temperature. *Geophysical Research Letters*, **47**(15), e2020GL088887.

Hua, H., Mirov, S., Vohra, Y. K. (1996). High-pressure and high-temperature studies on oxide garnets. *Physical Review B*, **54**(9), 6200–6209.

Jiang, F., Speziale, S., Duffy, T. S. (2004). Single-crystal elasticity of grossular-and almandine-rich garnets to 11 GPa by Brillouin scattering. *Journal of Geophysical Research*, **109**, B10210.

Kanzaki, M. (2010). Crystal structure of a new high-pressure polymorph of topaz-OH. *American Mineralogist*, **95**(8–9), 1349–1352.

Kim, D., Tracy, S. J., Smith, R. F., et al. (2021). Femtosecond X-ray diffraction of laser-shocked forsterite (Mg_2SiO_4) to 122 GPa. *Journal of Geophysical Research: Solid Earth*, **126**(1), e2020JB020337.

Langenhorst, F., Hornemann, U. (2005). Shock experiments on minerals: basic physics and techniques, in R. Miletich, ed., *Mineral Behaviour at Extreme Conditions*. Mineralogical Society, pp. 357–387.

Lei, L., He, D., Zou, Y., et al. (2008). Phase transitions of $LiAlO_2$ at high pressure and high temperature. *Journal of Solid State Chemistry*, **181**(8), 1810–1815.

Li, X., Kobayashi, T., Zhang, F., Kimoto, K., Sekine, T. (2004). A new high-pressure phase of $LiAlO_2$. *Journal of Solid State Chemistry*, **177**(6), 1939–1943.

Lin, C., Liu, J., Lin, J.-F., et al. (2013). Garnet-to-perovskite transition in $Gd_3Sc_2Ga_3O_{12}$ at high pressure and high temperature. *Inorganic Chemistry*, **52**(1), 431–434.

Liu, G., Liu, H. (2018). First principles study of $LiAlO_2$: new dense monoclinic phase under high pressure. *Journal of Physics: Condensed Matter*, **30**(11), 115401.

Liu, L.-G. (1978). High-pressure phase transformations of albite, jadeite and nepheline. *Earth and Planetary Science Letters*, **37**(3), 438–444.

Luo, S.-N., Akins, J. A., Ahrens, T. J., et al. (2004). Shock-compressed $MgSiO_3$ glass, enstatite, olivine, and quartz: optical emission, temperatures, and melting. *Journal of Geophysical Research*, **109**, B05205.

Mao, H. -k., Xu, J., Bell, P. M. (1986). Calibration of the ruby pressure gauge to 800 kbar under quasi-hydrostatic conditions. *Journal of Geophysical Research*, **91**(B5), 4673–4676.

Mao, Z., Dorfman, S. M., Shieh, S. R., et al. (2011). Equation of state of a high-pressure phase of $Gd_3Ga_5O_{12}$. *Physical Review B*, **83**(5), 054114.

Marsh, S. P. (1980). *LASL Shock Hugoniot Data*, University of California Press.

Mashimo, T., Chau, R., Zhang, Y., et al. (2006). Transition to a virtually incompressible oxide phase at a shock pressure of 120 GPa (1.2 Mbar): $Gd_3Ga_5O_{12}$. *Physical Review Letters*, **96**(10), 105504.

McQueen, R. G., Marsh, S. P., Fritz, J. N. (1967). Hugoniot equation of state of 12 rocks. *Journal of Geophysical Research*, **72**(20), 4999–5036.

Mookherjee, M., Tsuchiya, J., Hariharan, A. (2016). Crystal structure, equation of state, and elasticity of hydrous aluminosilicate phase, topaz-OH ($Al_2SiO_4(OH)_2$) at high pressures. *Physics of the Earth and Planetary Interiors*, **251**, 24–35.

Mosenfelder, J. L., Asimow, P. D., Ahrens, T. J. (2007). Thermodynamic properties of Mg_2SiO_4 liquid at ultra-high pressures from shock measurements to 200 GPa on forsterite and wadsleyite. *Journal of Geophysical Research-Solid Earth*, **112**(B6), B06208.

Mosenfelder, J. L., Asimow, P. D., Frost, D. J., Rubie, D. C., Ahrens, T. J. (2009). The $MgSiO_3$ system at high pressure: thermodynamic properties of perovskite, postperovskite, and melt from global inversion of shock and static compression data. *Journal of Geophysical Research-Solid Earth*, **114**, B01203.

Nellis, W. J. (2007). Adiabat-reduced isotherms at 100 GPa pressures. *High Pressure Research*, **27**(4), 393–407.

Nestola, F., Redhammer, G. J., Pamato, M. G., Secco, L., Negro, A. D. (2009). High-pressure phase transformation in $LiFeGe_2O_6$ pyroxene. *American Mineralogist*, **94** (4), 616–621.

Newman, M. G., Kraus, R. G., Akin, M. C., et al. (2018). In situ observations of phase changes in shock compressed forsterite. *Geophysical Research Letters*, **45**(16), 8129–8135.

O'Bannon, E., Beavers, C. M., Kunz, M., Williams, Q. (2018). High-pressure study of dravite tourmaline: Insights into the accommodating nature of the tourmaline structure. *American Mineralogist*, **103**(10), 1622–1633.

O'Bannon, E. F., Williams, Q. (2019). A Cr^{3+} luminescence study of natural topaz $Al_2SiO_4(F,OH)_2$ up to 60 GPa. *American Mineralogist*, **104**(11), 1656–1662.

Ohtani, E. (2020). The role of water in Earth's mantle. *National Science Review*, **7**(1), 224–232.

Ozaki, N., Nellis, W. J., Mashimo, T., et al. (2016). Dynamic compression of dense oxide ($Gd_3Ga_5O_{12}$) from 0.4 to 2.6 TPa: universal Hugoniot of fluid metals. *Scientific Reports*, **6**(1), 1–9.

Pamato, M. G., Myhill, R., Boffa Ballaran, T., Frost, D. J., Heidelbach, F., Miyajima, N. (2015). Lower-mantle water reservoir implied by the extreme stability of a hydrous aluminosilicate. *Nature Geoscience*, **8**(1), 75–79.

Panero, W. R., Caracas, R. (2020). Stability and solid solutions of hydrous aluminosilicates in the Earth's mantle. *Minerals*, **10**(4), 330.

Pommier, C. J. S., Denton, M. B., Downs, R. T. (2003). Raman spectroscopic study of spodumene ($LiAlSi_2O_6$) through the pressure-induced phase change from C2/c to $P2_1/c$. *Journal of Raman Spectroscopy*, **34**(10), 769–775.

Rubin, A. E., Ma, C. (2017). Meteoritic minerals and their origins. *Geochemistry*, **77**(3), 325–385.

Sano, A., Ohtani, E., Kondo, T., et al. (2008). Aluminous hydrous mineral δ-AlOOH as a carrier of hydrogen into the core-mantle boundary. *Geophysical Research Letters*, **35** (3), L03303.

Schoelmerich, M. O., Tschentscher, T., Bhat, S., et al. (2020). Evidence of shock-compressed stishovite above 300 GPa. *Scientific Reports*, **10**(1), 10197.

Sekine, T., Ahrens, T. J. (1992). Shock-induced transformations in the system $NaAlSiO_4$-SiO_2: a new interpretation. *Physics and Chemistry of Minerals*, **18**(6), 359–364.

Sharp, T. G., DeCarli, P. S. (2006). Shock effects in meteorites, in D. S. Lauretta, H. Y. McSween, eds., *Meteorites and the Early Solar System II*. University of Arizona Press, pp. 653–677.

Shen, G., Mao, H. K. (2017). High-pressure studies with X-rays using diamond anvil cells. *Reports on Progress in Physics*, **80**(1), 1–53.

Shieh, S. R., Dorfman, S. M., Kubo, A., Prakapenka, V. B., Duffy, T. S. (2011). Synthesis and equation of state of post-perovskites in the $(Mg,Fe)_3Al_2Si_3O_{12}$ system. *Earth and Planetary Science Letters*, **312**(3–4), 422–428.

Simakov, G. V., Pavlovskiy, M. N., Kalashnikov, N. G., Trunin, R. F. (1974). Shock compressibility of twelve minerals. *Izvestiya Akademii Nauk SSSR Fizika Zemli*, **10**, 488–492.

Simakov, G. V., Trunin, R. F. (1980). Minerals compression by shock waves (in Russian). *Izvestiya Akademii Nauk SSSR Fizika Zemli*, **2**, 77–81.

Soga, N. (1967). Elastic constants of garnet under pressure and temperature. *Journal of Geophysical Research*, **72**(16), 4227–4234.

Stan, C. V., Wang, J., Zouboulis, I. S., Prakapenka, V., Duffy, T. S. (2015). High-pressure phase transition in $Y_3Fe_5O_{12}$. *Journal of Physics: Condensed Matter*, **27**(40), 405401.

Svendsen, B., Ahrens, T. J. (1983). Dynamic compression of diopside and salite to 200 GPa. *Geophysical Research Letters*, **10**(7), 501–504.

Svendsen, B., Ahrens, T. J. (1990). Shock-induced temperatures of $CaMgSi_2O_6$. *Journal of Geophysical Research*, **95**(B5), 6943–6953.

Takazawa, E., Sekine, T., Kobayashi, T., Zhu, Y. (1998). Hugoniot equation of state and high-pressure transformation of jadeite. *Journal of Geophysical Research: Solid Earth*, **103**(B6), 12261–12268.

Telegin, G. S., Antoshev, G. V., Bugayeva, V. A., Simakov, G. V., Trunin, R. F. (1980). Calculated determination of Hugoniot curves of rocks and minerals. *Izvestiya: Physics of the Solid Earth, English Translation*, **16**, 319–324.

Tracy, S. J., Smith, R. F., Wicks, J. K., et al. (2019). *In situ* observation of a phase transition in silicon carbide under shock compression using pulsed X-ray diffraction. *Physical Review B*, **99**(21), 214106.

Tracy, S. J., Turneaure, S. J., Duffy, T. S. (2018). In situ X-ray diffraction of shock-compressed fused silica. *Physical Review Letters*, **120**(13), 135702.

Tracy, S. J., Turneaure, S. J., Duffy, T. S. (2020). Structural response of alpha-quartz under plate-impact shock compression. *Science Advances*, **6**(35), eabb3913.

Trunin, R. F. (2005). *Shock Compression of Condensed Materials*. Cambridge University Press.

Tutti, F., Dubrovinsky, L. S., Saxena, S. K. (2000). High pressure phase transformation of jadeite and stability of $NaAlSiO_4$ with calcium-ferrite type structure in the lower mantle conditions. *Geophysical Research Letters*, **27**(14), 2025–2028.

Ullrich, A., Schranz, W., Miletich, R. (2009). The nonlinear anomalous lattice elasticity associated with the high-pressure phase transition in spodumene: a high-precision static compression study. *Physics and Chemistry of Minerals*, **36**(10), 545.

Wackerle, J. (1962). Shock-wave compression of quartz. *Journal of Applied Physics*, **33**(3), 922–937.

Wentzcovitch, R., Stixrude, L., eds. (2010). *Theoretical and Computational Methods in Mineral Physics: Geophysical Applications,* Vol. 71. Mineralogical Society of America.

Wicks, J. K., Duffy, T. S. (2016). Crystal structures of minerals in the lower mantle, in H. Terasaki, R. A. Fischer, eds., *Deep Earth: Physics and Chemistry of the Lower Mantle and Core*. American Geophysical Union, pp. 69–87.

Xu, J., Kuang, Y., Zhang, B., et al. (2016). Thermal equation of state of natural tourmaline at high pressure and temperature. *Physics and Chemistry of Minerals*, **43**(5), 315–326.

Yusa, H., Yagi, T., Shimobayashi, N. (1995). A new unquenchable high-pressure polymorph of $Ca_3Al_2Si_3O_{12}$. *Physics of the Earth and Planetary Interiors*, **92**(1), 25–31.

Zhang, L., Ahsbahs, H., Kutoglu, A., Geiger, C. A. (1999). Single-crystal hydrostatic compression of synthetic pyrope, almandine, spessartine, grossular and andradite garnets at high pressures. *Physics and Chemistry of Minerals*, **7**(1), 52–58.

7

Scaling Relations for Combined Static and Dynamic High-Pressure Experiments

RAYMOND JEANLOZ

Waste heat – the pressure-volume area between the Rayleigh line and Hugoniot – offers a simple means of quantifying energy dissipation upon dynamic compression, confirming that (i) maximum compression on shock loading corresponds to the conditions at which all the shock energy goes into heating rather than compression; (ii) breaking a single shock into two shocks reduces heating, an effect optimized by the intermediate compression being about half the final compression; and (iii) static precompression further reduces heating upon shock loading to a given final compression. Combined static-dynamic experiments can thus maximize material compression by tuning dissipation.

7.1 Introduction

The highest pressures to which materials are well characterized in the laboratory are generated through dynamic loading involving shock, multiple-shock, or ramp compression and are already exceeding 10 TPa. Especially under shock loading, however, the sample is heated due to viscous dissipation, thereby limiting the compressions that can be achieved. In contrast, static high-pressure methods in principle allow near-dissipation-free (reversible), isothermal compression of materials, but only to pressures below 1 TPa.

As the terapascal range is of interest for imposing forces comparable to those determining the structure of atoms – the atomic unit of energy density is $E_H/a_0{}^3 = 29.4$ TPa (e.g., Hartree, 1928), with engagement of core electrons in interatomic bonding – static precompression can be used to enhance atomic *compression* in dynamic loading experiments (Jeanloz et al., 2007). Specifically, the combined use of static and dynamic methods makes it possible to tune the final pressure–density–temperature (P–ρ–T) states achieved in samples, with the reversible nature of isentropic compression leading to minimum final temperatures and therefore maximum compression under dynamic loading. Low-temperature compression is also valuable for characterizing crystalline matter, without melting, which is useful for extending phase diagrams to higher pressures and for comparison with theory.

The present chapter summarizes scaling relations for evaluating the internal energy (E) dissipated upon dynamic loading,

$$E_H(V_H) - E_S(V_H) = \int_{S_0}^{S_H} T_H \, dS_H, \tag{7.1}$$

where V, T, and S are volume, temperature, and entropy, and subscripts 0, S, and H indicate zero-pressure ($P = 0$, $V = V_0$), isentropic, and shock-compressed states, respectively. Dynamic states considered here include Hugoniot (single-shock), as well as multiply shocked and precompressed shock states (i.e., states off the principal Hugoniot). The internal energy difference in (7.1) represents the heating due to entropy production (e.g., viscous dissipation) in the shock front, and our interest is in quantifying this heating so as to control and perhaps minimize it.

In order to focus on scaling relations, both material strength and phase transformations are ignored in the present work. That is, the sample is treated as a fluid with a simple, single-phase equation of state. Also, the discussion is entirely one-dimensional and assumes a steady shock-loading front, with compression along the Rayleigh line, as suited to many equation of state and related measurements under dynamic loading.

The current analysis is not intended to compete with simulations based on realistic equations of state and hydrodynamics: It is most suitable for low pressures, and the final pressure under dynamic loading is not treated correctly except for the case of single-shock loading (i.e., the principal Hugoniot). Nevertheless, the approach used here can provide useful scaling information for energetics to high compression, as described later (e.g., Sections 7.2 and 7.3).

7.2 Waste Heat

Waste heat is given by the area between the Hugoniot and Rayleigh line in a pressure–volume plot, and corresponds to the dissipated energy (7.1) in the approximation that the P–V isentrope is indistinguishable from the Hugoniot starting at the same initial state, P_0, V_0 (Figure 7.1) (Duvall, 1962; Gault and Heitowit, 1963). Though crude, this approximation provides insight into the roles of precompression and multiple-shock loading in tuning the final states achieved in dynamic experiments.

Waste heat is determined from the Hugoniot relations that describe conservation of mass, momentum, and energy in terms of compression $\eta = 1 - \frac{V}{V_0}$, shock velocity U, material (particle) velocity u, shock pressure p_H, and specific internal energy e_H as

$$\eta = \frac{u}{U}, \tag{7.2}$$

$$p_H - p_0 = Uu = \frac{u^2}{\eta}, \tag{7.3}$$

$$e_H - e_0 = \frac{\eta}{2}(p_H + p_0) = \frac{u^2}{2} + p_0\eta. \tag{7.4}$$

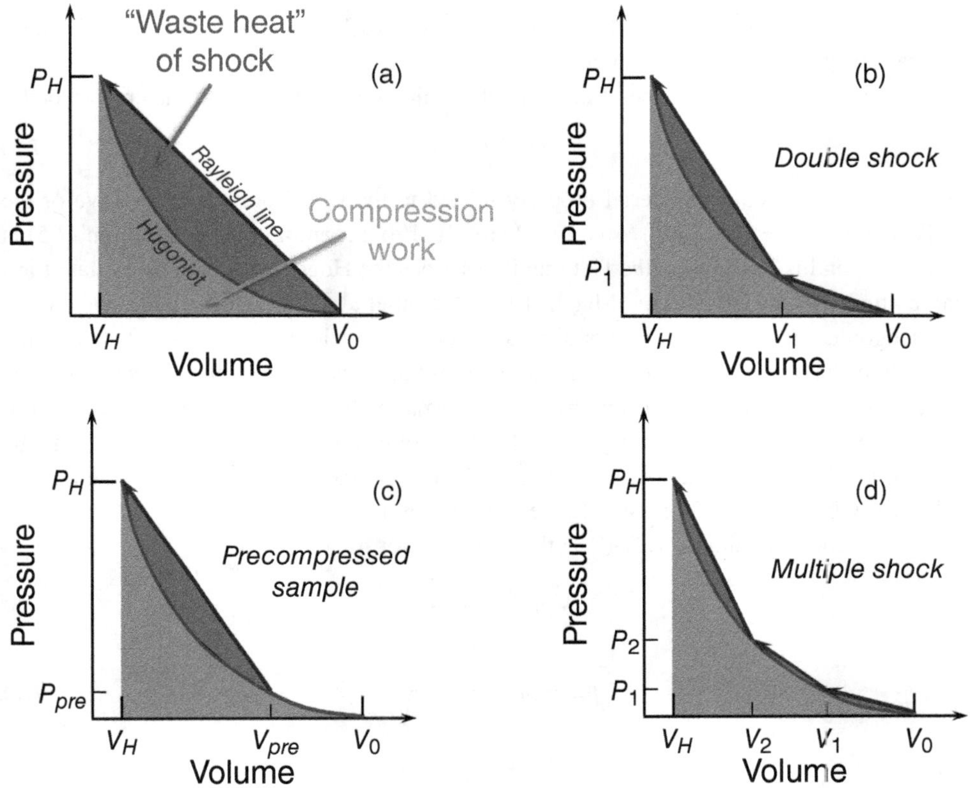

Figure 7.1 (a) Waste heat in single-shock compression (area in red) is the difference between the Hugoniot energy (7.4), given by the triangle under the Rayleigh line (dark blue arrow) connecting zero-pressure ($P_0 = 0$, V_0) and final (P_H, V_H) states, and the reversible energy of compression estimated as the area (orange) under the Hugoniot (blue curve). The Hugoniot is given by a linear relationship between shock velocity (*Us*) and material velocity (*up*), ignoring effects of both strength and phase transformation. The waste-heat approximation ignores the difference between the isentrope and Hugoniot. (b) Double shock reduces waste heat (red areas) because the Rayleigh lines (dark blue arrows) to the first (P_1, V_1) and final (P_H, V_H) shock states are closer to the Hugoniot (blue curve) than the Rayleigh line for a single shock (cf. (a)). (c) Static precompression to an initial pressure-volume condition (P_{pre}, V_{pre}) also reduces waste heat (red area) because the Rayleigh line (dark blue arrow) is now shortened relative to a single shock. Precompression is along an isotherm, which lies at pressures slightly below the isentrope, but is approximated here by the Hugoniot. (d) Multiple-shock compression, illustrated by a three-step shock, also reduces the waste heat (red areas) relative to a single shock and approximates continuous ramp compression as the loading approaches an isentrope. For the color version, refer to the plate section.

The variables are made nondimensional by dividing velocities, pressure, and specific energy by the zero-pressure bulk sound velocity c_0, bulk modulus $K_{0S} = \rho_0 c_0^2$, and sound velocity squared c_0^2, respectively ($U = Us/c_0$, $u = up/c_0$, $p_H = P_H/K_{0S}$, $e_H = E_H/(K_{0S}V_0)$, $e_0 = E_0/(K_{0S}V_0)$). Here, $\rho = 1/V$ is density, and subscripts S and 0 again indicate isentropic and initial conditions. Although $p_0 = 0$ for a single shock from zero pressure (i.e., for the principal Hugoniot), the initial states are kept in (7.3) and (7.4) because nonzero values are considered later.

The present analysis assumes a linear relationship between shock and material velocity:

$$U = 1 + su, \tag{7.5}$$

with the slope s being a material property related to the initial pressure derivative of the bulk modulus: $s = (K_{0S}{}' + 1)/4 \approx 1$ to 2 (Ruoff, 1967; Jeanloz, 1989). Equation (7.5) is empirical, and is justified by the fact that it describes the Hugoniot equations of state for a large number of materials (e.g., Marsh, 1980; Trunin et al., 2001). It is mainly violated at the Hugoniot elastic limit (i.e., conditions of dynamic yielding) and at solid-state transformations having large volume changes, with multiple linear segments characterizing the Hugoniot; in the latter case, the energy of transformation has to be taken into account. Still, electronic excitation and melting typically show only modest influence on (7.5) at the present level of approximation, as shown by the nearly linear relation for the Hugoniot of Al to the Gbar (100 TPa) pressure range (Avrorin et al., 1993).

The Hugoniot relations (7.2)–(7.4) then take the form

$$\eta = \frac{u}{1 + su}. \tag{7.6}$$

$$p_H - p_0 = \frac{\eta}{(1 - s\eta)^2}. \tag{7.7}$$

$$e_H - e_0 = \frac{\eta^2}{2(1 - s\eta)^2} + p_0\eta \tag{7.8}$$

and also

$$U = \frac{1}{(1 - s\eta)}. \tag{7.9}$$

$$u = \frac{\eta}{(1 - s\eta)}. \tag{7.10}$$

The material velocity u is a unique function of compression η in the waste-heat approximation, as given by (7.6) and (7.10), and in the case of multiple shocks it is the changes in u that determine successive states, as with impedance matching (e.g., Forbes, 2012).

Taking the reversible (isentropic) compression work to be approximately given by the integral over the Hugoniot pressure, integrating $de_S = -p_S\, dV/V_0 \approx p_H d\eta$ using (7.7), and subtracting this from the Hugoniot energy (7.8) gives the normalized waste heat (with $p_0 = 0$):

$$g(\eta, s) = \frac{1}{2}\left(\frac{\eta}{1 - s\eta}\right)^2 - \frac{1}{s^2}\left[\frac{s\eta}{1 - s\eta} + \ln(1 - s\eta)\right]. \tag{7.11}$$

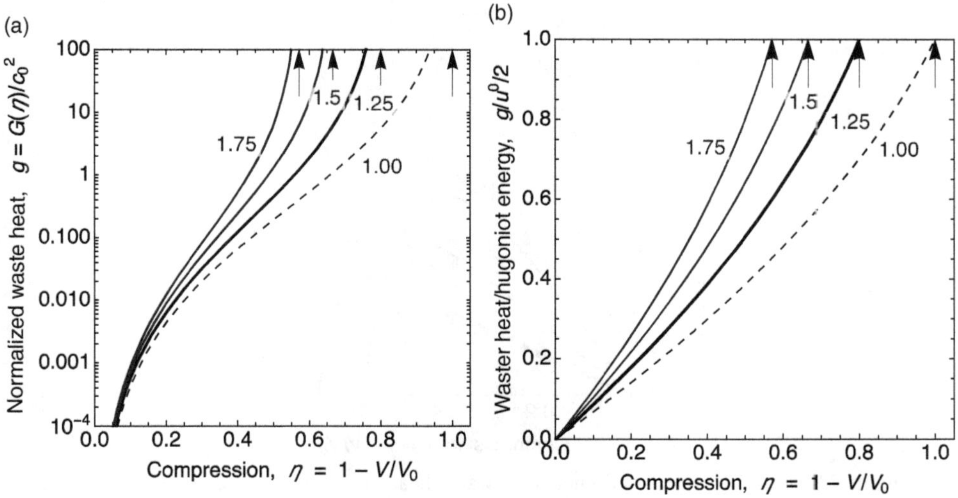

Figure 7.2 (a) Waste heat (G) normalized by sound velocity squared $\left(c_0^2 = K_{0S}/\rho_0\right)$, (7.11), and (b) ratio of normalized waste heat g to Hugoniot energy $u^2/2$, shown as a function of compression $(\eta = 1 - V/V_0)$ for a single shock to a Hugoniot state (Figure 7.1a). Results are plotted for typical values of the slope of the linear relationship between shock and material velocities (7.5): $s = (K_{0S}' + 1)/4 = 1.0$ (dashed), 1.25 (bold solid curve), 1.5 (solid curve), and 1.75 (solid curve). The baseline model is $s = 1.25$ $(K_{0S}' = 4)$ (Jeanloz, 1989), and arrows indicate the limiting compression values $\eta_{Max} = 1/s$ for the linear U–u relation. Thus, the maximum density ratio $\rho/\rho_0)_{Max} = s/(s-1) = (K_{0S}'+1)/(K_{0S}'-3) = 5$ and 3 for $K_{0S}' = 4$ and 5, respectively.

This result is equivalent to Gault and Heitowit's (1963) equation (17) with (7.10) substituted for the material velocity and (dimensional) waste heat $G = gc_0{}^2$.

The linear U–u relation (7.5) implies a maximum density or compression $\eta_{Max} = 1/s$, at which point the Hugoniot pressure (7.7) and energy (7.8) become infinite, the reason being that by maximum compression (minimum volume) all of the energy is going into waste heat rather than compression (Figure 7.2). Note that when half the shock energy is going into waste heat ($\eta \approx 0.3$–0.6 depending on s: Figure 7.2b), the actual magnitude of waste heat is still relatively small compared to values achieved at higher compression (Figure 7.2a).

This interpretation of the maximum compression achieved on shock loading illustrates the utility of the waste-heat analysis, and motivates checking its validity. One way to do so is to compare the waste heat (7.11) with the corresponding dissipated energy (7.1), the energy difference between Hugoniot and isentrope energies as a function of compression. Using (7.8) for e_H and the Eulerian finite-strain equation of state to calculate the isentrope energy e_S (Jeanloz, 1989), both as a function of K_{OS}' (hence s) and compression, the waste heat $g(\eta, s)$ from (7.11) is found to be within 10–20% of the dissipated energy $e_H - e_S$ in (7.1) (Figure 7.3).

To be specific, the isentrope energy is given by

$$e_S = \frac{9}{2} f^2 \left[1 + 3\left(K_{0S}' - 4\right) f + \cdots\right] \tag{7.12}$$

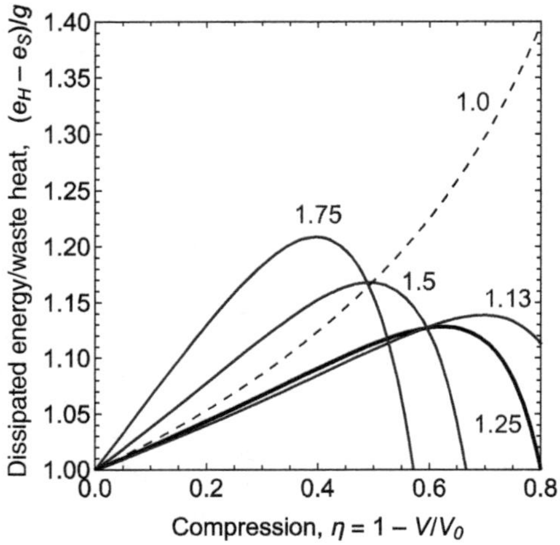

Figure 7.3 Ratio of dissipated energy to waste heat in single-shock compression to the Hugoniot state, shown as a function of the *U–u* slope $s = 1.0$ (dashed curve), 1.13 (solid curve), 1.25 (bold solid curve), 1.5 (solid curve), and 1.75 (solid curve). The dissipated heat is estimated from the difference between the Hugoniot energy (e_H) and the compression energy for a third-order Eulerian finite-strain isentrope (e_S), and the range of *U–u* slopes for most materials is $s = 1.13 - 1.75$ ($K_{0S}{}' = 3.5 - 6$), with $s = 1.25$ ($K_{0S}{}' = 4$) being the baseline model (Jeanloz, 1989).

with $f = [(V/V_0)^{-2/3} - 1]/2 = [(1-\eta)^{-2/3} - 1]/2$ being the spatial (Eulerian) measure of volume strain. The waste heat shows comparable agreement with respect to the dissipated shock energy as indicated in Figure 7.3, if the Rydberg–Vinet equation of state and isentrope energy are used instead of the Eulerian finite-strain function (7.12). The agreement at low compression is because the Hugoniot and isentrope deviate only at third order in volume (Zel'dovich and Raizer, 1967, I §18). At high compression, the waste heat is dominated by the maximum compression ($\eta = 1/s$) implied by the linear *U–u* relation, so it is insensitive to the detailed functional form used for the isentropic compressional energy, e_S.

A further application of waste heat is to double-shock compression, which is expected to greatly reduce dissipation relative to a single shock: That is, the area between Rayleigh lines and Hugoniot is significantly decreased for double relative to single shocks to a given final compression (Figure 7.1b). The Hugoniot energies for two shocks are obtained from (7.8), with the first shock (to compression η_1) starting from $p_0 = 0$, and the second shock (to final compression $\eta \geq \eta_1$) starting from p_1. The resulting waste heat for the double shock is

$$g_2(\eta, \eta_1, s) = \frac{\eta}{2}\left[\frac{\eta}{(1-s\eta)^2} - \frac{\eta_1}{(1-s\eta)^2} + \frac{\eta_1}{(1-s\eta_1)^2}\right] - \frac{1}{s^2}\left[\frac{s\eta}{1-s\eta} + \ln(1-s\eta)\right], \tag{7.13a}$$

$$g_2(\eta, \eta_1, s) = g_1(\eta, s) - \frac{\eta}{2}\left[\frac{\eta_1}{(1-s\eta)^2} - \frac{\eta_1}{(1-s\eta_1)^2}\right]. \tag{7.13b}$$

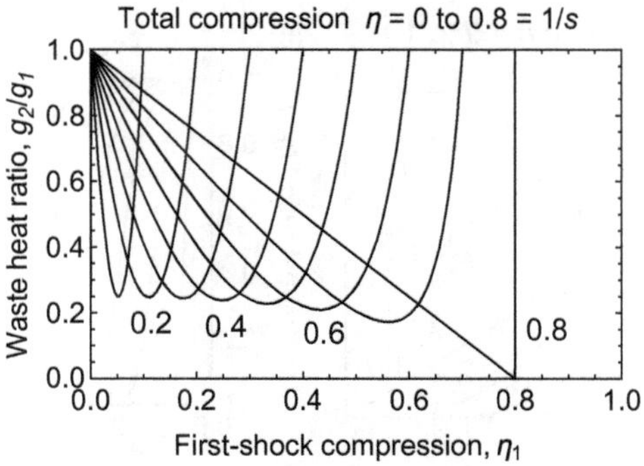

Figure 7.4 Ratio of waste heat for double-shock relative to single-shock g_2/g_1 obtained from (7.11) and (7.13) is plotted as a function of first-compression η_1 for total compressions of $\eta = 0.1$ to 0.8 (curves from left to right, shown in increments of 0.1 and labeled for $\eta = 0.2$, 0.4, 0.6, and 0.8; $\eta = 0$ lies along the left vertical axis). Here, $s = 1.25$ ($K_0{}' = 4$) such that maximum compression is $\eta_{Max} = 1/s = 0.8$. Minimum values of g_2/g_1 are obtained for $\eta_1/\eta \approx 0.5$ at small compressions.

Because $\eta_1 \leq \eta$, the term in square brackets in (7.13b) is positive, thus confirming that the waste heat for the double shock is less than that for the single shock $g_2(\eta) \leq g_1(\eta)$.

Calculations for typical values of s, show that the waste heat can be reduced four- to fivefold through double- relative to single-shock loading, and that there is an optimal value of η_1/η for minimizing dissipation (Figure 7.4). That is, both small and large initial shocks result in compression with waste heat differing little from that of a single shock. Optimal values of η_1/η are near 0.5 for small total compression η and drift upward for larger values of η. Evidently, breaking a single shock into two equal-compression shocks dramatically reduces the heating; the corresponding optimal value of first-shock pressure can be derived from (7.7).

7.3 Shock Loading Statically Precompressed Samples

Static precompression further diminishes heating for dynamic loading because there is no first shock and associated dissipation. The precompressed state lies on an isotherm starting at $P_0 = 0$, $V = V_0$ (e. g., at $T = 300$ K), but for simplicity it is taken to be on the isentrope starting from the same initial state (see Figure 7.1c) (Jeanloz et al., 2007). In the waste-heat approximation used here, the isentrope is approximated by the Hugoniot (7.7).

Extending the argument leading to (7.13), the waste heat for a sample precompressed to an initial state η_I, and then shock-loaded to final compression η is

$$g_{pre}(\eta, \eta_1, s) = \frac{(\eta - \eta_1)}{2}\left[\frac{\eta}{(1 - s\eta)^2} + \frac{\eta_1}{(1 - s\eta_1)^2}\right] - \frac{1}{s^2}\left[\frac{s\eta}{1 - s\eta} - \frac{s\eta_1}{1 - s\eta_1} + \ln\left(\frac{1 - s\eta}{1 - s\eta_1}\right)\right], \tag{7.14a}$$

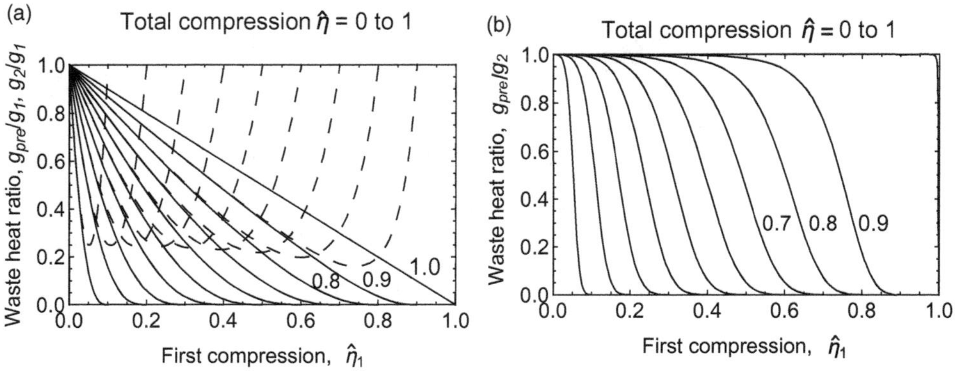

Figure 7.5 (a) Ratio of waste heat for precompressed (solid curves) and double-shocked sample (dashed curves) relative to single shock as a function of first compression, $\widehat{\eta}_1 = s\eta_1$, plotted for total compressions of $\widehat{\eta} = s\eta = 0$ to 1 in increments of 0.1 (values of $\widehat{\eta} = 0.8$, 0.9, and 1.0 are labeled for g_{pre}/g_1). (b) Ratio of waste heat for a precompressed to double-shocked sample as a function of first compression, $\widehat{\eta}_1 = s\eta_1$, plotted for total compressions of $\widehat{\eta} = s\eta = 0$ to 1 in increments of 0.1 (values of $\widehat{\eta} = 0.7$, 0.8, and 0.9 are labeled for g_{pre}/g_2; curves for 0 and 1 lie along the left and right vertical axes, respectively). First compression refers either to static precompression for g_{pre} or to a first shock (i.e., dynamic precompression) for g_2.

$$g_{pre}(\eta,\eta_1,s) = g_2(\eta,\eta_1,s) - \frac{1}{2}\left(\frac{\eta_1}{1-s\eta_1}\right)^2 + \frac{1}{s^2}\left[\frac{s\eta_1}{1-s\eta_1} + \ln(1-s\eta_1)\right], \tag{7.14b}$$

$$g_{pre}(\eta,\eta_1,s) = g_1(\eta,s) - \frac{1}{2}\left[\frac{\eta\eta_1}{(1-s\eta)^2} - \frac{\eta\eta_1}{(1-s\eta_1)^2} + \left(\frac{\eta_1}{1-s\eta_1}\right)^2\right] + \frac{1}{s^2}\left[\frac{s\eta_1}{1-s\eta_1} + \ln(1-s\eta_1)\right]. \tag{7.14c}$$

The second term on the right of (7.14b) is larger in magnitude than the third term, showing that the waste heat for precompressed samples is less than that for a double shock with identical first and final compression (η_1 and η), in accord with expectations from Figures 7.1b and 7.1c.

Plots of waste heat based on (7.11), (7.13), and (7.14) confirm that precompression significantly reduces dissipation for a single shock, and also reduces dissipation relative to that for a double shock (Figure 7.5). The way to read these plots is that waste heat vanishes as precompression η_1 approaches the final compression η (Figure 7.5a), and static precompression reduces waste heat more than double-shock loading, especially for initial compression η_1 half (or slightly more than half) the final compression η (Figure 7.5b).

The waste-heat approximation places the precompressed state on the Hugoniot, at initial compression η_1, so it is worth checking the results against values of energy dissipation $e_H - e_S$ calculated from a different model, such as the Eulerian finite-strain Mie–Grüneisen approach of Jeanloz et al. (2007) (Figure 7.6). The Mie–Grüneisen

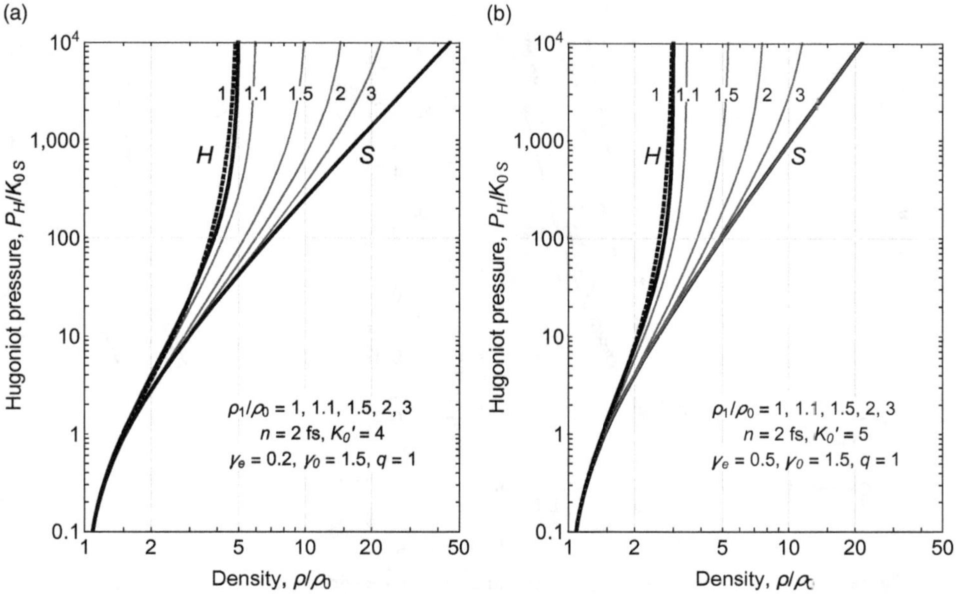

Figure 7.6 Hugoniots for samples precompressed to initial densities $\rho_1/\rho_0 = 1.1$, 1.5, 2, and 3 are compared with the principal Hugoniot ($\rho_1/\rho_0 = 1.0 : H$) and isentrope (*S*) for samples having bulk-modulus pressure derivative $K_{0S}' = 4$ (a) and 5 (b). These are calculated using the Eulerian finite-strain ($n = 2$) Mie–Grüneisen model of Jeanloz et al. (2007), with Grüneisen parameter given by $\gamma(V) = \gamma_0(V/V_0)^q + \gamma_e$. Typical values of $\gamma_0 = 1.5$, $q = 1.0$, and $\gamma_e = 0.2$ to 0.5 are used here and yield principal Hugoniots matching those obtained from the corresponding linear *U–u* relation (7.5) (i.e., with $s = 1.25$ (a) and 1.50 (b): dotted curves).

equation of state has long been used to model Hugoniots (e.g., Zel'dovich and Raizer, 1967, XI §8). It exhibits a maximum compression

$$\eta_{MaxMG} = \frac{2}{\gamma(\eta_{MaxMG}) + 2} \tag{7.15}$$

that depends on the value of the Grüneisen parameter at maximum density, $\gamma(\eta_{MaxMG})$, and in general differs from the maximum compression of the linear *U–u* relation, $\eta_{Max} = 1/s = 4/(K_{0S}' + 1)$. In the present chapter, the Grüneisen parameter is modeled as having a vibrational component (γ_0 at zero pressure) that decreases with compression, plus a constant electronic contribution $(\gamma_e) : \gamma = \gamma_0(V/V_0)^q + \gamma_e$.

Both Mie–Grüneisen and waste-heat approaches show that dissipated energy increases rapidly with final compression, but it is significantly reduced by precompression (Figure 7.7). For a given final density, increased precompression systematically pushes the dynamically loaded state toward an isentrope (Figure 7.8). Naturally, precompression produces no heating, and the Mie–Grüneisen and waste-heat models are in good accord in showing the initial increase in dissipation as the final, shock-induced density (ρ/ρ_0) rises past the initial precompression value, ρ_1/ρ_0.

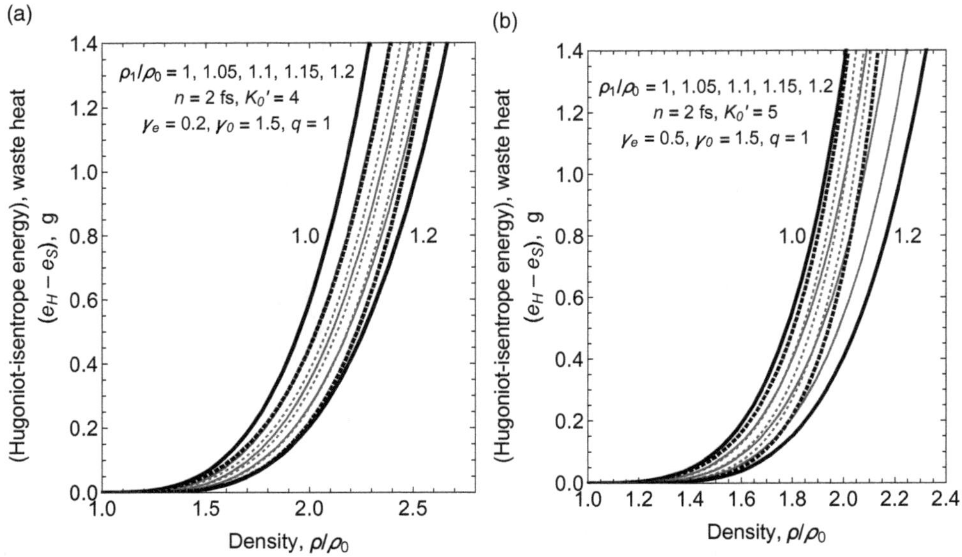

Figure 7.7 Dissipated energy $e_H - e_S$ (solid curves: Mie–Grüneisen model) and waste heat g_{pre} (dotted curves) for samples precompressed to densities $\rho_1/\rho_0 = 1.05$, 1.10, 1.15, and 1.20 (bold curve) are compared as a function of final density (ρ/ρ_0) with values for the principal Hugoniot ($\rho_1/\rho_0 = 1.0$: bold curve) for samples having bulk-modulus pressure derivative $K_{0S}' = 4$ (a) and 5 (b). The energies are normalized by c_0^2, meaning that a value of 1 is of order 20 MJ/kg ~ 20 eV, depending on atomic mass.

The estimates of heating at higher compressions diverge somewhat because of different assumptions about the limiting compressions in the Mie–Grüneisen and waste-heat (linear U–u) models, but there is still broad agreement about the effectiveness of precompression. For example, 10% precompression ($\rho_1/\rho_0 = 1.10$) reduces the shock dissipation to about 50–60% of the Hugoniot value at twofold compression ($\rho/\rho_0 = 2.0$ for $K_{0S}' = 4$ or 5); and 20% precompression ($\rho_1/\rho_0 = 1.20$) brings the energy of the shock-loaded state to within 20–50% of the isentrope energy at $\rho/\rho_0 = 2.0$ (Figure 7.8).

This comparison highlights one of the advantages of the waste-heat approximation, in that it requires knowledge only of the Hugoniot equation of state (ρ_0, c_0, and s); neither the isentrope p_S nor the thermal properties (γ_0, q, and γ_e) need to be known at high pressures and temperatures, as in the Mie–Grüneisen approach. In fact, the latter often needs adjustment due to added electronic contributions to γ (increased effective γ_e) at high compressions, bringing results closer to those of the waste-heat model. In any case, the waste-heat model gives an upper bound for the dissipation relative to the principal Hugoniot (Figure 7.8), with the maximum dissipation decreasing systematically with increasing precompression, as indicated by the solid diagonal line ($s\eta = 1$) in Figure 7.5a:

$$\left.\frac{g_{pre}}{g_1}\right|_{max} = 1 - s\eta_1 = 1 - \frac{s}{\rho_1/\rho_0}\left(\frac{\rho_1}{\rho_0} - 1\right) \tag{7.16}$$

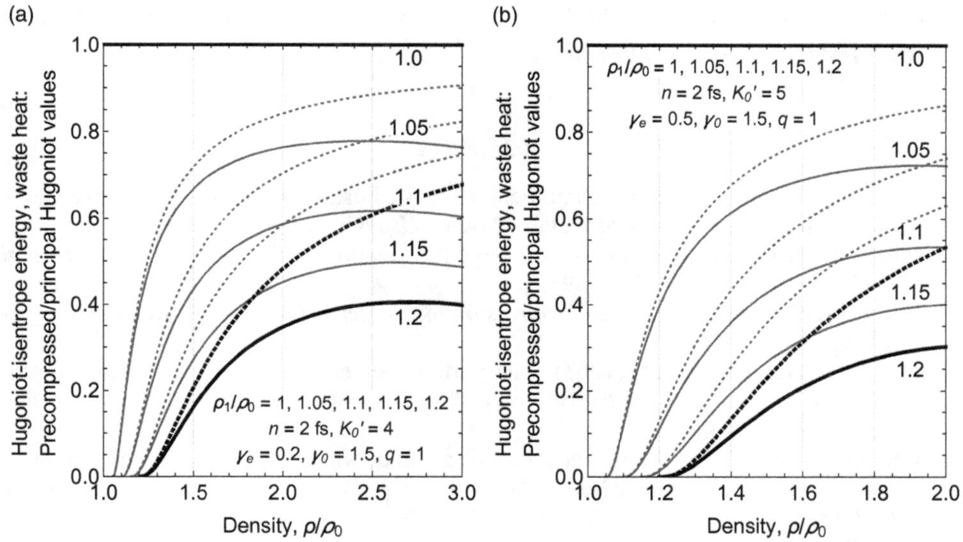

Figure 7.8 Ratio of dissipated energy $e_H - e_S$ (solid curves) and waste heat g_{pre} (dotted curves) for precompressed samples relative to the principal (single-shock) Hugoniot are shown as a function of final density (ρ/ρ_0) for samples precompressed to $\rho_1/\rho_0 = 1.0$ (no precompression; bold curve), 1.05, 1.10, 1.15, and 1.20 (bold curve): solid curves are labeled. Values of 0 and 1 on the vertical axis correspond to the isentrope and principal Hugoniot, respectively. Material properties are as in Figure 7.7, with $K_{0S}' = 4$ (a) and 5 (b).

with $s\eta_1 \leq s\eta < 1$. For the baseline model, $s = 1.25$ ($K_{0S}' = 4$), this yields suppression of dissipation and an approach to the isentrope, with g_{pre}/g_1) less than 79%, 64%, and 53% for precompression values of $\rho_1/\rho_0 - 1 = 20$, 40% and 60%, respectively.

7.4 Conclusion

The waste-heat approximation offers a simple scaling for the approach of shock-compressed states to an isentrope, either due to double-shock loading (Figure 7.1b) or to shock loading of a statically precompressed sample (Figure 7.1c). Multiple-shock loading further reduces dissipation relative to a single shock (Figure 7.1d), and approaches isentropic ramp compression in the limit of numerous steps. When combined with static precompression, multiple-shock or ramp compression offer the quickest approach to isentropic conditions (minimal dissipation), hence to the highest compressions of materials.

Acknowledgments

I thank G. W. Collins, B. K. Godwal, B. Militzer, and T. Smart for discussions, and T. S. Duffy and R. S. McWilliams for helpful comments on the manuscript. This work was supported by the National Nuclear Security Administration (NNSA) Center for Matter under Extreme Conditions and the National Science Foundation (NSF) Physics Frontier Center for Matter at Atomic Pressure.

References

Avrorin, E. N, Vodolaga., B. K., Simonienko, V. A., Fortov, V. E. (1993). Intense shock waves and extreme states of matter. *Physics Uspekhi*, 36, 337–364.

Duvall, G. E. (1962). Concepts of shock wave propagation. *Bulletin of the Seismological Society of America*, 52, 869–893.

Forbes, J. W. (2012). *Shock Wave Compression of Condensed Matter, a Primer*. Springer-Verlag.

Gault, D. E., Heitowit, E. D. (1963). The partition of energy for hypervelocity impact craters formed in rock, in *Proceedings of the 6th Hypervelocity Impact Symposium,* Vol. **2**, pp. 419–456.

Hartree, D. R. (1928). The wave mechanics of an atom with a non-Coulomb central field. Part I. Theory and methods. *Mathematical Proceedings of the Cambridge Philosophical Society*, 24, 89–110.

Jeanloz, R. (1989). Shock wave equation of state and finite strain theory. *Journal of Geophysical Research*, 94, 5873–5886.

Jeanloz, R., Celliers, P. M., Collins, G. W., et al. (2007). Achieving high-density states through shock-wave loading of precompressed samples. *Proceedings of the National Academy of Sciences*, 104, 9172–9177.

Marsh, S. P., ed. (1980). *LASL Shock Hugoniot Data*. University of California Press.

Ruoff, A. L. (1967). Linear shock-velocity–particle-velocity relationship. *Journal of Applied Physics*, 38, 4976–4980.

Trunin, R. F., Gudarenko, L. F., Zhernokletov, M. V., Simakov, G. V. (2001). *Experimental Data on Shock Compression and Adiabatic Expansion of Condensed Matter*. RFNC-VNIIEF.

Zel'dovich, Y. B., Raizer, Y. P. (1967). *Physics of Shock Waves and High-Temperature Hydrodynamic Phenomena*. Academic Press.

8

Equations of State of Selected Solids for High-Pressure Research and Planetary Interior Density Models

YINGWEI FEI

Our ability to determine the density (specific volume) as a function of pressure and temperature has drastically improved in the last several decades, with the combination of synchrotron X-ray diffraction and high-pressure techniques such as laser-heated diamond-anvil cell and large-volume multi-anvil press. The improvements are in both pressure–temperature range and data quality, and obtaining high-resolution 2D angle-dispersive diffraction data at over a megabar pressure and above 2,500 K is now routine. In parallel, dynamic compression techniques, such as laser-driven shock wave and magnetically accelerated flyer plate-impact experiments, have provided new ways to measure density at extreme conditions. The combination of static and dynamic compression data allows us to examine internal consistency in pressure determination and establish reliable pressure scales. Internally consistent pressure scales for several pressure standards are emerging through extensive comparison of compression data over a large pressure range and simultaneous measurements of elasticity and density. A concerted effort is needed to further expand and improve measurements under simultaneous high pressure and temperature, particularly at temperatures above 2,500 K, in order to accurately model the thermal pressure. To decipher the compositions of the Earth's interior based on density observations from seismology requires high accuracy in measuring the subtle compositional effects on the density of mantle and core materials. For a universal understanding of the thermal equations of state of solids, the emphasis should be on reconciling the static and dynamic data of well-studied materials that have substantial overlap in pressure–temperature ranges.

8.1 Introduction

Equations of state (EoS) for solids are fundamental functions to describe density (or specific volume) variations with pressure and temperature. Accurate determination of the equations of state of Earth materials is essential for developing density models of the Earth and planetary interiors and is also fundamental for understanding material behavior at extreme conditions. High pressures can be generated by static methods in which the force is directly applied to the surface. The combination of high-pressure techniques such as diamond-anvil cell (DAC) (e.g., Weir et al., 1959; Shen and Mao, 2017) and multi-anvil device (e.g., Kawai et al., 1973; Walker et al., 1990) with synchrotron X-ray diffraction has provided the opportunity for extensive study of the EoS of solids over a wide pressure and temperature range. The compression behavior of a vast number of solids has been investigated by in situ X-ray diffraction, particularly in the diamond-anvil cell. The pressures generated between two gem-quality diamond anvils are usually determined from the unit cell volumes of known equations of state of the pressure standards (pressure scales). These pressure scales were primarily established by direct measurements of pressure and density through shock wave techniques.

Shock waves generate dynamic compression that is characterized by the shock and particle velocities. With known initial density of the shocked material, the pressure and density can be directly measured along a Hugoniot. The reduced shock wave equations of state (pressure–density relation at room temperature) for many simple metals form the basis for pressure calibrations in static compression experiments. However, the reduction of Hugoniot measurements to the room temperature EoS relies on knowledge of the thermodynamic properties of the materials. The consistency in pressure prediction among different pressure scales is not guaranteed.

With advances in synchrotron X-radiation and its applications to high-pressure research, an increasing number of high-pressure experiments have been conducted at synchrotron facilities where X-ray diffraction data can be directly obtained for the standards and samples of interest. It has been demonstrated that different pressure standards subjected to the same pressure condition often result in significantly different calculated pressures (e.g., Dewaele et al., 2004; 2008a; Fei et al., 2004; 2007a). These comparisons have been mostly limited to static data below 150 GPa. Now that static compression data have expanded into the multimegabar pressure range and dynamic compression data to over a terapascal, there is a renewed urgency to examine the consistency among different pressure standards over a large pressure range and to understand the validity of extrapolation and the sources of discrepancy. In this chapter, I review the methods for EoS measurements, data analyses, and internal consistency among different data sources for some key solids, including platinum (Pt), gold (Au), neon (Ne), MgO, and NaCl, that are commonly used as pressure standards. I also summarize EoS data for mantle phases such as (Mg,Fe)O-ferropericlase, $CaSiO_3$-perovskite, and bridgmanite, and discuss the effect of light elements on iron (Fe) alloys relevant to Earth's core density models.

8.2 Methods

8.2.1 Shockwave Experiments

Chapters 5 and 6 in this volume provide a description for standard Hugoniot measurements by planar impact using a conventional gas gun. In recent years, laser-driven shock compression has extended shock pressure to well over 1,000 GPa (Tracy, Chapter 5, this volume).

Magnetically accelerated flyers launched using Sandia National Laboratories' Z Pulsed Power Facility are capable of generating shock pressures over 1,000 GPa (e.g., Knudson and Desjarlais, 2013; McCoy et al., 2017; Fei et al., 2021). The plate-impact experiments on the Z machine utilize the impedance-matching technique that allows high-precision determination of the Hugoniot states of the sample through the shock response of a known standard (Knudson and Desjarlais, 2013). Recently, Root et al. (2018) reported the principal Hugoniot of forsterite (Mg_2SiO_4) to 950 GPa, and Fei et al. (2021) reported melting and density of $MgSiO_3$ up to 1,254 GPa using the Z Pulsed Power Facility. Details of the experimental facility and capability are described in recent publications (Desjarlais et al., 2017; Tomlinson et al., 2018).

8.2.2 Static Compression Experiments

8.2.2.1 In Situ X-Ray Diffraction in Laser-Heated Diamond Anvil Cell

In recent years, significant advances have been made in laser-heated diamond anvil cell (LHDAC) techniques at synchrotron user facilities such as the Advanced Photon Sources (APS) (USA), European Synchrotron Radiation Facility (ESRF) (France), and Super Photon ring-8 GeV (SPring-8) (Japan) (e.g., Shen et al., 2001; Schultz et al., 2005; Meng et al., 2006; Ohishi et al., 2008; Prakapenka et al., 2008; Fukui et al., 2013; Hirao et al., 2020). It is now routine to obtain high-quality X-ray diffraction data at simultaneous high pressure and temperature by combining a tightly focused X-ray beam (<5 μm) with double-sided laser heating in the diamond anvil cell. The system at the APS consists of a double-sided laser heating system that consists of two fiber lasers and two spectroradiometric systems with optics to heat the sample and collect thermal emissions for temperature measurement on both sides (Prakapenka et al., 2008). Figure 8.1 shows the general

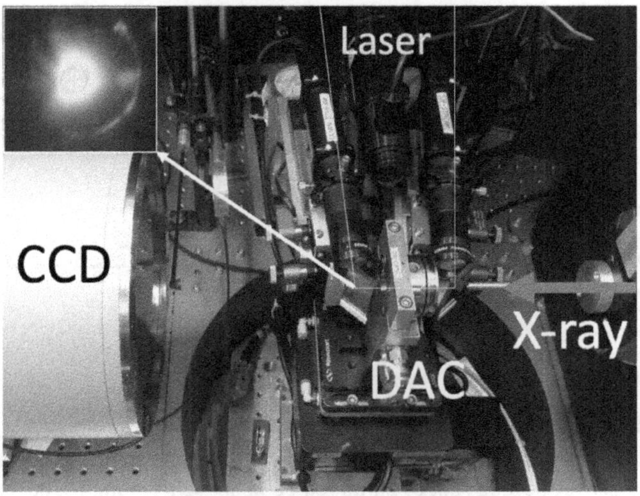

Figure 8.1 Experimental configuration for an in situ X-ray diffraction study in the laser-heated diamond anvil cell at a synchrotron facility (APS). For the color version, refer to the plate section.

experimental setup for in situ X-ray diffraction measurements at the APS synchrotron facility. The fiber laser and optics have improved temperature uniformity and stability. The system is designed to generate a large heating spot (25 μm in diameter) compared to the X-ray beam size (5 μm), which minimizes the sample temperature gradients both radially and axially in the diamond anvil cell and maximizes heating stability. These new features greatly enhance the data quality of pressure (*P*)-volume (*V*)-temperature (*T*) measurements.

For EoS studies, it is important to load samples under conditions as close to hydrostatic as possible so that the measured *P–V–T* relations represent the hydrostatic EoS of materials. Helium, neon, and argon are commonly used as quasihydrostatic pressure media. For gas-medium loading, typically a thin prepressed or sintered sample is loaded in a preindented metal (commonly Re) gasket hole with less than one third of the chamber volume filled (e.g., Fei, 1999). The sample chamber is then filled with gas to 1.5–2 kbar in a high-pressure gas-loading device and subsequently sealed at pressures >0.3 GPa. For solid-media (such as NaCl) loading, a thin sample is sandwiched between layers of the solid pressure-transmitting medium. NaCl as a pressure medium provides an efficient thermal insulator for laser heating. The compression behavior of some solids is sensitive to deviatoric stress. Annealing the sample at high temperature is an effective way to reduce the deviatoric stress (Meng et al., 1993), and it is especially critical to reduce the effect of deviatoric stress in measuring the thermal pressure. If the EoS of the solid is sensitive to deviatoric stress, the deviatoric stress changes with increasing temperature. With a possible contribution from the stress effect, the measured thermal pressure can be uncertain. Figure 8.2 shows an angle-dispersive X-ray diffraction pattern of Fe and Ne at simultaneous high pressure and temperature during laser heating in the DAC. Such high-quality diffraction data are critical for developing thermal EoS of solids (Fei et al., 2016).

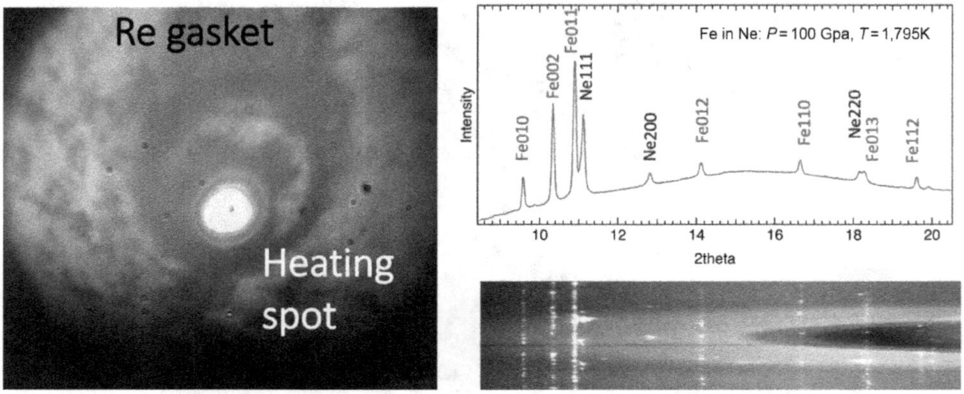

Figure 8.2 Representative angle-dispersive X-ray diffraction patterns of Fe and Ne at high temperature. Left: A heating spot of 20 μm on Fe in an Ne medium. Right: The integrated diffraction pattern with labeled hcp-Fe and Ne peaks (top) and the corresponding 2D cake pattern (bottom). The 2D image shows a spotty diffraction pattern for Ne. For the color version, refer to the plate section.

8.2.2.2 In Situ X-Ray Diffraction in the Multi-Anvil Press

While the diamond anvil cell is a versatile tool for high-pressure research, the large-volume multi-anvil press has advantages in stable heating with accurate temperature measurements and large sample volume for complex chemical systems. In addition to phase equilibrium studies, the large-volume presses at SPring-8 and APS synchrotron facilities have produced complementary *P–V–T* measurements of pressure standards and Earth materials (e.g., Fei et al., 2004; Katsura et al., 2009; Tange et al., 2012; Litasov et al., 2013a; Yamazaki et al., 2019). Figure 8.3 shows the experimental setup at beamline BL04B1 (SPring-8) for in situ X-ray diffraction measurements. A similar setup at the APS Geo Soil Enviro Consortium for Advanced Radiation Sources (GSECARS) beamline is also available for general users. A hydraulic press with an MA8 double-stage system is capable of generating pressures typically up to 30 GPa using tungsten carbide (WC) anvils and over 100 GPa using sintered diamond (SD) anvils (e.g., Yamazaki et al., 2014; Yamazaki and Ito, 2020). The second-stage assembly consists of eight corner-truncated WC or SD cubes that form an octahedral cavity for the cell assembly. The synchrotron X-ray is accessed through the gap between the cubes. A detailed description of the facility is given by Utsumi et al. (2002).

The insert in Figure 8.3 shows an X-ray radiograph image of the sample chamber through the X-ray path between the anvil gaps. The sample temperature can be directly measured with a thermocouple. Multiple sample chambers are possible (Figure 8.3 insert) because of the large sample volume, but sample volume becomes limited with the use of sintered diamond anvils for achieving higher pressure. An energy-dispersive diffraction technique is typically used for unit cell parameter determination, using a single-element (Ge) solid-state detector and polychromatic (white) synchrotron radiation. The incident X-ray beam is collimated to form a thin beam with dimensions of 50×100 μm (Katsura et al., 2004). Figure 8.4 shows representative energy-dispersive X-ray diffraction patterns of MgO and Pt at high pressure, collected using the multi-anvil setup at APS GSECARS beamline.

Different anvil materials have been tested for achieving higher pressure. Ultrahard 26 mm WC anvils (Fuji Die, TF-05) with a truncated edge length of 2.0 mm are capable of reaching simultaneous high *P–T* of 33 GPa and 1,673 K (Litasov et al., 2013b). It is possible to generate pressures as high as 65 GPa at room temperature using tapered WC anvils (Fuji Die, TJS01), but pressure drops significantly at high temperature, down to 48 GPa at 2,000 K (Ishii et al., 2017). There is considerable trade-off between pressure generation efficiency and assembly stability, especially at high temperature.

The use of SD anvils has extended pressures to above 50 GPa and up to 120 GPa at room temperature (e.g., Tange et al., 2008; Ito et al., 2010; Yamazaki et al., 2014; Arimoto et al., 2019; Yamazaki et al., 2019; Yamazaki and Ito, 2020). We have also tested large sintered diamond anvils (1 inch cubes) to optimize sample volume above 30 GPa (Figure 8.4). The development has led to successful synthesis of large $MgSiO_3$-bridgmanite samples for dynamic compression experiments (Fei et al., 2021). However, it is challenging to reach simultaneous high pressure and temperature above 65 GPa and 1,800 K (e.g. Tange et al., 2009a; Irifune et al., 2010; Liu et al., 2016; Arimoto et al., 2019; Yamazaki and Ito, 2020). Nevertheless, routine in situ X-ray diffraction measurements in the

Figure 8.3 Experimental configuration for in situ X-ray diffraction study in the large-volume multi-anvil press at synchrotron facility (SPring-8). A 1,500 ton press (top) is shown with X-ray path and detector, with the top view of the second stage (bottom-left) and radiograph image of the sample chambers at ambient pressure and high pressure (632 tons). For the color version, refer to the plate section.

multi-anvil press up to 60 GPa and 1,500 K (Tange et al., 2012; Yamazaki et al., 2014) would be complementary to measurements by using the laser-heated diamond anvil cell.

8.3 Equation of State at Room Temperature

Practical pressure scales such as the ruby fluorescence pressure gauge at room temperature are usually established using the reduced shock wave equations of state of several metals

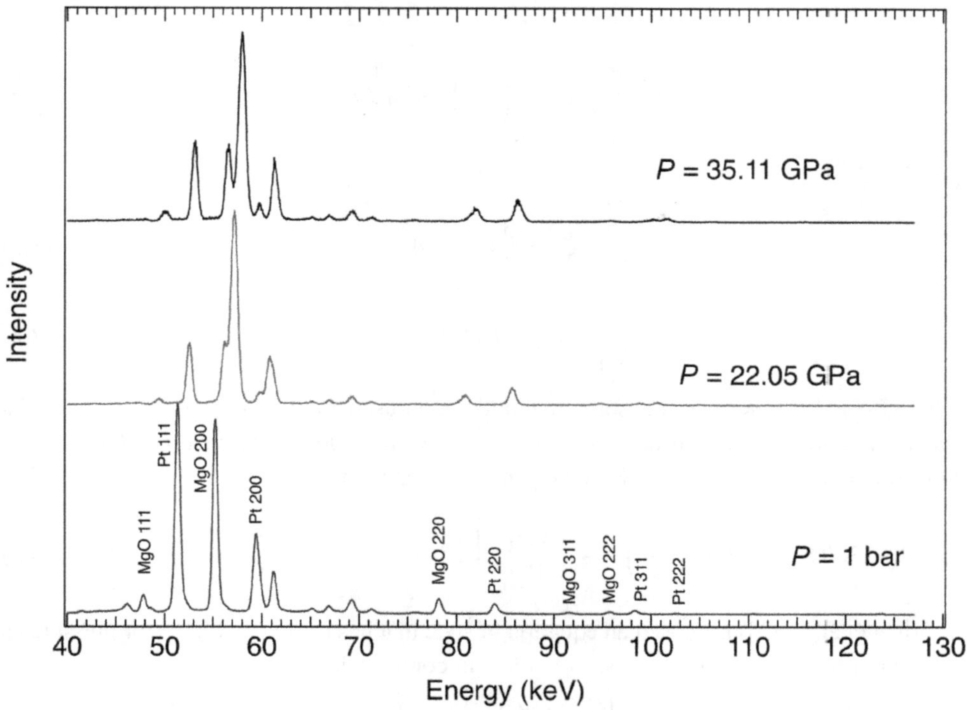

Figure 8.4 Representative X-ray diffraction patterns of MgO and Pt collected at different pressures, with energy-dispersive technique at $2\theta = 6.1092°$. The pressures were generated with large sintered diamond-anvils (1 inch cubes).

(Mao et al., 1986; Dewaele et al., 2004), assuming that they are close to the absolute pressure scales. Comparison of different pressure scales using in situ X-ray diffraction measurements in the multi-anvil press (Fei et al., 2004) and diamond anvil cell (Dewaele et al., 2004) has demonstrated that large discrepancies among different pressure scales exist, and these discrepancies can lead to significantly different pressure determinations (e.g., Fei et al., 2004). There have been several approaches to improve pressure scales and achieve self-consistency. These include the following: (1) simultaneous measurements of elasticity and density to independently solve for pressure, (2) optimization of pressure-scale-free data sets to derive the best-fitted EoS parameters, (3) combination of first principles quasi-harmonic approximation calculations with correction using low-pressure experimental data, (4) refinement of pressure scales by finding the global minimum between the static and dynamic data for many well-studied solids, and (5) establishment of internally consistent pressure scales by optimizing the EOS parameters using compression data with common references.

All pressure scales adopt a form that describes the pressure–volume relation at a reference temperature. Derived from finite-strain theory (Birch, 1978), pressure (P) as a function of volume (V) at 300 K is expressed as follows:

$$P(V, 300K) = 3f(1+2f)^{5/2}K_0\left(1+\xi_1 f+\xi_2 f^2+\cdots\right), \tag{8.1}$$

where

$$f = \frac{1}{2}\left[\left(\frac{V_0}{V}\right)^{2/3} - 1\right] \tag{8.2}$$

and

$$\xi_1 = \frac{3}{2}(K_0' - 4), \tag{8.3}$$

$$\xi_2 = \frac{3}{2}\left[K_0 K_0'' + K_0'(K_0' - 7) + \frac{143}{9}\right], \tag{8.4}$$

where K_0, K_0', and K_0'' are the isothermal bulk modulus, its first pressure derivative, and its second pressure derivative at 300 K, respectively. Equation (8.1) becomes the popular third-order Birch–Murnaghan (B–M) equation of state when $\xi_2 = 0$,

$$P(V, 300\,\mathrm{K}) = \frac{3}{2}\mathrm{K}_0\left[\left(\frac{V_0}{V}\right)^{7/3} - \left(\frac{V_0}{V}\right)^{5/3}\right]\left\{1 - \frac{3}{4}(4 - K_0')\left[\left(\frac{V_0}{V}\right)^{2/3} - 1\right]\right\}. \tag{8.5}$$

Vinet et al. (1986) proposed an equation of state that has a universal form for pressure as a function of volume for all classes of solids in compression,

$$P(V, 300\,\mathrm{K}) = 3K_0\left(\frac{V}{V_0}\right)^{-2/3}\left[1 - \left(\frac{V}{V_0}\right)^{1/3}\right]\exp\left\{\frac{3}{2}(K_0' - 1)\left[1 - \left(\frac{V}{V_0}\right)^{1/3}\right]\right\}. \tag{8.6}$$

K_0 and K_0' in the B–M or the Vinet EoS are the fitted parameters, and their covariance within a 90% confidence region depends on the *P*–*V* data range. A better constrained K_0' value requires a large compression range. For a given set of *P*–*V* data, the fitted K_0 and K_0' values are unique to the EoS used in the fit, and therefore the values should not be interchanged between the Vinet and B–M EoS. In general, the Vinet EoS gives a better representation of the data over a large compression range (Cohen et al., 2000), particularly for very compressible solids. However, the extrapolation of the Vinet form tends to predict lower pressure than that of B–M EoS at multimegabar pressures.

The adiabatic bulk modulus (K_S) and its pressure derivative (K_S') can be independently measured by techniques such as Brillouin scattering and ultrasonic interferometry. K_S relates to the isothermal bulk modulus (K_T) through the thermodynamic relation:

$$K_S = (1 + \alpha\gamma T)K_T, \tag{8.7}$$

where α and γ are the thermal expansion coefficient and the Grüneisen parameter, respectively. Simultaneous measurement of K_S and density (ρ) at high pressure could lead to an absolute pressure scale. Using the Brillouin scattering and ultrasonic interferometry techniques, the adiabatic bulk modulus (K_S) and the shear modulus (G_S) are determined from the measured longitudinal sound velocity (V_P) and shear sound velocity (V_s) by

$$K_S = \left(V_P^2 - \frac{4}{3}V_S^2\right)\rho \tag{8.8}$$

and

$$G_S = V_S^2 \rho. \tag{8.9}$$

The absolute pressure scale established by this method is applicable to the pressure range of the experiment. The extrapolation is still dependent on the chosen form of the EoS. In principle, it is possible to assess the validity of an EoS fit if the experimental data cover a sufficiently large compression range to provide a tight constraint on the K_0' value.

MgO with a simple cubic structure is often used as a prototype for a primary pressure scale. Zha et al. (2000) established the MgO primary pressure scale by measuring the elastic properties of MgO using the Brillouin scattering technique, combined with specific volume measurements by X-ray diffraction (XRD) measurements up to 55 GPa. The MgO primary pressure scale is also proposed by measuring the sound velocities of MgO up to 11 GPa (Li et al., 2006) and 22 GPa (Kono et al., 2010) using a combination of ultrasonic interferometry and in situ X-ray diffraction techniques. More recently, Fan et al. (2019) measured the elasticity of MgO at simultaneous high P–T conditions up to 30 GPa and 900 K by combining Brillouin scattering and XRD. There is excellent agreement on the isothermal bulk modulus, where $K_0 = 160.5 \pm 0.5$ GPa. However, the measured pressure derivative K_0' values range from 4.03 to 4.35, which would translate into large pressure differences at pressures above 50 GPa.

In order to ensure consistency in pressure prediction over a large pressure range and to understand the causes of the discrepancies in literature, I compiled the EOS parameters for several common pressure scales (Table 8.1), and compared the differences in pressure extrapolation up to 365 GPa in the following section.

8.3.1 Common Pressure Standards

8.3.1.1 Neon

Neon (Ne) is often used as a quasihydrostatic pressure medium. In situ X-ray diffraction measurements in Ne also produce unit cell parameters of Ne along with those of the sample of interest. Therefore, Ne is often used as a secondary pressure standard, and a reliable EoS of Ne would be useful for pressure determination. Fei et al. (2007a) proposed the Ne pressure scale based on the compression data of Hemley et al. (1989) up 110 GPa and additional compression data for solid Ne up to 61 and 115 GPa using Pt and Au, respectively, as the internal pressure standards. It was found that the pressures of Hemley et al. (1989), recalculated with the ruby scale of Dewaele et al. (2004), were more consistent with the additional compression data using the Pt and Au pressure scales (Fei et al., 2007a).

Since 2007, Ne compression data have been extended to pressures over 240 GPa using different pressure scales. Figure 8.5 plots data published since 2007, compared with the calculated compression curve of Ne using the third-order B–M and the Vinet EoS (Fei et al., 2007a). The compression data of Dewaele et al. (2008b) using either the ruby or W pressure scales are in excellent agreement with the B–M EoS of Fei et al. (2007a). The

Table 8.1 *EOS parameters for MgO, Ne, NaCl, Pt, and Au at 300 K.*

		Vinet EoS		B–M EoS			
References	V_0, Å^3	K_0, GPa	K_0'	K_0, GPa	K_0'	Methods	Max P, GPa
MgO							
Zha et al. (2000)				160.2	4.03	Brillouin	55
Speziale et al. (2001)	74.71			160.2	3.99	Combination	52
Tange et al. (2009b)	74.698	160.6	4.37	160.6	4.22	Combination	
Kono et al. (2010)	74.71			160.9	4.35	Ultrasonic	23
Ne							
Fei et al. (2007a)	88.967	1.16	8.23	1.43	8.02	Diffraction	115
Dewaele et al. (2008b)	88.936	1.07	8.4	1.42	8.03	Diffraction	209
Dorfman et al. (2012)	88.967	1.04	8.48			Diffraction	245
Ye et al. (2018)	88.967	1.28	8	2.01	6.48	Diffraction	128
NaC1–B2							
Fei et al. (2007a)	41.35	26.86	5.25	30.69	4.33	Diffraction	140
Sakai et al. (2011)	37.73	41.36	4.96	47.8	4.06	Diffraction	302
Dorfman et al. (2012)	41.35	24.2	5.76			Diffraction	245
Dewaele et al. (2019)	42.3	22.664	5.735			Diffraction	155
Pt							
Fei et al. (2007a)	60.38	277	5.08	277	4.95	Diffraction	90
Dorogokupets and Dewaele (2007)		277.3	5.12			Combination	
Matsui et al. (2009)		273	5.2			Combination	
Yokoo et al. (2009)		276.4	5.48	276.4	5.12	Shock	
Dorfman et al. (2012)		277	5.43			Diffraction	228
Ye et al. (2017)		277.3	5.23			Diffraction	140
Kamada et al. (2019)		273	5.3	273.3	5.23	IXS	20
Au							
Tsuchiya (2003)	67.85	166.7	6.12			Theory	
Fei et al. (2007a)	67.85	167	6	167	5.77	Diffraction	140
Takemura and Dewaele (2008)	67.85	167	5.9			Diffraction	131
Yokoo et al. (2009)	67.72	167.5	5.94	167.5	5.79	Shock	
Dorfman et al. (2012)	67.85	167	5.88			Diffraction	253
Ye et al. (2017)	67.85	167	5.9			Diffraction	140

data of Takemura et al. (2010) and Dorfman et al. (2012), which used Pt and Au as pressure standards, respectively, showed a systematic shift to higher pressures but are within 2% error even at 245 GPa. The data of Ye et al. (2018) and Yang et al. (2019), using MgO and Fe for pressure determination respectively, are also within 2% uncertainty. Note that the Vinet EoS always predicts lower pressures when it is extrapolated to higher pressure (Figure 8.5). The pressure difference between the B–M and the Vinet EoS exceeds 2% beyond 180 GPa and reaches 5.7% at 350 GPa.

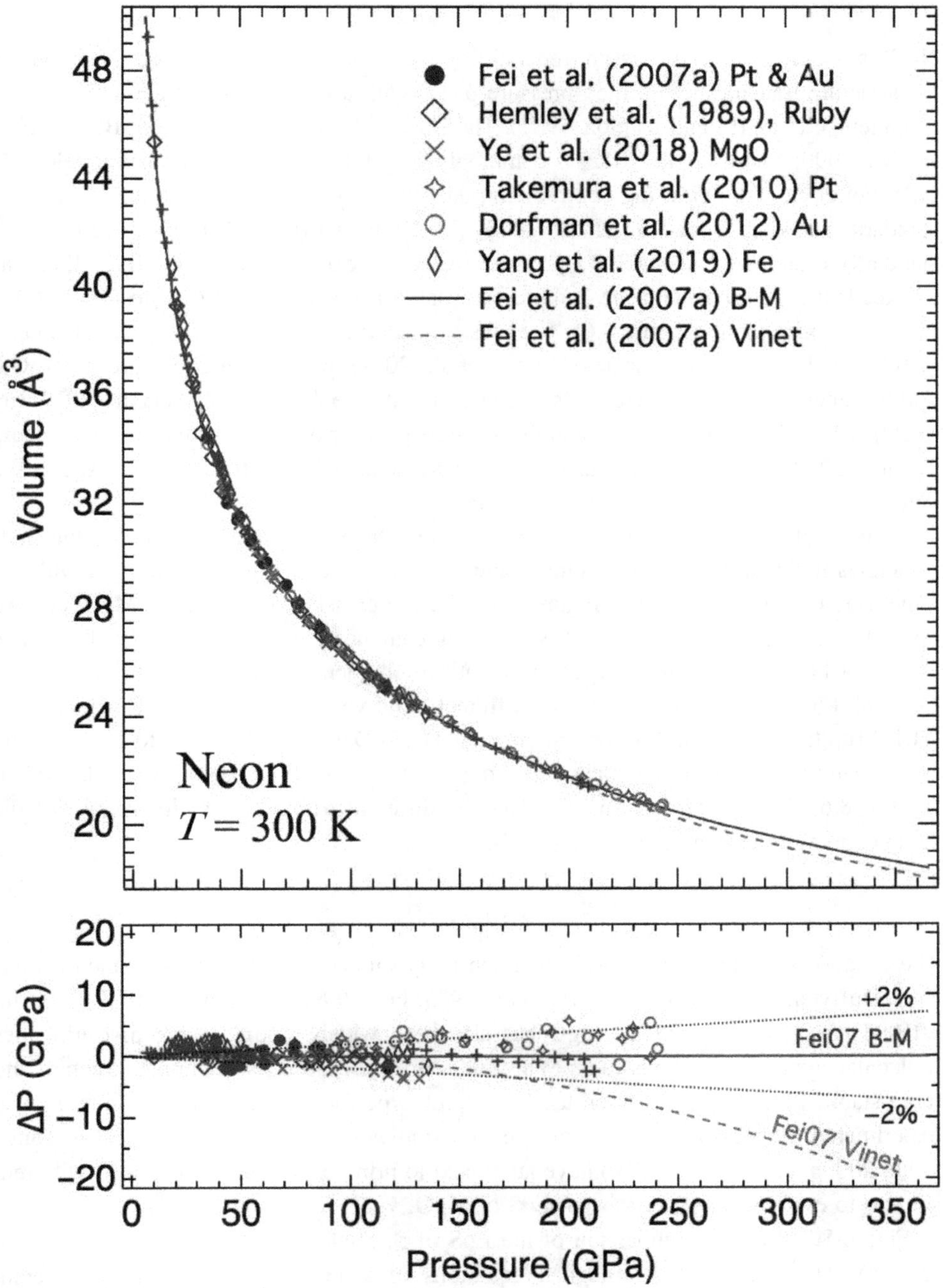

Figure 8.5 Compression data of Ne at 300 K, compared with the B–M EoS (Fei07 B–M) of Fei et al. (2007a). The Vinet EoS is also plotted for comparison. Different symbols represent various studies of Ne compression with different pressure scales (see discussion in the text). The pressure deviations from the Fei07 B–M are compared in the bottom graph.

8.3.1.2 NaCl

NaCl is widely used as a pressure transmitting media and pressure standard. NaCl with a B1 structure transforms to a high-pressure phase with a B2 structure at about 27 GPa at room temperature (Li and Jeanloz, 1987). The NaCl–B1 scale (Decker, 1965; Birch, 1978; Brown, 1999; Matsui et al., 2012) is commonly used for pressure determination below 27 GPa. Fei et al. (2007a), using compression data up to 107 GPa based on the Pt pressure standard, and data from Sata et al. (2002) (MgO standard) and Ono et al. (2006) (Au standard), established an NaCl–B2 pressure scale. The extrapolation of the B–M EoS was in excellent agreement with the compression data of Sakai et al. (2011) up to 304 GPa, within 2% uncertainty (Figure 8.6). The pressure determination in the study of Sakai et al. (2011) was based on the Pt scale of Matsui et al. (2009), which is in good agreement with that of Fei et al. (2007a). There is also agreement with the data set of Sata et al. (2010) (up to 180 GPa and a Fei07 Pt scale), and the absolute primary pressure scale of Murakami and Takata (2019) based on simultaneous measurements of densities and acoustic wave velocities of NaCl–B2 phase up to ~120 GPa.

Ueda et al. (2008) and Ono (2010) calculated the EoS of NaCl–B2 using the first-principles molecular dynamics method, and their results are in good agreement with the Vinet EoS of Fei et al. (2007a) (Figure 8.6). The compression data of NaCl, MgO, and Au up to 140 GPa (Hirose et al., 2008) showed agreement when the MgO scale of Speziale et al. (2001) was used, but predicted systematically higher pressures when the Au scale of Fei et al. (2007a) was used. There is a different trend with the data sets of Dorfman et al. (2012) (up to 265 GPa and Ne pressure medium) and Dewaele (2019) (up to 155 GPa and He pressure medium), showing about 6% higher pressures than those of Fei et al. (2007a) (Figure 8.6). The difference could reflect the different degrees of hydrostaticity of the sample environment on the measured EoS.

8.3.1.3 MgO

The equation of state of MgO has been extensively studied by static and dynamic methods (e.g., Duffy and Ahrens, 1993; Duffy et al., 1995; Fei, 1999; Zha et al., 2000; Kono et al., 2010; Fat'yanov et al., 2018). MgO has also been widely studied using first-principles molecular dynamics methods (e.g., Wu et al., 2008). Because of its simple structure and large stability field, MgO is often used as a prototype for calculations and a reference in experiments for pressure comparison. Several studies (e.g., Speziale et al., 2001; Tange et al., 2009b; Kono et al., 2010) have attempted to bring experimental data from different sources to establish a self-consistent EoS of MgO.

Figure 8.7 shows a comparison of the EoS of Speziale et al. (2001) and Tange et al. (2009b). The two Birch–Murnaghan equations of state predict significantly different pressures at high pressure (>100 GPa) because of the different pressure derivatives of the bulk modulus used ($K_0' = 3.99$ vs. 4.221), even though a similar bulk modulus (K_0 ~ 160.5 GPa) was adopted. Various experimental studies (e.g., Jackson and Neisler, 1982; Fei, 1999; Zha et al., 2000; Speziale et al., 2001; Li et al., 2006; Jacobsen et al., 2008) have converged to K_0 ~ 160.5 GPa based on experimental pressures below 100 GPa.

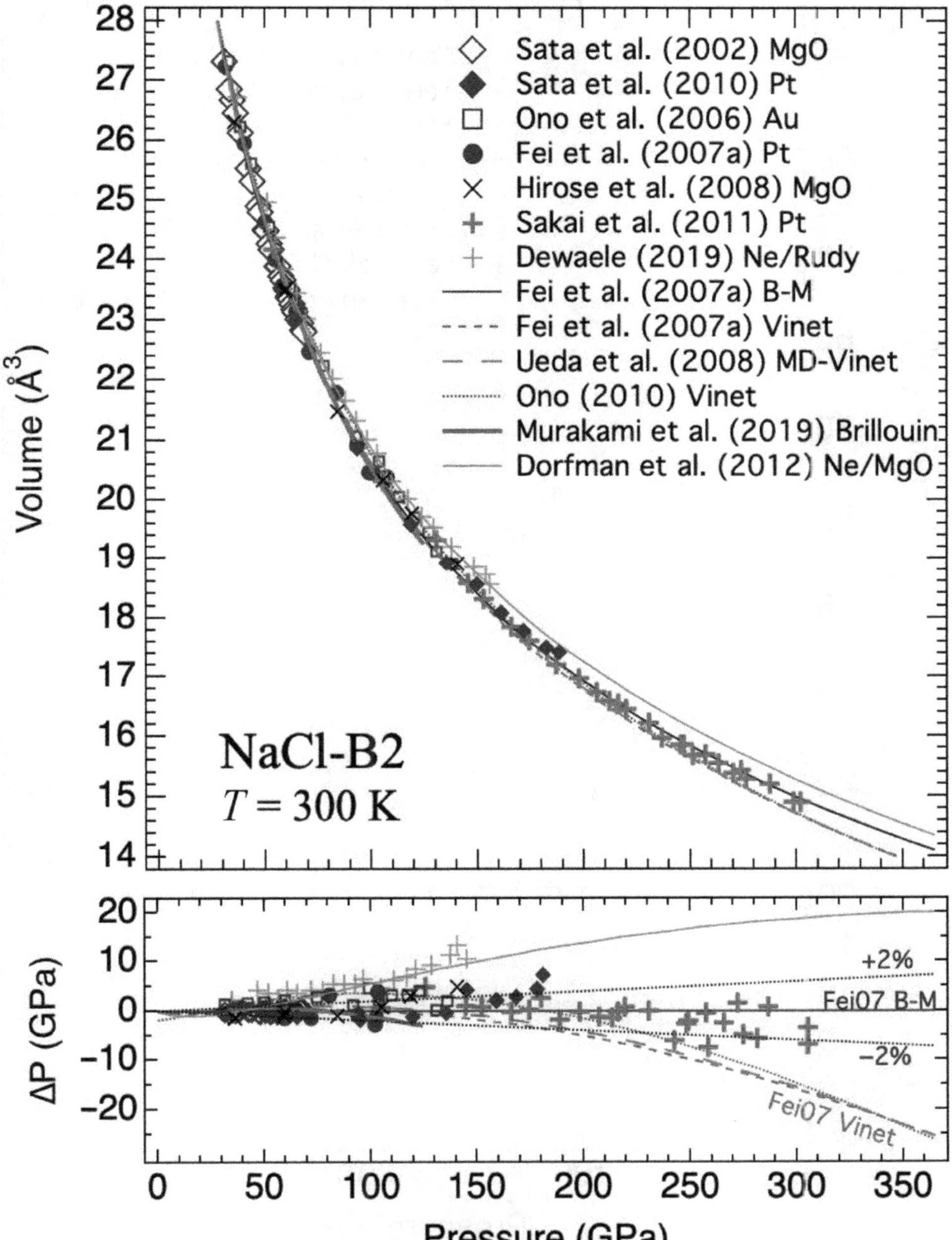

Figure 8.6 Compression data of NaCl at 300 K, compared with the B–M EoS of Fei et al. (2007a). The Vinet EoS of Fei et al. (2007a), Ueda et al. (2008), Ono (2010), and Dorfman et al. (2012) are also plotted for comparison.

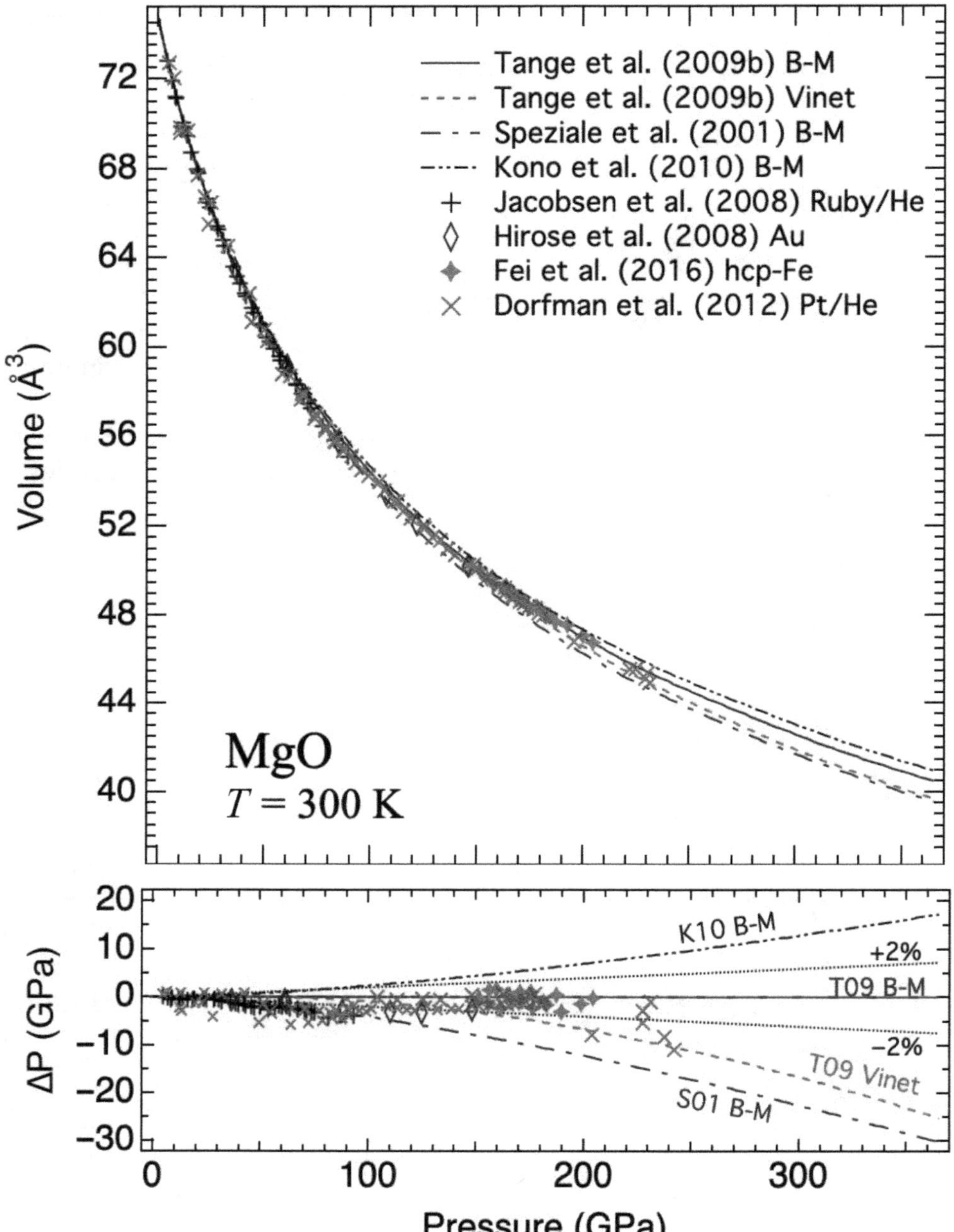

Figure 8.7 Compression data of MgO at 300 K compared with the B–M EoS of Tange et al. (2009b) (T09 B-M). The Vinet EoS of Tange et al. (2009b) (T09 Vinet) and the B–M EoS of Speziale et al. (2001) (S01 B–M) are also plotted for comparison.

The pressure derivative of the bulk modulus can be best evaluated by data with a large compression range. However, those data are often influenced by the choice of pressure scales and the nonhydrostatic nature of the compression.

Tange et al. (2009b) used pressure-scale-free data sets to establish an MgO pressure scale. The data sets include thermal expansion data and high-temperature adiabatic bulk modulus at ambient pressure, high-pressure adiabatic bulk modulus up to 55 GPa at room temperature, and shock compression data. The shock compression data played a critical role in optimizing K_0' because these data extend to high simultaneous pressure and temperature. The K_0' value cannot be uniquely constrained because of the trade-off with the fitted Grüneisen parameter when only Hugoniot data are available. The degree of hydrostaticity under compression also plays an important role in determining the EoS. It has been demonstrated that compression of MgO under nonhydrostatic condition tends to make MgO less compressible (higher K_0' or higher K_0 value) (Duffy et al., 1995; Fei, 1999; Jacobsen et al., 2008). The quasihydrostatic data (Jacobsen et al., 2008) based on the ruby scale of Dewaele et al. (2004) agree better with the EoS of Speziale et al. (2001). On the other hand, the very high-pressure data in the study of hcp–Fe (Fei et al., 2016) agree well with the EoS of Tange et al. (2009b) in the pressure range of 150–204 GPa, where the EoS of hcp–Fe is consistent with multiple pressure scales.

Evaluation of the EoS of MgO at multimegabar pressure strongly depends on the pressure scales used in the experiments. Dorfman et al. (2012) compared the pressure standards between MgO and NaCl up to 265 GPa between MgO and Mo up to 213 GPa, and between MgO and Pt up to 228 GPa, in either Ne or He pressure media. They established self-consistent pressure scales using the Vinet EoS of MgO (Tange et al., 2009b) as a reference. It should be noted that the difference between the Vinet EoS of Tange et al. (2009b) and the B–M EoS of Speziale et al. (2001) is small when extrapolated to multimegabar pressures, but they both underestimate pressures compared with the B–M EoS of Tange et al. (2009b) (Figure 8.7).

8.3.1.4 Au

Experimental and theoretical studies of Au compression behavior have converged to a bulk modulus (K_0) of 167 GPa (e.g., Shim et al., 2002; Tsuchiya, 2003; Greeff and Graf, 2004; Dorogokupets and Dewaele 2007; Fei et al., 2007a; Takemura and Dewaele, 2008; Yokoo et al., 2009; Dorfman et al., 2012; Ye et al., 2017). However, the reported pressure derivative of the bulk modulus (K_0') varies from 5.0 to 6.2, which results in very large difference in pressure extrapolation. Figure 8.8 shows the compression curves of Au up to 365 GPa, using EoS parameters of Fei et al. (2007a) and Dorfman et al. (2012). The Vinet EoS (F07 Vinet) predicts lower pressures than the B–M EoS (F07 B–M) when it is extrapolated to very high pressure. Comparing the parameters used for the Vinet EoS of Dorfman et al. (2012) (D12 Vinet) and F07 Vinet, even a small difference in K_0' (Table 8.1) could result in significant pressure difference at multimegabar pressures (Figure 8.8). Therefore, it requires a high level of accuracy in the fitted parameters such as K_0' when the pressure scale is intended for experiments in the multimegabar pressure range.

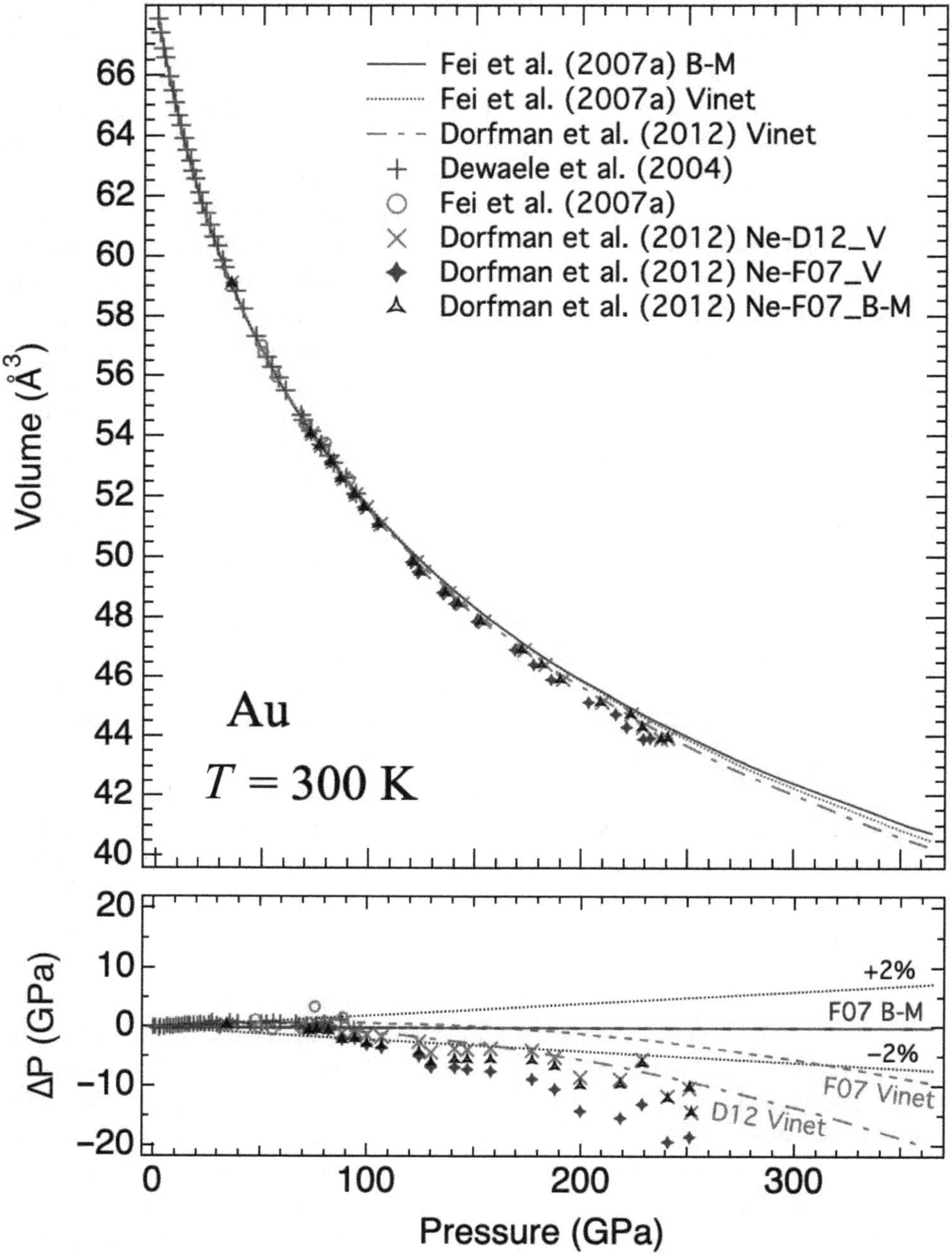

Figure 8.8 Compression data of Au at 300 K, compared with the B–M EoS of Fei et al. (2007a). The Vinet EoS of Fei et al. (2007a) is also plotted for comparison.

The remeasured shock compression data at pressures up to 580 GPa produced a K_0' of 5.94 using the Vinet EoS (Yokoo et al., 2009).

8.3.1.5 Pt

Several studies (e.g., Dewaele et al., 2004; Fei et al., 2007a; Matsui et al., 2009; Kamada et al., 2019) suggested that the Pt scales of Jamieson et al. (1982) and Holmes et al. (1989) derived from shock wave data overestimate pressure. Figure 8.9 shows the comparison among different equations of state of Pt. Using the B–M EoS of Fei et al. (2007a) as a reference, the Vinet EoS of Fei et al. (2007a), Matsui et al. (2009) and Kamada et al. (2019) showed similar negative deviation from these early studies. The parameters of the EoS of Matsui et al. (2009) were optimized using both shock wave and thermal expansion data along with static compression data at simultaneous high P and T, whereas the EoS of Kamada et al. (2019) was derived from phonon velocity and density measurements based on inelastic X-ray scattering and X-ray diffraction up to 20 GPa. The Vinet EoS of Ye et al. (2017), optimized from compression data up to 140 GPa using the MgO scale, shows similar behavior as the B–M EoS of Fei et al. (2007a) when extrapolated to higher pressure, but predicts systematically higher pressures than the Vinet EoS (Figure 8.9).

There is still a significant discrepancy among Pt EoS. Notably, two compression data sets in Ne and He pressure media (Dorfman et al., 2012) plotted at pressures higher than those predicted by the pressure scales discussed previously. These data also showed systematically higher pressures even at relatively low pressures (Figure 8.9). The compression data of Sata et al. (2010) showed the largest deviation at higher pressure, agreeing with the EoS of Holmes et al. (1989). The same data set showed agreement in pressure comparison between the NaCl–B2 EoS and the Pt EoS of Fei et al. (2007a) (Figure 8.6). All data together seem to show two divergent trends for the EoS of Pt at high pressure, which reflect the difference in K_0'. With the K_0 value centered at 277 GPa, the lower value, $K_0' = 5.08$ (Fei et al., 2007a), is more consistent with the results from the comprehensive data analysis of Matsui et al. (2009) and from the combined measurements of velocity and density (Kamada et al., 2019). On the other hand, the higher value, $K_0' = 5.43$ (Dorfman et al., 2012), agrees with results of the reanalysis of the shock wave data (Yokoo et al., 2009) and compression data (Sata et al., 2010) using the MgO pressure scale.

8.3.1.6 Other Pressure Standards

For diamond anvil cell experiments, the ruby fluorescence pressure gauge has been adopted as a practical pressure scale for pressure determination at room temperature. The widely cited Mao ruby scale (Mao et al., 1986) was based on the reduced shock wave equations of state of Cu and Ag (Carter et al., 1971). The wavelength (λ) shift of the ruby R-line emission as a function of pressure up to 80 GPa under quasihydrostatic Ar pressure medium is expressed as $P(\text{GPa}) = 1{,}904/7.665[(\lambda/\lambda_0)^{7.665}-1]$. Copper and aluminum are commonly used as flyers for dynamic compression experiments, and their EoS have been well calibrated in the shock wave community. There have been extensive recalibrations of

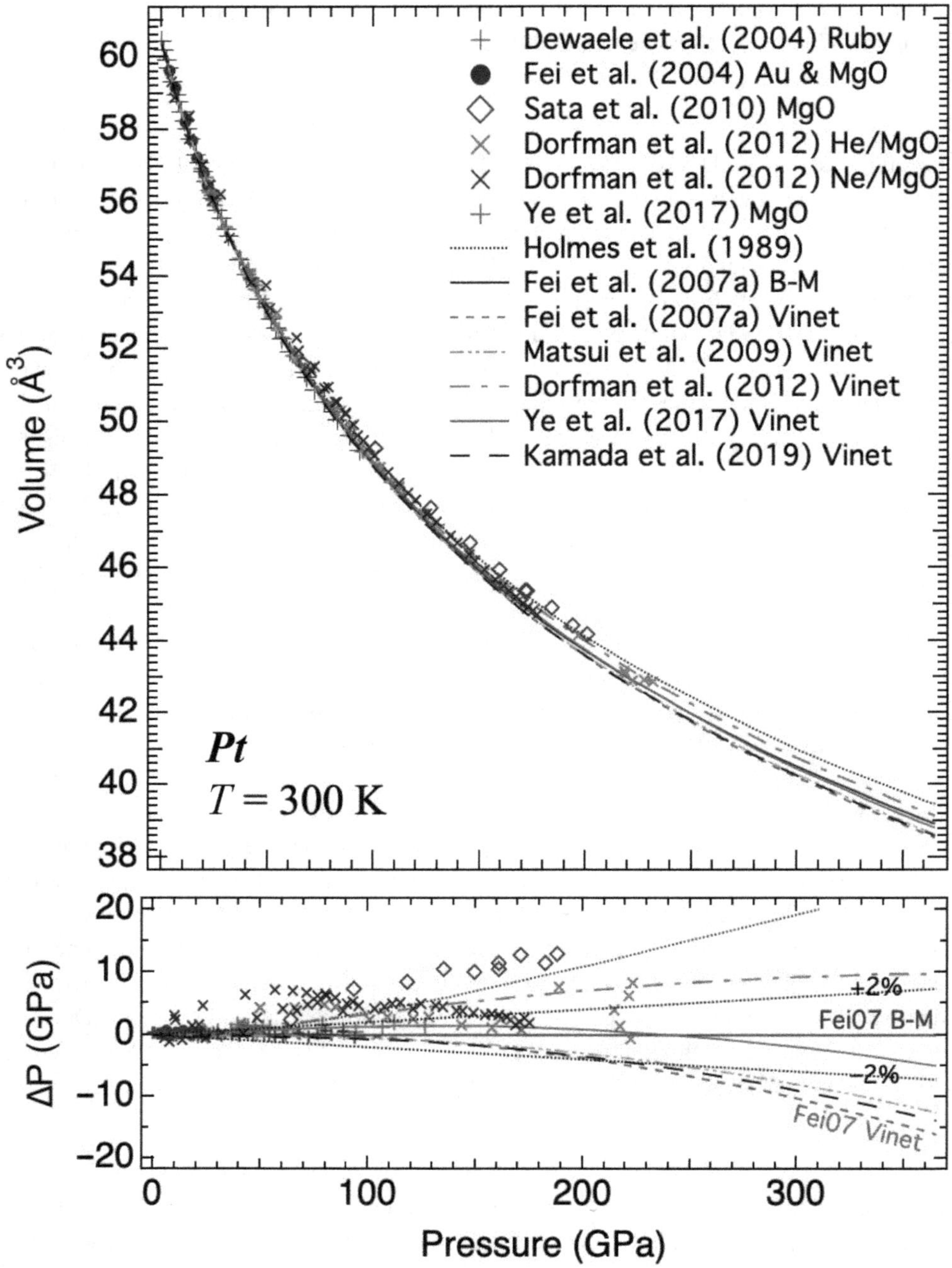

Figure 8.9 Compression data of Pt at 300 K, compared with the B–M EoS of Fei et al. (2007a). The Vinet EoS of Holmes et al. (1989), Fei et al. (2007a), Matsui et al. (2009), Dorfman et al. (2012), Ye et al. (2017), and Kamada et al. (2019) are also plotted for comparison.

the ruby pressure scale to higher pressures against X-ray diffraction data for metal standards such as Au, Pt, Ta, W, Cu, Mo, and Al (Dewaele et al. 2004, 2008a; Chijioke et al., 2005; Dorogokupets and Oganov, 2006). The ruby scale by Dewaele et al. (2004), P (GPa) = 1,904/9.5[$(\lambda/\lambda_0)^{9.5}$–1], is best supported by the reduced equations of state of W and Mo (Nixson and Fritz, 1992), Cu (Wang et al., 2002), Al (Nellis et al., 1988), and Ni (Kennedy and Keeler, 1972). The ruby scale was further refined by including the compression data of the transition metals (Dewaele et al., 2008a). The refinement was based on the reduced shock wave EoS of 10 metals (Ag, Al, Au, Co, Cu, Mo, Ni, Pt, Ta, and W), assuming the global minimum between the static and dynamic data would produce the best pressure calibration. Using this approach, Dewaele et al. (2008a) produced a new ruby scale, P(GPa) = 1,920/9.61[$(\lambda/\lambda_0)^{9.61}$–1], that is consistent with the reduced EoS of Al (Nellis et al., 1988). However, there is still a large number of metal standards that do not produce internally consistent equations of state when they are compared with static compression data (Dewaele et al. 2004, 2008a; Fei et al., 2007a).

The reduction of shock wave data to the room temperature compression curve relies on the model of the Grüneisen parameter and the electronic contribution for metals. The reduced shock wave EoS of different metals are not converging to internally consistent pressure scales when they are checked by static experiments. It is evident that there is still inconsistency between the Pt and the Au scale (Figures 8.8 and 8.9). Increasing the K_0' value for Au and decreasing the K_0' value for Pt from the reduced shock wave EoS could produce consistent pressure predictions in the static experiments (Fei et al., 2007a). In order to further understand the cause of the discrepancy, it would be useful to directly compare the compression behavior with other well-established metal standards such as Al, Cu, and W, at pressures between 150–300 GPa.

8.4 Thermal Pressure

8.4.1 Models of Thermal Equation of State

Once the equation of state of a solid is established, the pressure(P)–volume(V)–temperature (T) relation can be described by

$$P(V,T) = P(V, 300\,\mathrm{K}) + P_{th}, \tag{8.10}$$

where $P(V,\ 300\ \mathrm{K})$ describes the compression behavior at 300 K and P_{th} is the thermal pressure. The thermal pressure may consist of the phonon ($P_{th,vib}$) and electronic ($P_{th,el}$) contributions of metals:

$$P_{th} = P_{th,vib} + P_{th,el}. \tag{8.11}$$

The phonon contribution to the thermal pressure is calculated from the Mie–Grüneisen formulation with the Debye model,

$$P_{th,vib} = \frac{\gamma_{vib}}{V}[E(V,T) - E(V, 300\,\mathrm{K})], \tag{8.12}$$

where γ_{vib} is the vibrational Grüneisen parameter, which is assumed to be a function of volume only. The Grüneisen parameter is commonly described with a simple function,

$$\gamma_{vib} = \gamma_0 \left(\frac{V}{V_0} \right)^q, \tag{8.13}$$

where q is the volume dependence. A constant q usually cannot sufficiently describe the complex volume dependence of the Grüneisen parameter. Tange et al. (2009b) introduced a new function with two adjustable parameters,

$$\gamma_{vib} = \gamma_0 \left\{ 1 + a \left[\left(\frac{V}{V_0} \right)^b - 1 \right] \right\}. \tag{8.14}$$

Equation (8.14) is identical to (8.13) with $b = q$ when $a = 1$. The harmonic internal energy, $E(V, T)$, can be calculated from Debye model by using

$$E(V, T) = 9nRT \left(\frac{\theta_D}{T} \right)^{-3} \int_0^{\theta_D/T} \frac{x^3}{e^x - 1} dx. \tag{8.15}$$

The Debye temperature (θ_D) is related to the Grüneisen parameter by

$$\theta_D = \theta_0 \left(\frac{V}{V_0} \right)^{-(1-a)\gamma_0} \exp \left[\frac{\gamma_0 - \gamma_{vib}}{b} \right]. \tag{8.16}$$

The specific heat capacity at constant volume (C_V) from the Debye model is given by

$$C_V = 9nR \left(\frac{\theta_D}{T} \right)^{-3} \int_0^{\theta_D/T} \frac{x^4 e^x}{(e^x - 1)^2} dx. \tag{8.17}$$

For metal pressure standards, the electronic contribution to the thermal pressure could be significant, especially at high temperature. In shock wave data analysis, the $P_{th,el}$ is often estimated by

$$P_{th,el} = \frac{\gamma_e}{V} \int_{300}^{T} C_{V,el} \, dT, \tag{8.18}$$

where γ_e is the electronic part of the Grüneisen parameter. The electronic specific heat capacity, $C_{v,el}$, may be approximated by

$$C_{V,el} = \beta_0 \left(\frac{V}{V_0} \right)^{\kappa} T \tag{8.19}$$

to account for the anharmonic and electronic contributions (Boness et al., 1986).

Thermal expansivity (α) and heat capacity (C_P) as a function of temperature can be experimentally measured, and they are important data to assess the anharmonic contribution. The C_P and C_v are related by

$$C_P = (1 + \gamma \alpha T) C_V. \tag{8.20}$$

The average Debye temperature at ambient pressure (θ_0) can be optimized by fitting the measured C_P at 1 atm.

8.4.2 Data Analysis and Thermal Pressure Calculations

There are multiple data sources that can be used to establish a reliable thermal equation of state of a solid. The best way to establish a pressure-scale-free EoS is to simultaneously obtain elasticity and density measurements. These measurements are often performed in a limited pressure range at room temperature. Although shock wave experiments provide direct density-pressure measurements over a large pressure range, the thermal state has to be modeled using thermodynamic relations. Tange et al. (2009b) used pressure-scale-free data sets to develop a *P–V–T* equation of state for MgO that could be a reference for pressure determination at high pressure and temperature. To optimize the parameters of thermal EoS of MgO, the comprehensive analyses utilize thermal expansion data up to 2,990 K, high-temperature adiabatic bulk modulus (K_S) data up to 1,800 K, high-pressure K_S data up to 55 GPa, and shock compression data up to 196 GPa and 3,663 K. There is excellent agreement among different data at pressure below 50 GPa. However, the extrapolation to multimegabar bar pressures depends on the EoS forms to be adopted. Using the same data set for optimization, the pressure predicted by the Vinet EoS is about 7% lower than that by the B–M EoS at 365 GPa (Figure 8.7).

It is hard to judge which form of EoS is better for extrapolation. Laser-driven ramp compression of MgO to very high pressure (Coppari et al., 2013) may provide some constraint. Figure 8.10a shows compression data between 300–450 GPa that seems to indicate a better match with the Vinet EoS extrapolation without the thermal pressure correction. On the other hand, the combination of first-principles local density quasiharmonic approximation (QHA) calculations with experimental low-pressure data (Wu et al., 2008) agrees better with the B–M EoS extrapolation (Figure 8.10a).

Volume (or density) measurements at simultaneous high pressure and temperature are limited. The static compression data for MgO were limited to 1,100 K up to 24 GPa in the externally heated DAC (Fei, 1999) and to 1,500 K and 24 GPa in the multi-anvil press (Kono et al., 2010). The thermal pressures derived from these experiments cover a limited temperature range, and they are not sensitive enough to resolve the dependence of V/V_0 on the thermal pressure (Figure 8.10b). In the model of Tange et al. (2009b), the optimized thermal parameters were mostly influenced by the shock Hugoniot data up to 196 GPa and 3,663 K (Duffy and Ahrens, 1993). Recent shock wave data with preheated MgO samples (Fat'yanov et al., 2018) provide an independent check of the model at very high temperature. The calculated thermal pressures are generally higher than the measurements. The difference originates from the discrepancy of shock temperature measurements at ~200 GPa between the two shock wave studies (Figure 8.10b). Effective assessment of the thermal pressure from static experiments requires measurements at much higher temperatures. Additional static *P–V–T* measurements at temperatures above 2,500 K are preferred to further evaluate the thermal EoS of MgO.

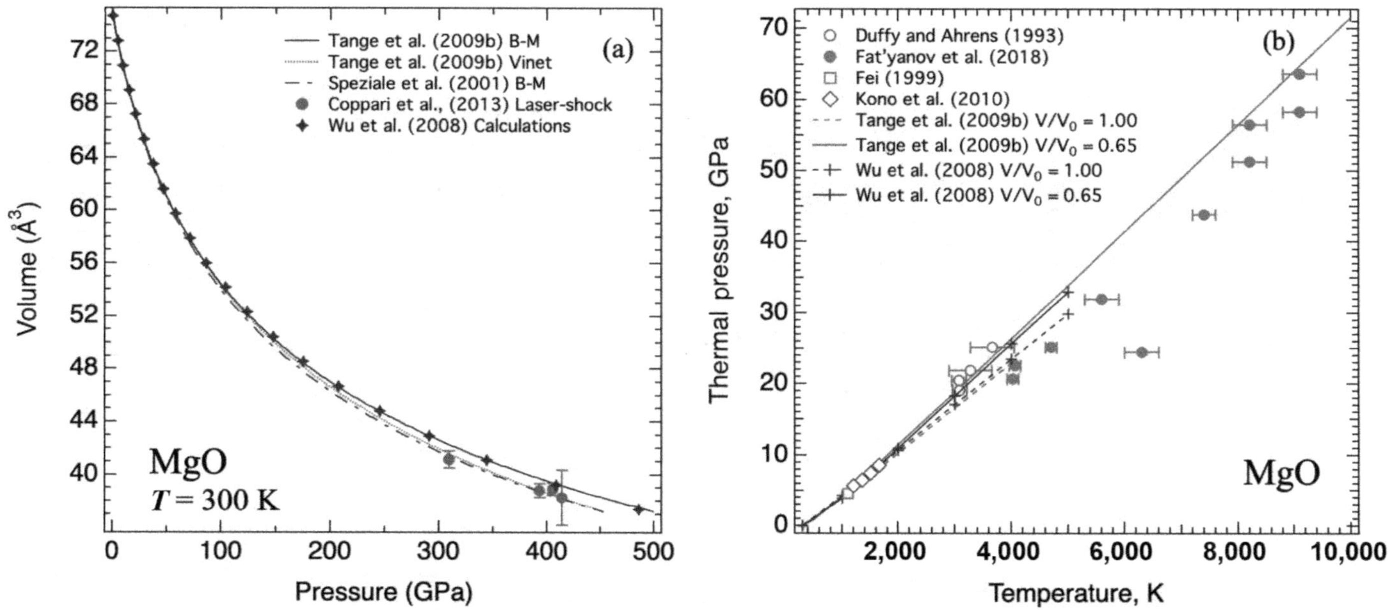

Figure 8.10 (a) Compression curves of MgO up to 500 GPa compared with laser-driven ramp-compression data (Coppari et al., 2013) and theoretical calculations (Wu et al., 2008). (b) Thermal pressure of MgO as a function of temperature, compared with shock wave Hugoniot data (Duffy and Ahrens, 1993; Fat'yanov et al., 2018), calculations (Wu et al., 2008), and static P–V–T data (Fei, 1999; Kono et al., 2010).

Table 8.2 lists the commonly used thermal EoS for MgO, Pt, Au, Ne, and NaCl–B2 for pressure calibration. The consistency in pressure determination at high temperature using these models needs to be further verified because limited *P–V–T* data were used to determine the thermal pressure. Figure 8.11 shows the calculated thermal pressures for the Au thermal EoS (Tsuchiya, 2003; Fei et al., 2007a; Yokoo et al., 2009). The analyses of Yokoo et al. (2009) yielded a reduced Au EoS similar to that of Fei et al. (2007a) at room temperature, by combining the thermal expansion, bulk modulus, and specific heat data as a function of temperature at ambient pressure with the shock wave data. The calculated thermal pressures have a very weak V/V_0 dependence, even at 60% compression (Figure 8.11). In comparison, Fei et al. (2007a) just fitted the measured static *P–V–T* data to obtain the Grüneisen parameter with a single volume dependence parameter, $q = 0.6$. A large q value ($q = 2.15$) in the model of Tsuchiya (2003), based on the first-principles electronic structure calculation method, results in substantially smaller thermal pressures.

Although the calculated thermal pressures from Fei et al. (2007a) and Yokoo et al. (2009) are identical at $V/V_0 = 1.0$ because the same γ_0 is used in the calculation, the calculated thermal pressures at high pressure (e.g., $V/V_0 = 0.6$) from Fei et al. (2007a) are generally higher than those from Yokoo et al. (2009) (Figure 8.11). The deviation reflects how the Grüneisen parameter is modeled as a function of the volume (Figure 8.11 insert). The two-parameters function of the Grüneisen parameter (Tange et al., 2009b) is necessary because a strong volume dependence is needed to fit the thermal expansion data at ambient pressure. The functional form has the flexibility to fit both high-temperature thermal properties and the Hugoniot data at high compression. However, it is not clear if the function is the right expression to fit both static compression data at moderate compression. To effectively resolve the volume dependence of the Grüneisen parameter, static compression data to very high temperature (>2,000 K) are required. The caveat is that the measured thermal pressures from static high–*P–T* data are also tied to a reference pressure scale, but they should provide some internally consistent checks among different scales at simultaneous high pressure and temperature.

The thermal EoS of Pt has been reassessed by Yokoo et al. (2009), whose reduced EoS at 300 K agrees better with that of Holmes et al. (1989). On the other hand, Matsui et al. (2009) developed a thermal EoS of Pt using both the shock wave and static data, and its reduced EoS is more consistent with that of Fei et al. (2007a) (Figure 8.9). Matsui et al. (2009) and Yokoo et al. (2009) used (8.13) and (8.14) to model the Grüneisen parameter, respectively. The calculated thermal pressures are similar (Figure 8.12). In contrast, the derived thermal pressures from ab initio molecular dynamics simulations (Ono et al., 2011) show much lower values. Comparison of static compression data among different pressure scales (e.g., Fei et al., 2007a; Dewaele et al., 2008a) indicate that the high K_0' value of the shock wave–derived EoS led to overestimated pressures in the multimegabar pressure range. There is an unresolved inconsistency between the shock wave and static compression data. The static compression data clearly show a divergence in pressure prediction between the reduced shock wave EoS of Pt and Au (Dewaele et al., 2008a). One way to resolve the difference is to either increase the thermal pressure for the Pt EoS or decrease the thermal pressure of the Au EoS. Recently, static compression data in the laser-heated

Table 8.2 *Parameters for thermal equations of state of selected solids.*

	B–M EoS			Mie–Grüneisen relation			
References	V_0, Å^3	K_0, GPa	K_0'	θ_0	γ_0	a	b (q)
MgO							
Speziale et al. (2001)	74.71	160.2	3.99	773	1.524	1	1.65*
Tange et al. (2009b)	74.698	160.6	4.22	761	1.431	0.29	3.5
Kono et al. (2010)	74.71	160.9	4.35	773	1.53	1	1.5
Ne							
Fei et al. (2007a)	88.967	1.43	8.02	75.1	2.05	1	0.6
Dewaele et al. (2008b)	88.936	1.42	8.03	75.1	2.442‡	1	0.97
NaCl–B2							
Fei et al. (2007a)	41.35	30.69	4.33	290	1.7	1	0.5
Pt							
Fei et al. (2007a)	60.38	277	4.95	230	2.72	1	0.5
Matsui et al. (2009)	60.38	273(v)	5.2(v)	230	2.7	1	1.1
Yokoo et al. (2009)	60.38	276.4	5.12	230	2.63	0.39	5.2
Au							
Fei et al. (2007a)	67.85	167	5.77	170	2.97	1	0.6
Yokoo et al. (2009)	67.72	167.5	5.79	170	2.96	0.45	4.2
Tsuchiya (2003)	67.85	166.7	6.12	180	3.16	1	2.15
Bridgmanite							
Tange et al. (2012), Mg-br	162.373	256.7	4.09	950	1.54	1	1.5
Wolf et al. (2015), MgFe-br	163.5	238.4	4	1,000	1.4	1	0.56
Sun et al. (2018), MgFeAl-br	164.4	247	4	1,000	1.52	1	0.58
Ferropericlase							
Mao et al. (2011), High-spin	76.1	159	4	763	1.4	1	0.16
Mao et al. (2011), Low-spin	74.4	165	4	763	1.5	1	0.01
$CaSiO_3$-perovskite							
Sun et al. (2016)	45.4	249	4	1,000	1.8	1	1.1
Shim et al. (2000)	46.6	248	4	1,000	0.96	1	1
Shim et al. (2000)	46.1	208	4	1,000	2.3	1	0.97

Notes:

*: $q = 1.65(V/V_0)^{11.8}$.

‡: $\gamma = 2.446(V/V_0)^{0.97}+1/2$.

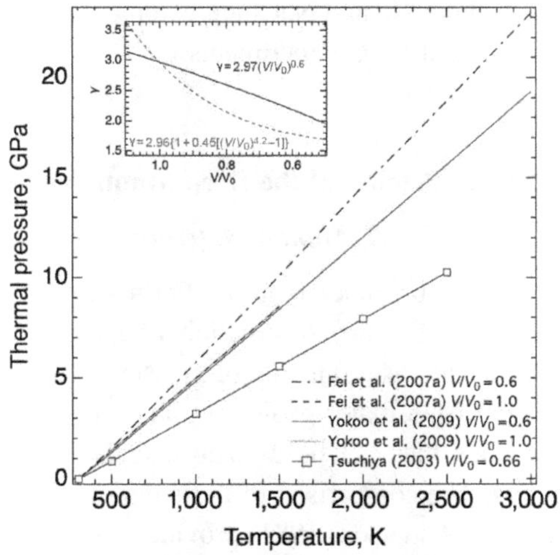

Figure 8.11 Comparison of the calculated thermal pressures of Au as a function of temperature from different models (Tsuchiya, 2003; Fei et al., 2007a; Yokoo et al., 2009). The insert shows the comparison of the Grüneisen parameter as a function of V/V_0 (solid curve = Fei et al., 2007a; dashed curve = Yokoo et al., 2009).

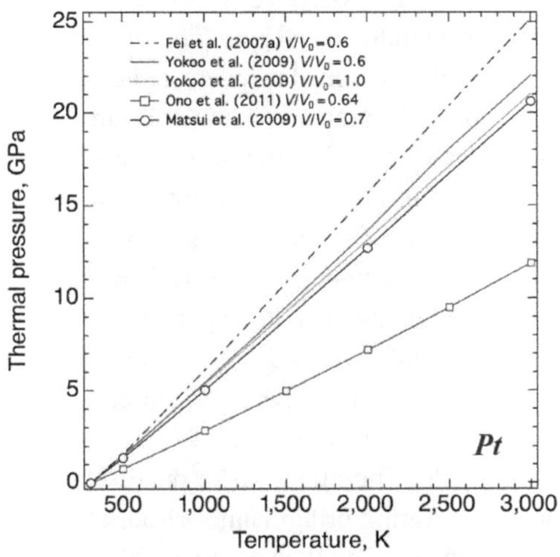

Figure 8.12 Comparison of the calculated thermal pressures of Pt as a function of temperature from different models (Fei et al., 2007a; Matsui et al., 2009; Yokoo et al., 2009; Ono et al., 2011).

DAC up to 140 GPa and 2,500 K provided some comparisons among different pressure scales (Ye et al., 2017). Additional *P–V–T* measurements of Au and Pt at high temperatures, particularly above 2,500 K, are needed to resolve the discrepancy.

The thermal EoS of other standards such as Ne and NaCl-B2, have even less experimental data to evaluate the thermal pressure. The thermal EoS of Ne and NaCl-B2 were based on *P–V–T* data up to 1,000 K in the externally heated DAC (Fei et al., 2007a; Dewaele et al., 2008b). Further evaluation of thermal pressures to higher temperature would be beneficial because both Ne and NaCl are commonly used as pressure media and thermal insulation in laser-heated DAC experiments.

8.5 Density Profiles of the Deep Mantle and Core

8.5.1 Mantle Materials

For a peridotitic mantle, the major minerals in the Earth's lower mantle are bridgmanite, ferropericlase, and Ca-perovskite. The construction of mantle density profiles requires accurate measurements of density as a function of pressure, temperature, and composition for the mantle phases. One of the major goals in mineral physics is to establish thermal equations of state of minerals that can be used to calculate mantle density profiles for mantle model compositions. The comparison of these density profiles with seismic observations (e.g., Dziewonski and Anderson, 1981) provides insights into mantle composition and possible chemical heterogeneity in the deep mantle.

The combination of high-pressure techniques and synchrotron X-ray diffraction has revolutionized the study of EoS of materials over a wide *P–T* range, with high precision. The first comprehensive study of the EoS of (Mg,Fe)SiO_3-bridgmanite was performed at the Cornell High Energy Synchrotron Source (CHESS) with a neon pressure medium using an external heater to achieve simultaneous high pressure and temperature (Mao et al., 1991). High-resolution X-ray diffraction data were collected with angle-dispersive and classical film techniques. Table 8.3 summarizes the measurements of the EoS of (Mg,Fe) SiO_3-bridgmanite. Further advances in EoS measurements came from the combination of in situ laser heating and imaging-plate detector for angle-dispersive X-ray diffraction (Fiquet et al., 2000), expanding the *P–T* conditions to 57 GPa and 2,500 K. The use of an imaging plate with a fast scan technique was critical for collecting high-quality *P–V–T* data. The implement of the double-sided laser-heating system (Shen et al., 2001) at synchrotron beamlines with a highly focused X-ray beam (down to 3–5 μm) has further improved the capability of *P–V–T* measurements in subsequent years.

Table 8.3 lists EoS parameters from studies that focused on the compositional effect on the bulk modulus of bridgmanite. The effect of Fe on EoS is further complicated by the electronic spin transition in Fe-bearing bridgmanites (Badro et al. 2004; Li et al. 2004) and the Fe^{3+} distribution in the structure (Catalli et al. 2010, 2011; Wolf et al., 2015; Mao et al., 2017). The effect of Al substitutions in the structure on the elastic properties of bridgmanite has been evaluated in the Fe-free samples (Andrault et al., 2001; Walter et al., 2004, 2006; Catalli et al., 2011). It has been a challenge to deconvolute the combination effect of cation substitution (Al, Fe) in the structure on EoS because of multiple possible coupled substitutions (Andrault et al., 2001; Walter et al., 2006; Boffa Ballaran et al., 2012; Mao et al., 2017; Sun et al., 2018).

There is covariance between the isothermal bulk modulus (K_0) and its pressure derivative (K'_0). The derived parameters are also influenced by the pressure scales used in the data analysis and hydrostaticity of the sample environment. With these uncertainties and trade-offs, it is generally difficult to determine the compositional effect of the EoS

Table 8.3 *Parameters for Birch–Murnaghan equations of state of bridgmanites.*

References	X_{Fe}, X^{Al}	V_0, Å^3	K_0, GPA	K_0'	max P, GPa	P standard	Methods	P medium	Heating
Mao et al. (1991)	0, 0	162.49	261	4	30	Ruby-M86, Au-A89	XRD	Ne	External
	0.1, 0	162.79							
	0.2, 0	163.53							
Fiquet et al. (1998)	0, 0	162.3	261	4	57	Ruby-M86, Pt-J82	XRD	Ar	Laser
Fiquet et al. (2000)	0, 0	162.3	259	3.7	90	Pt-J82	XRD	Ar	Laser
Andrault et al. (2001)	0.05, 0	162.7	255	4	57	Ruby-M86	XRD	Ar	
	0, 0.05	163.3	265	4	57			Ar	
	0, 0.22	163.5	277	4	69			Ar	
	0.05, 0.05		267	4	60			Ar	
Walter et al. (2004)	0, 0	162.47	267.3	4	59	KBr-K97	XRD	KBr	Laser
	0, 0.1	162.95	268.9	4	101			KBr	
	0, 0.2	163.42	262.6	4	70			KBr	
	0, 0.4	164.37	254.2	4	90			KBr	
	0, 0.5	164.85	253.4	4	83			KBr	
Lundin et al. (2008)	0, 0	162.3	261	4	93	Au-T03	XRD	Ar	
	0.09, 0	163.18	259	4	86			Ar or NaCl	
	0.15, 0	163.3	259	4	108			Ar	
Katsura et al. (2009)	0, 0	162.39	256	3.8	48	MgO-S01, M00	MA-XRD		TiB_2 heater
Catalli et al. (2010)	0.085, 0	165.78	237	4	55	Au-T03	XRD	Ar	
Catalli et al. (2011)	0, 0.10	163.83	244	4	94	Au-T03	XRD	Ar	Laser
Tange et al. (2012)	0, 0	162.373	256.7	4.09	100	MgO–T09	MA-, DAC-XRD	$MgSiO_3$ glass	La CrO_3/ Laser

Table 8.3 (*cont.*)

References	X_{Fe}, X^{Al}	V_0, Å^3	K_0, GPA	K_0'	max P, GPa	P standard	Methods	P medium	Heating
Boffa Ballaran et al. (2012)	0, 0	162.36	251	4.11	77	Ruby-J08	Single-XRD	He	
	0.04, 0	163.09	253	3.99	74	Ruby-J08	Single-XRD	He	
	0.41, 0.36	168.93	240	4.12	74	Ruby-J08	Single-XRD	He	
Wolf et al. (2015)	0.13, 0	163.16	243.8	4.16	120	Ne-D08	XRD	Ne	
Z. Mao et al. (2017)	0.06, 0.01	162.96	255	4	85	Pt-F07	Single-XRD	He	
	0.12, 0.11	164.05	264	4	110	Pt-F07	Single-XRD	He	
Sun et al. (2018)	0.21, 0.07	164.4	247	4	120	Au-F07	XRD	NaCl	Laser

References for the P standard column: M86: Mao et al. (1986); A89: Anderson et al. (1989); J82: Jamieson et al. (1982); K97: Kohler et al. (1997); T03: Tsuchiya (2003); S01: Speziale et al. (2001); M00: Matsui et al. (2000); T09: Tange et al. (2009b); J08: Jacobsen et al. (2008); D07: Dewaele et al. (2008b); F07: Fei et al. (2007a).

parameters. Mao et al. (2017) provided an assessment of the confidence ellipses of the bulk modulus (K_0) and its pressure derivative (K_0') of bridgmanite at the 1σ level and found that $K_0 = 253 \pm 3$ GPa at $K_0' = 4$ is a good representative value for all Fe-bearing bridgmanites. While the effect of Fe in bridgmanites on the bulk modulus is negligible, the effect of Al has not reached a consensus (Table 8.3) (e.g., Andrault et al., 2001; Walter et al., 2004; Nishio-Hamane and Yagi, 2009; Catalli et al., 2011; Mao et al., 2017; Sun et al., 2018).

Assessment of the compositional effect on the thermal EoS parameters is even more challenging. The best approach is to establish a robust thermal EoS for $MgSiO_3$-bridgmanite and then assess any deviation from the reference that could be attributed to the compositional effect in additional to the V_0 as a function of composition. Tange et al. (2012) used *P–V–T* data from both multi-anvil pressure (up to 63 GPa and 1,500 K) and LHDAC (up to 108 GPa and 2,400 K) to derive the thermal EoS parameters (Table 8.2). The compositional effect on the thermal pressure is expected to be small, comparing to the uncertainties in the bulk modulus and its pressure derivative for bridgmanites with a range of compositions. Recent *P–V–T* data for Fe-bearing and (Fe,Al)-bearing bridgmanites (e.g., Wolf et al., 2015; Sun et al., 2018) provided an assessment of these effects by direct comparison of the *P–V–T* data.

The second most abundant mineral in the lower mantle is $(Mg_{1\text{-}x}Fe_x)O$-ferropericlase. The Fe content is determined by the bulk iron content of the mantle and the Mg–Fe partitioning between coexisting bridgmanite and ferropericlase. The thermal EoS of the MgO endmember is well established, and it has been used as a reference for pressure calibration. The effect of Fe on EoS is complicated by the Fe spin transition at high pressure (Badro et al., 2003). The effect of the spin transition on the compression behavior of $(Mg_{1\text{-}x}Fe_x)O$-ferropericlase has been well documented (e.g., Lin et al., 2005, 2006; Fei et al., 2007b; Komabayashi et al., 2010; Mao et al., 2011; Solomatova et al., 2016). Mantle density profiles must incorporate the equations of state of high-spin and low-spin ferropericlase as a function of Fe content (Mao et al., 2011; Sun et al., 2016).

The lower mantle contains about 8 vol% $CaSiO_3$-perovskite for a peridotitic composition. Because $CaSiO_3$-perovskite cannot be quenched to ambient condition, the fitted bulk modulus and its derivative usually have a large uncertainty without an accurate determination of V_0. However, the compression data covering the entire mantle pressure range (Shim et al., 2000; Ricolleau et al., 2009; Sun et al., 2016) have converged to $K_0 = 246 \pm 3$ GPa at $K_0' = 4$ (Table 8.2). The experimental *P–V–T* measurements (Shim et al., 2000; Ricolleau et al., 2009; Noguchi et al., 2013; Sun et al., 2016) yielded different thermal pressure models. These differences are small compared to the variations in the room temperature compression curves from different studies.

Ricolleau et al. (2009) determined *P–V–T* relations up to 112 GPa and 2,500 K using a natural KLB-1 peridotite as the starting material. In situ diffraction data provided direct density measurements of mantle bridgmanite, ferropericlase, and $CaSiO_3$-perovskite over the entire mantle conditions. Figure 8.13 compares density profiles for bridgmanites, ferropericlase, and $CaSiO_3$-perovskite along a mantle adiabat with a potential temperature of 1,990 K at the top of the lower mantle (Katsura et al., 2010), against a Preliminary Reference Earth Model (PREM) density profile (Dziewonski and Anderson, 1981). The

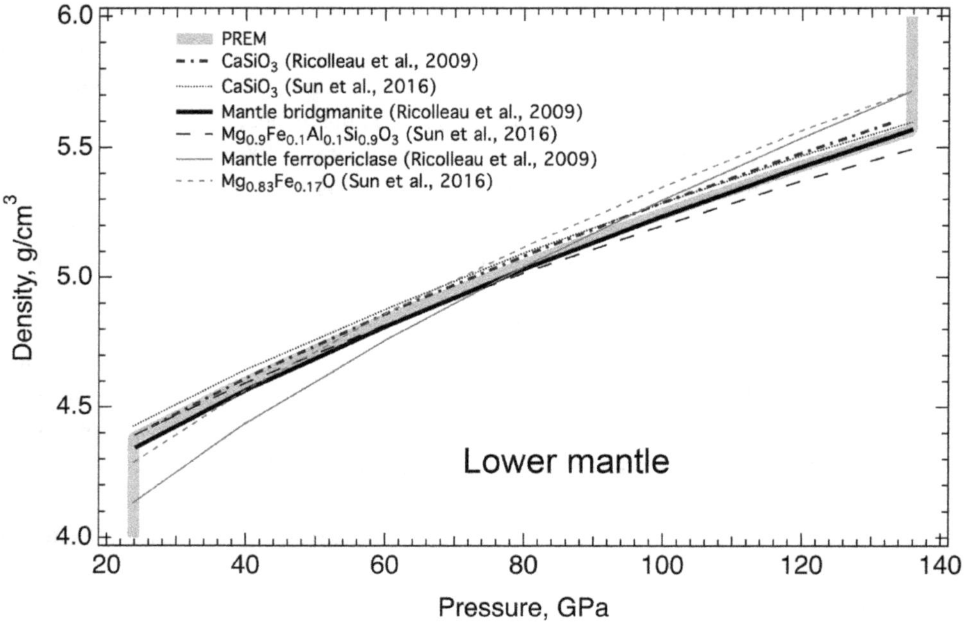

Figure 8.13 Density profiles for bridgmanite, ferropericlase, and $CaSiO_3$–perovskite along adiabatic geotherm (Katsura et al., 2010), compared with PREM model.

chemical compositions of the minerals reflect the coexisting equilibrium composition at the synthesis conditions. Their density profiles were compared with measurements of the individual phase. The adiabatic bulk modulus of ferropericlase is substantially smaller than the PREM bulk modulus. Adding any ferropericlase component to the mantle would lower the mantle bulk modulus. The adiabatic bulk modulus of the mantle bridgmanite from Ricolleau et al. (2009) is smaller than the model parameter used in Sun et al. (2016), but it is comparable to the PREM bulk modulus (Figure 8.13). This difference is important because it determines if a peridotitic composition or a bridgmanite composition would provide a better fit to the mantle density profile.

8.5.2 Core Materials

The Earth's core mainly consists of iron with ~5 wt.% Ni (McDonough and Sun, 1995) based on cosmochemical constraints. The observed density of the core is smaller than that of pure Fe or Fe–Ni alloy. The core density deficit can be explained by incorporating light elements, such as S, Si, O, C, and H in the core (Birch, 1952; Li and Fei, 2014). The density jump from the liquid outer core to the solid inner core reflects different amounts of light elements incorporated in the liquid and solid core, controlled by the element partitioning between liquid and solid metal at the inner-core boundary (ICB). It is important to establish a reference density profile for pure Fe as accurately as possible so that the deviation from the reference can be attributed to the compositional effect.

The inner-core pressure starts at 329 GPa. If the uncertainty in pressure determination is about 10%, the inferred compositional effect from density difference would be problematic because the corresponding uncertainty in density is more than 1.5%. Fei et al. (2016) assessed the hcp–Fe compression data and reconciled the discrepancies among different studies (Mao et al., 1990; Dubrovinsky et al., 2000; Yamazaki et al., 2012; Sakai et al., 2014) by using internally consistent pressure scales. Reanalysis of the room temperature compression data has brought the agreement among different studies within 2% up to 280 GPa.

The effect of Ni on the density has been investigated for $Fe_{0.8}Ni_{0.2}$ alloy to 260 GPa (Mao et al., 1990) and $Fe_{0.9}Ni_{0.1}$ alloy to 279 GPa (Sakai et al., 2014). Adding 10 wt.% Ni would make the alloy slightly more compressible, resulting in about 1% density increase at the ICB conditions (Sakai et al., 2014).

How carbon affects the density of the solid inner core has focused on EoS measurements of iron carbides, such as Fe_3C (Li et al., 2002; Sata et al., 2010; Litasov et al., 2013c; Hu et al., 2019; Takahashi et al., 2019) and Fe_7C_3 (Chen et al., 2012, 2014; Prescher et al., 2015; Lai et al., 2018). For limited carbon content in the core, the inner core is likely crystalized as Fe with interstitial carbon. Yang et al. (2019) examined carbon substitution in hcp–Fe and its effect on the density up to 109 GPa, by direct comparison of the density difference between pure Fe and Fe with 1.37 wt.% C. The results show that the incorporation of carbon in hcp–Fe leads to the expansion of the lattice. Using the EoS of hcp–Fe with dissolved C, the estimated carbon content to explain the density deficit in Earth's inner core is about 1 wt.%, assuming an ICB temperature of 6,000 K (Yang et al., 2019).

At least 2 wt.% sulfur (S) is expected in the core based on the cosmochemical arguments (McDonough and Sun, 1995). The amount of S that can be incorporated into the inner core depends on the S solubility in solid iron and partitioning between solid and liquid. Significant amounts of S would be partitioned into the liquid outer core. For a model Fe–S core, 10 wt.% S is required to account for the outer core density deficit, based on direct density measurements of liquid Fe with 11.8 wt.% S by shock compression (Huang et al., 2013). Although S solubility in metallic iron increases with pressure, up to 4.8 wt.% at 120 GPa and eutectic temperature (Li et al., 2001; Kamada et al., 2012), the EoS of Fe–S alloy has not been measured because of the nonquenchable nature of hcp–Fe. The estimated S content in the inner core reaches 7 wt.% assuming an ICB temperature of 6,000 K (Kamada et al., 2014) and even higher values from recent study of thermal EoS (Thompson et al., 2020), based on density interpolation between hcp–Fe and Fe_3S. This could be an overestimated value if the Fe–S alloy is less compressible than Fe_3S.

The effect of Si on the hcp–Fe EoS has been recently investigated by static compression method (Asanuma et al., 2011; Fischer et al., 2014; Tateno et al., 2015) and shock wave compression (Zhang et al., 2016, 2018; Huang et al., 2019). All studies indicate that Fe–Si alloy is less compressible than pure hcp–Fe. Under the conditions of Earth's core, adding 3.8 wt.% Si to the inner core would reduce the density enough to match the observed inner-core density assuming an ICB temperature of 6,000 K (Huang et al., 2019).

Figure 8.14 shows the composition space allowed in the inner core in the Fe–S–Si–C system for given ICB temperatures, based on *P–V–T* measurements in the Fe-bearing

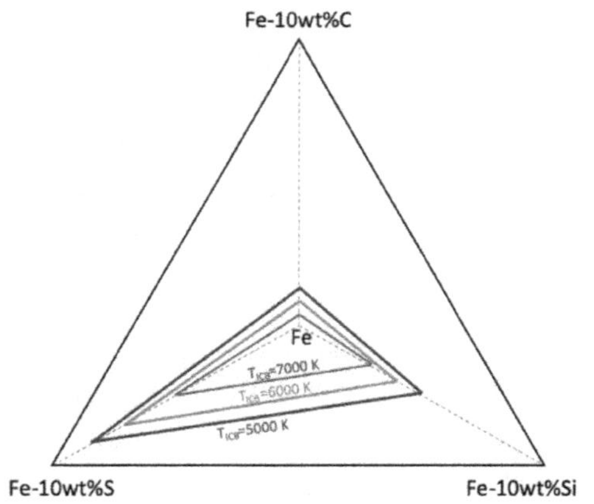

Figure 8.14 Possible inner-core compositions that would explain the core density deficit. Assuming an ICB temperature of 6,000 K, any compositions within Fe 0.9 wt.% C, Fe 3.8 wt.% Si, and Fe 6.9 wt.% S would satisfy the inner-core density requirement.

binary systems. The core composition cannot be uniquely defined just from the density measurements. Sound velocity measurements and element partitioning data between solid and liquid at core conditions are critical to further constrain the core composition.

8.6 Perspectives

Tremendous progress in determining the equations of state of materials has been made in the last 50 years. Synchrotron-based X-ray sources opened a new way to conduct X-ray diffraction measurements in the diamond anvil cell and large-volume multi-anvil press, and the use of the intense X-ray source has significantly increased the efficiency of data collection. More importantly, the highly collimated beam has allowed for analysis of smaller sample size, resulting in high-quality data at extreme pressure and at simultaneous high pressure and temperature. The introduction of gas-pressure media in DAC experiments and an area detector for angle-dispersive diffraction has significantly improved the data quality for EoS measurements. The combination of a tightly focused X-ray beam (<5 μm) and double-sided laser heating in the DAC with fast data acquisition using a charged coupled device (CCD) area detector made high-quality *P–V–T* data possible over a wide *P–T* range. It has become routine to obtain *P–V* measurements up to 200 GPa at room temperature and *P–V–T* measurements over 120 GPa and 2,500 K.

Static EoS measurements over 200 GPa are still challenging, but an increasing number of studies have reached these pressures. With development of two-stage anvils (Dubrovinsky et al., 2012; Dubrovinskaia et al., 2016; Kuramochi et al., 2020) and toroidal DAC (Dewaele et al., 2018; Jenei et al., 2018), it is possible to reach terapascal static pressures. The accuracy in pressure determination in this unchartered territory needs to be examined by multiple approaches (Rech et al., 2019). The critical part of studying EoS in the multimegabar pressure range is to examine the validity of extrapolation of the established EoS. The advances in dynamic compression that routinely reaches over a terapascal pressure, and the improved capability of theoretical predictions, provide independent checks for pressure determination under extreme conditions.

There are still limited measurements under simultaneous high pressure and temperature. It is critical to obtain volume or density data at very high temperature to effectively model the thermal equations of state of solids. As demonstrated, the effect of the compression rate (V/V_0) on the thermal pressure is sensitive only at very high temperature. The consistency among MgO, Au, and Pt pressure scales should be cross-checked at temperatures over 3,000 K even at moderate pressures. Improved laser-driven ramp-compression data for MgO should provide a critical test for the validity of extrapolation of the current MgO pressure scale.

With improved pressure scales and internal consistency among different scales, the accuracy in density measurements on the iron-alloy core compositions should further improve. However, density data alone can only provide upper bounds for each light element in the solid inner core. Additional constraints have to come from other observations and experiments such as element partitioning at the ICB and differentiation processes. Looking beyond the Earth, applications of EoS measurements at extreme conditions will play an increasingly important role in modeling the internal structures and properties of super-Earths.

Acknowledgments

The project is supported by NSF grant EAR-1619868 and the Carnegie Institution for Science. I thank Lily Thompson and Zhu Mao for their thoughtful and constructive reviews.

References

Anderson, O. L., Isaak, D. G., Yamamoto, S. (1989). Anharmonicity and the equation of state for gold. *Journal of Applied Physics*, **65**, 1534–1543.

Andrault, D., Bolfan-Casanova, N., Guignot, N. (2001). Equation of state of lower mantle (Al, Fe)-$MgSiO_3$ perovskite. *Earth and Planetary Science Letters*, **193**, 501–508.

Arimoto, T., Irifune, T., Nishi, M., Tange, Y., Kunimoto, T., Liu, Z. (2019). Phase relations of $MgSiO_3$-$FeSiO_3$ system up to 64 GPa and 2300 K using multianvil apparatus with sintered diamond anvils. *Physics of the Earth and Planetary Interiors*, **295**, 106297.

Asanuma, H., Ohtani, E., Sakai, T., et al. (2011). Static compression of $Fe_{0.83}Ni_{0.09}Si_{0.08}$ alloy to 374 GPa and $Fe_{0.93}Si_{0.07}$ alloy to 252 GPa: implications for the Earth's inner core. *Earth and Planetary Science Letters*, **310**(1–2), 113–118.

Badro, J., Fiquet, G., Guyot, F., et al. (2003). Iron partitioning in Earth's mantle: toward a deep lower mantle discontinuity. *Science*, **300**, 789–791.

Badro, J., Rueff, J.-P., Vanko, G., Monaco, G., Fiquet, G., Guyot, F. (2004). Electronic transitions in perovskite: possible nonconvecting layers in the lower mantle. *Science*, **305**, 383–386.

Birch, F. (1952). Elasticity and constitution of the Earth's interior. *Journal of Geophysical Research*, **57**, 227–286.

Birch, F. (1978). Finite strain isotherm and velocities for single-crystal and polycrystalline NaCl at high pressures and 300 K. *Journal of Geophysical Research*, **83**, 1257–1268.

Boffa Ballaran, T., Kurnosov, A., Glazyrin, K., et al. (2012). Effect of chemistry on the compressibility of silicate perovskite in the lower mantle. *Earth and Planetary Science Letters*, **333**, 181–190.

Boness, D. A., Brown, J. M., McMahan, A. K. (1986). The electronic thermodynamics of iron under Earth core conditions. *Physics of the Earth and Planetary Interiors*, **42**, 227–240.

Brown, J. M. (1999). The NaCl pressure standard. *Journal of Applied Physics*, **86**(10), 5801–5808.

Carter, W. J., Marsh, S. P., Fritz, J. N., McQueen, R. G. (1971). The equation of state of selected materials for high-pressure references, in E. C. Lloyd, ed., *Accurate Characterization of the High-Pressure Environment*. National Bureau of Standards, Spec. Publ. 326, pp. 147–158.

Catalli, K., Shim, S. H., Prakapenka, V. B., et al. (2010). Spin state of ferric iron in $MgSiO_3$ perovskite and its effect on elastic properties. *Earth and Planetary Science Letters*, **289**, 68–75.

Catalli, K., Shim, S. H., Dera, P., et al. (2011). Effects of the Fe^{3+} spin transition on the properties of aluminous perovskite – new insights for lower-mantle seismic heterogeneities. *Earth and Planetary Science Letters*, **310**, 293–302.

Chen, B., Gao, L., Lavina, B., et al. (2012). Magneto-elastic coupling in compressed Fe_7C_3 supports carbon in Earth's inner core. *Geophysical Research Letters*, **39**(18), L18301.

Chen, B., Li, Z., Zhang, D., et al. (2014). Hidden carbon in Earth's inner core revealed by shear softening in dense Fe_7C_3. *Proceedings of the National Academy of Sciences*, **111**(50), 17755–17758.

Chijioke, A. D., Nellis, W. J., Silvera, I. F. (2005). High-pressure equations of state of Al, Cu, Ta, and W. *Journal of Applied Physics*, **98**(7), 073526.

Cohen, R. E., Gulseren, O., Hemley, R. J. (2000). Accuracy of equation of state formulations. *American Mineralogist*, **85**(2), 338–344.

Coppari, F., Smith, R. F., Eggert, J. H., et al. (2013). Experimental evidence for a phase transition in magnesium oxide at exoplanet pressures. *Nature Geoscience*, **6**, 926–929.

Decker, D. L. (1965). Equation of state of NaCl and its use as a pressure gauge in high-pressure research. *Journal of Applied Physics*, **36**(1), 157–161.

Desjarlais, M. P., Knudson, M. D., Cochrane, K. R. (2017). Extension of the Hugoniot and analytical release model of α-quartz to 0.2–3 TPa. *Journal of Applied Physics*, **122**, 035903.

Dewaele, A. (2019). Equations of state of simple solids (including Pb, NaCl and LiF) compressed in helium or neon in the Mbar range. *Minerals*, **9**, 684.

Dewaele, A., Datchi, F., Loubeyre, P., Mezouar, M. (2008b). High pressure–high temperature equations of state of neon and diamond. *Physical Review B*, **77**(9), 094106.

Dewaele, A., Loubeyre, P., Occelli, F., Marie, O., Mezouar, M. (2018). Toroidal diamond anvil cell for detailed measurements under extreme static pressure. *Nature Communications,* **9**, 2913.

Dewaele, A., Loubeyre, P., Mezouar, M. (2004). Equations of state of six metals above 94 GPa. *Physical Review B*, **70**(9), 094112.

Dewaele, A., Loubeyre, P., Occelli, F., Mezouar, M., Dorogokupets, P. I., Torrent, M. (2006). Quasihydrostatic equation of state of iron above 2 Mbar. *Physical Review Letters,* **97**(21), 215504.

Dewaele, A., Torrent, M., Loubeyre, P., Mezouar, M. (2008a). Compression curves of transition metals in the Mbar range: Experiments and projector augmented-wave calculations. *Physical Review B*, **78**(10), 104102.

Dorfman, S. M., Prakapenka, V. B., Meng, Y., Duffy, T. S. (2012). Intercomparison of pressure standards (Au, Pt, Mo, MgO, NaCl and Ne) to 2.5 Mbar. *Journal of Geophysical Research*, **117**, B08210.

Dorogokupets, P. I., Dewaele, A. (2007). Equations of state of MgO, Au, Pt, NaCl-B1, and NaCl-B2: internally consistent high-temperature pressure scales. *High Pressure Research,* **27**(4), 431–446.

Dorogokupets, P. I., Oganov, A. R. (2006). Equations of state of Al, Au, Cu, Pt, Ta, and W and revised ruby pressure scale. *Doklady Earth Sciences,* **410**(7), 1091–1095.

Dubrovinskaia, N., Dubrovinsky, L., Solopova, N. A., et al. (2016). Terapascal static pressure generation with ultrahigh yield strength nanodiamond. *Science Advances*, **2**(7), e1600341.

Dubrovinsky, L., Dubrovinskaia, N., Prakapenka, V., Abakumov, A. M. (2012). Implementation of micro-ball nanodiamond anvils for high-pressure studies above 6 Mbar. *Nature Communications,* **3**, 1163.

Dubrovinsky, L. S., Saxena, S. K., Tutti, F., Rekhi, S., Le Bihan, T. (2000). In situ X-ray study of thermal expansion and phase transition of iron at multimegabar pressure. *Physical Review Letters,* **84**, 1720–1723.

Duffy, T. S., Ahrens, T. J. (1993). Thermal expansion of mantle and core materials at very high pressure. *Geophysical Research Letters*, **20**, 1103–1106.

Duffy, T. S., Hemley, R. J., Mao, H.-K. (1995). Equation of state and shear strength at multimegabar pressures: magnesium oxide to 227 GPa. *Physical Review Letters,* **74**, 1371–1374.

Dziewonski, A. M., Anderson, D. L. (1981). Preliminary reference Earth model. *Physics of the Earth and Planetary Interiors*, **25**(4), 297–356.

Fan, D., Fu, S., Yang J., Tkachev, S. N., Prakapenka, V. B., Lin, J. (2019). Elasticity of single-crystal periclase at high pressure and temperature: the effect of iron on the elasticity and seismic parameters of ferropericlase in the lower mantle. *American Mineralogist*, **104**, 262–275.

Fat'yanov, O. V., Asimow, P. D., Ahrens, T. J. (2018). Thermodynamically complete equation of state of MgO from true radiative shock temperature measurements on samples preheated to 1850 K. *Physical Review B*, **97**, 024106.

Fei, Y. (1999). Effects of temperature and composition on the bulk modulus of (Mg,Fe)O. *American Mineralogist*, **84**, 272–276.

Fei, Y., Li, J., Hirose, K., et al. (2004). A critical evaluation of pressure scales at high temperatures by in situ X-ray diffraction measurements. *Physics of the Earth and Planetary Interiors*, **143–144**, 515–526.

Fei, Y., Murphy, C., Shibazaki, Y., Shahar, A., Huang, H. (2016). Thermal equation of state of hcp-iron: constraint on the density deficit of Earth's solid inner core. *Geophysical Research Letters*, **43**, L069456.

Fei, Y., Ricolleau, A., Frank, M., Mibe, K., Shen, G., Prakapenka, V. B. (2007a). Toward an internally consistent pressure scale. *Proceedings of the National Academy of Sciences*, **104**(22), 9182–9186.

Fei, Y., Seagle, C. T., Townsend, J. P., et al. (2021). Melting and density of $MgSiO_3$ determined by shock compression of bridgmanite to 1254GPa. *Nature Communications,* **12,** 876.

Fei, Y., Zhang, L., Corgne, A., et al. (2007b). Spin transition and equations of state of (Mg, Fe)O solid solutions. *Geophysical Research Letters*, **34**, L17307.

Fiquet, G., Andrault, D., Dewaele, A., Charpin, T., Kunz, M., Hauesermann, D. (1998). P–V–T equation of state of $MgSiO_3$ perovskite. *Physics of the Earth and Planetary Interiors*, **105**, 21–31.

Fiquet, G., Dewaele, A., Andrault, D., Kunz, M., Le Bihan, T. (2000). Thermoelastic properties and crystal structure of $MgSiO_3$ perovskite at lower mantle pressure and temperature conditions. *Geophysical Research Letters*, **27**, 21–24.

Fischer, R. A., Campbell, A. J., Caracas, R., et al. (2014). Equations of state in the Fe–FeSi system at high pressures and temperatures. *Journal of Geophysical Research: Solid Earth*, **119**, 2810–2827.

Fukui, H., Sakai, T., Sakamaki, T., et al. (2013). A compact system for generating extreme pressures and temperatures: an application of laser-heated diamond anvil cell to inelastic X-ray scattering. *Review of Scientific Instruments*, **84**, 113902.

Greeff, C. W., Graf, M. J. (2004). Lattice dynamics and the high- pressure equation of state of Au. *Physical Review B*, **69**(5), 054107.

Hemley, R. J., Zha, C.-S., Jephcoat, A. P., Mao, H.-K., Finger, L. W., Cox, D. E. (1989). X-ray diffraction and equation of state of solid neon to 110 GPa. *Physical Review B*, **39**(16), 11820–11827.

Hirao, N., Kawaguchi, S. I., Hirose, K., Shimizu, K., Ohtani, E., Ohishi, Y. (2020). New developments in high-pressure X-ray diffraction beamline for diamond anvil cell at SPring-8. *Matter and Radiation at Extremes*, **5**, 018403.

Hirose, K., Sata, N., Komabayashi, T., Ohishi, Y. (2008). Simultaneous volume measurements of Au and MgO to 140 GPa and thermal equation of state of Au based on the MgO pressure scale. *Physics of the Earth and Planetary Interiors*, **167**(3–4), 149–154.

Hixson, R. S., Fritz, J. N. (1992). Shock compression of tungsten and molybdenum. *Journal of Applied Physics*, **71**, 1721–1728.

Holmes, N. C., Moriarty, J. A., Gathers, G. R., Nellis, W. J. (1989). The equation of state of platinum to 660 GPa (6.6 Mbar). *Journal of Applied Physics,* **66**(7), 2962–2967.

Hu, X., Fei, Y., Yang, J., et al. (2019). Phase stability and thermal equation of state of iron carbide Fe_3C to 245 GPa. *Geophysical Research Letters*, **46**, L084545.

Huang, H., Wu, S., Hu, X., Wang, Q., Wang, X., Fei, Y. (2013). Shock compression of Fe-FeS mixture up to 204 GPa. *Geophysical Research Letters*, **40**, 687–691.

Huang, H., Leng, C., Wang, Q., et al. (2019). Equation of state for shocked Fe-8.6 wt% Si up to 240 GPa and 4,670 K. *Journal of Geophysical Research: Solid Earth*, **124**, B017983.

Irifune, T., Shinmei, T., McCammon, C. A., Miyajima, N., Rubie, D. C., Frost, D. J. (2010). Iron partitioning and density changes of pyrolite in Earth's lower mantle. *Science*, **327**, 193–195.

Ishii, T., Yamazaki, D., Tsujino, N., et al. (2017). Pressure generation to 65 GPa in a Kawai-type multi-anvil apparatus with tungsten carbide anvils. *High Pressure Research*, **37**(4), 507–515.

Ito, E., Yamazaki, D., Yoshino, T., et al. (2010). Pressure generation and investigation of the post-perovskite transformation in $MgGeO_3$ by squeezing the Kawai-cell equipped with sintered diamond anvils. *Earth and Planetary Science Letters*, **293**, 84–89.

Jackson, I., Niesler, H. (1982). The elasticity of periclase to 3 GPa and some geophysical implications, in S. Akimoto, M. H. Manghnani, eds., *High Pressure Research in Geophysics*. Center for Academic Publishing, pp. 93–113.

Jacobsen, S. D., Holl, C. M., Adams, K. A., et al. (2008). Compression of single-crystal magnesium oxide to 118 GPa and a ruby pressure gauge for helium pressure media. *American Mineralogist*, **93**, 1823–1828.

Jamieson, J. C., Fritz, J. N., Manghnani, M. H. (1982). Pressure measurement at high temperature in X-ray diffraction studies: gold as a primary standard, in S. Akimoto, M. H. Manghnani, eds., *High Pressure Research in Geophysics*. Center for Academic Publishing, pp. 27–48.

Jenei, Z., O'Bannon, E. F., Weir, S. T., Cynn, H., Lipp, M. J., Evans, W. J. (2018). Single crystal toroidal diamond anvils for high pressure experiments beyond 5 megabar. *Nature Communications,* **9**, 3563.

Kamada, S., Ohtani, E., Terasaki, H., et al. (2012). Melting relationships in the Fe–Fe_3S system up to the outer core conditions. *Earth and Planetary Science Letters*, **359–360**, 23–33.

Kamada, S., Ohtani, E., Terasaki, H., et al. (2014). Equation of state of Fe_3S at room temperature up to 2 megabars. *Physics of the Earth and Planetary Interiors*, **228**, 106–113.

Kamada, S., Fukui, H., Yoneda, A., et al. (2019). Elastic constants of single-crystal Pt measured up to 20 GPa based on inelastic X-ray scattering: Implication for the establishment of an equation of state. *Comptes Rendus Geoscience*, **351**(2–3), 236–242.

Katsura, T., Funakoshi, K., Kubo, A., et al. (2004). A large-volume high P-T apparatus for in situ X-ray observation, "SPEEDMkII". *Physics of the Earth and Planetary Interiors*, **143–144**, 497–506.

Katsura, T., Yokoshi, S., Kawabe, K., et al. (2009). P–V–T relations of $MgSiO_3$ perovskite determined by in situ X-ray diffraction using a large-volume high-pressure apparatus. *Geophysical Research Letters*, **36**, L01305.

Katsura, T., Yoneda, A., Yamazaki, D., Yoshino, T., Ito, E. (2010). Adiabatic temperature profile in the mantle. *Physics of the Earth and Planetary Interiors*, **183**, 212–218.

Kawai, N., Togaya, M., Onodera, A. (1973). A new device for high pressure vessels. *Proceedings of the Japan Academy*, **49**, 623–626.

Kennedy, G., Keeler, R. (1972). Compressibility, in D. E. Gray, ed., *American Institute of Physics Handbook*, 3rd edition. McGraw-Hill, pp. 4–39.

Knudson, M. D., Desjarlais, M. P. (2013). Adiabatic release measurements in α-quartz between 300 and 1200 GPa: characterization of α-quartz as a shock standard in the multimegabar regime. *Physical Review B*, **88**, 184107.

Kohler, U., Johannsen, P. G., Holzapfel, W. B. (1997). Equation-of-estate data for CsCl-type alkali halides. *Journal of Physics: Condensed Matter*, **9**, 5581–5592.

Komabayashi, T., Hirose, K., Nagaya, Y., Sugimura, E., Ohishi, Y. (2010). High temperature compression of ferropericlase and the effect of temperature on iron spin transition. *Earth and Planetary Science Letters*, **297**, 691–699.

Kono, Y., Irifune, T., Higo, Y., Inoue, T., Barnhoorn, A. (2010). PVT relation of MgO derived by simultaneous elastic wave velocity and in situ X-ray measurements: a new pressure scale for the mantle transition region. *Physics of the Earth and Planetary Interiors*, **183**(1–2), 196–211.

Kuramochi, K., Ishimatsu, N., Sakai, T., Kawamura, N., Irifune, T. (2020). An application of NPD to double-stage diamond anvil cells: XAS spectra of rhenium metal under high pressures above 300 GPa. *High Pressure Research*, **40**(1), 119–129.

Lai, X., Zhu, F., Liu, J., et al. (2018). The high-pressure anisotropic thermoelastic properties of a potential inner core carbon-bearing phase, Fe_7C_3, by single-crystal X-ray diffraction. *American Mineralogist,* **103**, 1568–1574.

Li, B., Woody, K., Kung, J. (2006). Elasticity of MgO to 11 GPa with an independent absolute pressure scale: Implications for pressure calibration. *Journal of Geophysical Research*, **111**, B11206.

Li, J., Fei, Y. (2014). Experimental constraints on core composition, in R. W. Carlson, ed., *Treatise on Geochemistry*. Elsevier, pp. 527–557.

Li, J., Fei, Y., Mao, H. K., Hirose, K., Shieh, S. R. (2001). Sulfur in the Earth's inner core. *Earth and Planetary Science Letters*, **193**, 509–514.

Li, J., Mao, H. K., Fei, Y., Gregoryanz, E., Eremets, M., Zha, C. S. (2002). Compression of Fe_3C to 30 GPa at room temperature. *Physics and Chemistry of Minerals,* **29**, 166–169.

Li, J., Struzhkin, V. V., Mao, H. -K., et al. (2004). Electronic spin state of iron in lower mantle perovskite. *Proceedings of the National Academy of Sciences*, **101**, 14027–14030.

Li, X., Jeanloz, R. (1987). Measurement of the B1–B2 transition pressure in NaCl at high temperatures. *Physical Review B: Condensed Matter and Materials Physics,* **36**(1), 474–479.

Lin, J. F., Gavriliuk, A. G., Struzhkin, V. V., et al. (2006). Pressure-induced electronic spin transition of iron in magnesiowüstite-(Mg,Fe)O. *Physical Review B*, **73**, 113107.

Lin, J. F., Struzhkin, V. V., Jacobsen, S. D., et al. (2005). Spin transition of iron in magnesiowüstite in the Earth's lower mantle. *Nature*, **436**, 377–380.

Litasov, K. D., Dorogokupets, P. I., Ohtani, E., et al. (2013a). Thermal equation of state and thermodynamic properties of molybdenum at high pressures. *Journal of Applied Physics,* **113**, 093507.

Litasov, K. D., Gavryushkin, P. N., Dorogokupets, P. I., et al. (2013b). Thermal equation of state to 33.5 GPa and 1673 K and thermodynamic properties of tungsten. *Journal of Applied Physics,* **113**, 133505.

Litasov, K. D., Sharygin, I. S., Dorogokupets, P. I., et al. (2013c). Thermal equation of state and thermodynamic properties of iron carbide Fe_3C to 31 GPa and 1473 K. *Journal of Geophysical Research*, **118**, 5274–5284.

Liu, Z., Irifune, T., Nishi, M., Tange, Y., Arimoto, T., Shinmei, T. (2016). Phase relations in the system $MgSiO_3$–Al_2O_3 up to 52 GPa and 2000 K. *Physics of the Earth and Planetary Interiors*, **257**, 18–27.

Lundin, S., Catalli, K., Santillan, J., et al. (2008). Effect of Fe on the equation of state of mantle silicate perovskite over 1 Mbar. *Physics of the Earth and Planetary Interiors*, **168**, 97–102.

Mao, H. K., Hemley, R. J., Fei, Y., et al. (1991). Effect of pressure, temperature, and composition on the lattice parameters and density of three (Fe, Mg)SiO_3 perovskites up to 30 GPa. *Journal of Geophysical Research*, **96**, 8069–8079.

Mao, Z., Lin, J. F., Liu, J., Prakapenka, V. B. (2011). Thermal equation of state of lower-mantle ferropericlase across the spin crossover. *Geophysical Research Letters,* **38**, L23308.

Mao, Z., Lin, J. F., Yang, J., Inoue, T., Prakapenka, V. B. (2015). Effects of the Fe^{3+} spin transition on the equation of state of bridgmanite. *Geophysical Research Letters,* **42**, 4335–4342.

Mao, Z., Wang, F., Lin, J.-F., et al. (2017). Equation of state and hyperfine parameters of high-spin bridgmanite in the Earth's lower mantle by synchrotron X-ray diffraction and Mössbauer spectroscopy. *American Mineralogist*, **102**, 357–368.

Mao, H. K., Wu, Y., Chen, L. C., Shu, J. F., Jephcoat, A. P. (1990). Static compression of iron to 300 GPa and $Fe_{0.8}Ni_{0.2}$ Alloy to 260 GPa: implications for composition of the core. *Journal of Geophysical Research*, **95**, 21737–21742.

Mao, H. K., Xu, J., Bell, P. M. (1986). Calibration of the ruby pressure gauge to 800 kbar under quasi-hydrostatic conditions. *Journal of Geophysical Research: Solid Earth*, **91**(B5), 4673–4676.

Matsui, M., Higo, Y., Okamoto, Y., Irifune, T., Funakoshi, K.-I. (2012). Simultaneous sound velocity and density measurements of NaCl at high temperatures and pressures: Application as a primary pressure standard. *American Mineralogist*, **97**(10), 1670–1675.

Matsui, M., Ito, E., Katsura, T., et al. (2009). The temperature–pressure–volume equation of state of platinum. *Journal of Applied Physics,* **105**(1), 013505.

Matsui, M., Parker, S. C., Leslie, M. (2000). The MD simulation of the equation of state of MgO: application as a pressure calibration standard at high temperature and high pressure, *American Mineralogist*, **85**, 312– 316.

McCoy, C. A., Knudson, M. D., Root, S. (2017). Absolute measurement of the Hugoniot and sound velocity of liquid copper at multimegabar pressures. *Physical Review B*, **96**, 174109.

McDonough, W. F., Sun, S.-S. (1995). The composition of the Earth. *Chemical Geology*, **120**, 223–253.

Meng, Y., Shen, G., Mao, H. K. (2006). Double-sided laser heating system at HPCAT for in situ X-ray diffraction at high pressures and high temperatures. *Journal of Physics: Condensed Matter*, **18**, S1097–S1103.

Meng, Y., Weidner, D. J., Fei, Y. (1993). Deviatoric stress in a quasi-hydrostatic diamond anvil cell: Effect on the volume-based pressure calibration. *Geophysical Research Letters,* **20**, 1147–1150.

Murakami, M., Takata, N. (2019). Absolute primary pressure scale to 120 GPa: toward a pressure benchmark for Earth's lower mantle. *Journal of Geophysical Research*, **124**, 6581–6588.

Nellis, W. J., Moriarty, J. A., Mitchell, A. C., et al. (1988). Metals physics at ultrahigh pressure: aluminum, copper, and lead as prototypes. *Physical Review Letters,* **60**, 1414–1417.

Nishio-Hamane, D., Yagi, T. (2009). Equations of state for postperovskite phases in the $MgSiO_3$–$FeSiO_3$–$FeAlO_3$ system. *Physics of the Earth and Planetary Interiors*, **175**, 145–150.

Noguchi, M., Komabayashi, T., Hirose, K., Ohishi, Y. (2013). High-temperature compression experiments of $CaSiO_3$ perovskite to lowermost mantle conditions and its thermal equation of state. *Physics and Chemistry of Minerals*, **40**, 81–91.

Ohishi, Y., Hirao, N., Sata, N., Hirose, K., Takata, M. (2008). Highly intense monochromatic X-ray diffraction facility for high-pressure research at SPring-8. *High Pressure Research*, **28**, 163–173.

Ono, S. (2010). The equation of state of B2-type NaCl. *Journal of Physics: Conference Series,* **215**, 012196.

Ono, S., Brodholt, J. P., Price, G. D. (2011). Elastic, thermal and structural properties of platinum. *Journal of Physics and Chemistry of Solids*, **72**(3), 169–175.

Ono, S., Kikegawa, T., Ohishi, Y. (2006). Structural property of CsCl-type sodium chloride under pressure, *Solid State Communications*, **137**(10), 517–521.

Prakapenka, V. B., Kubo, A., Kuznetsov, A., et al. (2008). Advanced flat top laser heating system for high pressure research at GSECARS: application to the melting behavior of germanium. *High Pressure Research*, **28**, 225–235.

Prescher, C., Dubrovinsky, L., Bykova, E., et al. (2015). High Poisson's ratio of Earth's inner core explained by carbon alloying. *Nature Geoscience*, **8**(3), 220–223.

Rech, G. L., Zorzi, J. E., Perottoni, C. A. (2019). Equation of state of hexagonal-close-packed rhenium in the terapascal regime. *Physical Review B*, **100**, 174107.

Ricolleau, A., Fei, Y., Cottrell, E., et al. (2009). Density profile of pyrolite under the lower mantle conditions. *Geophysical Research Letters*, **36**, L06302.

Root, S., Townsend, J. P., Davies, E., et al. (2018). The principal Hugoniot of forsterite to 950 GPa. *Journal Geophysical Research Letters,* **45**, 3865–3872.

Sakai, T., Ohtani, E., Hirao, N., Ohishi, Y. (2011), Equation of state of the NaCl-B2 phase up to 304 GPa. *Journal of Applied Physics,* **109**(8), 084912.

Sakai, T., Takahashi, S., Nishitani, N., Mashini, I., Ohtani, E., Hirao, N. (2014). Equation of state of pure iron and $Fe_{0.9}Ni_{0.1}$ alloy up to 3 Mbar. *Physics of the Earth and Planetary Interiors*, **228**, 114–126.

Sata, N., Hirose, K., Shen, G., Nakajima, Y., Ohishi, Y., Hirao, N. (2010). Compression of FeSi, Fe_3C, $Fe_{095}O$, and FeS under the core pressures and implication for light element in the Earth's core. *Journal of Geophysical Research*, **115**, B09204.

Sata, N., Shen, G., Rivers, M. L., Sutton, S. R. (2002). Pressure-volume equation of state of the high-pressure B2 phase of NaCl. *Physical Review B*, **65**(10), 104114.

Schultz, E., Mezouar, M., Crichton, W., et al. (2005). Double-sided laser heating system for in situ high pressure and high temperature monochromatic X-ray diffraction at the ESRF. *High Pressure Research*, **25**, 71–83.

Shen, G., Mao, H. K. (2017). High-pressure studies with X-rays using diamond anvil cells. *Reports on Progress in Physics,* **80**, 1–53.

Shen, G., Rivers, M. L., Wang, Y., Sutton, S. R. (2001). A laser heated diamond cell system at the Advanced Photon Source for in situ X-ray measurements at high pressure and temperature. *Review of Scientific Instruments*, **72**, 1273–1282.

Shim, S. H., Duffy, T. S., Shen, G. (2000). The stability and P–V–T equation of state of $CaSiO_3$ perovskite in the Earth's lower mantle. *Journal of Geophysical Research*, **105**, 25955–25968.

Shim, S. H., Duffy, T. S., Takemura, K. (2002). Equation of state of gold and its application to the phase boundaries near 660 km depth in Earth's mantle. *Earth and Planetary Science Letters*, **203**(2), 729–739,

Solomatova, N. V., Jackson, J. M., Sturhahn, W., et al. (2016). Equation of state and spin crossover of (Mg, Fe)O at high pressure, with implications for explaining topographic relief at the core-mantle boundary. *American Mineralogist*, **101**, 1084–1093.

Speziale, S., Zha, C.-S., Duffy, T. S., Hemley, R. J., Mao, H.-K. (2001). Quasi-hydrostatic compression of magnesium oxide to 52 GPa: implications for the pressure–volume–temperature equation of state. *Journal of Geophysical Research*, **106**(B1), 515–528.

Sun, N., Mao, Z., Yan, S., Wu, X., Prakapenka, V. B., Lin, J.-F. (2016). Confirming a pyrolitic lower mantle using self-consistent pressure scales and new constraints on $CaSiO_3$ perovskite. *Journal of Geophysical Research*, **121**, 4876–4894.

Sun, N., Wei, W., Han, S., et al. (2018). Phase transition and thermal equation of state of (Fe,Al)-bridgmanite and post-perovskite: implication for the chemical heterogeneity at the lowermost mantle. *Earth and Planetary Science Letters*, **490**, 161–169.

Takahashi, S., Ohtani, E., Ikuta, D., et al. (2019). Thermal equation of state of Fe_3C to 327 GPa and carbon in the core. *Minerals*, ***9***, 744.

Takemura, K., Dewaele, A. (2008). Isothermal equation of state for gold with a He-pressure medium. *Physical Review B*, **78**(10), 104119.

Takemura, K., Watanuki, T., Ohwada, K., Machida, A., Ohmura, A., Aoki, K. (2010). Powder X-ray diffraction study of Ne up to 240 GPa. *Journal of Physics: Conference Series,* **215**, 012017.

Tange, Y., Irifune, T., Funakoshi, K. (2008). Pressure generation to 80 GPa using multi-anvil apparatus with sintered diamond anvils. *High Pressure Research*, **28**, 245–254.

Tange, Y., Kuwayama, Y., Irifune, T., Funakoshi, K., Ohishi, Y. (2012). P–V–T equation of state of $MgSiO_3$ perovskite based on the MgO pressure scale: a comprehensive reference for mineralogy of the lower mantle. *Journal of Geophysical Research*, **117**, L08988.

Tange, Y., Nishihara, Y., Tsuchiya, T. (2009b). Unified analyses for P-V-T equation of state of MgO: a solution for pressure-scale problems in high P–T experiments. *Journal of Geophysical Research*, **114**, B03208.

Tange, Y., Takahashi, E., Nishihara, Y., Funakoshi, K., Sata, N. (2009a). Phase relations in the system MgO-FeO-SiO_2 to 50 GPa and 2000 °C: an application of experimental techniques using multianvil apparatus with sintered diamond anvils. *Journal of Geophysical Research*, **114**, B02214.

Tateno, S., Kuwayama, Y., Hirose, K., Ohishi, Y. (2015). The structure of Fe–Si alloy in Earth's inner core. *Earth and Planetary Science Letters*, **418**, 11–19.

Thompson, S., Komabayashi, T., Breton, H., et al. (2020). Compression experiments to 126 GPa and 2500 K and thermal equation of state of Fe_3S: implications for sulphur in the Earth's core. *Earth and Planetary Science Letters*, **534**, 116080.

Tomlinson, K., Seagle, C. T., Huang, H., Smith, G. E., Taylor, J. L., Paguio, R. R. (2018). Enhanced dual confocal measurement system. *Fusion Science and Technology,* **73**, 139–148.

Tsuchiya, T. (2003). First-principles prediction of the P–V–T equation of state of gold and the 660-km discontinuity in Earth's mantle. *Journal of Geophysical Research*, **108** (B10), 2462.

Ueda, Y., Matsui, M., Yokoyama, A., Tange, Y., Funakoshi, K. (2008). Temperature–pressure–volume equation of state of the B2 phase of sodium chloride. *Journal of Applied Physics,* **103**(11), 113513.

Utsumi, W., Funakoshi, K., Katayama, Y., Yamakata, M., Okata, T., Shimomura, O. (2002). High-pressure science with a multi-anvil apparatus at SPring-8. *Journal of Physics: Condensed Matter*, **14**, 10497–10504.

Vinet, P., Ferrante, J., Smith, J. R., Rose, J. H. (1986). A universal equation of state for solids. *Journal of Physics C: Solid State Physics,* **19**(20), L467–L473.

Walker, D., Carpenter, M. A., Hitch, C. M. (1990). Some simplifications to multianvil devices for high pressure experiments. *American Mineralogist*, **75**, 1020–1028.

Walter, M., Kubo, A., Yoshino, T., Brodholt, J., Koga, K. T., Ohishi, Y. (2004). Phase relations and equation-of-state of aluminous Mg–silicate perovskite and implications for Earth's lower mantle. *Earth and Planetary Science Letters*, **222**, 501–516.

Walter, M., Trønnes, R. G., Armstrong, L. S., Lord, O., Caldwell, W. A., Clark, A. M. (2006). Subsolidus phase relations and perovskite compressibility in the system MgO–$AlO_{1.5}$–SiO_2 with implications for Earth's lower mantle. *Earth and Planetary Science Letters*, **248**, 77–89.

Wang, Y., Ahuja, R., Johansson, B. (2002). Reduction of shock-wave data with mean-field potential approach. *Journal of Applied Physics,* **92**, 6616.

Weir, C. E., Lippincott, E. R., Valkenburg, A. V., Bunting, E. N. (1959). Infrared studies in the 1-to 15-micron ragion to 30,000 atmospheres. *Journal of Research NBS*, **63A**(1), 55–62.

Wolf, A. S., Jackson, J. M., Dera, P., Prakapenka, V. B. (2015). The thermal equation of state of (Mg, Fe)SiO_3 bridgmanite (perovskite) and implications for lower mantle structures. *Journal of Geophysical Research*, **120**, 7460–7489.

Wu, Z., Wentzcovitch, R. M., Umemoto, K., Li, B., Hirose, K., Zheng, J.-C. (2008). Pressure–volume–temperature relations in MgO: an ultra- high pressure-temperature scale for planetary sciences applications. *Journal of Geophysical Research*, **113**, B06204.

Yamazaki, D., Ito, E. (2020). High pressure generation in the Kawai-type multianvil apparatus equipped with sintered diamond anvils. *High Pressure Research*, **40,** 3–11.

Yamazaki, D., Ito, E., Yoshino, T., (2012). P–V–T equation of state for ε-iron up to 80 GPa and 1900 K using the Kawai-type high pressure apparatus equipped with sintered diamond anvils. *Geophysical Research Letters*, **39**, L20308.

Yamazaki, D., Ito, E., Yoshino, Y., et al. (2014). Over 1 Mbar generation in the Kawai-type multianvil apparatus and its application to compression of $(Mg_{0.92}Fe_{0.08})SiO_3$ perovskite and stishovite. *Physics of the Earth and Planetary Interiors*, **228**, 262–267.

Yamazaki, D., Ito, E., Yoshino, T., et al. (2019). High-pressure generation in the Kawai-type multianvil apparatus equipped with tungsten-carbide anvils and sintered- diamond anvils, and X-ray observation on $CaSnO_3$ and $(Mg,Fe)SiO_3$. *Comptes Rendus Geoscience*, **351**(2–3), 253–259.

Yang, J., Fei, Y., Hu, X., Greenberg, E., Prakapenka, V. B. (2019). Effect of carbon on the volume of solid iron at high pressure: implications for carbon substitution in iron structures and carbon content in the Earth's inner core. *Minerals*, **9**, 720.

Ye, Y., Prakapenka, V., Meng, Y., Shim, S.-H. (2017). Intercomparison of the gold, platinum, and MgO pressure scales up to 140 GPa and 2500 K. *Journal of Geophysical Research*, **122**, 3450–3464.

Ye, Y., Shim, S. H., Prakapenka, V., Meng, Y. (2018). Equation of state of solid Ne intercalibrated with the MgO, Au, Pt, NaCl-B2, and ruby pressure scales up to 130 GPa. *High Pressure Research,* **38**, 377–395.

Yokoo, M., Kawai, N., Nakamura, K. G., Kondo, K. (2009). Ultrahigh-pressure scales for gold and platinum at pressures up to 550 GPa. *Physical Review B*, **80**, 104114.

Zha, C.-S., Mao, H.-K., Hemley, R. J. (2000). Elasticity of MgO and a primary pressure scale to 55 GPa. *Proceedings of the National Academy of Sciences*, **97**(25), 13,494–13,499.

Zha, C.-S., Mibe, K., Bassett, W. A., Tschauner, O., Mao, H.-K., Hemley, R. J. (2008). P–V–T equation of state of platinum to 80 GPa and 1900 K from internal resistive heating/X-ray diffraction measurements. *Journal of Applied Physics,* **103**(5), 054908.

Zhang, Y., Sekine, T., He, H., Yu, Y., Liu, F., Zhang, M. (2016). Experimental constraints on light elements in the Earth's outer core. *Scientific Reports*, **6**(1), 22473.

Zhang, Y., Sekine, T., Lin, J. F., et al. (2018). Shock compression and melting of an Fe-Ni-Si alloy: implications for the temperature profile of the Earth's core and the heat flux across the core–mantle boundary. *Journal of Geophysical Research*, **123**, 1314–1327.

9

Elasticity at High Pressure with Implication for the Earth's Inner Core

SEIJI KAMADA, TATSUYA SAKAMAKI, AND EIJI OHTANI

Elastic wave velocities and densities of iron and candidate iron alloys are important properties for understanding the seismological observations of Earth's core. Several methods have been applied to measure the elastic wave velocities of iron and iron alloys at room temperature. Recently, measurements have been extended to simultaneous high-pressure and high-temperature conditions. Birch's law, which is the linearity between density and compressional wave velocity (V_P), is applicable to the experimental results of density and V_P at high pressure and room temperature. The effect of temperature on Birch's law is discussed, and it is not negligible at temperatures greater than 1,000–2,000 K. The V_P and density of hcp Fe are extrapolated to pressure and temperature conditions of the inner core. V_P of hcp Fe at 330–360 GPa is higher than the inner core seismic velocity, thus suggesting that iron should be alloyed with other elements so as to reduce not only its density, but also its velocity at inner core conditions. The V_P of Fe–Si, Fe–H, and Fe–C alloys is slower than that of Fe at the pressure of the inner core. If the temperature effect on Birch's law is taken into account, Si and H can be candidates for the major light elements in the inner core, while C, O, and S may not be included or exist as minor constituents.

9.1 Introduction

Elastic wave velocity is crucial for understanding the interior of the Earth (e.g., Dziewonski and Anderson, 1981; Kennett and Engdahl, 1991; Kennett et al., 1995). Compressional wave velocity (V_P), shear wave velocity (V_S), and density (ρ) profiles show discontinuous changes in the Earth's interior. In particular, the Earth's core consists of a liquid outer core, confirmed by the disappearance of the shear waves, and a solid inner core. The pressure and density distributions were estimated from seismic data based on the Williamson–Adams model (e.g., Dziewonski and Anderson, 1981). In order to reveal the main constituents of the two core sections, the relationships between the densities and seismic wave velocities of the core were compared with those of candidate materials at high pressure and high temperature (e.g., Birch, 1961). The Earth's core densities are lower than those of pure iron (Fe) (e.g., Birch, 1952); thus, some light elements, such as hydrogen (H), carbon (C), oxygen (O), silicon (Si), and sulfur (S), have been considered as additional core components (e.g., Poirier, 1994; McDonough, 2017).

In order to account for seismic observations, the V_P, V_S, and density of Fe and Fe alloys with light elements have been investigated. Recent studies revealed that the density deficits of the outer and inner cores were 7.5–7.6% (Kuwayama et al., 2020) and 1.4–6.4% (Dewaele et al., 2006; Fei et al., 2016), respectively. Kuwayama et al. (2020) reported a density jump of 0.32 g/cm^3 between liquid and solid Fe at 330 GPa, smaller than the seismically observed density jump, 0.60–0.82 g/cm^3, across the inner-core boundary (Dziewonski and Anderson, 1981; Masters and Gubbings, 2003). This additional density jump must be caused by preferential partitioning of light elements between the liquid outer core and the solid inner core. In addition to the density constraint, the incorporation of light elements in the core should also explain the observed seismic wave velocity in this region.

The observed seismic wave velocity profiles in the Earth's interior provide important constraints on the physical and chemical properties of constituent materials. Specific to the Earth's core, the link between the seismological observations and the geophysical models is provided by measurements, or calculations, of pertinent physical properties of candidate materials, such as pure Fe and Fe light element alloys. The focus of mineral physics research has been providing accurate measurements of those properties at conditions relevant to the core. Tateno et al. (2010) experimentally determined that the hexagonal close-packed (hcp) Fe is the stable phase up to the pressure and temperature of the inner core. Accurate determination of the equation of state (EoS) of pure Fe provides an estimate of the density deficit of the inner core. Upgraded measurements and pressure calibration have improved the estimates from ~ 8% (Mao et al., 1990) to a commonly accepted value of 4% at 6,000 K (Dewaele et al., 2006; Fei et al., 2016). Many studies have focused on measurements of the densities of Fe alloys to explain the core density deficit by incorporating light elements in the core. In this chapter, we focus on elastic wave velocity measurements at high pressure. The data are used to directly compare with the seismic observations to provide constraints on the composition of the inner core.

9.2 Summary of Elastic Wave Velocity Data for hcp Fe and Fe Light Element Alloys

The elastic wave velocities of core materials have been measured at high pressures using various techniques, including inelastic X-ray scattering (IXS), nuclear inelastic scattering (NIS, NRIXS), and shock wave experiments. Figure 9.1 summarizes the measurements of the elastic wave velocities of pure Fe and Fe light element alloys up to 170 GPa from room temperature to high temperatures using IXS, NIS, and shock wave methods.

Figure 9.1a compares the V_P-density data of Fe and Fe–Ni alloys at room and high temperature obtained by different techniques. V_P from IXS and shock wave experiments were directly measured, while V_P and V_S from NIS measurements were calculated from Debye sound velocities (V_{Debye}) and the EoS of Fe and Fe alloys. Out of the shock wave experimental data presented in Brown and McQueen (1986), we only considered those estimated in a pressure–temperature range where the hcp structure is expected to be stable based on the phase diagram of Fe reported by Anzellini et al. (2013). The V_P of hcp Fe from the shock data monotonically increases to 12 g/cm^3 along the Hugoniot. The V_P decreases at approximately 12 g/cm^3, likely caused by the onset of melting (Brown and McQueen, 1986; Nguyen and Holmes, 2004). The compressional velocities of Fe at room temperature were discussed by Antonangeli and Ohtani (2015). They proposed reference relations for the pressure and density dependence of the compressional velocity of hcp Fe at room temperature. Measurements of V_P for Fe–Ni alloy shows that $Fe_{92}Ni_8$ has slower V_P compared to that of Fe at inner-core conditions (Lin et al., 2003b; Sakamaki et al., 2016). Decremps et al. (2014) and Wakamatsu et al. (2018) measured V_P of hcp Fe and Fe–Ni alloys using picosecond acoustics and femtosecond laser pulses in a diamond anvil cell, respectively. V_P of hcp Fe by Decremps et al. (2014) showed similar values to those by IXS. V_P of Fe–Ni alloys becomes slower with increasing Ni contents (Wakamatsu et al., 2018).

Figures 9.1b and 9.1c show V_P as a function of density of Fe light element alloys. For FeHX, Shibazaki et al. (2012) and Mao et al. (2004) measured V_P by IXS up to 70 GPa and by NRIXS up to 52 GPa, respectively. The results are consistent below 30 GPa, whereas the values by NRIXS are systematically lower than those by IXS at pressures above 30 GPa. There is a general agreement on V_P measurements of Fe_3C up to 50 GPa at room temperature among previous studies (Fiquet et al., 2009; Gao et al., 2011; Takahashi et al., 2019a). The V_P at 68 GPa reported by Fiquet et al. (2009) is significantly faster than those of Gao et al. (2008, 2011) and Takahashi et al. (2019a) (Figure 9.1b). Although Fiquet et al. (2009) showed a clear softening at the highest measurement pressure (~83 GPa), Takahashi et al. (2019a) did not observe such a phenomenon up to 86 GPa. The NRIXS data by Gao et al. (2008) are systematically lower than other measurements. In the case of Fe_3S, the V_P measurements by IXS (Kamada et al., 2014a) is in agreement with those based on the NRIXS method (Lin et al., 2004) (Figure 9.1c).

When the amount of Si in Fe is small, the V_P measurements of Fe–Si alloys are consistent with each other (Figure 9.1c). For $Fe_{85}Si_{15}$ alloy, Mao et al. (2012) obtained V_P by IXS coupled with density measurements by XRD. The measured V_P values are faster than those by Lin et al. (2003b), who used the NRIXS method. The discrepancy may be partially caused by the hydrostatic conditions of the samples. The pressure medium (NaCl or Ne) was used in Mao et al. (2012), whereas Lin et al. (2003b) did not use any pressure medium. Furthermore, the choice of EoS in the NRIXS method introduces additional uncertainty. Antonanglei et al. (2018) studied V_P of a similar alloy, $Fe_{84}Si_{16}$, using the IXS method, and their results were consistent with those by Mao et al. (2012).

The analysis of the combined experimental data sets at room temperature indicates that the V_P values of pure Fe are faster than that of the inner core under core pressure conditions (e.g., Antonangeli and Ohtani, 2015). The V_P and V_S values of most Fe light element alloys at room temperature also appear to be faster than that of the inner core by extrapolation (Figure 9.1c).

Elastic wave velocities of iron have also been studied by first-principles calculations. Vočadlo et al. (2009) and Martorell et al. (2013) showed a temperature effect on V_P and V_S of hcp Fe (Figures 9.1a and 9.1d). Martorell et al. (2013) calculated V_P and V_S from 0 to 7,340 K at 360 GPa. Above 5,000 K, the V_P of pure Fe became close to that of the inner core, while the V_S of pure Fe was faster than that of V_S of the inner core below

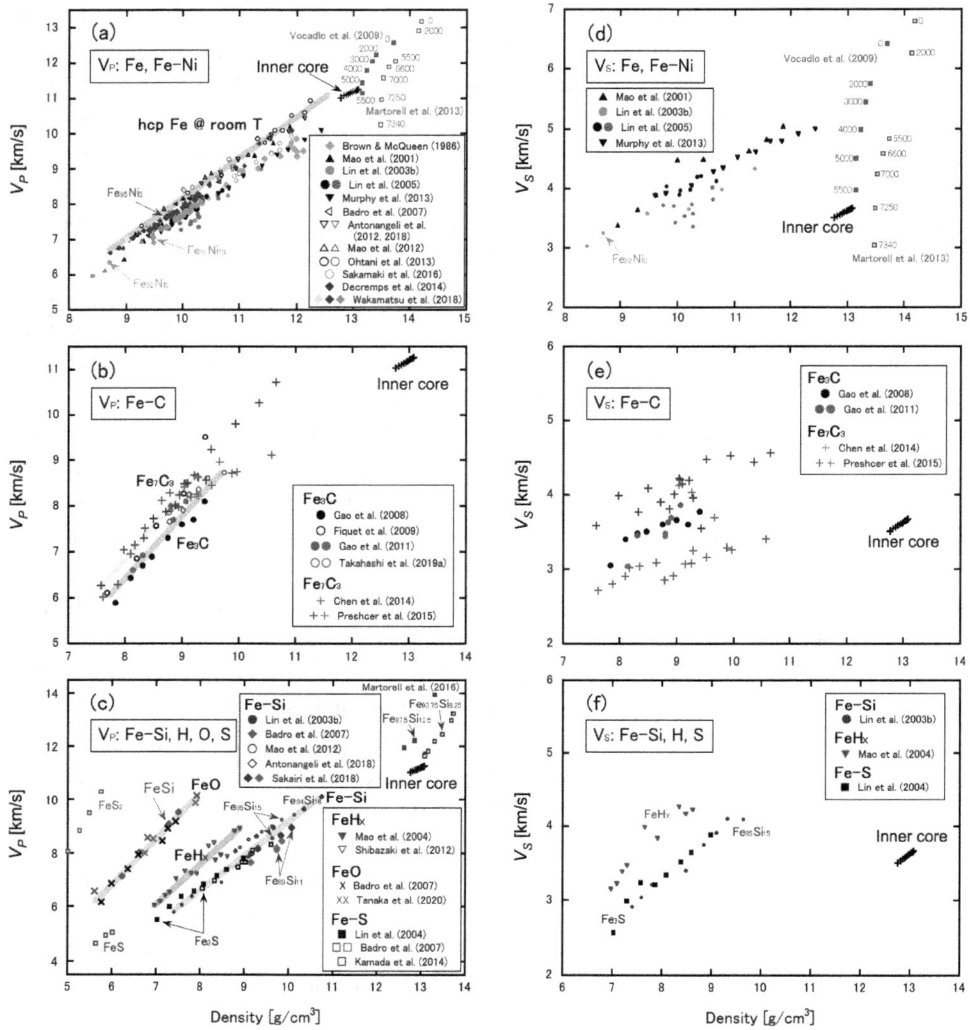

Figure 9.1 V_P and V_S as a function of density at both room and high temperature. Relationships between densities and V_P of (a) Fe and Fe–Ni alloy, (b) Fe–C alloys, and (c) Fe–Si, Fe–H, Fe–S alloys. Plus marks represent V_P of the inner core. Solid symbols represent V_P based on the NIS method and open symbols represent those based on the IXS method. Light blue diamonds represent V_P of hcp Fe based on shock wave experiments (Brown and McQueen, 1986; also see Table 9.1). Black, gray, and green symbols in Figure 9.1a represent V_P of pure hcp iron (Mao et al., 2001; Lin et al., 2003b, 2005; Badro et al., 2007; Antonangeli et al., 2012, 2018; Mao et al., 2012; Murphy et al., 2013; Ohtani et al., 2013; Sakamaki et al., 2016), V_P of $Fe_{92}Ni_8$ (Lin et al., 2003b), and calculated V_P of pure hcp iron (Vočadlo et al., 2009; Martorell et al., 2013), respectively. Open and solid symbols represent Martorell et al. (2013) at 360 GPa, and Vočadlo et al. (2009) at pressure conditions from 295–316 GPa, respectively. The numbers in green represent temperature conditions of their calculations. Red and orange symbols in Figure 9.1a represent V_P of pure iron at high temperature conducted by external-heated diamond anvil cell (Lin et al., 2005; Antonangeli et al., 2012; Mao et al., 2012; Ohtani et al., 2013), and

6,000 K. They reported that the temperature effect on V_P and V_S becomes large when approaching the melting temperature of hcp Fe ($T/T_M > 0.9$). Martorell et al. (2016) also examined the effect of Si content on V_P by ab initio molecular dynamics simulations of hcp–Fe–Si alloys at 360 GPa and at temperatures up to melting. The results showed a decrease of V_P with increasing Si contents in hcp Fe and a clear temperature dependence on V_P of Fe–Si alloys (Figure 9.1c).

The effects of temperature on V_P and V_S are still hotly debated. Lin et al. (2005) reported a relatively large temperature effect on V_P and V_S on the basis of the NRIXS method under pressure and temperature conditions between 36–73 GPa, and 300–1,700 K. Sakamaki et al. (2016) reported V_P of pure Fe at high pressure (90–170 GPa) and temperature (2,300 and 3,000 K). Those results showed a temperature effect on Birch's law. On the other hand, Antonangeli et al. (2012) and Ohtani et al. (2013) reported that the V_P of pure Fe at high temperature (up to 1,100 K) is close to that at room temperature. However, the V_P of pure

Figure 9.1 *(cont.)* those of pure iron at high temperature conducted by laser-heated diamond anvil cell (Sakamaki et al., 2016), respectively. Black, green, and blue symbols in Figure 9.1b are V_P at room temperature (Fe_3C: Gao et al., 2008; Fiquet et al., 2009; Gao et al., 2011; Takahashi et al., 2019a, Fe_7C_3: Chen et al., 2014; Prescher et al., 2015), and red symbols are V_P at high temperatures (Fe_3C: Gao et al., 2008; Fiquet et al., 2009; Gao et al., 2011; Takahashi et al., 2019a, Fe_7C_3: Chen et al., 2014; Prescher et al., 2015). Blue circles, green right-pointing triangles, and blue left-pointing triangles in Figure 9.1c are V_P of Fe–Si alloys at room temperature (FeSi: Badro et al., 2007, $Fe_{84}Si_{16}$: Antonangeli et al., 2018; $Fe_{85}Si_{15}$: Lin et al., 2003b; Mao et al., 2012, $Fe_{89}Si_{11}$: Sakairi et al., 2018) and red left-pointing triangles are V_P of $Fe_{89}Si_{11}$ at high temperature (Sakairi et al., 2018). V_P of $Fe_{93.75}Si_{6.25}$ (360 GPa and 0–7350 K) and $Fe_{87.5}Si_{12.5}$ (360 GPa and 0–6550 K) are from Martorell et al. (2016). Purple downward triangles in Figure 9.1c are V_P of FeH_X at room temperature (Mao et al., 2004; Shibazaki et al., 2012). Black and green outline plus marks in Figure 9.1c are V_P of FeO at room temperature (Badro et al., 2007; Tanaka et al., 2020), and red outline plus signs are V_P of FeO at high temperature (Tanaka et al., 2020). Black solid and open squares in Figure 9.1c are V_P of Fe_3S (Lin et al., 2004; Kamada et al., 2014a). Blue open squares represent V_P of FeS (Badro et al., 2007), and green open squares represent V_P of FeS_2 (Badro et al., 2007). The relationships between density and V_S of alloys are shown for (d) Fe and Fe–Ni alloys, (e) Fe–C alloys, and (f) Fe–Si, Fe–H, and Fe–S alloys. Plus signs represent V_S of the inner core. Green open and solid squares represent V_S of pure hcp Fe from Martorell et al. (2013) and Vočadlo et al. (2009), respectively. The calculated conditions are the same as those for V_P. Black upward and downward triangles in Figure 9.1d represent V_S at room temperature of Mao et al. (2001) and Murphy et al. (2013), respectively. Black and red circles are V_S of pure iron at room temperature and high temperature (Lin et al., 2005), respectively. Gray symbols represent those of $Fe_{92}Ni_8$ (Lin et al., 2003b). Black and green triangles in Figure 9.1e represent V_S at room temperature of Fe_3C of Gao et al. (2008) and Gao et al. (2011), respectively. Red triangles are V_S at high temperature of Fe_3C (Gao et al., 2011). Green and blue triangles in Figure 9.1e are V_S of Fe_7C_3 at room temperature from Chen et al. (2014) and Prescher et al. (2015), respectively and red triangles are V_S of Fe_7C_3 at high temperature from Prescher et al. (2015). Blue, black, and purple symbols in Figure 9.1f represent V_S at room temperature of $Fe_{85}Si_{15}$ alloys (Lin et al., 2003b), Fe_3S (Lin et al., 2004), and FeH_X (Mao et al., 2004), respectively. For the color version, refer to the plate section.

Table 9.1 *Hugoniot data from shock wave experiments reported by Brown and McQueen (1986), compared with other data.*

P [GPa][a]	T_H [K][a]	ρ [kg/m^3][a]	V_P [m/s][a]	T_H/T_M	T^b [K]	T^b/T_M	ρ^c [kg/m^3]	V_P [m/s]	$V_{P\ HT}/V_{P\ RT}$	For fit[d]
119(2)	2,250(90)	11,020(70)	8,950(30)	0.5650(220)	2,580(370)	0.6487(928)	11,380(30)	9,850(670)	0.9089(618)	Used
120(2)	2,270(90)	11,040(70)	8,960(30)	0.5693(222)	2,570(370)	0.6446(930)	11,400(30)	9,870(670)	0.9081(617)	Used
142(2)	2,830(120)	11,320(70)	9,040(120)	0.6632(274)	3,160(370)	0.7405(876)	11,760(30)	10,270(680)	0.8802(594)	Used
177(2)	3,770(170)	11,710(70)	9,390(130)	0.8056(374)	4,120(370)	0.8811(798)	12,280(30)	10,850(700)	0.8652(567)	
204(3)	4,520(230)	11,980(80)	9,860(70)	0.9081(467)	4,850(420)	0.9744(854)	12,640(40)	11,260(710)	0.8754(554)	
207(3)	4,600(240)	12,010(80)	9,500(70)	0.9190(479)	4,920(420)	0.9826(849)	12,680(40)	11,310(710)	0.8402(530)	
77(1)	1,300(50)	10,330(40)	8,130(40)	0.3814(143)	1,760(190)	0.5149(568)	12,840(30)	11,490(710)	0.9077(655)	Used
102(1)	1,840(70)	10,760(40)	8,550(40)	0.4908(186)	2,240(210)	0.5964(546)	13,120(20)	11,810(720)	0.8992(625)	Used
151(1)	3,060(130)	11,440(40)	9,310(50)	0.7007(298)	3,330(220)	0.7611(493)	13,490(90)	12,220(740)	0.8929(587)	Used
193(1)	4,210(210)	11,870(40)	9,670(70)	0.8672(427)	4,570(210)	0.9409(437)	10,590(20)	8,960(640)	0.8712(555)	
196(1)	4,290(210)	11,900(40)	9,760(70)	0.8785(438)	4,650(210)	0.9506(434)	11,080(20)	9.510(660)	0.8758(556)	
201(1)	4,430(230)	11,950(40)	9,570(50)	0.8971(456)	4,770(210)	0.9658(429)	11,900(20)	10,430(680)	0.8530(538)	
206(1)	4,570(240)	12,000(40)	9,380(30)	0.9154(475)	4,900(210)	0.9799(425)	12,490(10)	11,100(700)	0.8307(521)	

[a] Values were taken from Brown and McQueen (1986).
[b] Temperature are calculated using EoS of Fe (Dewaele et al., 2006) to reproduce P and ρ reported in Brown and McQueen (1986).
[c] ρ for room temperature are calculated using EoS of Fe (Dewaele et al., 2006) to reproduce P reported in Brown and McQueen (1986).
[d] The data used for analysis of the temperature effect on Birch's law are denoted as "used" (see the text).
Numbers in parentheses are errors in two or three last digits.
T_M represents a melting temperature of iron based on Anzellini et al. (2013).
T_H represents a Hugoniot temperature reported by Brown and McQueen (1986).

Fe could not explain the observed V_P of the inner core even if the high-temperature effect was taken into account.

Gao et al. (2011) reported a temperature effect of Birch's law in Fe_3C for V_P and V_S in the temperature range of 740–1,450 K, whereas Takahashi et al. (2019a) could not detect a clear temperature effect at high temperatures up to 1,400 K at 40 GPa and 2,300 K at 67 GPa. In the case of Fe_7C_3 and Fe–Si alloys, Prescher et al. (2015) reported a slight temperature effect on Birch's law for V_P and V_S, and Sakairi et al. (2018) reported a clear temperature effect on Birch's law for V_P in $Fe_{89}Si_{11}$. The temperature effects on Birch's law for V_P and V_S of Fe and Fe light element alloys were measurable when the temperature was higher than 1,800 K.

Figure 9.1d–f compares the observed V_S of the inner core with measurements of pure Fe and Fe light element alloys at room temperature and high temperature. The trend of the temperature effect on Birch's law is not clear based on limited high-temperature data (e.g., Lin et al., 2005; Antonangeli et al., 2012; Ohtani et al., 2013; Prescher et al., 2015). High-precision measurements of V_P and V_S at high temperature are needed to further determine the temperature effect, which is critical for discussion of the physical properties of the Earth's inner core.

9.3 Methods of Elastic Wave Velocity Measurements

9.3.1 Ultrasonic Interferometry

The ultrasonic echo method is one of the oldest techniques used to measure the elastic wave velocities of specimens. The frequencies of the ultrasonic waves used in this method are typically in the range of MHz to GHz, depending on specimen size. The elastic wave velocity can be determined using the specimen length and the travel time. The travel time can be determined by detecting the ultrasonic wave reflected on the top and bottom surfaces of the specimen. The elastic wave velocity can be calculated as follows:

$$V_{elas} = \frac{2\Delta L}{\Delta t}, \tag{9.1}$$

where V_{elas}, ΔL, and Δt represent the elastic wave velocity, sample thickness, and travel time, respectively. The sample thickness can be estimated by assuming hydrostatic compression ($(\Delta L/\Delta L_0)^3 = (V/V_0)$, where ΔL_O, V_O, and V represent the initial sample thickness and the initial sample volume, and sample volume at high pressure, respectively) or by Cook's method (Cook, 1957). The sample thickness can also be measured using X-radiography at synchrotron facilities. The ultrasonic echo method has been applied to elastic wave velocity measurements of samples at high pressure and temperature (e.g., Spetzler et al., 1996; Mao et al., 1998; Higo et al., 2006; Jacobsen et al., 2006; Nishida et al., 2013; Gréaux et al., 2019; Thomson et al., 2019; Yoneda et al., 2019). This technique allows us to measure the V_P and V_S of single crystals, polycrystalline samples, liquids, and glass of mantle and core materials. It is crucial to measure the sample thickness precisely because the error in thickness is usually the main source of the error in elastic wave velocities.

By combining XRD technique, the V_P, V_S, and density can be directly measured, leading to determination of the bulk modulus. Then pressure can be calculated by a numerical integration of the bulk modulus and volume relation. For example, Kono et al. (2010) examined the *P–V–T* relations of MgO by using XRD and ultrasonic measurements and proposed an equation of state of MgO that can be used as a primary pressure scale. Details of techniques and recent studies of ultrasonic interferometry are reviewed by Li and Liebermann (2014).

9.3.2 Brillouin Scattering

The Brillouin scattering method is used to measure the elastic wave velocities (V_P and V_S) (see review by Speziale et al., 2014). Since Weidner et al. (1975) measured V_P and V_S on quartz, this technique has been applied to mantle minerals from ambient to high pressure using a diamond anvil cell. Brillouin scattering, which is inelastic scattering, occurs because of the interaction between light and acoustic phonons in the materials. The incident light is scattered due to the interaction, and the frequency of the scattered light is changed from that of the incident light. The scattered light is detected through a Fabry–Perot interferometer. A Brillouin shift ($\Delta\omega$) between incident and scattered light is then obtained. If all the surfaces in diamond anvils and a sample are parallel, V_P or V_S can be calculated as follows (Sinogeikin and Bass, 2000):

$$V_{P\,or\,S} = \frac{\Delta\omega\lambda}{2\sin\left(\frac{\theta}{2}\right)}, \tag{9.2}$$

where θ is the external scattering angle and λ is the wavelength of the excitation laser. Because the transmission method is compatible with the diamond anvil window, the V_P and V_S of many transparent mantle minerals have been measured under high pressures (e.g., Weidner et al., 1975; Zha et al., 1994; Zha et al., 1996, 2000; Sinogeikin and Bass, 2000). Zha et al. (2000) proposed a primary pressure scale by measuring the V_P and V_S of MgO, which was established without any pressure scale.

There are a few limitations in the Brillouin scattering technique, although it is a very versatile way to measure the V_P and V_S of mantle minerals. Because the excitation laser goes through diamond anvils in the DAC, the Brillouin peak of the longitudinal and transverse phonon mode of the diamonds can be observed, which overlaps with the Brillouin peak of samples above 10 km/s. This makes it difficult to measure V_P and V_S under lower mantle conditions. In the case of the transmission Brillouin scattering method, another problem is the color of the samples. When the samples contain certain amounts of Fe, the color of the samples usually becomes darker and visible light cannot pass through the samples. Although there are some disadvantages as described, there are also advantages in this technique. V_P and V_S can be measured from single crystals, polycrystals, glass, and liquids. More importantly for high-pressure samples, Brillouin scattering has the ability to measure V_P and V_S of a small sample at a few tens of microns.

There are some notable studies for the elastic wave velocities of lower mantle minerals at high pressure. For instance, Murakami et al. (2012) measured V_S of aluminous bridgmanite (Al–$MgSiO_3$) and ferropericlase ($(Mg_{0.92}Fe_{0.08})O$) up to 120 GPa and at 2,700 K using this technique with a laser-heating system. They reported an approximately 4% drop of V_S in bridgmanite at around 90 GPa and a 7% drop of V_S in ferropericlase and concluded that 93 vol% of silicate perovskite explained seismic wave velocities of the lower mantle, which is a volume fraction much higher than that of the peridotitic mantle model. Moreover, Kurnosov et al. (2017) examined V_P and V_S of single-crystal (Fe, Al) bearing silicate perovskite, and concluded that a high Fe^{3+} contents in silicate perovskite was able to explain thr seismic data of the lower mantle shallower than 1,200 km. Their velocity model was not able to explain seismic data of the lower mantle deeper than 1,200 km, suggesting either a change in silicate perovskite cation ordering or in the ferric iron content in mantle perovskite.

9.3.3 Inelastic X-Ray Scattering

The inelastic X-ray scattering method has been used to measure the V_P of Fe (e.g., Fiquet et al., 2001; Antonangeli et al., 2004a; Mao et al., 2012; Ohtani et al., 2013) and V_P and V_S of single crystals (e.g., Antonangeli et al., 2004b, 2011; Fukui et al., 2008; Yoneda et al., 2017; Kamada et al., 2019). The frequency shifts of X-rays due to the interaction between X-ray photons and acoustic phonons in the sample are detected using IXS spectrometers. The obtained momentum and energy transfers show a phonon dispersion relationship. The acoustic phonons can be regarded as elastic waves in the long wave limit. The phonon dispersion relation is linear near the Γ point, i.e., a group velocity equal to a phase velocity. In this case, the elastic wave velocity can be obtained as the slope of the linear relation. The slope of the phonon dispersion near the Γ point, which is the elastic wave velocity, can be obtained using a sine dispersion relationship as follows:

$$E = 4.192 \times 10^{-4} V_P \times Q_{Max} \sin\left(\frac{\pi}{2}\frac{Q}{Q_{Max}}\right), \tag{9.3}$$

where E [meV] and Q [nm^{-1}] represent the energy and momentum transfers, respectively. V_P and Q_{Max} are the elastic wave velocity and the position of the edge of the first Brillouin zone, respectively, which are free parameters during the fits.

Because this method requires X-rays instead of visible light as used in Brillouin scattering, measurements for opaque samples are possible. Measurements for core materials such as Fe and Fe light element alloys have been performed (e.g., Badro et al., 2007; Fiquet et al., 2009; Antonangeli et al., 2010; Shibazaki et al., 2012; Ohtani et al., 2013; Kamada et al., 2014a). Moreover, it is also possible to measure V_P at high temperatures (Antonangeli et al., 2010; Fukui et al., 2013; Liu et al., 2014; Sakamaki et al., 2016; Sakairi et al., 2018; Takahashi et al., 2019a). Recently, the V_P of liquid Fe and Fe–Ni–S alloys have also been measured (e.g., Kawaguchi et al., 2017; Kuwayama et al., 2020). Because IXS is a synchrotron-based X-ray technique, it is typically possible to collect X-ray diffraction patterns in parallel, and hence the density is directly measured in addition to the velocities. For example, Kuwayama et al. (2020) measured V_P and density of liquid

Fe directly. They compared the results with those of the Earth's outer core and concluded that the outer core was 7.5% less dense and 4% faster in V_P than liquid pure Fe.

9.3.4 Nuclear Inelastic Scattering

The nuclear inelastic scattering technique allows the determination of the partial phonon density of states of samples containing a Mössbauer isotope. Incoherent inelastic X-ray scattering measured from detectors placed radially around the samples is analyzed to derive the Debye sound velocity (V_D), while electronic information is derived from the nuclear forward scattering collected downstream from the sample. The samples are usually enriched in ^{57}Fe in order to effectively perform NIS measurements. The Debye sound velocity is determined by a parabolic to the low-energy part of the partial phonon density of states, typically lower than 20 meV. V_D is related to V_P and V_S as follows:

$$\frac{3}{V_D^3} = \frac{1}{V_P^3} + \frac{2}{V_S^3}, \tag{9.4}$$

where V_D, V_P, and V_S represent the Debye sound velocity and compressional and shear wave velocities, respectively. V_P and V_S are expressed using adiabatic bulk moduli (K_S), shear moduli (G), and density (ρ) as follows:

$$\rho V_P^2 = K_S + \frac{4}{3}G, \tag{9.5}$$

$$\rho V_S^2 = G. \tag{9.6}$$

If the EoS of the sample is known, the density and K_S can be calculated. Since V_P and V_S can be obtained in addition to the electron spin states of Fe, this technique has been applied to mantle and core materials (e.g., Mao et al., 2001, 2004; Lin et al., 2005; Sinmyo et al., 2014; Prescher et al., 2015). The sample density can be directly obtained by using the XRD technique under the same conditions as the NIS measurements. If the XRD technique was not used, the EoS would be needed to calculate the sample density at a certain pressure. Therefore, the accuracy of V_P and V_S depends on the accuracy of the EoS used. The accuracy of EoS is usually not very high if there is a lack of a primary pressure scale, and some of the EoS parameters were fixed to simplify EoS fitting, especially with high-temperature data.

By the NIS method, the time domain Mössbauer spectrum can be obtained, which provides information on the electron state in an atom. The Mössbauer spectrum measurements in the energy domain using a nuclear Bragg monochromator have also been made at high pressure by using the synchrotron X-ray combined with the DAC (e.g., Mitsui et al., 2009; Hirao et al., 2020). A center shift (CS) related to s-electron density at the nucleus and a quadrupole splitting (QS) related to an electric field gradient acting on the nucleus can be obtained from Mössbauer spectra. When the nucleus is placed in a magnetic field, Zeeman splitting can be observed, which suggests that the mineral has a magnetic field. In particular, ^{57}Fe is the most important Mössbauer nuclide for geoscientists because Fe is a fundamental element in mantle minerals and core materials. There have been many investigations on mantle minerals and core materials based on Mössbauer spectroscopy.

Fe undergoes an electron spin transition in mantle minerals, such as ferropericlase (e.g., Badro et al., 2003; Lin et al., 2006; Lin et al., 2009) and bridgmanite (e.g., Badro et al., 2004; Jackson et al., 2005; Li et al., 2006). Antonangeli et al. (2011) reported an anomaly in C_{44} of ferropericlase, although it did not affect V_P and V_S. The valence state of Fe in bridgmanite is also of importance, which can also be revealed by Mössbauer spectroscopy. Fukui et al. (2016) measured the elastic moduli (C_{ij}) and discussed the effect of chemical composition. Only Fe^{2+} was found in their samples. Yoshino et al. (2016) discussed the effect of Fe valence on the electrical conductivity of Al-bearing bridgmanite. Moreover, Maeda et al. (2017) indicated no electron spin transition of Fe^{2+} in a basaltic glass, and that the distortion of Fe^{2+} polyhedra inferred from the QS pressure dependence might gradually stabilize Fe^{2+} in basaltic glass. Hamada et al. (2016) examined the spin transition of Fe^{2+} in FeO up to 200 GPa, suggesting that the Fe–O bonding became shorter due to the spin transition. Hence, dense Fe–O liquids might exist in the liquid outer core.

Fe in metallic Fe and Fe alloys undergoes an electronic topological transition (ETT) (e.g., Lifshitz, 1960; Glazyrin et al., 2013). The transition in hcp metals is accompanied by anomalies in the lattice parameter ratio of c/a at finite temperature and in the second-order Doppler shifts of the Mössbauer spectra, which are related to the heat capacity (Glazyrin et al., 2013). Thus, these anomalies affect the sound velocities. In the case of Fe–Si alloys, Kamada et al. (2018) showed anomalies in the pressure dependence of the center shift, suggesting the occurrence of ETT.

9.3.5 Shock Wave

Density and compressional velocity can also be measured by shock wave compression experiments such as explosive-driven, two-stage light gas-gun, and high-power pulse laser shock experiments (e.g., Brown and McQueen, 1986; Sakaiya et al., 2014). The shock wave velocity and particle velocity are measured in the shock wave experiments to derive the pressure and density along the Hugoinot using the Rankin–Hugoniot equations that express the conservation of mass, momentum, and energy across the shock front as follows:

$$P_H - P_0 = \rho_0 (U_S - u_0)(u_p - u_0), \tag{9.7}$$

$$\rho_H = \rho_0 \frac{U_S - u_0}{U_S - u_p}, \tag{9.8}$$

$$E_H - E_0 = \frac{1}{2}(P_H + P_0)\left(\frac{1}{\rho_0} - \frac{1}{\rho_H}\right), \tag{9.9}$$

where P_H, ρ_H, and E_H are the pressure, density, and internal energy of the sample on the Hugoniot, and P_O, ρ_O, and E_O are the pressure, density, and internal energy of the sample at initial conditions. U_S and u_P represent the shock velocity and particle speed after the shock wave, respectively, and u_O is the particle speed before impact and is usually zero. If there is no phase transition, the Hugoniot temperature (T_H) can be calculated by numerical integration along the Hugoniot using the following differential equation (e.g., Brown and McQueen, 1986):

$$dT_H = -\gamma T_H \frac{dV}{V_H} + \frac{(V_0 - V_H)dP + (P_H - P_0)dP}{2C_V}, \tag{9.10}$$

where γ and C_V represent the Grüneisen parameter and the specific heat at constant volume, respectively. The Grüneisen parameter and C_V are related by the following formula:

$$\gamma = \frac{K_T \alpha V_H}{C_V}. \tag{9.11}$$

The calculated T_H depends on γ, which has several empirical expressions as a function of volume (e.g., Sakai et al., 2014).

In addition to pressure and density determination, the compressional velocity in the shock wave experiments is determined either using the overtaking rarefaction method with multisample technique (e.g., Brown and McQueen, 1986; Barker and Hollenbach, 1972; Huang et al., 2011) or using the reverse-impact method (Duffy and Ahrens, 1995; Huang et al., 2011). The multisample technique leverages the different sample thicknesses to relate the window overtake time with the sample thickness. The Lagrangian sound velocity is determined from the thickness and time at which overtake would occur in an infinitely thick sample. The uncertainty in the velocity measurements is directly related to the error in the fitted the catch-up thickness of the sample. The reverse-impact method uses the sample as the flyer, which may limit the maximum impact velocity achievable in the shock wave experiments. The method provides direct measurements of both compressional and bulk sound velocities, allowing calculations of the shear sound velocity (Huang et al., 2018).

Shock wave dynamic compression data are the sources of equations of state of many materials (see Chapter 6, this volume). Simple metals such as Au and Pt were chosen as pressure standards based on dynamic compression data (e.g., Jamieson et al., 1982; Holmes et al., 1989; Yokoo et al., 2009). However, these pressure standards contain uncertainty associated with the estimation of the Hugoniot temperature and the reduction from Hugoniot to the isothermal condition. The technique also provides unique sound velocity measurements at simultaneous high pressure and temperature along the Hugoniot. Measurements on the Earth's core materials provide a direct comparison to the observed seismic velocity profiles in the core. Huang et al. (2011) reported V_P of liquid Fe–O–S alloys up to 207 GPa and concluded that oxygen was depleted in the outer core. In the case of an Fe–S system, Huang et al. (2018) measured V_P of liquid Fe-11.8 wt.%S up to 211 GPa, and their results implied an upper limit of 10wt.% S to be able to explain V_P and densities of the outer core.

9.3.6 Pulsed Laser

The ultrashort pulsed laser (pulse width from subnanoseconds to femtoseconds) enables us to measure the elastic wave velocities of thin samples in a DAC under high pressure. The pulse width is in the range between picoseconds and femtoseconds (e.g., Chigarev et al.,

2008; Decremps et al., 2014). A pump laser excites elastic waves in the sample when the laser hits its surface. When the elastic waves reach where a probe laser hits, intensity of the probe laser changes. By measuring time delays of the intensity changes, the travel time can be calculated. As shown in (9.1), the travel time can be estimated. When V_P and ΔL are 10 km/s and 10 μm, Δt is 1 ns. This suggests that the elastic wave excited by the pulsed laser with a pulse width of subnanoseconds, picoseconds, or femtoseconds does not overlap. Therefore, the travel time is simply obtained as a time lag between when the pulsed laser hits the sample surface and when the compressional wave reaches the opposite surface (e.g., Decremps et al., 2014). Chigarev et al. (2008) observed reflective P and S waves, which are converted from the P waves due to a reflection at the interface of the diamond anvil and the sample.

For higher-pressure conditions, the femtosecond pulsed laser is preferred to distinguish the time lag. Decremps et al. (2014) measured V_P of Fe up to 150 GPa, and Wakamatsu et al. (2018) reported V_P of Fe–Ni alloys up to 60 GPa. Using this technique, the sample thickness must be estimated using the EoS of the sample. Therefore, the error in V_P depends on the accuracy of the EoS. Although the thickness cannot be measured directly, the travel time can be precisely measured. In previous studies, only V_P was measured. Decremps et al. (2015) improved this method in order to record two-dimensional reflectivity of the sample surface. The sample thickness and elastic wave velocity can be calculated by analyzing the propagation of the compressional wave in the plane perpendicular to the incident axis of the pulse laser. The disadvantage of the two-dimensional recording method is that it needs relatively large sample dimensions (~100 μm in diameters and ~50 μm in thickness) and takes more than a few hours for the measurements. The experimental pressures have to be measured in order to estimate the sample densities and thicknesses for the V_P calculations.

Edmund et al. (2019) measured V_P of Fe-5wt.%Si using this technique at room temperature in addition to XRD data. They suggested the hcp Fe–Si alloy might have a texture along the compression axis of the DAC, and therefore the measured V_P were slightly higher than those based on the IXS method. Although they showed the texture effect on V_P, they also indicated that $dV_P/d\rho$ decreased with increasing Si contents in the alloy.

9.3.7 Radial X-Ray Diffraction

The elastic properties, such as elastic constants (C_{ij}s), can be estimated from X-ray diffraction patterns. Mao et al. (1998) reported the C_{ij}s of hcp Fe at 211 GPa. In this method, X-rays hit the sample through an X-ray transparent gasket in the direction perpendicular to the compression axis. The method of data analysis was described by Singh (1993) for the cubic system, Singh and Balasingh (1994) for the hexagonal system, and Singh et al. (1998) for all crystal systems. The observed d spacing, $d_{obs}(hkl)$, can be expressed as follows:

$$d_{obs}(hkl) = d_P(hkl)\left[1 + \left(1 - 3\cos^2\psi\right)Q(hkl)\right], \tag{9.12}$$

$$Q(hkl) = \frac{t}{3}\left\{\frac{\alpha}{2G_R(hkl)} + \frac{(1-\alpha)}{2G_V}\right\}, \tag{9.13}$$

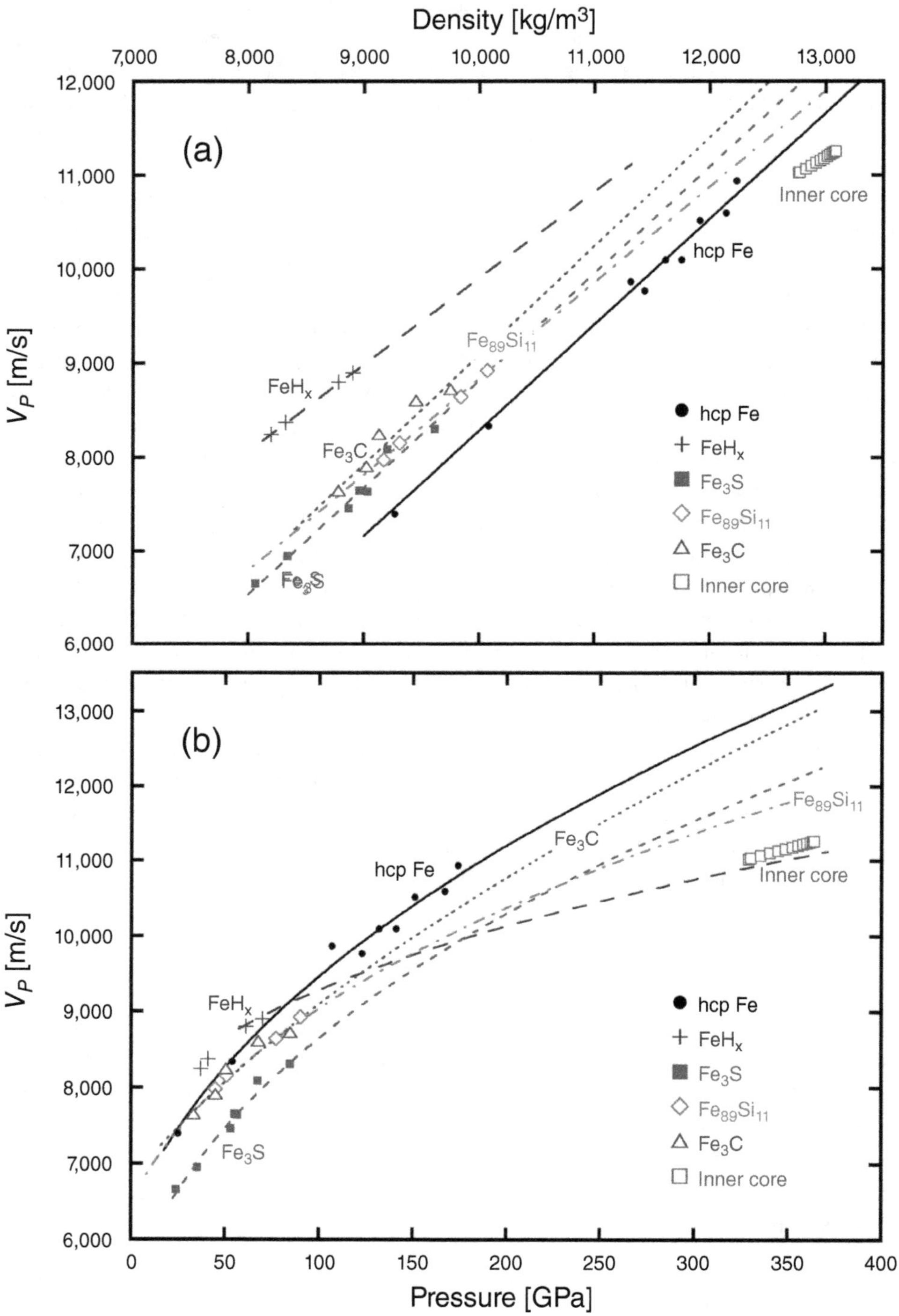

Figure 9.2 V_P of the inner core and core materials at room temperature and high temperature as a function of (a) density and (b) pressure. Solid and open symbols represent V_P at

where $d_P(hkl)$ and ψ represent the d spacing under hydrostatic conditions and the angle between the compression axis and the diffraction plane normal, respectively. The uniaxial stress component, t, is the difference between the maximum and minimum stresses. G_R and G_V are the aggregate shear moduli calculated under the Reuss (iso-stress) and Voigt (iso-strain) conditions, respectively. The factor α in (9.13) varies from 0 to 1 corresponding to the fraction of the Ruess and Voigt conditions and is difficult to define from experimental data. Therefore, the factor α is usually fixed. Merkel et al. (2005) reported the C_{ij}s of hcp Fe under pressure with α value of 1 and 0.5.

This technique assumes a pure elastic deformation in the sample. However, there is also plastic deformation (e.g., Antonangeli et al., 2006; Merkel et al., 2006, 2009). Antonangeli et al. (2006) discussed C_{ij}s based on plastic deformation and purely elastic deformation. The obtained V_P anisotropies in hcp cobalt showed a similar result among radial X-ray diffraction studies. On the other hand, the results were different from those based on IXS single crystal measurements, theoretical results including generalized gradient approximation (GGA), and local density approximation (LDA) calculations. They explained that this discrepancy was due to inhomogeneous stresses by a grain boundary interaction. The C_{44} obtained based on the radial XRD method differed from that based on the IXS single-crystal measurement and GGA or LDA calculations because C_{44} was strongly linked to plastic shear strain on the basal plane (Antonangeli et al., 2006).

9.4 Elastic Wave Velocity at High Pressure

9.4.1 Room Temperature Data

Several studies have reported the V_P of Fe and Fe light element alloys at room temperature (Shibazaki et al., 2012; Ohtani et al., 2013; Kamada et al., 2014a; Sakamaki et al., 2016; Sakairi et al., 2018; Takahashi et al., 2019a). Figure 9.2a shows the relationship between V_P and density. The empirical rule between V_P and density, Birch's law (Birch, 1961), seems to hold. The relationships between the pressure based on the EoS of Fe (Dewaele et al., 2006) and V_P are shown in Figure 9.2b. We used the EoS of Fe by Dewaele et al. (2006) because they compressed Fe under quasihydrostatic conditions and established their EoS based on the ruby and tungsten scales, which were also supported

Figure 9.2 (*cont.*) room temperature and high temperature, respectively. The V_P values of the core materials are cited from previous studies (Ohtani et al., 2013; Kamada et al., 2014a; Sakamaki et al., 2016; Sakairi et al., 2018; Takahashi et al., 2019a; Tanaka et al., 2020). Temperature conditions for Fe are shown in different colors (blue: 400 K, purple: 700 K, green: 1,000 K, red: 2,300 K, and orange: 3,000 K). Lines represent Birch's law at room temperature. V_P of Fe below 1,000 K shows similar V_P at room temperature. In contrast, V_P of Fe above 2,300 K shows slower V_P than those below 1,000 K. V_P of $Fe_{89}Si_{11}$ at 1,800 K are slower than those at room temperature. As for Fe_3C, V_P at high temperature are similar to those at room temperature.

by available shock waves, ultrasonic, X-ray, and thermochemical data (Dorogokupets and Oganov, 2007). Figure 9.2 also compares the compressional velocity, V_P, of various Fe alloys with that of pure Fe. Pressures were calculated using the available EoS (FeH_X: Hirao et al., 2004; Fe_3S: Kamada et al., 2014b; $Fe_{89}Si_{11}$: Tateno et al., 2015; Fe_3C: Takahashi et al., 2019a). The V_P of the inner core is slower than that of Fe, Fe_3S, $Fe_{89}Si_{11}$, and Fe_3C. However, the extrapolated V_P of FeH_X is slower than that of the inner core. Therefore, the dissolution of H into hcp Fe can explain the V_P of the inner core, although the effect of temperature must be taken into account to quantify the amount of H in the inner core.

9.4.2 High-Temperature Data

High-temperature and high-pressure measurements of IXS have been performed using external heated or laser-heated diamond anvil cell. A portable laser-heating system, such as the COMPAT system (Fukui et al., 2013), has been developed in order to install in the IXS beamline. Using the portable laser-heating system, high-temperature IXS has been measured at BL35XU and BL43LXU of SPring-8. The effect of temperature on the compressional velocities of Fe is not significant between 400–1,000 K (Ohtani et al., 2013). On the other hand, those of Fe above 2,300 K are slower than those below 1,000 K (Sakamaki et al., 2016), indicating a clear temperature effect. As for Fe–Si alloy, the similar temperature effect was observed at 1,800 K (Sakairi et al., 2018). In contrast, the effect of temperature on V_P was very small in Fe_3C based on measurements up to 2,300 K (Takahashi et al., 2019a). Precise measurements of the V_P of Fe_3C at higher temperatures are needed to further understand the temperature effect.

The effect of temperature on the V_P of Fe is discussed in this section. The relationships between the normalized temperature and V_P are shown in Figure 9.3. The experimental temperatures are normalized by the melting temperature of Fe (Anzellini et al., 2013) at the same experimental pressures. The measured V_P data by IXS are normalized by that at room temperature based on Birch's law for Ohtani et al. (2013) and the power law for Mao et al. (2012). The V_P data from shock experiments (Brown and McQueen, 1986) are normalized by V_P at room temperature based on Birch's law reported by Ohtani et al. (2013). Temperatures on the Hugoniot of the shock experiments are recalculated using the EoS of hcp Fe (Dewaele et al., 2006) in order to reproduce pressures and densities reported by Brown and McQueen (1986). The calculated Hugoniot temperatures vary from 4,100–5300 K at 200 GPa depending on the model, indicating a large uncertainty. The recalculated temperatures are listed in Table 9.1, showing only the shock compression data from Brown and McQueen (1986) under the stable conditions of hcp Fe (Anzellini et al., 2013), which are denoted as "used" in Table 9.1. As shown in Figure 9.3, the temperature effect on V_P seems linear below T/T_M of approximately 0.85. Shock wave data with T/T_M values larger than 0.85 are excluded from the fitting because these data may represent mixed phases (Brown and McQueen, 1986). As shown in Figure 9.4, the V_P of Fe at room temperature of Ohtani et al. (2013) was fitted by Birch's law ($V_P = 1.126(\pm 0.042) \times \rho - 2970(\pm 470)$)

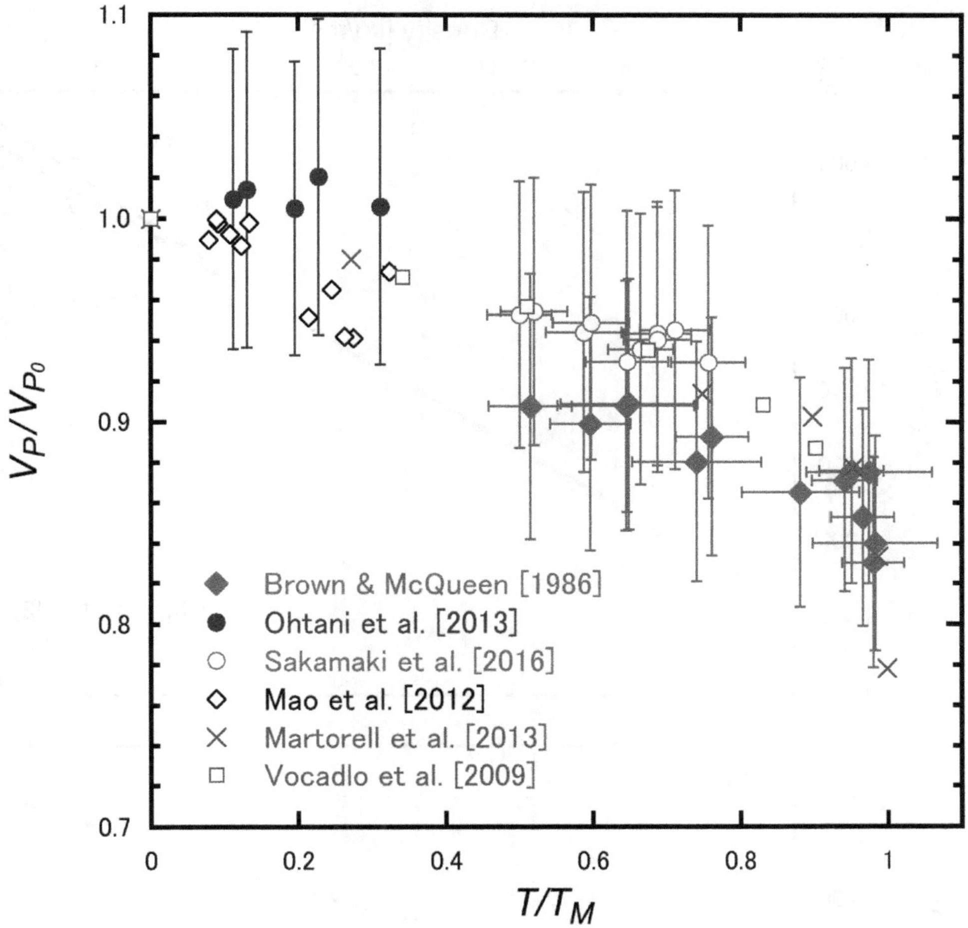

Figure 9.3 Relationships between normalized temperature and V_P. The temperature is normalized by T_M, the melting temperature of Fe (Anzellini et al., 2013) at the same experimental pressures. V_P determined by calculation (Vočadlo et al., 2009; Martorell et al., 2013) is normalized by that at 0 K, whereas that determined experimentally is normalized by that at room temperature (Brown and McQueen, 1986; Mao et al., 2012; Ohtani et al., 2013). The details are given in the text.

and power law ($V_P = 0.0293(\pm 0.0001) \times \rho^{1.3622(\pm 0.0004)}$). The two fitting methods showed a maximum V_P difference of 2 % at the pressure corresponding to the center of the Earth. Therefore, Birch's law is considered to discuss the temperature effect. The following expressions can empirically express the temperature effect on Birch's law (e.g., Sakamaki et al., 2016):

$$V_P = (a_0 + a_1 T)\rho + (b_0 + b_1 T), \tag{9.14}$$

or

$$V_P = M\rho + B + A(T - T_0)(\rho - \rho^*), \tag{9.15}$$

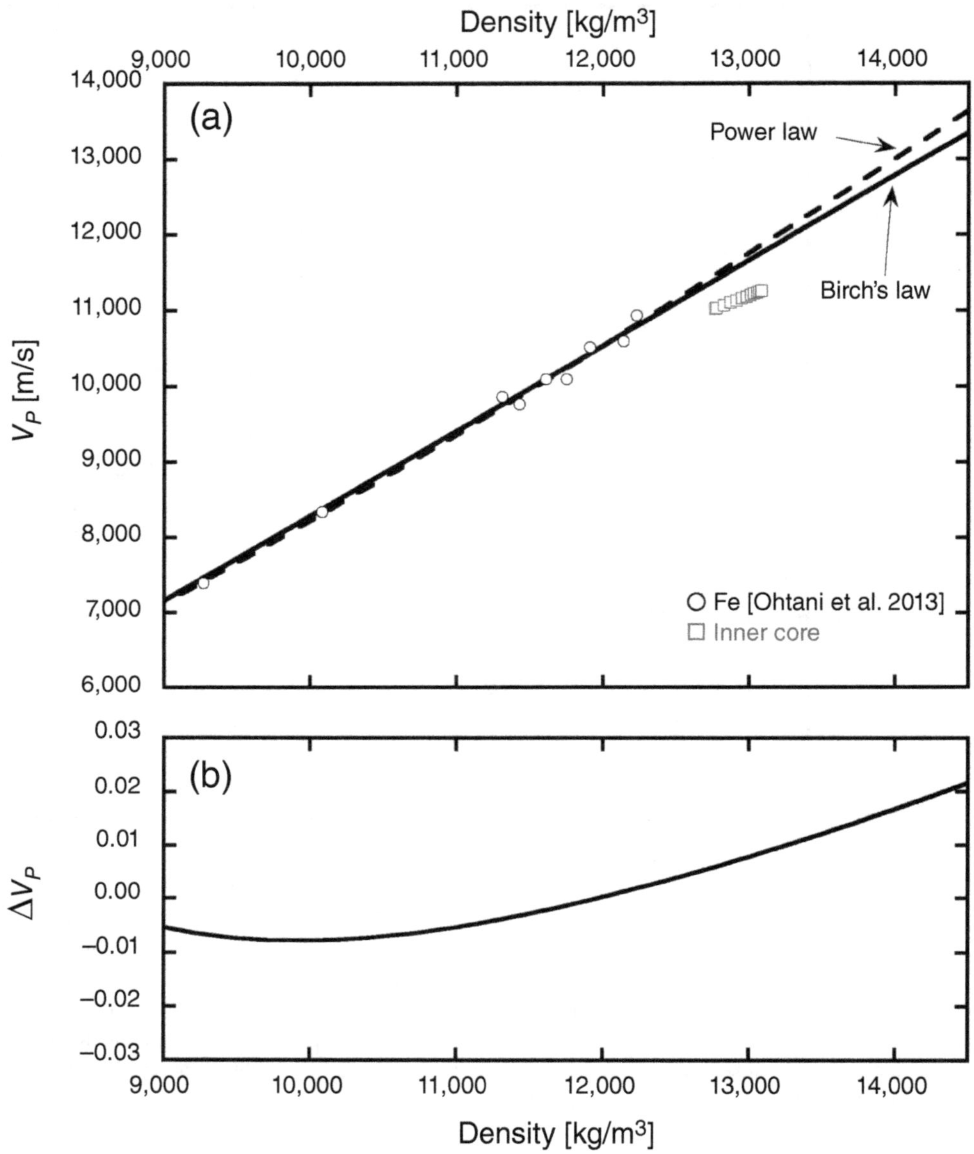

Figure 9.4 (a) Density–V_P data of Fe at high pressure and room temperature. Symbols are fitted by Birch's law (solid line) and power law (broken line), respectively. (b) V_P difference between Birch's law and power law. ΔV_P represents $(V_{P\ Power\ law} - V_{P\ Birch\ law})/V_{P\ Birch\ law}$. The difference is at most 2%.

where V_P is in m/s, ρ is in kg/m^3, $M = a_0 + a_1 T_0$, $B = b_0 + b_1 T_0$, $A = a_1$, $\rho^* = -\frac{b_1}{a_1}$, and T_0 = 300. The data at 300, 400, 700, and 1,000 K from Ohtani et al. (2013), at 2,300 and 3,000 K of Sakamaki et al. (2016), and at around 2,000 and 3,000 K of Brown and McQueen (1986) were fitted using (9.14). We obtained a_0 of 1.071(±0.029), a_1 of

5.56(±0.09)×10^{-5}, b_0 of −2250(±320), and b_1 of −0.872(±0.018). Using these values, we calculated M of 1.088(±0.029), B of −2,510(±320), A of 5.56(±0.09)×10^{-5}, and ρ^* of 15,690(±410), which represents the density at 510 GPa at room temperature based on the EoS of Fe (Dewaele et al., 2006). The fitting results are shown in Figure 9.5a,b, and they reproduce the experimental data well. Figure 9.5b shows the relationships between V_P and pressure. The fitted parameters here are slightly different from those of Sakamaki et al. (2016), because data at 400, 700, and 1,000 K from Ohtani et al. (2013) are also included in the fit. These velocities' data are slightly faster than those at room temperature, making the slope of Birch's law smaller.

The V_S of hcp Fe can be obtained using its EoS, thermal expansion, and Grüneisen parameter. The EoS of hcp Fe and the Grüneisen parameter reported by Dewaele et al.

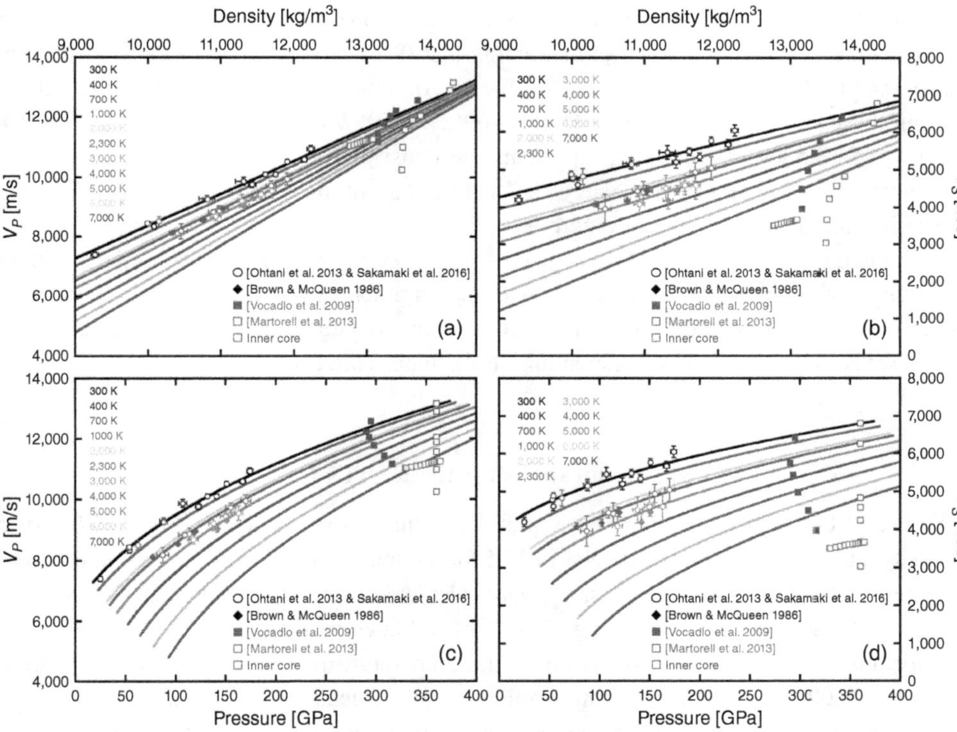

Figure 9.5 Temperature effect on Birch's law. Relationships between V_P and (a) density, and (b) pressure. Relationships between V_S and (c) density and (d) pressure. Temperature conditions are expressed using different colors as shown in the figures. Green solid and green open squares represent calculated V_P and V_S reported by Vočadlo et al. (2009) and Martorell et al. (2013), respectively. Red diamonds are V_P from Brown and McQueen (1986) between 1,700–2,600 K, and orange diamonds are V_P from Brown and McQueen (1986) between 3,100–3,400 K. The shock temperatures are recalculated as shown in Table 9.1. Black, blue, purple, and pink circles are V_P at temperatures from 300–1,000 K by Ohtani et al. (2013), and red and orange circles are V_P at 2,300–3,000 K by Sakamaki et al. (2016). For the color version, refer to the plate section.

(2006) are used, and the thermal expansion reported by Yamazaki et al. (2012) is used to calculate V_S utilizing the following relations:

$$K_T = -V\left(\frac{\partial P}{\partial V}\right)_T, \tag{9.16}$$

$$K_S = (1 + \alpha\gamma T)K_T, \tag{9.17}$$

$$V_S = \frac{3}{4}\sqrt{\left({V_P}^2 - \frac{K_S}{\rho}\right)}, \tag{9.18}$$

where K_T and K_S represent the isothermal and adiabatic bulk moduli, respectively. The effect of pressure and temperature on thermal expansion (α) and the Grüneisen parameter (γ) were taken from Yamazaki et al. (2012) and Dewaele et al. (2006), respectively. The calculated V_S values are shown in Figures 9.5c and 9.5d and fitted to (9.14) to reveal the temperature dependencies on V_S by assuming that (9.14) is applicable to V_S. We obtained a_0 of 0.454(±0.003), a_1 of 4.81(±0.16)×10^{-5}, b_0 of 330(±35), and b_1 of −0.891(±0.018). Using these values, we calculated M of 0.469(±0.003), B of 64(±35), A of 4.81(±0.16), and ρ^* of 18,530(±710), which represents the density at 930 GPa at room temperature based on the EoS of Fe (Dewaele et al., 2006). The fitting results of V_S reproduce the experimental data, as shown in Figure 9.5c–d.

The extrapolated V_P and V_S of hcp Fe to the pressure and density conditions corresponding to the inner core do not agree with measurements of the inner core. The discrepancy of V_S between hcp Fe and the inner core is notably large, ~ 37% at 4,000 K, and 33% at 5,000 K, at the pressure corresponding to the inner-core boundary.

9.5 Implications for the Earth's Core

We use (9.14) to extrapolate the V_P of pure Fe to inner-core conditions. Figure 9.6 shows density and V_P comparisons between PREM for the inner core and pure hcp Fe at different isothermal temperatures. The densities were calculated using the EoS of hcp Fe (Dewaele et al., 2006) and V_P calculated using (9.14). Dashed lines are densities and V_P calculated at temperatures corresponding to the adiabatic temperature profiles (T_{Ad}), with T_{ICB} from 4,000 to 6,000 K. The temperature profiles can be calculated using $T_{IC} = T_{ICB}\left(\frac{\rho_{IC}}{\rho_{ICB}}\right)^{\gamma}$, where T_{IC} and T_{ICB} represent the temperature of the inner core at a certain depth and the inner-core boundary (ICB), and ρ_{IC} and ρ_{ICB} are the density of the inner core at a certain depth and the ICB, respectively. The Grüneisen parameter (γ) is assumed to be 1.5 (e.g., Brown and Shankland, 1981). As shown in Figure 9.6a, the densities of hcp Fe are larger than those of the inner core. The density deficit of the inner core ($\Delta\rho = \frac{-(\rho_{IC}-\rho_{Fe})}{\rho_{Fe}}$), which is considered as the density at the adiabatic temperature profiles, varies from 6.5% (T_{Ad} with T_{ICB} of 4,000 K) to 4.5% (T_{Ad} with T_{ICB} of 6,000 K). The difference of V_P ($\Delta V = \frac{-(V_{IC}-V_{Fe})}{V_{Fe}}$) varies from 8.4% ($T_{Ad}$ with T_{ICB} of 4,000 K) to 3.4% (T_{Ad} with T_{ICB} of 6,000 K). While the pressure dependencies of $\Delta\rho$ are negligible, those of ΔV are not

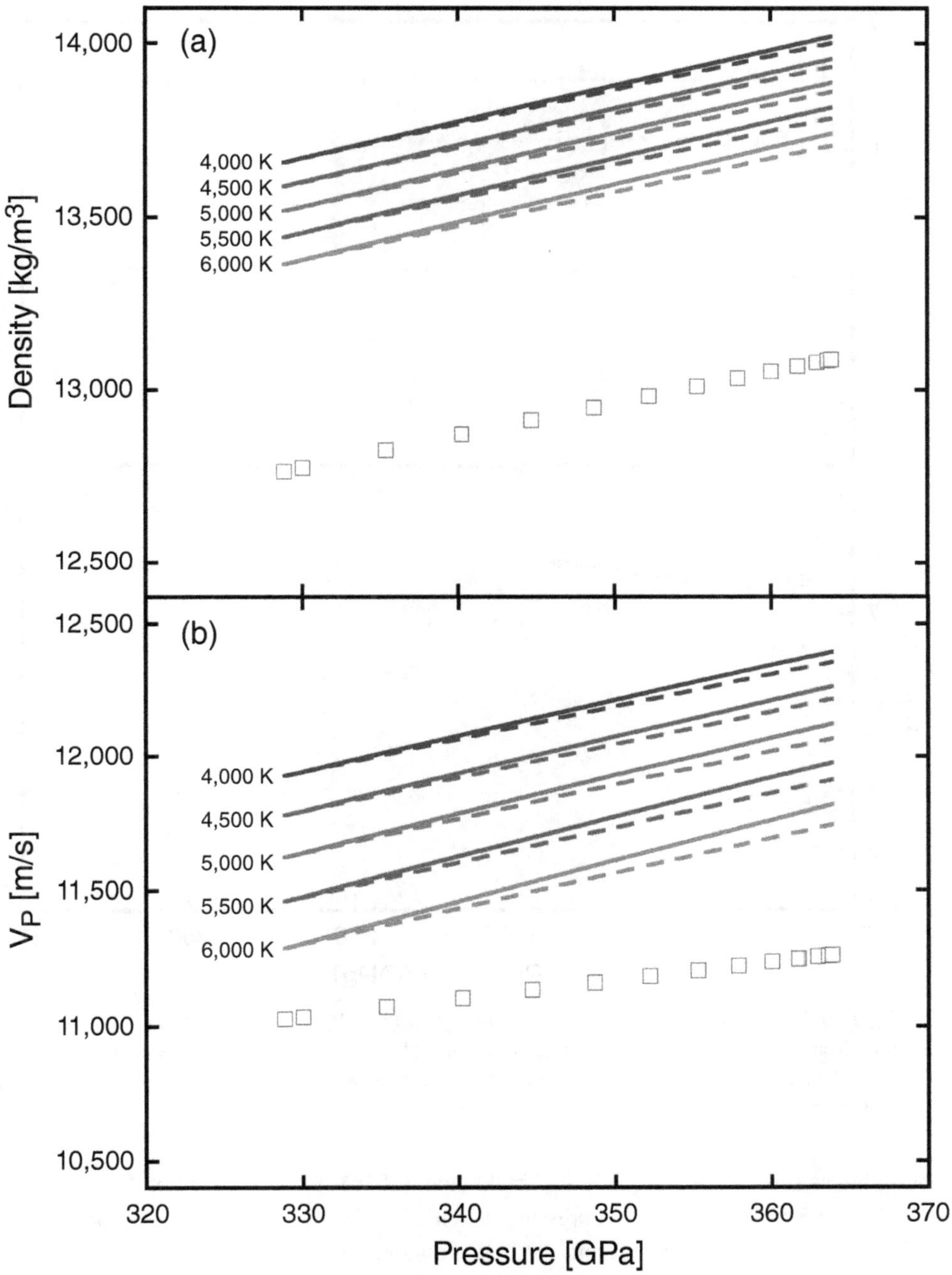

Figure 9.6 Comparison between (a) densities and (b) V_P of pure iron and the PREM inner core. Solid lines represent densities and V_P at isothermal conditions from 4,000–6,000 K. Dashed lines are densities and V_P calculated along the adiabatic temperature profiles, with T_{ICB} = 4,000, 4,500, 5,000, 5,500, and 6,000 K. Red squares represent densities and V_P of the PREM inner core. For the color version, refer to the plate section.

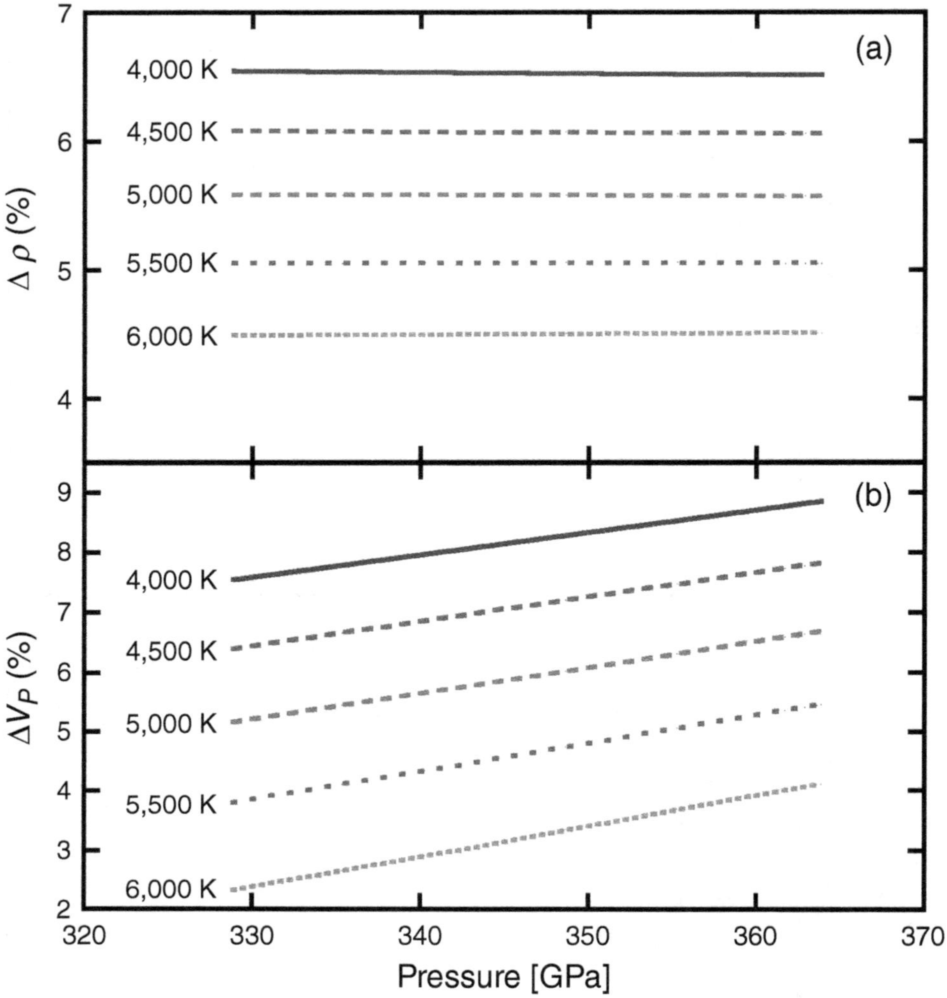

Figure 9.7 Density deficit of the inner core ($\Delta\rho$) (a) and V_P difference (ΔV_P) (b) between pure iron and the PREM inner core, as a function of pressure. Density and V_P of pure iron are calculated along the adiabatic temperature profiles with T_{ICB} = 4,000, 4,500, 5,000, 5,500, and 6,000 K.

negligible as shown in Figure 9.7. Sakamaki et al. (2016) suggested a $\Delta\rho$ of 4–5% and ΔV of 4–10% at 5,500 K and the ICB pressure. Birch's law for hcp Fe reported here shows a $\Delta\rho$ of 5% and ΔV of 4% at 5,500 K and the ICB pressure. Since the present temperature effect on the compressional velocity is larger than that of Sakamaki et al. (2016), the estimated ΔV is smaller than that of Sakamaki et al. (2016). This suggests that a smaller amount of light elements may explain the V_P and density of the inner core than previously reported on the basis of V_P and density (e.g., Badro et al., 2007). Antonangeli et al. (2018) measured V_P of hcp Fe and hcp Fe-9wt.%Si alloy, and discussed the temperature effect on elastic wave velocities based on Martorell et al. (2013) or Sakamaki et al. (2016). Since the anharmonic effect due to the atomic vibrations becomes significant approaching the melting temperature (Martorell et al., 2013), the temperature effects on V_P and V_S are large. Antonangeli et al. (2018) estimated elasticity of hcp Fe up to the melting temperature.

Light elements in the inner core are needed to lower both V_P and density of pure hcp Fe because the V_P and density of pure Fe at inner-core conditions are larger than those of the inner core, as shown in Figure 9.6. Most Fe light element alloys have a faster V_P than those of the inner core as shown in Figure 9.2b. Only FeH_X has a slower V_P at room temperature. Therefore, hydrogen is the only candidate light element that would lower the V_P to match that of the inner core if the temperature effect is neglected. However, the temperature effect on the V_P of Fe and Fe–Si alloys shown in Figures 9.3 and 9.6 is clearly measurable. Therefore, the temperature effect on V_P must be considered for other Fe light elements. Matching of V_S is also problematic because there is no direct determination. Therefore, it is difficult to discuss the V_S of the inner core. The reported V_S values were calculated using the EoS of pure hcp Fe and Fe light element alloys, which introduces a large uncertainty in the calculated V_S.

Ideal mixing between Fe and Fe light element alloys is often assumed in order to discuss the inner core. The density of the inner core (ρ_{IC}) and elastic wave velocity (V_{IC}) can be expressed as follows:

$$\rho_{IC} = x\rho_{Fe} + (1-x)\rho_{FeX}, \tag{9.19}$$

$$V_{IC} = \frac{V_{Fe}V_{FeX}}{(1-x)V_{Fe} + xV_{FeX}}, \tag{9.20}$$

where x, ρ_{Fe}, and ρ_{FeX} represent the volume ratio of Fe in the inner core and densities of Fe and Fe light element alloys, respectively. V_{IC}, V_{Fe}, and V_{FeX} represent elastic wave velocities (i.e., V_P or V_S) of the inner core, Fe, and Fe light element alloys, respectively. Sakairi et al. (2018) reported that 3–6 wt.% Si can explain the V_P and density of the inner core. By using the present temperature effect on Fe, the amount of Si in the inner core can be estimated to be in the range of 4.2 wt.% at 5,000 K to 4.7 wt.% at 6,000 K, based on a combination of matching the density and V_P to the PREM profiles in the inner core.

Ni is also an important component in the inner core. The reported V_P of hcp Fe–Ni alloys are slower than that of hcp Fe (Lin et al., 2003b; Decremps et al., 2014; Wakamatsu et al., 2018; Morrison et al., 2019). Wakamatsu et al. (2018) reported V_P of Fe, Fe-5wt.%Ni, and Fe-15wt.%Ni. The Ni effect on V_P decreased by about 2% for Fe-5wt.% Ni and 4% for Fe-15 wt.%Ni at 30 GPa compared to the V_P of pure hcp Fe. Morrison et al. (2019) compared V_P of hcp Fe and hcp $Fe_{91}Ni_9$. The drop of V_P was about 2% at around 75 GPa because of the existence of Ni in hcp Fe. On the other hand, Antonangeli et al. (2018) suggested that up to 5 wt.% Ni does not affect significantly the V_P of hcp Fe based on study of Fe–Ni–Si and Fe–Si alloys. Liu et al. (2016) reported elastic wave velocities (V_P, V_S, and V_D) of hcp Fe alloyed with Ni and Si. The combination of Ni and Si slightly increases V_P but significantly decreases V_S. These studies highlight the importance of V_P and V_S measurements in ternary systems, such as Fe–Ni–Si. The V_S of the inner core is much slower than that of hcp Fe, even at high temperatures. In order to solve this discrepancy, the light element should significantly lower the V_S of hcp Fe. Moreover, the premelting effect on elastic properties (Martorell et al., 2013) might be needed to be investigated experimentally.

The temperature effect on the V_P of Fe_3C is probably negligible up to 2,000 K. Carbon may not be a good candidate for the light elements in the inner core because V_P of both hcp Fe and Fe_3C are faster than those of the inner core at the pressures corresponding to the inner core. As for FeH_X, there is no report on its temperature effect. However, V_P at room temperature extrapolated to the inner core pressures are slower than those of the inner core. H can therefore be a candidate light element in the inner core. If temperature would further decrease the V_P of FeH_X, the amount of H in the inner core would be more than that estimated using V_P at room temperature.

In the case of S, the V_P of Fe_3S at room temperature is faster than that of the inner core, hence incorporating S alone in the core would not explain the observed V_P unless high temperature could reduce the V_P of Fe_3S. However, there are no data on V_P and V_S of Fe sulfides at high temperatures. As for O, the temperature effect on the V_P of $Fe_{0.92}O$ was observed (Tanaka et al., 2020). The V_P of FeO at the pressure corresponding to the inner core boundary can be calculated to be 13,630 m/s at 5,000 K, which is too fast to explain the V_P of the inner core by ideal mixing between Fe and FeO. In addition, it has been reported that FeO transforms from B1 to B2 under core conditions (Ozawa et al., 2011). The elastic properties of B2 FeO need to be investigated to further understand the possibility of O as a light element in the inner core, although Alfè et al. (2002) estimated the O content in the inner core to be 0.2 at.%, suggesting O is not a major light element in the inner core.

Based on the elastic properties of core materials alone, Si and H are viable candidates for the light elements in the inner core, while C and O are unlikely the light element candidates in the inner core. S is not a good candidate for the light elements in the inner core according to the V_P data at room temperature. To assess the possibility of S as a light element in the inner core, V_P measurements at high temperature and pressure are required.

The V_S of the inner core is much slower than that of hcp Fe and Fe light element alloys. The temperature effect on V_S needs to be greater than that of V_P in order to explain the V_S of the inner core. The lack of experimental measurements of V_S makes the assessment more difficult. On the other hand, theoretical calculations may provide some insights into this difficult issue. Vočadlo et al. (2009) reported 11% and 38% reductions in V_P and V_S of pure Fe at $T/T_M \sim 0.902$, respectively, compared to those at 0 K. Martorell et al. (2013) reported 23% and 53% reduction of V_P and V_S at $T/T_M \sim 0.999$, respectively, compared to those at 0 K. However, Martorell et al. (2016) did not predict a strong drop in the elastic velocities of Fe–Si alloys. According to their results, Si as a sole light element in the inner core was able to explain the observed inner-core density and V_P, but not V_S. Expanding the calculations to ternary systems, Li et al. (2018) showed that hcp-$Fe_{30}Si_1C_1$ alloy can explain the density and seismic wave velocity without considering anelastic or premelting phenomena.

9.6 Concluding Remarks

In summary, the V_P of pure hcp-Fe at 330–360 GPa is faster than the observed V_P in the inner core. Therefore, light elements in the inner core need to play a role in lowering the sound velocity in addition to lowering the density. The V_P of Fe–Si, Fe–H, and Fe–C alloys are below V_P of pure Fe compared at the inner-core pressure and room temperature, but only the extrapolated V_P value of FeH_X is below that of the inner core. If the temperature effect on Birch's law is taken into account, Si and H can be considered as the major light elements candidates for the inner core, while C, O, and S are not likely present as major light elements. To explain the V_S of the inner core, the premelting effect on the elastic wave velocities of the core materials must be further investigated. Due to the experimental difficulty, experimental measurements of V_S of core materials are limited, especially at simultaneous high temperature and pressure. In addition to expanding experimental pressure and temperature range, we need to focus on the high-precision velocity measurements of hcp structure of iron alloys instead of the endmember compositions. Because multiple light elements in the core have been estimated on the basis of geochemical and cosmochemical arguments (e.g., McDonough, 2017), measurements in multiple light elements system would provide a direct test of the inner-core composition models.

Acknowledgments

The authors thank Dr. Naohisa Hirao, Dr. Hiroshi Fukui, and Dr. Alfred Q. R. Baron for experimental help at SPring-8. S. K. thanks Dr. Fumiya Maeda for experimental help and discussion. S. K. was supported by JSPS KAKENHI Grant Numbers JP25800291, JP15H05831, and JP16K13902. E. O. was supported by JSPS KAKENHI Grant Numbers JP22000002, JP15H05748, and JP20H00187. T. S. was supported by JSPS KAKENHI Grant Numbers JP25800292 and JP16H01112. The synchrotron X-ray radiation experiments were performed at SPring-8.

References

Alfè, D., Gillan, M. J., Price, G. D. (2002). Composition and temperature of the Earth's core constrained by combining ab initio calculations and seismic data. *Earth and Planetary Science Letters*, **195**, 91–8.

Antonangeli, D., Komabayashi, T., Occelli, F. (2012). Simultaneous sound velocity and density measurements of hcp iron up to 93 GPa and 1100 K: an experimental test of the Birch's law at high temperature. *Earth and Planetary Science Letters*, **331–332**, 210–214.

Antonangeli, D., Krisch, M., Fiquet, G., et al. (2004b). Elasticity of cobalt at high pressure studied by inelastic X-ray scattering. *Physical Review Letters*, **93**(21), 215505.

Antonangeli, D., Merkel, S., Farber, D. L. (2006). Elastic anisotropy in hcp metals at high pressure and the sound wave anisotropy of the Earth's inner core. *Geophysical Research Letters*, **33**, L24303.

Antonangeli, D., Morard, G., Paolasini, L. et al. (2018). Sound velocities and density measurements of solid hcp–Fe and hcp–Fe–Si (9 wt.%) alloy at high pressure:

constraints on the Si abundance in the Earth's inner core. *Earth and Planetary Science Letters*, **482**, 446–453.

Antonangeli, D., Occelli, F., Requardt, H. (2004a). Elastic anisotropy in textured hcp–iron to 112 GPa from sound wave propagation measurements. *Earth and Planetary Science Letters*, **225**, 243–251.

Antonangeli, D., Ohtani, E. (2015). Sound velocity of hcp–Fe at high pressure: experimental constrants, extrapolations and comparison with seismic models. *Progress in Earth and Planetary Science*, **2**, 3.

Antonangeli, D., Siebert, J., Aracne, C. M. (2011). Spin crossover in ferropericlase at high pressure: a seismologically transparent transition? *Science*, **331**, 64–67.

Antonangeli, D., Siebert, J., Badro, J. (2010). Composition of the Earth's inner core from high-pressure sound velocity measurements in Fe–Ni–Si alloys. *Earth and Planetary Science Letters*, **295**, 292–296.

Anzellini, S., Dewaele, A., Mezouar, M., et al. (2013). Melting of iron at Earth's inner core boundary based on fast X-ray diffraction. *Science*, **340**, 464–466.

Badro, J., Fiquet, G., Guyot, F., et al. (2003). Iron partitioning in Earth's mantle: toward a deep lower mantle discontinuity. *Science*, **300**, 789–791.

Badro, J., Fiquet, G., Guyot, F. et al. (2007). Effect of light elements on the sound velocities in solid iron: implications for the composition of Earth's core. *Earth and Planetary Science Letters*, **254**, 233–238.

Badro, J., Rueff, J.-P., Vankó, G., et al. (2004). Electronic transitions in perovskite: possible nonconvecting layers in the lower mantle. *Science*, **305**, 383–386.

Barker, L. M., Hollenbach, R. E. (1972). Laser interferometer for measuring velocities of any reflecting of surface. *Journal of Applied Physics*, **43**, 4669–4675.

Birch, F. (1952). Elasticity and constitution of the Earth's interior. *Journal of Geophysical Research*, **57**, 227–286.

Birch, F. (1961). Composition of the Earth's mantle. *Geophysical Journal of the Royal Astronomical Society*, **4**(S0), 295–311.

Brown, M. J., Shankland, T. J. (1981). Thermodynamic parameters in the Earth as determined from seismic profiles. *Geophysical Journal of the Royal Astronomical Society*, **66**, 579–596.

Brown, M. J., McQueen, R. G. (1986). Phase transitions, Grüneisen parameter, and elasticity for shocked iron between 77 GPa and 440 GPa. *Journal of Geophysical Research*, **91**(B7), 7485–7494.

Chen, B., Li, Z., Zhang, D., et al. (2014). Hidden carbon in Earth's inner core revealed by shear softening in dense Fe_7C_3. *Proceedings of the National Academy of Sciences of the United States of America*, **111**(50), 17755–17758.

Chigarev, N., Zinin, P., Ming, L.-C., et al. (2008). Laser generation and detection of longitudinal and shear acoustic waves in a diamond anvil cell. *Applied Physics Letters*, **93**, 181905.

Cook, R. K. (1957). Variation of elastic constants and static strains with hydrostatic pressure: a method for calculation from ultrasonic measurements. *Journal of the Acoustical Society of America*, **29**, 445–449.

Decremps, F., Antonangeli, A., Gauthier, M., et al. (2014). Sound velocity of iron up to 152 GPa by picosecond acoustics in diamond anvil cell. *Geophysical Research Letters*, **41**, 1459–1464.

Decremps, F., Gauthier, M., Ayrinhac, S., et al. (2015). Picosecond acoustics method for measuring the thermodynamical properties of solids and liquids at high pressure and high temperature. *Ultrasonics*, **56**, 129–140.

Figure 1.0 Dave Mao at the Geophysical Laboratory, Washington DC, 2018 (photos courtesy of the Carnegie Institution for Science).

Figure 1.3 Dave Mao at HPCAT, the Advanced Photon Source.

Figure 1.4 Attendees at the symposium to honor Ho Kwang "Dave" Mao and 50 years of high-pressure science at the Geophysical Laboratory, held in October 2018, at the Broad Branch Road Campus of the Carnegie Institution for Science
(photo courtesy of the Carnegie Institution for Science).

Figure 2.3 "Kawai-type apparatus" using a 5,000 ton press and E. Ito at the Institute of Thermal Springs, Okayama University, Misasa.

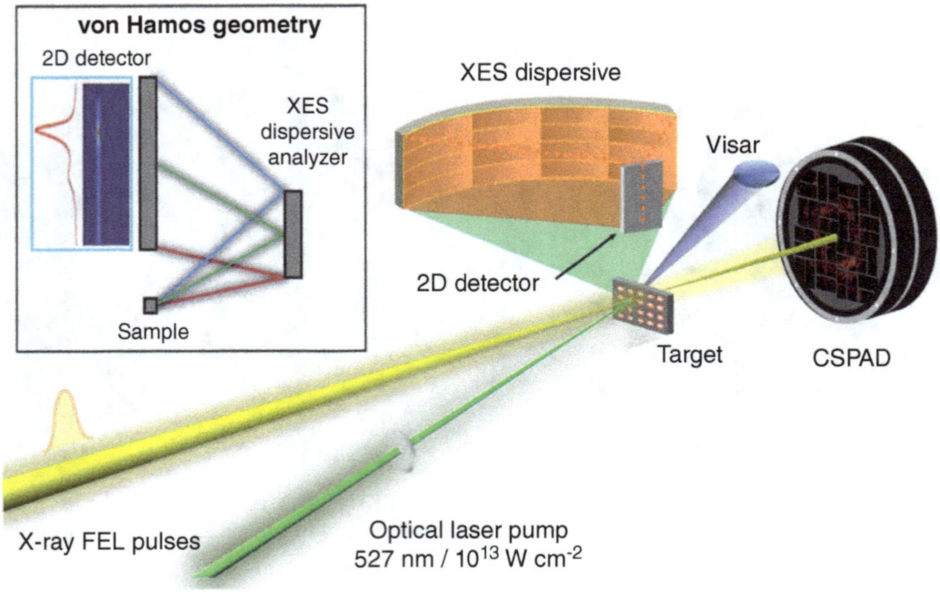

Figure 3.5 Schematic of an ultrafast XES setup at MEC, LCLS. An optical laser pump generates a shock wave in a target, and an X-ray FEL pulse probes the sample at extreme conditions using an XES spectrometer combined with XRD detectors (CSPAD).

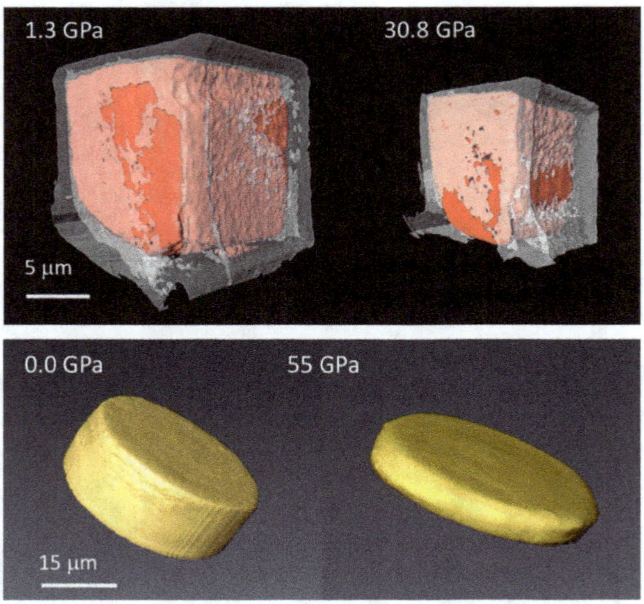

Figure 3.6 Top panel: representative 3D renderings from a high-pressure nano- transmission X-ray microscopy (TxM) experiment on a Pt-coated SiO_2 cube. The Pt coating is shown in gray and the "hollow" volume representing the SiO_2 cube is shown in red. The volume has decreased by a factor of two from 1.3 to 30.8 GPa. Bottom panel: representative 3D renderings from high-pressure nano-TxM experiment on an uncoated SiO_2 cylinder.
Adapted from Mao et al. (2019).

Figure 4.1 In 2010, EFree Partners focused on SNAP are attending a meeting at the SNS hosted by Dean Myles (ORNL). The meeting was attended by (in order from left to) Yang Ding (EFree/APS), Jamie Molaison (ORNL), Dave Mao (EFree Director), Maria Baldini (EFree), Jeff Yarger (ASU), Maddury Somayazulu (EFree), Malcolm Guthrie (EFree Chief Scientist), John Parise (Stonybrook, SNAP PI), Chris Tulk (ORNL), Reini Boehler (EFree Chief Scientist), Xiaojia Chen (EFree), Neelam Pradhan (ORNL), and Antonio dos Santos (ORNL). Note that affiliations given are those held in 2010.

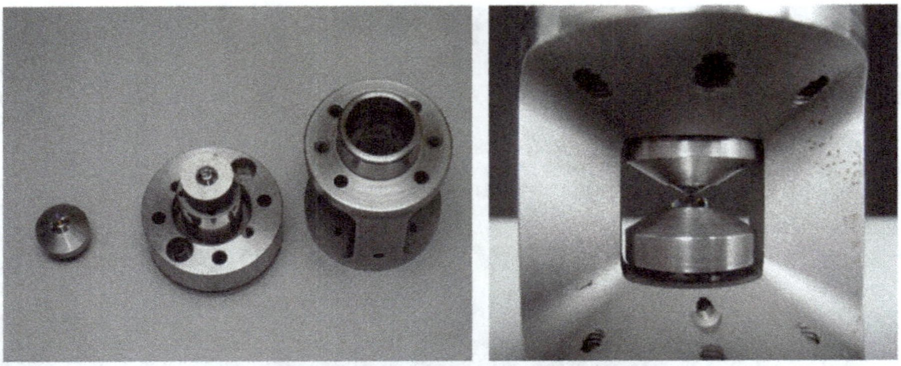

Photo credit: Carlos Jones

Photo credit: Mark Loguillo

Figure 4.2 Pictures of the various generations of ORNL DACs. Top: the panoramic DAC shown disassembled as well as with a close-up view of anvils in steel seats. Middle: the new SNAP DAC with CVD anvils mounted into seats. Bottom: an assembly of two clamped CuBe DAC with Versimax^{TM} anvils and gasket components.

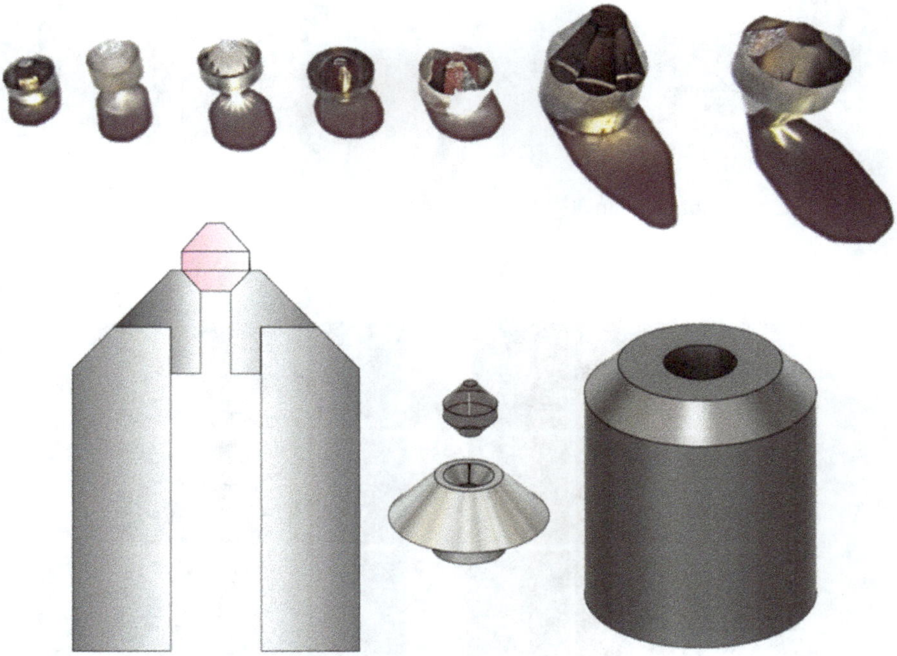

Figure 4.3 Top: a photo of various synthetic diamonds grown by chemical vapor deposition lined up together. For reference of scale, the anvil on the right is a 10 mm diameter CVD anvil. Bottom: schematic images of precision-machined pistons, steel seats, and conical CVD anvils used on the current generation of SNAP DACs with a typical anvil diameter of 6 mm.

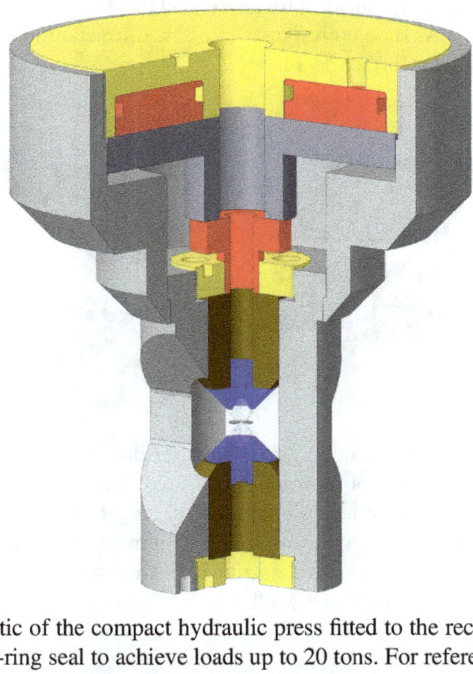

Figure 4.4 Schematic of the compact hydraulic press fitted to the recent neutron DACs. It employs a double O-ring seal to achieve loads up to 20 tons. For reference of size, the outer diameter of the top can, i.e., the largest horizontal extent of the press, is 115 mm.

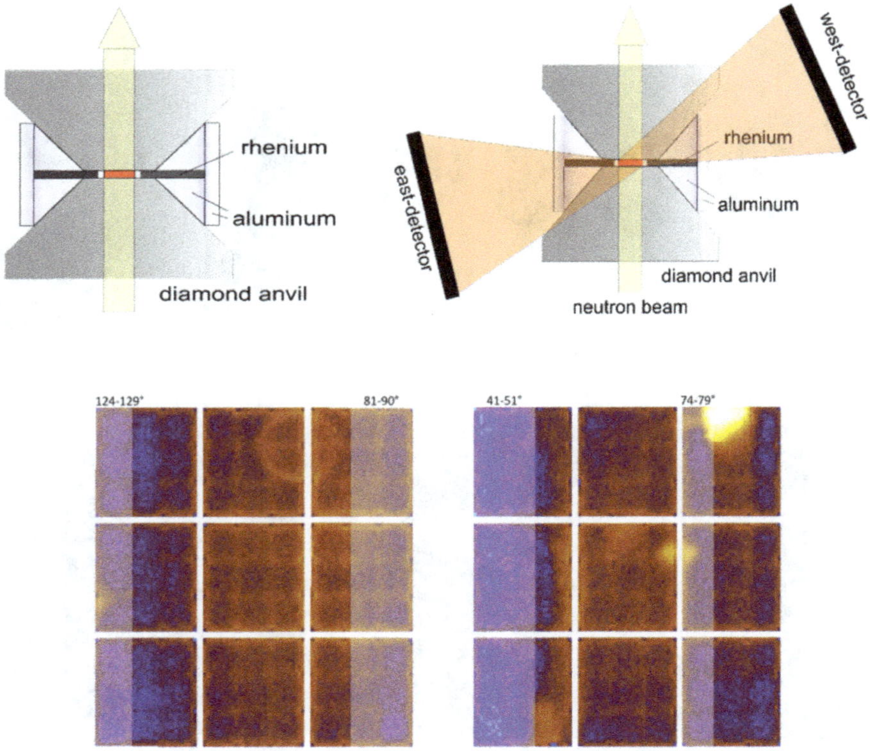

Figure 4.5 Top-left: close-up of the Re gasket, Al wedges, and ring assembled around the diamond anvils. Top-right: schematics of the diffraction geometry: incoming neutron beam (yellow), illuminated sample (red), and scattered beam (orange) as captured by the typical detector setting on SNAP. Bottom: 2D detector intensity map of the two SNAP detector banks West (left) and East (right) as seen in Mantid. Each detector is made up of 3×3 modules, and the black lines correspond to their boundaries. The data shown are obtained from ice at ~45 GPa contained inside a Re gasket and the diffraction signal is constricted to a d-range of 1.98–2.10 Å. Blue corresponds to a low diffraction intensity and yellow to a high intensity. See text for further detail.

Figure 5.5 (a) Experimental configuration for laser-driven shock experiments at MEC showing the set of detectors positioned about direct beam and the translatable target cassette. Two arms of the drive laser impinge on the target at 20°to the target normal. (b) Experimental configuration at SACLA's BL3-EH5 showing the detector positioned at an angle to intercept diffracted X-rays in a grazing incidence refection geometry. Insets show schematics of typical target packages with a polycarbonate (CH) ablator, a sample, and LiF window.

After Schoelmerich et al. (2020).

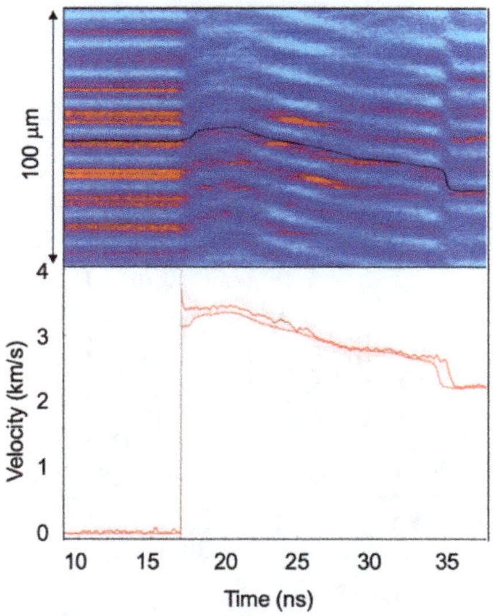

Figure 5.6 Line VISAR data showing the 2D interferogram from the streak camera with the wave profiles below. The two traces are from independent VISARs with different sensitivities.

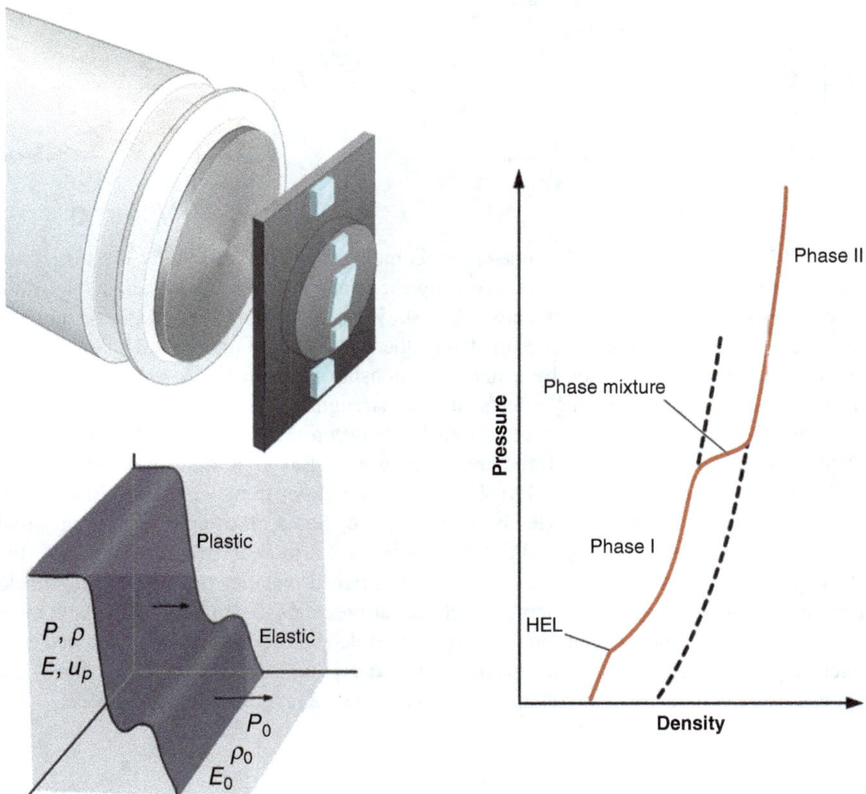

Figure 6.1 Upper left: Schematic illustration of traditional plate-impact shock wave experiment. The projectile carrying the impactor is shown on the left. The target assembly on the right schematically illustrates one common measurement approach used for silicates in which transit times and velocities are determined using flat and inclined mirrors. For details, see Ahrens (1987). Lower left: Illustration of a two-wave structure of a shock wave. Right: Schematic illustration of the major regions of a Hugoniot curve for a material undergoing a phase transformation. HEL = Hugoniot elastic limit.

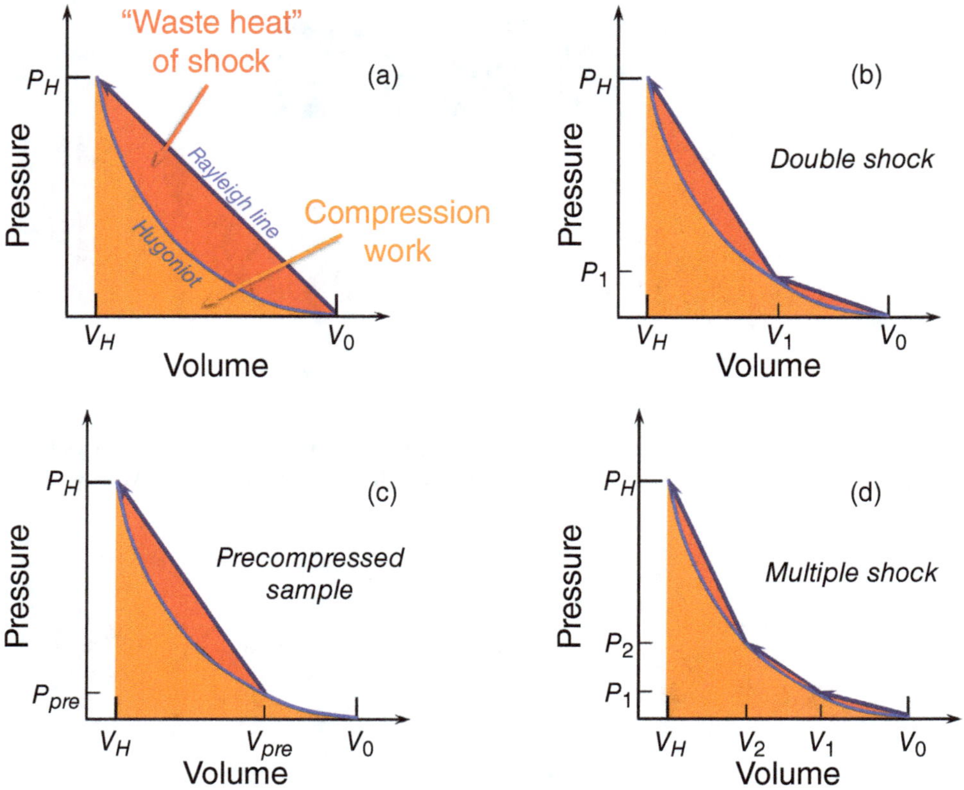

Figure 7.1 (a) Waste heat in single-shock compression (area in red) is the difference between the Hugoniot energy (7.4), given by the triangle under the Rayleigh line (dark blue arrow) connecting zero-pressure ($P_0 = 0$, V_0) and final (P_H, V_H) states, and the reversible energy of compression estimated as the area (orange) under the Hugoniot (blue curve). The Hugoniot is given by a linear relationship between shock velocity (*Us*) and material velocity (*up*), ignoring effects of both strength and phase transformation. The waste-heat approximation ignores the difference between the isentrope and Hugoniot. (b) Double shock reduces waste heat (red areas) because the Rayleigh lines (dark blue arrows) to the first (P_1, V_1) and final (P_H, V_H) shock states are closer to the Hugoniot (blue curve) than the Rayleigh line for a single shock (cf. (a)). (c) Static precompression to an initial pressure-volume condition (P_{pre}, V_{pre}) also reduces waste heat (red area) because the Rayleigh line (dark blue arrow) is now shortened relative to a single shock. Precompression is along an isotherm, which lies at pressures slightly below the isentrope, but is approximated here by the Hugoniot. (d) Multiple-shock compression, illustrated by a three-step shock, also reduces the waste heat (red areas) relative to a single shock and approximates continuous ramp compression as the loading approaches an isentrope.

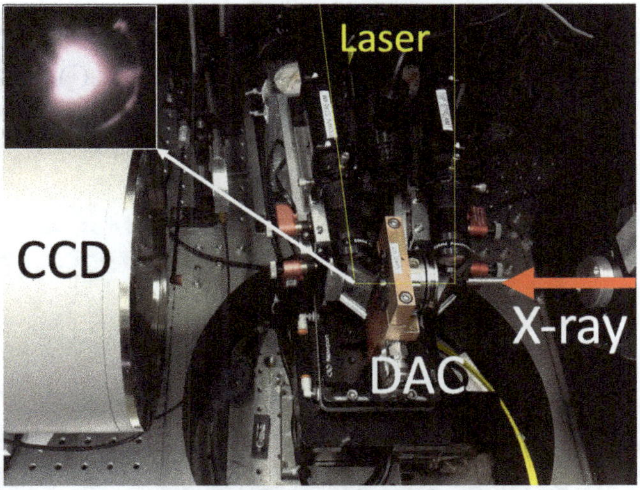

Figure 8.1 Experimental configuration for an in situ X-ray diffraction study in the laser-heated diamond anvil cell at a synchrotron facility (APS).

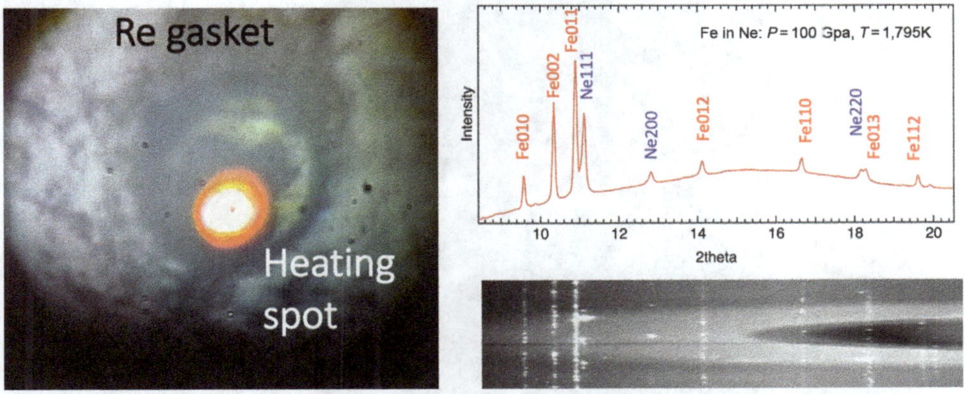

Figure 8.2 Representative angle-dispersive X-ray diffraction patterns of Fe and Ne at high temperature. Left: A heating spot of 20 μm on Fe in an Ne medium. Right: The integrated diffraction pattern with labeled hcp-Fe and Ne peaks (top) and the corresponding 2D cake pattern (bottom). The 2D image shows a spotty diffraction pattern for Ne.

Figure 8.3 Experimental configuration for in situ X-ray diffraction study in the large-volume multi-anvil press at synchrotron facility (SPring-8). A 1,500 ton press (top) is shown with X-ray path and detector, with the top view of the second stage (bottom-left) and radiograph image of the sample chambers at ambient pressure and high pressure (632 tons).

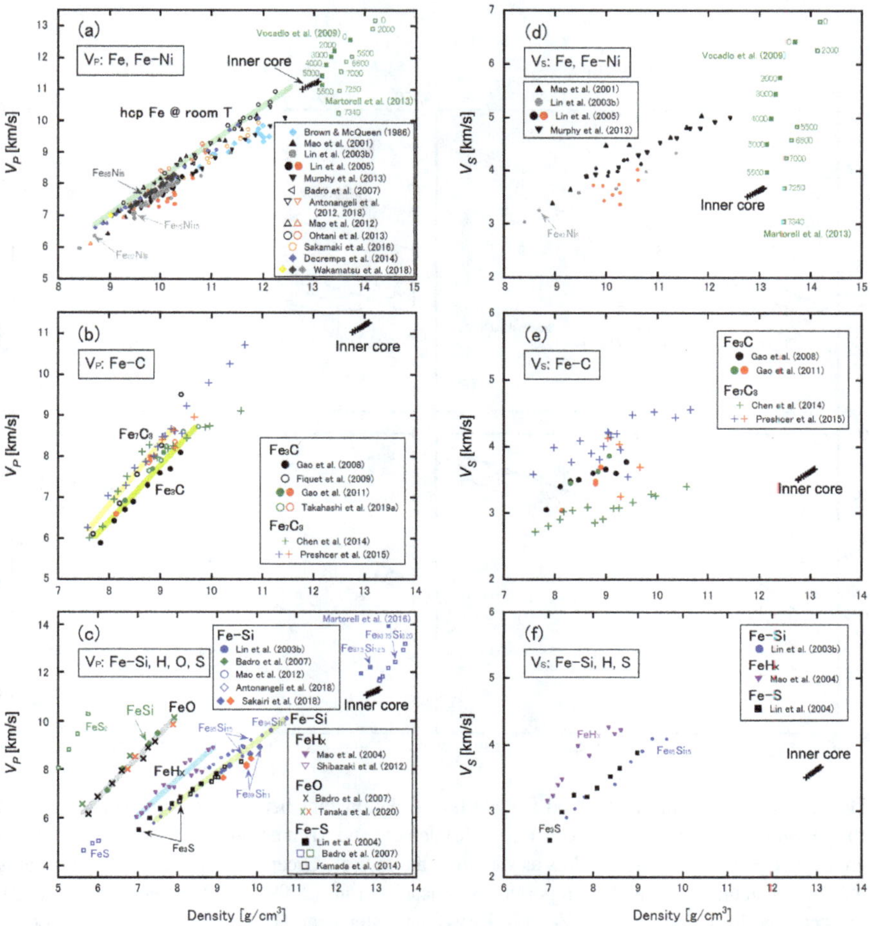

Figure 9.1 V_P and V_S as a function of density at both room and high temperature. Relationships between densities and V_P of (a) Fe and Fe–Ni alloy, (b) Fe–C alloys, and (c) Fe-Si, Fe–H, Fe–S alloys. Plus marks represent V_P of the inner core. Solid symbols represent V_P based on the NIS method and open symbols represent those based on the IXS method. Light blue diamonds represent V_P of hcp Fe based on shock wave experiments (Brown and McQueen, 1986; also see Table 9.1). Black, gray, and green symbols in Figure 9.1a represent V_P of pure hcp iron (Mao et al., 2001; Lin et al., 2003b, 2005; Badro et al., 2007; Antonangeli et al., 2012, 2018; Mao et al., 2012; Murphy et al., 2013; Ohtani et al., 2013; Sakamaki et al., 2016), V_P of $Fe_{92}Ni_8$ (Lin et al., 2003b), and calculated V_P of pure hcp iron (Vočadlo et al., 2009; Martorell et al., 2013), respectively. Open and solid symbols represent Martorell et al. (2013) at 360 GPa, and Vočadlo et al. (2009) at pressure conditions from 295–316 GPa, respectively. The numbers in green represent temperature conditions of their calculations. Red and orange symbols in Figure 9.1a represent V_P of pure iron at high temperature conducted by external-heated diamond anvil cell (Lin et al., 2005; Antonangeli et al., 2012; Mao et al., 2012; Ohtani et al., 2013), and those of pure iron at high temperature conducted by laser-heated diamond anvil cell (Sakamaki et al., 2016), respectively. Black, green, and blue symbols in Figure 9.1b are V_P at room temperature (Fe_3C: Gao et al., 2008; Fiquet et al., 2009; Gao et al., 2011; Takahashi et al., 2019a, Fe_7C_3: Chen et al., 2014; Prescher et al., 2015), and red symbols are V_P at high temperatures (Fe_3C: Gao et al., 2008; Fiquet et al., 2009; Gao et al., 2011; Takahashi et al., 2019a, Fe_7C_3: Chen et al., 2014; Prescher et al., 2015). Blue circles, green right-pointing triangles, and blue left-pointing triangles in Figure 9.1c are V_P of Fe–Si alloys at room temperature (FeSi: Badro et al., 2007, $Fe_{84}Si_{16}$: Antonangeli et al., 2018; $Fe_{85}Si_{15}$: Lin et al., 2003b; Mao et al., 2012, $Fe_{89}Si_{11}$: Sakairi et al., 2018) and red left-pointing triangles are V_P of $Fe_{89}Si_{11}$ at high temperature (Sakairi et al., 2018). V_P of $Fe_{93.75}Si_{6.25}$ (360 GPa and 0–7350 K) and $Fe_{87.5}Si_{12.5}$ (360 GPa and 0–6550 K) are from Martorell et al. (2016). Purple downward triangles in Figure 9.1c are V_P of FeH_X at room temperature (Mao et al., 2004; Shibazaki et al.,

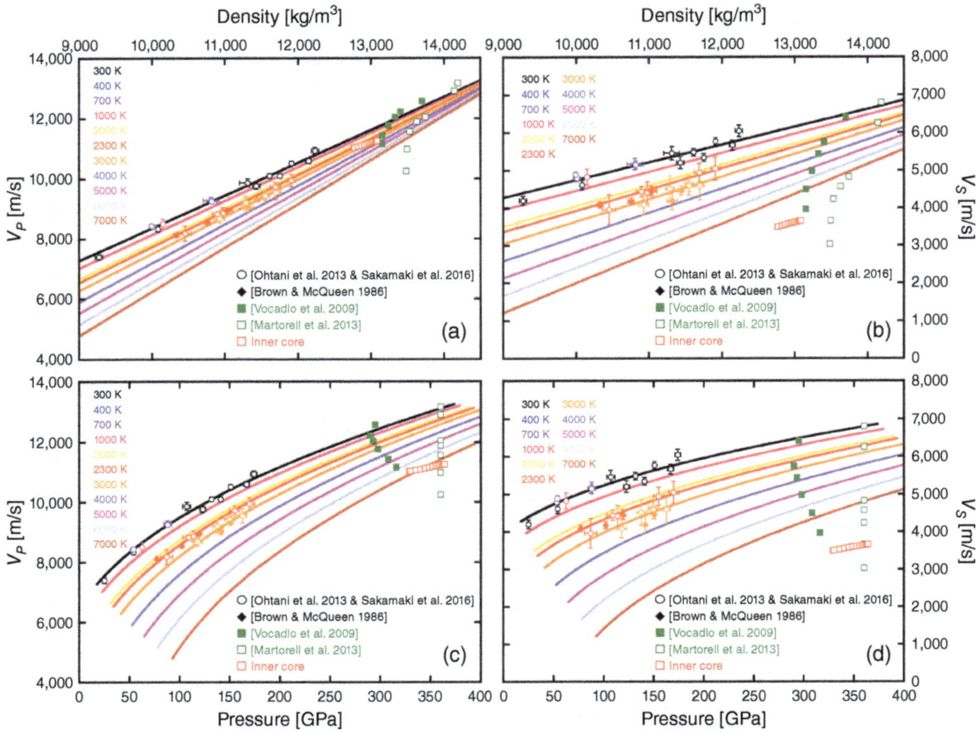

Figure 9.5 Temperature effect on Birch's law. Relationships between V_P and (a) density, and (b) pressure. Relationships between V_S and (c) density and (d) pressure. Temperature conditions are expressed using different colors as shown in the figures. Green solid and green open squares represent calculated V_P and V_S reported by Vočadlo et al. (2009) and Martorell et al. (2013), respectively. Red diamonds are V_P from Brown and McQueen (1986) between 1,700–2,600 K, and orange diamonds are V_P from Brown and McQueen (1986) between 3,100–3,400 K. The shock temperatures are recalculated as shown in Table 9.1. Black, blue, purple, and pink circles are V_P at temperatures from 300–1,000 K by Ohtani et al. (2013), and red and orange circles are V_P at 2,300–3,000 K by Sakamaki et al. (2016).

Figure 9.1 (*cont.*) 2012). Black and green outline plus marks in Figure 9.1c are V_P of FeO at room temperature (Badro et al., 2007; Tanaka et al., 2020), and red outline plus signs are V_P of FeO at high temperature (Tanaka et al., 2020). Black solid and open squares in Figure 9.1c are V_P of Fe_3S (Lin et al., 2004; Kamada et al., 2014a). Blue open squares represent V_P of FeS (Badro et al., 2007), and green open squares represent V_P of FeS_2 (Badro et al., 2007). The relationships between density and V_S of alloys are shown for (d) Fe and Fe–Ni alloys, (e) Fe–C alloys, and (f) Fe–Si, Fe–H, and Fe–S alloys. Plus signs represent V_S of the inner core. Green open and solid squares represent V_S of pure hcp Fe from Martorell et al. (2013) and Vočadlo et al. (2009), respectively. The calculated conditions are the same as those for V_P. Black upward and downward triangles in Figure 9.1d represent V_S at room temperature of Mao et al. (2001) and Murphy et al. (2013), respectively. Black and red circles are V_S of pure iron at room temperature and high temperature (Lin et al., 2005), respectively. Gray symbols represent those of $Fe_{92}Ni_8$ (Lin et al., 2003b). Black and green triangles in Figure 9.1e represent V_S at room temperature of Fe_3C of Gao et al. (2008) and Gao et al. (2011), respectively. Red triangles are V_S at high temperature of Fe_3C (Gao et al., 2011). Green and blue triangles in Figure 9.1e are V_S of Fe_7C_3 at room temperature from Chen et al. (2014) and Prescher et al. (2015), respectively and red triangles are V_S of Fe_7C_3 at high temperature from Prescher et al. (2015). Blue, black, and purple symbols in Figure 9.1f represent V_S at room temperature of $Fe_{85}Si_{15}$ alloys (Lin et al., 2003b), Fe_3S (Lin et al., 2004), and FeH_X (Mao et al., 2004), respectively.

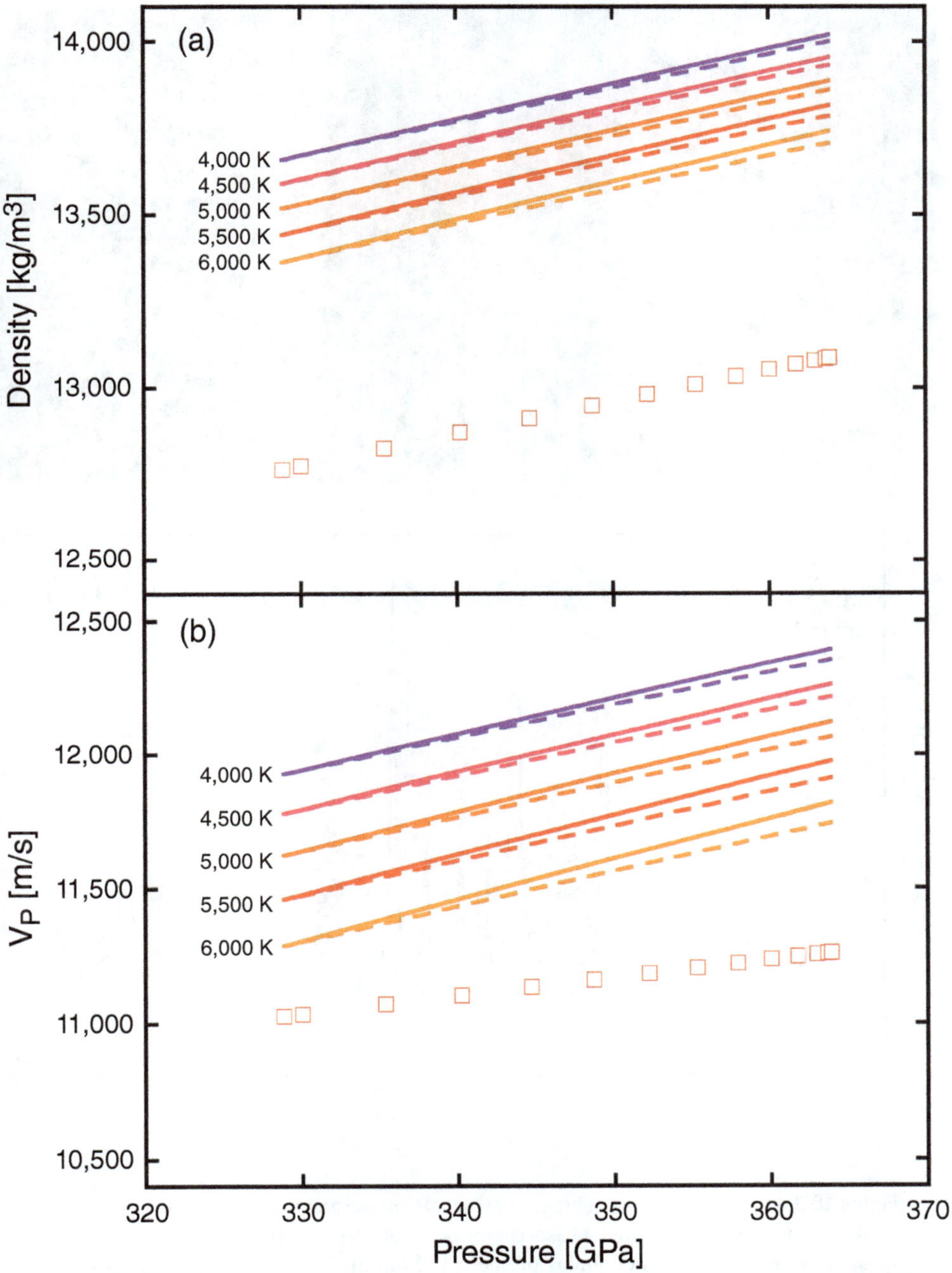

Figure 9.6 Comparison between (a) densities and (b) V_P of pure iron and the PREM inner core. Solid lines represent densities and V_P at isothermal conditions from 4,000–6,000 K. Dashed lines are densities and V_P calculated along the adiabatic temperature profiles, with T_{ICB} = 4,000, 4,500, 5,000, 5,500, and 6,000 K. Red squares represent densities and V_P of the PREM inner core.

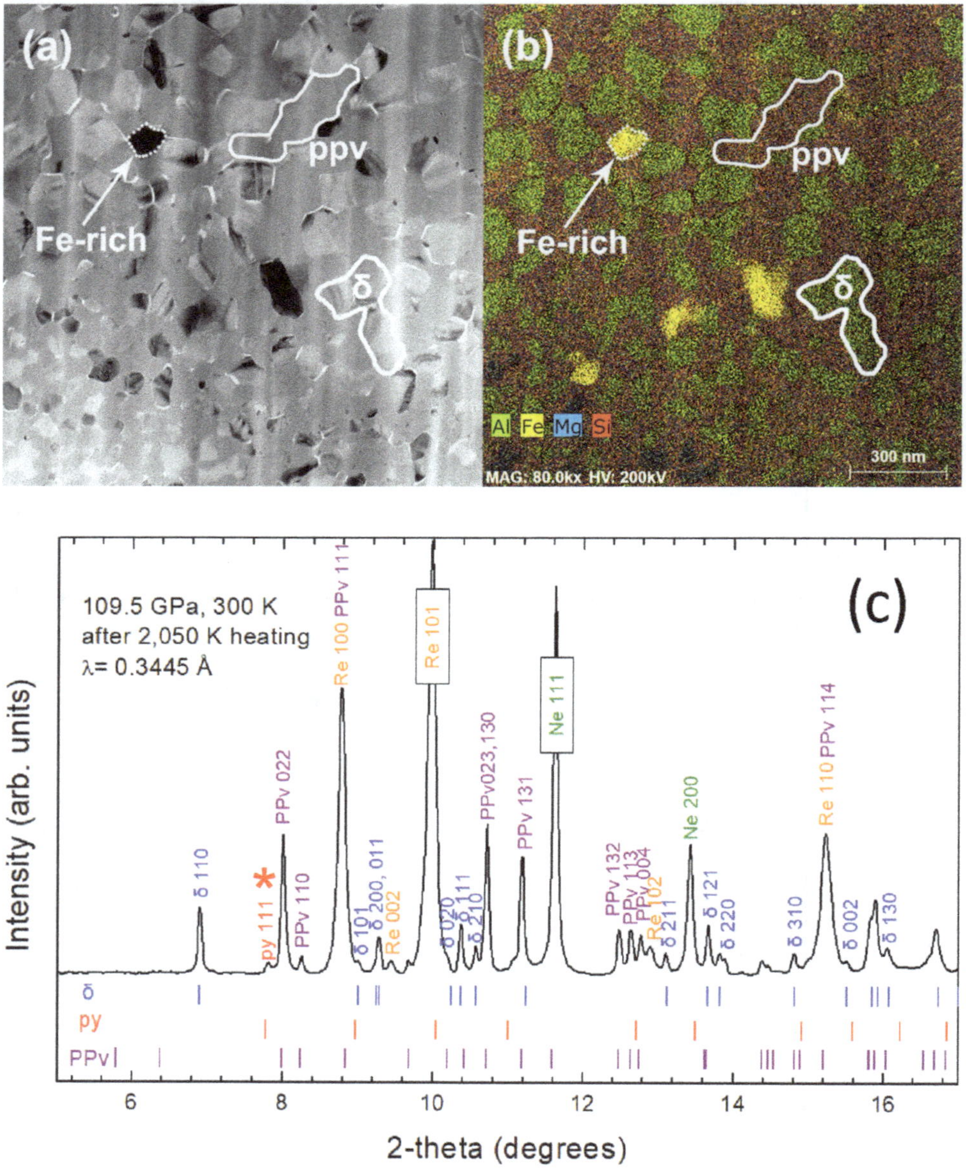

Figure 10.3 A multiphase assemblage contains pPv phase and hydrous δ-phase, coexisting with a minor pyrite-structured phase (py-phase) FeOOH at 110 GPa. (a and b) TEM mapping shows a few Fe-rich grains of about 100 nm in dimensions, which is consistent with the existence of a minor py-phase suggested by (c) in situ XRD where only the (111) peak is visible, marked by a red asterisk. A list of 14 reflections of one selected grain of the py-phase is shown in Table 10.1.

Reproduced from Zhang et al. (2019).

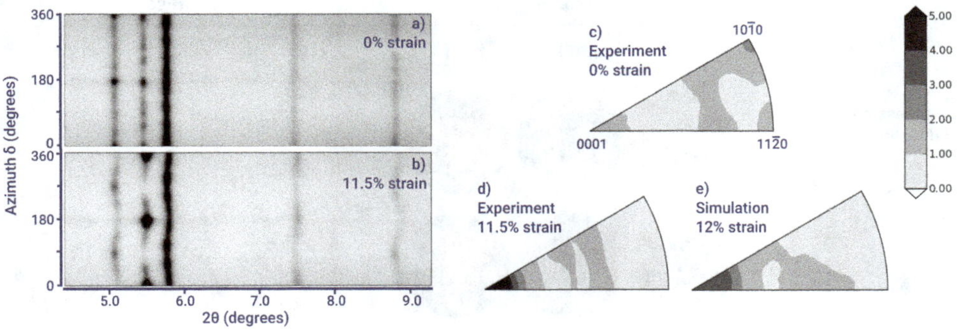

Figure 11.4 Diffraction images for ε-Fe at 17 GPa and 400 K (Merkel et al., 2012) measured at the start of the deformation (a) and after 11.5% axial strain in a D-DIA (b). Variations of peak intensities and positions with azimuth are indicative of the sample LPO and stress, respectively. Measured inverse pole figure of the compression direction at the start of deformation (c) and after 11.5% axial strain (d). (e) Results of a self-consistent model after 12% axial strain reproducing the measured sample strain and textures and constraining the strength of basal, prismatic, and pyramidal slip as well as tensile twinning in ε-Fe at those conditions.

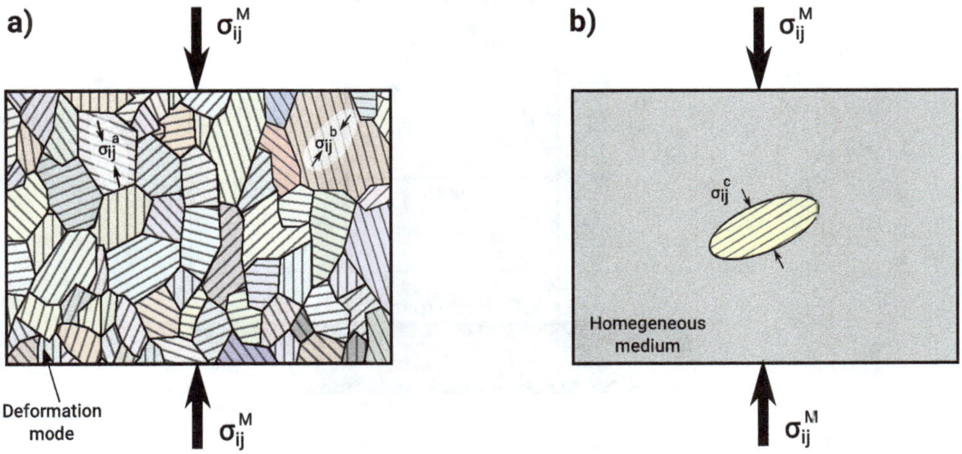

Figure 11.5 Polycrystalline simulation (after Amodeo et al., 2018). (a) In a polycrystal, stress and strain distributions are heterogeneous. The behavior of each grain depends on its local environment, elastic, and plastic properties. (b) In self-consistent models, each grain is treated as an elliptical inclusion inside a homogeneous medium.

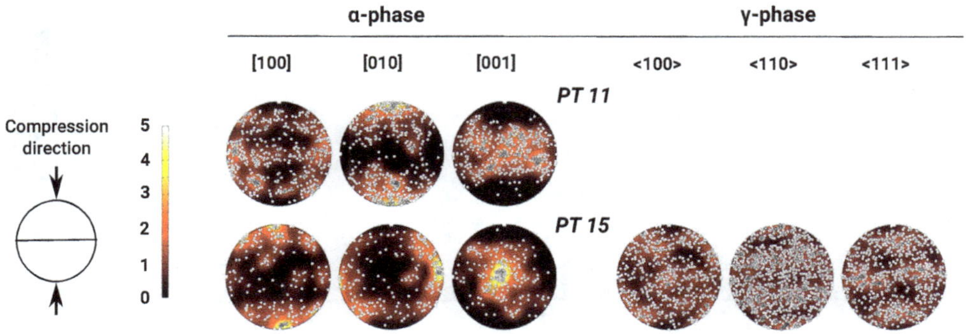

Figure 11.9 Sample results from multigrain crystallography in the DAC (Rosa et al., 2016). Pole figures presenting the orientations of individual olivine and ringwoodite grains during a phase transformation at ≈18 GPa and 880 K. White circles are individual grain orientations. Background color scale is recalculated based on an orientation distribution function fitted to the sample. Scale in multiples of a random distribution (m.r.d.).

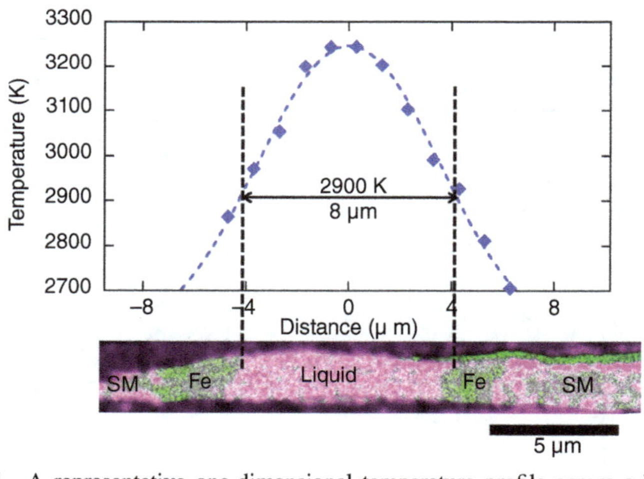

Figure. 13.1 A representative one-dimensional temperature profile across a laser-heated hot spot and its relation to composite X-ray maps for an Fe-C sample recovered from 138 GPa (Mashino et al., 2019). In melting experiments such as this one, by combining such a temperature profile with a sample cross section, one can obtain the temperature at the liquid–solid boundary, which indicates the liquidus temperature. This procedure is important for precisely determining melting temperatures because some temperature variations always exist in laser-heated DAC samples, even with a liquid portion, and therefore the temperature at the center of a hot spot may be higher than the melting temperature.

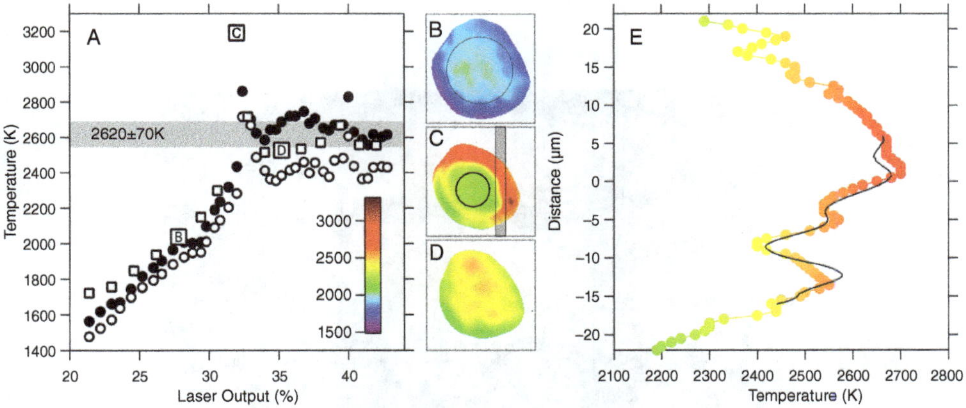

Figure 13.2 In situ temperature measurements and melting determination of Ni metal in a KCl pressure medium for run 77A at ~28 GPa (after Lord et al., 2014). (a) Temperature versus laser power plot showing spectroradiometric measurements from two sides of the sample (circles) together with multispectral imaging radiometry measurements (squares, made on the same side as the open circles). (b–d) Two-dimensional temperature maps determined by multispectral (four-color) imaging radiometry, color coded as a function of temperature. The black circle in (b) represents the approximate location of the ~20μm diameter incident laser beam. The gray bar in (c) represents the approximate location of the slit aperture (~3μm wide at the sample) used for 1D spectroradiometry, whereas the circle shows an example of the area interrogated using a 10μm diameter pinhole aperture. (e) A spectroradiometric line profile (circles) collected nearly simultaneously with the image in (c). The black line is an averaged line profile taken from the region in (c) denoted by the gray bar.

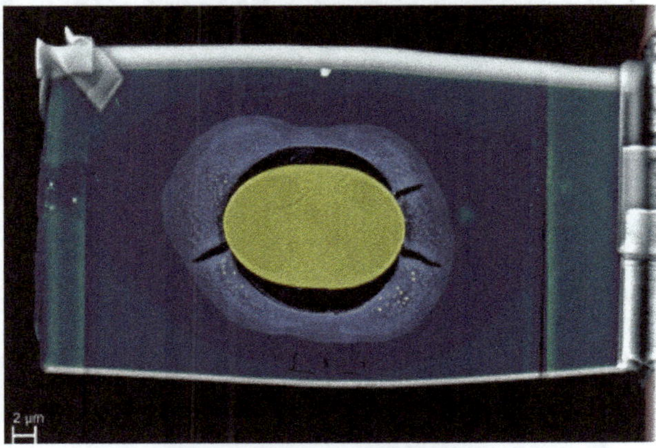

Figure 13.11 FIB thinned section removed from the center of the molten sample in Figure 13.4 showing the molten metal that has coalesced to the center (yellow) surrounded by molten silicate around it (light blue), itself surrounded by solidified silicate (dark blue) and with the untransformed starting silicate occupying the far field (turquoise). Such samples can be measured by traditional chemical analytical techniques (electron microprobe, scanning electron microscopy with EDX analysis) or more advanced techniques such as analytical transmission electron microscopy (Auzende et al., 2008; Piet et al., 2016) or ion probe analysis (Badro et al., 2007; Suer et al., 2017). It is noteworthy that loading samples by stacking monolithic disks of silicate of identical thickness and composition produces perfectly symmetrical samples at the end of the run, and one can note that the metal is far from the diamonds and isolated from them by the starting material, which bears no cracks or grain boundaries, hence efficiently blocking chemical reactions between the melt and the diamonds
(Huang and Badro, 2018).

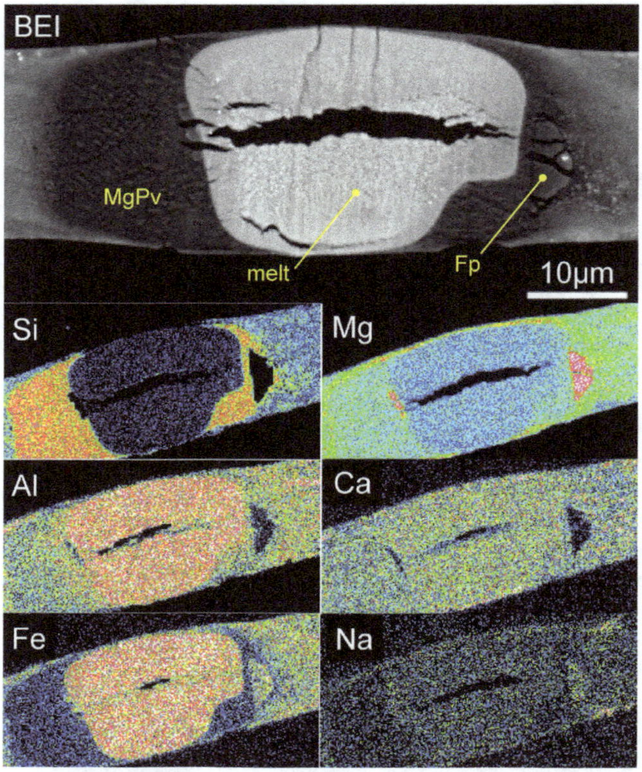

Figure 13.12 Melting experiment on the evolution of the magma ocean from Caracas et al. (2019). Backscattered electron images (top) and the X-ray maps for Si, Mg, Al, Ca, Fe, and Na obtained at 59 GPa. The Fe-rich melt crystallizes $MgSiO_3$-rich bridgmanite (Bdg) and a trace amount of (Mg,Fe)O ferropericlase (Fp) at 63% solidification from a fully molten pyrolite.

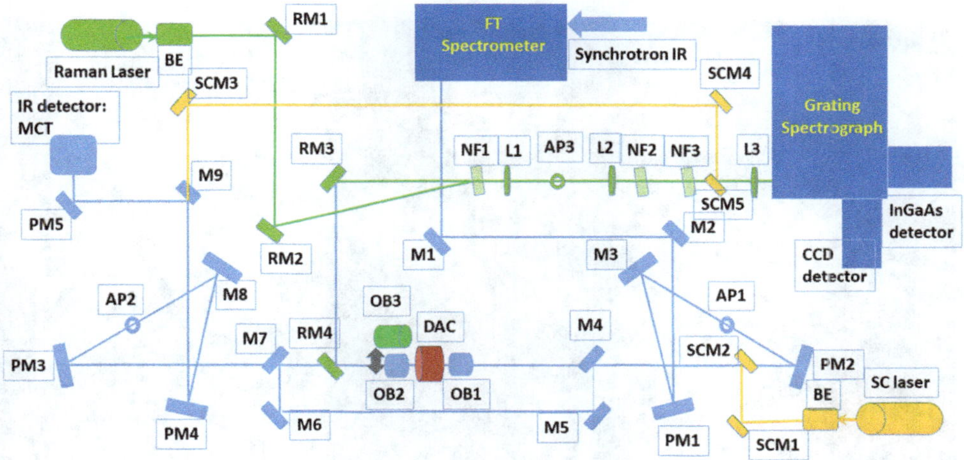

Figure 14.1 Combined synchrotron IR and Raman/optical spectroscopy system at the beamline BL01B of the SSRF (Shanghai). Left panel: Optical layout. Optical elements shown in blue correspond to synchrotron IR, while green and yellow indicate the optical elements that belong to Raman and visible/nearly IR parts of the system, respectively (Goncharov et al., 2019). Right panel: Photograph of the system with a continuous flow He cryostat.

Figure 14.1 (*cont.*)

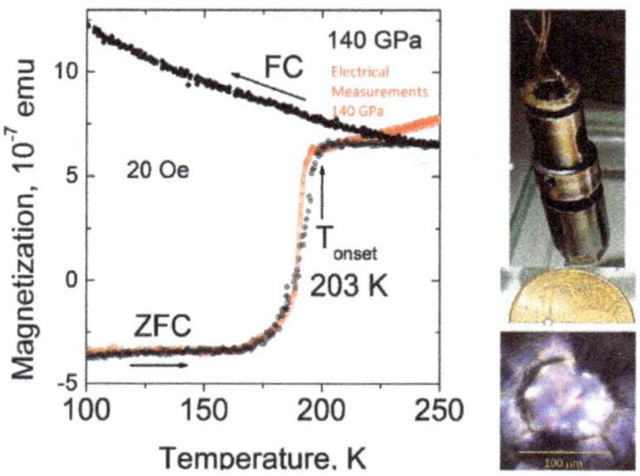

Figure 15.4 Left: Temperature dependence of the magnetization of sulfur hydride at a pressure of 155 GPa in zero-field-cooled (ZFC) and 20 Oe field-cooled (FC) modes (black circles). For comparison, the superconducting step obtained for sulfur hydride from electrical measurements at 145 GPa is shown (red circles). Right: Tiny DAC and view of the metallic hydrogen sulphide sample in the DAC at 150 GPa.

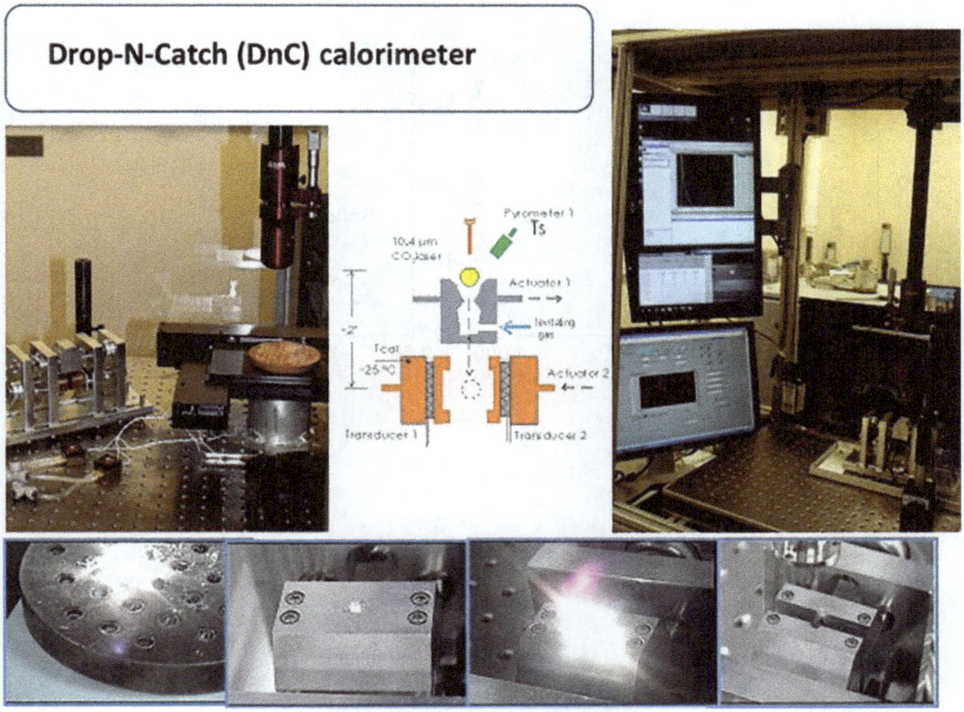

Figure 16.3 Top: Drop-n-catch calorimeter for measurements to 3,000°C. Bottom: Data obtained for melting of zirconia and hafnia. Fusion enthalpy from drop-n-catch calorimetry on HfO_2 (a) and ZrO_2 (b) in argon flow. T_s – surface temperature measured by FAR spectropyrometer
(Hong et al., 2018).

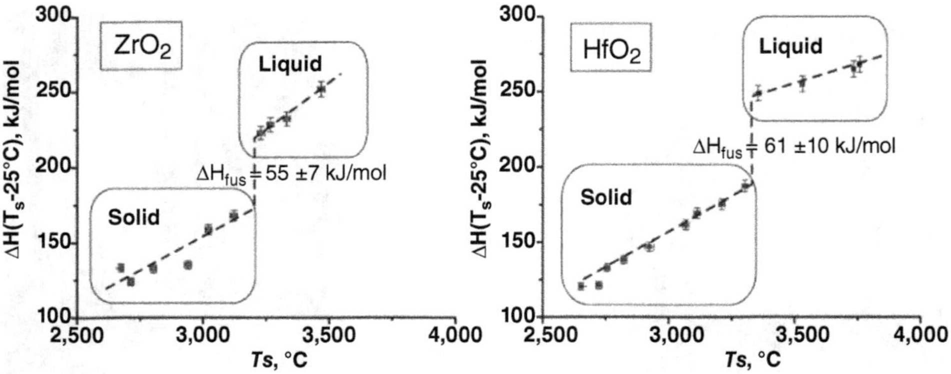

Figure 16.3 *(cont.)*

Dewaele, A., Loubeyre, P., Occelli, F., et al. (2006). Quasihydrostatic equation of state of iron above 2 Mbar. *Physical Review Letters*, **97**, 215504.

Dorogokupets, P. I., Oganov, A. R. (2007). Ruby, metals, and MgO as alternative pressure scales: a semiempirical description of shock-wave, ultrasonic, X-ray, and thermochemical data at high temperatures and pressures. *Physical Review B*, **75**, 024115.

Duffy, T., Ahrens, T. J. (1995). Compressional sound velocity, equation of state, and constitutive response of shock-compressed magnesium oxide. *Journal of Geophysical Research*, **100**, 529–542.

Dziewonski, A. M., Anderson, D. L. (1981). Preliminary reference Earth model. *Physics of the Earth and Planetary Interiors*, **25**, 297–356.

Edmund, E., Antonangeli, D., Decremps, F., et al. (2019). Velocity-density systematics of Fe-5wt%Si: constraints on Si content in the Earth's inner core. *Journal of Geophysical Research*, **124**, 3436–3447.

Fei, Y., Murphy, C., Shibazaki, Y., et al. (2016). Thermal equation of state of hcp–iron: constraint on the density deficit of Earth's solid inner core. *Geophysical Research Letters*, **43**, 6837–6843.

Fiquet, G., Badro, J., Gregoryanz, E., et al. (2009). Sound velocity in iron carbide (Fe_3C) at high pressure: implications for the carbon content of the Earth's inner core. *Physics of the Earth and Planetary Interiors*, **172**, 125–129.

Fiquet, G., Badro, J., Guyot, F., et al. (2001). Sound velocities in iron to 110 gigapascals. *Science*, **291**, 468–471.

Fischer, R. A., Campbell, A. J., Caracas, R., et al. (2014). Equations of state in the Fe–FeSi system at high pressures and temperatures. *Journal of Geophysical Research*, **119**, 2810–2827.

Fukui, H., Katsura, T., Kuribayashi, T., et al. (2008). Precise determination of elastic constants by high-resolution inelastic X-ray scattering. *Journal of Synchrotron Radiation*, **15**, 618–623.

Fukui, H., Sakai, T., Sakamaki, T., et al. (2013). A compact system for generating extreme pressures and temperatures: an application of laser-heated diamond anvil cell to inelastic X-ray scattering. *Review of Scientific Instruments*, **84**, 113902.

Fukui, H., Yoneda, A., Nakatsuka, A. et al. (2016) Effect of cation substitution on bridgmanite elasticity: A key to interpret sesmic anomalies in the lower mantle. *Scientific Reports*, **6**, 33337.

Gao, L., Chen, B., Wang, J., et al. (2008). Pressure-induced magnetic transition and sound velocities of Fe_3C: implications for carbon in the Earth's inner core. *Geophysical Research Letters*, **35**, L17306.

Gao, L., Chen, B., Zhao, J., et al. (2011). Effect of temperature on sound velocities of compressed Fe_3C, a candidate component of the Earth's inner core. *Earth and Planetary Science Letters*, **309**, 213–220.

Glazyrin, K., Pourovskii, L.V., Dubrovinsky, L., et al. (2013). Importance of correlation effects in hcp iron revealed by a pressure-induced electronic topological transition. *Physical Review Letters*, **110**, 117206.

Gréaux, S., Irifune, T., Higo, Y., et al. (2019). Sound velocity of $CaSiO_3$ perovskite suggests the presence of basaltic crust in the Earth's lower mantle. *Nature*, **565**, 218–221.

Hamada, M., Kamada, S., Ohtani, E., et al. (2016). Magnetic and spin transitions in wüstite: a synchrotron Mössbauer spectroscopic study. *Physical Review B*, **93**, 155165.

Higo, Y., Inoue, T., Li, B., et al. (2006). The effect of iron on the elastic properties of ringwoodite at high pressure. *Physics of the Earth and Planetary Interiors*, **159**, 276–285.

Hirao, N., Kawaguchi-Imada, S., Hirose, K., et al. (2020). New developments in high-pressure X-ray diffraction beamline for diamond anvil cell at SPring-8. *Matter and Radiation at Extremes*, **5**, 018403.

Hirao, N., Kondo, T., Ohtani, E., et al. (2004). Compression of iron hydride to 80 GPa and hydrogen in the Earth's core. *Geophysical Research Letters*, **31**, L06616.

Holmes, N. C., Moriarty, J. A., Gathers, G. R., et al. (1989). The equation of state of platinum to 660 GPa (6.6 Mbar). *Journal of Applied Physics*, **66**, 2962–2967.

Huang, H., Fei, Y., Cai, L., et al. (2011). Evidence for an oxygen-depleted liquid outer core of the Earth. *Nature*, **479**, 513–516.

Huang, H., Leng, C., Yang, G., et al. (2018). Measurements of sound velocity of liquid Fe-11.8 wt%S up to 211.4 GPa and 6150 K. *Journal of Geophysical Research*, **123**, 4730–4739.

Jackson, J. M., Sturhahn, W., Shen, G., et al. (2005). A synchrotron Mössbauser spectroscopy study of (Mg,Fe)SiO_3 perovskite up to 120 GPa. *American Mineralogist*, **90**, 199–205.

Jacobsen, S. D., Reichmann, H. J., Spetzler, H. A., et al. (2006). Structure and elasticity of single-crystal (Mg,Fe)O and a new method of generating shear waves for gigahertz ultrasonic interferometry. *Journal of Geophysical Research*, **107**(B2), 2037.

Jamieson, J. C., Fritz, J. N., Manghnani, M. H. (1982). Pressure measurement at high temperature in X-ray diffraction studies: gold as a primary standard, in S. Akimoto, M. H. Manghnani, eds., *High-Pressure Research in Geophysics*. Center for Academic Publications, pp. 27–48.

Kamada, S., Fukui, H., Yoneda, A., et al. (2019) Elastic constants of single-crystal Pt measured up to 20 GPa based on inelastic X-ray scattering: implication for the establishment of an equation of state. *Comptes Rendus Geoscience*, **351**, 236–242.

Kamada, S., Ohtani, E., Fukui, H., et al. (2014a). The sound velocity measurements of Fe_3S. *American Mineralogist*, **99**, 98–101.

Kamada, S., Ohtani, E., Terasaki, H., et al. (2014b). Equation of state of Fe_3S at room temperature up to 2 megabars. *Physics of the Earth and Planetary Interiors*, **228**, 106–113.

Kamada, S., Suzuki, N., Maeda, F., et al. (2018). Electronic properties and compressional behavior of Fe–Si alloys at high pressure. *American Mineralogist*, **103**, 1959–1965.

Kawaguchi, S. I., Nakajima, Y., Hirose, K., et al. (2017) Sound velocity of liquid Fe–Ni–S at high pressure. *Journal of Geophysical Research*, **122**, 3624–3634.

Kennett, B. L. N., Engdahl, E. R. (1991). Traveltimes for global earthquake location and phase identification. *Geophysical Journal International*, **105**(2), 429–465.

Kennett, B. L. N., Engdahl, E. R., Buland, R. (1995). Constraints on seismic velocities in the Earth from traveltimes. *Geophysical Journal International*, **122**(1), 108–124.

Kono, Y., Irifune, T., Higo, Y., et al. (2010). P–V–T relation of MgO derived by simultaneous elastic wave velocity and in situ X-ray measurements: a new pressure scale for the mantle transition region. *Physics of the Earth and Planetary Interiors*, **183**, 196–211.

Kurnosov, A., Marquardt, H., Frost, D., et al. (2017). Evidence for a Fe^{3+}-rich pyrolytic lower mantle from (Al,Fe)-bearing bridgmanite elasticity data. *Nature*, **543**, 543–546.

Kuwayama, Y., Morard, G., Nakajima, Y., et al. (2020). Equation of state of liquid iron under extreme conditions. *Physical Review Letters*, **124**, 165701.

Li, B., Liebermann, R. C. (2014). Study of the Earth's interior using measurements of sound velocities in minerals by ultrasonic interferometry. *Physics of the Earth and Planetary Interiors*, **233**, 135–153.

Lifshitz, I. M. (1960). Anomalies of electron characteristics of a metal in the high pressure region. *Soviet Physics JETP*, **11**(5), 1130–1135.
Li, J., Sturhahn, W., Jackson, J.M. (2006). Pressure effect on the electronic structure of iron in (Mg,Fe)(Si,Al)O_3 perovskite: a combined synchrotron Mössbauer and X-ray emission spectroscopy study up to 100 GPa. *Physics and Chemistry of Minerals*, **33**, 575–585.
Li, Y., Vočadlo, L., Brodhol, J. P. (2018). The elastic properties of hcp–Fe alloys under the conditions of the Earth's inner core. *Earth and Planetary Science Letters*, 493, 118–127. *Physics of the Earth and Planetary Interiors*, **233**, 135-153.
Lin, J.-F., Campbell, A. J., Heinz, D. L., et al. (2003a). Static compression of iron–silicon alloys: implications for silicon in the Earth's core. *Journal of Geophysical Research*, **108**(B1), 2045.
Lin, J.-F., Fei, Y., Sturhahn, W., et al. (2004). Magnetic transition and sound velocities of Fe_3S at high pressure: implications for Earth and planetary cores. *Earth and Planetary Science Letters*, **226**, 33–40.
Lin, J. F., Gavriliuk, A. G., Struzhkin, V. V., et al. (2006). Pressure-induced electronic spin transition of iron in magnesiowustite-(Mg.Fe)O. *Physical Review B*, **73**, 113107.
Lin, J. F., Gavriliuk, A. G., Sturhahn, W. et al. (2009). Synchrotron Mossbauer spectroscopic study of ferropericlase at high pressures and temperatures. *American Mineralogist*, **94**, 594–599.
Lin, J. F., Struzhkin, V. V., Sturhahn, W., et al. (2003b). Sound velocities of iron–nickel and iron–silicon alloys at high pressures. *Geophysical Research Letters*, **30**(21), 2112.
Lin, J. F., Sturhahn, W., Zhao, J., et al. (2005). Sound velocities of hot dense iron: Birch's law revisited. *Science*, **308**, 1892–1894.
Lin, J. F., Tsuchiya, T. (2008). Spin transition of iron in the Earth's lower mantle. *Physics of the Earth and Planetary Interiors*, **170**, 248–259.
Liu, J., Lin, J. F., Alatas, A., et al. (2014). Sound velocities of bcc–Fe and $Fe_{0.85}Si_{0.15}$ alloy at high pressure and temperature. *Physics of the Earth and Planetary Interiors*, **233**, 24–32.
Liu, J., Lin, J.-F., Alatas, A., et al. (2016). Seismic parameters of hcp–Fe alloyed with Ni and Si in the Earth's inner core. *Journal of Geophysical Research*, **121**, 610–623.
Maeda, F., Kamada, S., Ohtani, E., et al. (2017). Spin state and electronic environment of iron in basaltic glass in the lower mantle. *American Mineralogist*, **102**, 2106–2112.
Mao, H. K., Shu, J., Shen, G., et al. (1998). Elasticity and rheology of iron above 220 GPa and the nature of the Earth's inner core. *Nature*, **396**, 741–743.
Mao, H. K., Wu, Y., Shu, J. F., et al. (1990). Static compression of iron to 300 GPa and $Fe_{0.8}Ni_{0.2}$ alloy to 260 GPa: implications for composition of the core. *Journal of Geophysical Research*, **95**(B13), 21737–21742.
Mao, H. K., Xu, J., Struzhkin, V. V., et al. (2001). Phonon density of states of iron up to 153 gigapascals. *Science*, **292**, 914–916.
Mao, W. L., Sturhahn, W., Heinz, D. L., et al. (2004). Nuclear resonant X-ray scattering of iron hydride at high pressure. *Geophysical Research Letters*, **31**, L15618.
Mao, Z., Lin, J. F., Liu, J. et al. (2012). Sound velocities of Fe and Fe–Si alloy in the Earth's core. *Proceedings of the National Academy of Sciences of the United States of America*, **109**(26), 10239–10244.
Martorell, B., Vočadlo, L., Brodholt, J., et al. (2013). Strong premelting effect in the elastic properties of hcp–Fe under inner-core conditions. *Science*, **342**, 466–468.
Martorell, B., Wood, I. G., Vočadlo, L. (2016). The elastic properties of hcp–$Fe_{1-x}Si_x$ at Earth's inner-core conditions. *Earth and Planetary Science Letters*, **451**, 89–96.

Masters, G., Gubbins, D. (2003). On the resolution of density within the Earth. *Physics of the Earth and Planetary Interiors*, **140**, 159–167.

McDonough, M. F. (2017). Earth's core, in W. White, eds., *Encyclopedia of Geochemistry. Encyclpedia of Earth Sciences Series.* Springer, Cham. https://doi.org/10.1007/978-3-319-39193-9_258-1.

Merkel, S., Miyajima, N., Antonangeli, D., et al. (2006). Lattice preferred orientation and stress in polyscrytalline hcp–Co plastically deformed under high pressure. *Journal of Applied Physics*, **100**, 023510.

Merkel, S., Shu, J., Gillet, P., et al. (2005). X-ray diffraction study of the single-crystal elastic moduli of ε–Fe up to 30 GPa. *Journal of Geophysical Research*, **110**, B05201.

Merkel, S., Tomé, C., Wenk, H.-R. (2009). Modeling analysis of the influence of plasticity on high pressure deformation of hcp–Co. *Physical Review B*, **79**, 064110.

Mitsui, T., Hirao, N., Ohishi, Y., et al. (2009). Development of an energy-domain ^{57}Fe–Mössbauer spectrometer using synchrotron radiation and its application to ultrahigh-pressure studies with a diamond anvil cell. *Journal of Synchrotron Radiation*, **16**, 723–729.

Morrison, R. A., Jackson, J. M., Struhahn, W., et al. (2018). Equations of state and anisotropy of Fe–Ni–Si alloys. *Journal of Geophysical Research*, **123**, 4647–4675.

Morrison, R. A., Jackson, J. M., Struhahn, W., et al. (2019). High pressure thermoelasticity and sound velocities of Fe–Ni–Si alloys. *Physics of the Earth and Planetary Interiors*, **294**, 106268.

Murakami, M., Ohishi, Y., Hirao, N., et al. (2012). A perovskitic lower mantle inferred from high-pressure, high-temperature sound velocity data. *Nature*, **485**, 90–94.

Murphy, C. A., Jackson, J. M., Sturhahn, W. (2013). Experimental constraints on the thermodynamics and sound velocities of hcp–Fe to core pressures. *Journal of Geophysical Research*, **118**, 1999–2016.

Nguyen, J. H., Holmes, N. C. (2004). Melting of iron at the physical conditions of the Earth's core. *Nature*, **427**, 339–342.

Nishida, K., Kono, Y., Terasaki, H., et al. (2013). Sound velocity measurements in liquid Fe–S at high pressure: implications for Earth's and lunar cores. *Earth and Planetary Science Letters*, **362**, 182–186.

Ohtani, E., Shibazaki, Y., Sakai, T., et al. (2013). Sound velocity of hexagonal close-packed iron up to core pressures. *Geophysical Research Letters*, **40**, 5089–5094.

Ozawa, H., Takahashi, F., Hirose, K., et al. (2011). Phase transition of FeO and stratification in Earth's outer core. *Science*, **334**, 792–794.

Poirier, J. P. (1994). Light elements in the Earth's outer core: a critical review. *Physics of the Earth and Planetary Interiors*, **85**, 319–337.

Prescher, C., Dubrovinsky, L., Bykova, E., et al. (2015). High Poisson's ratio of Earth's inner core explained by carbon alloying. *Nature Geoscience*, **8**, 220–223.

Sakai, T., Takahashi, S., Nishitani, N., et al. (2014). Equation of state of pure iron and $Fe_{0.9}Ni_{0.1}$ alloy up to 3 Mbar. *Physics of the Earth and Planetary Interiors*, **228**, 114–126.

Sakairi, T., Sakamaki, T., Ohtani, E., et al. (2018). Sound velocity measurements of hcp Fe–Si alloy at high pressure and high temperature by inelastic X-ray scattering. *American Mineralogist*, **103**, 85–90.

Sakaiya, T., Takahashi, H., Kondo, T., et al. (2014). Sound velocity and density measurements of liquid iron up to 800 GPa: a universal relation between Birch's law coefficients for solid and liquid metals. *Earth and Planetary Science Letters*, **392**, 80–85.

Sakamaki, T., Ohtani, E., Fukui, H., et al. (2016). Constraints on Earth's inner core composition inferred from measurements of the sound velocity of hcp–iron in extreme conditions. *Science Advances*, **2**(2), e1500802.

Scott, H. P., Williams, Q., Knittle, E. (2001). Stability and equation of state of Fe_3C to 73 GPa: implications for carbon in the Earth's core. *Geophysical Research Letters*, **28** (9), 1875–1878.

Seagle, C. T., Campbell, A. J., Heinz, D. L., et al. (2006). Thermal equation of state of Fe_3S and implications for sulfur in Earth's core. *Journal of Geophysical Research*, **111**, B06209.

Shibazaki, Y., Ohtani, E., Fukui, H., et al. (2012). Sound velocity measurements in dhcp–FeH up to 70 GPa with inelastic X-ray scattering: implications for the composition of the Earth's cores. *Earth and Planetary Science Letters*, **313–314**, 79–85.

Singh, A. K. (1993). The lattice strains in a specimen (cubic system) compressed nonhydrostatically in an opposed anvil device. *Journal of Applied Physics*, **73**(9), 4278–4286.

Singh, A. K., Balasingh, C. (1994). The lattice strains in a specimen (hexagonal system) compressed nonhydrostatically in an opposed anvil high pressure setup. *Journal of Applied Physics*, **75**(10), 4956–4962.

Singh, A. K., Balasingh, C., Mao, H. K., et al. (1998). Analysis of lattice strains measured under nonhydrostatic pressure. *Journal of Applied Physics*, **83**(12), 7567–7575.

Sinmyo, R., Glazyrin, K., McCammon, C., et al. (2014). The influence of solid solution on elastic wave velocity determination in (Mg,Fe)O using nuclear inelastic scattering. *Physics of the Earth and Planetary Interiors*, **229**, 16–23.

Sinogeikin, S. V., Bass, J. D. (2000). Single-crystal elasticity of pyrope and MgO to 20 GPa by Brillouin scattering in the diamond cell. *Physics of the Earth and Planetary Interiors*, **120**, 43–62.

Spetzler, H., Shen, A., Chen, G., et al. (1996). Ultrasonic measurements in a diamond anvil cell. *Physics of the Earth and Planetary Interiors*, **98**, 93–9.

Speziale, S., Marquardt, H., Duffy, T. S. (2014). Brillouin scattering and its application in geosciences. *Reviews in Mineralogy & Geochemistry*, **78**, 543–603.

Takahashi, S., Ohtani, E., Ikuta, D., et al. (2019b) Thermal equation of state of Fe_3C to 327 GPa and carbon in the core. *Minerals*, **9**(12), 744.

Takahashi, S., Ohtani, E., Sakamaki, T., et al. (2019a). Sound velocity of Fe_3C at high pressure and high temperature determined by inelastic X-ray scattering. *Comptes Rendus Geoscience*, **351**, 190–196.

Tanaka, R., Sakamaki, T., Ohtani, E., et al. (2020). The sound velocity of wüstite at high pressures: implications for low-velocity anomalies at the base of the lower mantle. *Progress in Earth and Planetary Science*, **7**, 23.

Tateno, S., Hirose, K., Ohishi, Y., et al. (2010). The structure of iron in Earth's inner core. *Science*, **330**, 359–361.

Tateno, S., Kuwayma, Y., Hirose, K., et al. (2015). The structure of Fe–Si alloy in Earth's inner core. *Earth and Planetary Science Letters*, **418**, 11–19.

Thomson, A. R., Crchton, W. A., Brodholt, J. P., et al. (2019). Seismic velocities of $CaSiO_3$ perovskite can explain LLSVPs in Earth's lower mantle. *Nature*, **572**, 643–647.

Vočadlo, L., Dobson, D. P., Wood, I. G. (2009). Ab initio calculations of the elasticity of hcp–Fe as a function of temperature at inner-core pressure. *Earth and Planetary Science Letters*, **288**, 534–538.

Wakamatsu, T., Ohta, K., Yagi, T. (2018). Measurements of sound velocity in iron–nickel alloys by femtosecond laser pulses in a diamond anvil cell. *Physics and Chemistry of Minerals*, **45**, 589–595.

Weidner, D. J., Swyler, K., Carleton, H. R. (1975). Elasticity of microcrystals. *Geophysical Research Letters*, **2**(5), 189–192.

Yamazaki, D., Ito, E., Yoshino, T., et al. (2012). P–V–T equation of state for ε-iron up to 80 GPa and 1900 K using the Kawai-type high pressure apparatus equipped with sintered diamond anvils. *Geophysical Research Letters*, **39**, L20308.

Yokoo, M., Kawai, N., Nakamura, K. G., et al. (2009). Ultrahigh-pressure scales for gold and platinum at pressures up to 550 GPa. *Physical Review B*, **80**, 104114.

Yoneda, A., Fukui, H., Gomi, H., et al. (2017). Single crystal elasticity of gold up to ~20 GPa: bulk modulus anomaly and implication for a primary pressure scale. *Japanese Journal of Applied Physics*, **56**, 095801.

Yoneda, A., Kobayashi, S., Kamada, S. (2019). GHz ultrasonic velocity measurement in diamond anvil cell. *Review of High Pressure Science and Technology*, **29**(2), 144–155.

Yoshino, T., Kamada, S., Zhao, C., et al. (2016). Electrical conductivity model of Al-bearing bridgmanite with implications for the electrical structure of the Earth's lower mantle. *Earth and Planetary Science Letters*, **434**, 208–219.

Zha, C. S., Duffy, T.S., Downs, R. T., et al. (1996). Sound velocity and elasticity of single-crystal forsterite to 16 GPa. *Journal of Geophysical Research*, **101**(B8), 17535–17545.

Zha, C. S., Hemley, R. J., Mao, H. K., et al. (1994). Acoustic velocities and refractive index of SiO_2 glass to 57.5 GPa by Brillouin scattering. *Physical Review B*, **50**(18), 13105–13112.

Zha, C. S., Mao, H. K., Hemley, R. J. (2000). Elasticity of MgO and a primary pressure scale to 55 GPa. *Proceedings of the National Academy of Sciences of the United States of America*, **97**(25), 13494–13499.

10

Multigrain Crystallography at Megabar Pressures

LI ZHANG, JUNYUE WANG, AND HO-KWANG MAO

Applications of synchrotron X-ray diffraction techniques have enabled crystallographic characterization of pressure-induced phase transitions in diamond anvil cells (DACs) at megabar pressures. Accurate determination of high-pressure structures is crucial for understanding all other pressure-induced property changes. This chapter discusses current capabilities, technical challenges, and future perspectives of the multigrain techniques for high-pressure studies. Through single-crystal structure analysis of seifertite SiO_2 at 129 GPa, we conclude that single-crystal structure determination and refinement is possible in general cases at megabar pressures. A nearly full convergence of the structure can be achieved applying the multigrain method, and high-quality crystallographic data can then be obtained. In addition, multigrain indexation can be applied for fast online analysis of multiphase systems during synchrotron sessions. Future development of software will certainly promote wide application of the multigrain techniques. The multigrain capabilities can be further extended to multimegabar pressures. Combination of in situ X-ray powder diffraction, multigrain indexation, and single-crystal structure determination on individual grains provides new opportunities to characterize new phases at megabar pressures and beyond.

10.1 Introduction

Under the high-pressure and temperature (*P–T*) conditions that correspond to the deep Earth, the crystal structures of minerals that are stable under ambient conditions are replaced by denser structures through atomic rearrangement. The diamond anvil cell (DAC), where samples are compressed between two opposing diamonds, has been used to generate megabar pressures for over four decades (Mao and Bell, 1978). Recent developments have enabled static compression in the DAC to cover Earth's entire pressure range up to ~4 Mbar and several thousand K (Tateno et al., 2010; Li et al., 2018). To achieve the desired high *P–T* conditions, a near- or midinfrared laser is applied to heat the sample compressed between two anvils. Coupling of laser heating with synchrotron X-ray diffraction (XRD) measurements in DACs has been one of the major technical advancements (Prakapenka et al., 2008; Liermann et al., 2015; Meng et al., 2015; Shen and Mao,

2017; Hirao et al., 2020) and are responsible for numerous discoveries of new materials and high-pressure structures under high *P*–*T* conditions. Crystal structure determination of high-pressure phases can provide basic information for understanding all other pressure-induced properties and remains the dominant task at high-pressure synchrotron beamlines.

Structure characterization becomes increasingly challenging with increasing pressure. In situ powder XRD has been the most accessible technique in the past half century and is commonly used to detect structure changes under high *P*–*T* conditions. Some high-pressure structures are quenchable to ambient conditions and can be examined using conventional methods after recovery. Most of the structures formed at megabar pressures, however, are unquenchable, and therefore in situ XRD is the only available experimental approach for determination of these high-pressure structures. The powder diffraction technique has limitations in its application to high-pressure data. For instance, in a powder diffraction pattern from a multiphase assemblage, overlapping diffraction peaks from multiple phases often causes difficulties in phase identification and structure characterization. In addition, powder diffraction is often inadequate for structure determination and refinement of a high-pressure phase partly because reliable intensity measurements cannot be achieved due to the spottiness of XRD patterns and texturing of samples under extreme high-pressure conditions. Single-crystal X-ray diffraction is regarded as the ideal method for structure characterization, but single crystals, as required for conventional single-crystal diffraction measurements, are often unachievable after phase transitions under high *P*–*T* conditions.

High-pressure single-crystal diffraction techniques have been developed at the synchrotron beamlines (Dera et al., 2013; Merlini and Hanfland, 2013; Shen and Mao, 2017), and applied to characterize new structures and discover new materials at high pressures (Gregoryanz et al., 2008; Lavina et al., 2011; Bykova et al., 2016). In situ single-crystal diffraction measurements on a presynthesized single-crystal have also been made possible in the laser-heated DAC at high temperatures and pressures (Dubrovinsky et al., 2010). In megabar experiments, however, single crystals often break down to polycrystalline samples after phase transitions or reequilibrium under high *P*–*T* conditions, with only a few exceptions, such as sodium (Na) (Gregoryanz et al., 2008) and solid oxygen (O_2) (Lundegaard et al., 2006). Commonly, a polycrystalline multiphase sample is obtained in a phase equilibrium experiment simulating processes within the Earth's deep interior. Modifications of existing single-crystal methods have been used to successfully index several grains in a high-pressure data set (Lavina and Meng, 2015; Merlini et al., 2015, 2017; Cerantola et al., 2017). However, the conventional methods cannot be scaled to a large number of submicron-sized crystals in a multiphase polycrystalline sample and are far from a general solution for megabar experiments.

Synchrotron radiation X-ray sources have long been regarded as the ideal probe for studying materials in situ at high pressures (Mao and Bell, 1978), and the application of synchrotron X-ray probes is responsible for many discoveries in high-pressure research. Beveled diamond anvils with small culets (e.g., <150 μm) have made achieving megabar pressures routinely, and to probe an increasingly smaller sample with increasing pressure, the X-ray beam size has been gradually reduced from 30–50 μm three decades ago to the currently common 2–10 μm (Merlini and Hanfland, 2013; Liermann et al., 2015; Shen and

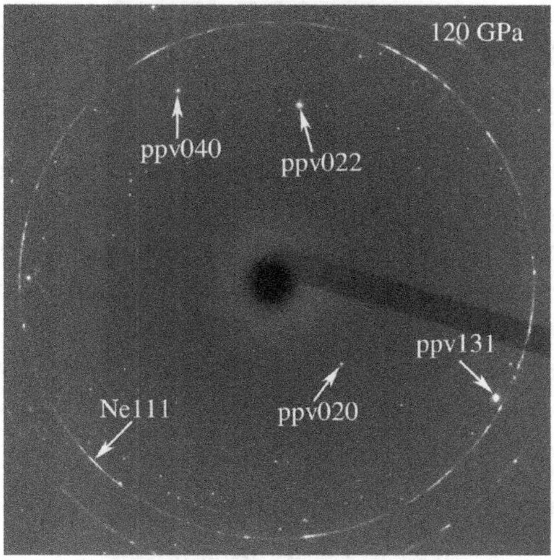

Figure 10.1 A two-dimensional XRD image of (Mg,Fe)SiO_3 pPv shows separation of the reflection spots. The diffraction image was collected at a fixed rotation angle at 120 GPa after temperature quenched from 2,400 K. Neon (Ne) served as the pressure medium. Reproduced from (Zhang et al., 2013).

Mao, 2017; Hirao et al., 2020). On the other hand, improved double-sided laser-heating techniques can be used to promote crystal growth in a polycrystalline sample under high *P*–*T* conditions. Concurrently, the sample thickness is reduced to only a few microns within a rhenium (Re) gasket above megabar pressures (Li et al., 2018). As a result, spotty XRD patterns are obtained when there exist a few thousand grains or less in the X-ray beam. Figure 10.1 shows a spotty XRD pattern obtained for the $MgSiO_3$ postperovskite (pPv) phase at 1.2 Mbar (Zhang et al., 2013).

Separated reflection spots on diffraction patterns have inspired a new crystallographic solution. Multigrain crystallography has been developed to handle samples in the form of a heterogeneous assembly consisting of a large number of crystalline grains by applying the FABLE (Fully Automatic BeamLine Experiments) package (Sørensen et al., 2012). Several hundred grains have been simultaneously indexed from each high-pressure data set collected at megabar pressures through determination of the crystallographic orientations for individual grains (Nisr et al., 2012; Zhang et al., 2013, 2016b). Each of the grains can then be treated as a single crystal. For example, the (Mg,Fe)SiO_3-pPv phase has been the focus of study since its discovery in 2004 as the dominant phase at the base of the lower mantle (Murakami et al., 2004). Pressure release to ambient conditions leads to amorphization of the structure, and therefore in situ crystal structure determination is required. Applying the multigrain indexation to a rotation dataset of spotty diffraction images (Figure 10.1), the single-crystal structure of the (Mg,Fe)SiO_3-pPv phase was obtained for the first time (Zhang et al., 2013). The progress suggested great potential for developing a general solution for single-crystal structure determination in coarse-grained polycrystalline samples at megabar pressures.

This chapter introduces the current capabilities of applying multigrain crystallography to high-pressure data sets, explains the technical limitations, and discusses future perspectives. Through demonstration of the single-crystal structure determination of seifertite SiO_2, we conclude that single-crystal structure determination and refinement has been made possible for general cases at megabar pressures. A nearly full convergence of the structure can be achieved using a combination of random crystallographic orientations from several grains in a merged data set, and we are able to obtain structures of a quality exceeding that of structures determined by conventional single-crystal methods. Lastly, multigrain indexing is particularly useful for fast online processing of DAC data at synchrotron beamlines, and some developments are suggested, with a particular emphasis on handling multiphase systems at high pressures.

10.2 Multigrain Indexation at High Pressures

A schematic setup for data collection in a DAC at the synchrotron beamline is shown in Figure 10.2. The sample is aligned to the rotation center, and a set of two-dimensional (2D) XRD images can be collected at small incremental steps similar to the rotation method used in conventional single-crystal crystallography. Small incremental steps are used to enhance the signal-to-noise ratio by reducing the background level, which is especially important for studying low-Z materials. The principles for conducting single-crystal diffraction in a DAC and requirements for the experimental setup are similar to those described previously (Miletich et al., 2000). A major concern when conducting single-crystal diffraction in a DAC is the limited X-ray access defined by the geometry of the opening cone of the DAC. The conically supported diamond anvils allow for relatively large optical apertures at megabar pressures (Boehler and De Hantsetters, 2004) so that they are widely used for single-crystal diffraction measurements.

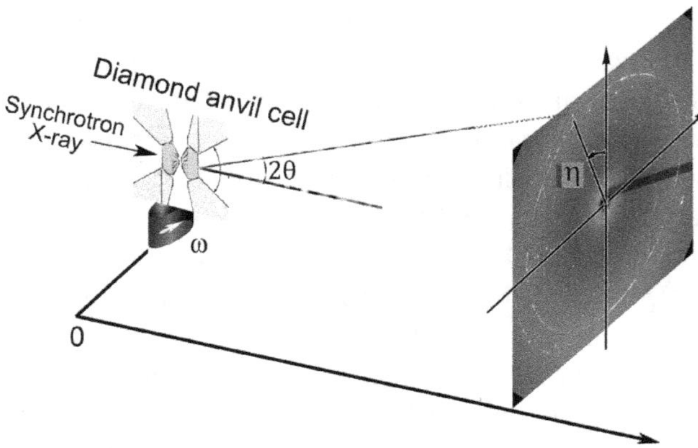

Figure 10.2 A schematic setup for high-pressure multigrain XRD data collection in a DAC. The sample is aligned to the rotation center, and a set of 2D XRD patterns are collected at small incremental steps, where ω, η, and θ represent the rotation, azimuth, and Bragg angles of each reflection, respectively.
Reproduced from Zhang et al. (2019).

High pressures produce complicated crystal structures, and phase identification remains as the most challenging step in high-pressure research. Powder diffraction is the most accessible X-ray diffraction technique and has been responsible for most discoveries of high-pressure structures in the past half-century, but powder diffraction methods encounter great difficulties when applied to multiphase assemblage at high pressures. First of all, overlapping peaks from multiple phases in a typical powder XRD pattern have prevented identification of new phases and weak phases in sample assemblages. Figure 10.3 shows transmission electron microscopy (TEM) images of a recovered multiphase sample from megabar pressure, where the grain size is about 100 nm on average and the minor Fe-rich phase makes up only ~3% by volume (Zhang et al., 2019). Accordingly, only the (111) peak of the cubic pyrite-structured phase (py-phase) was weakly visible in the integrated XRD pattern at 110 GPa (Figure 10.3c). Spotty XRD patterns were obtained using a synchrotron X-ray beam of approximately 3×4 μm^2 in size. Despite the extreme low proportion of the Fe-rich phase in the assemblage, 42 grains of the py-phase were indexed with nine to 18 reflections confirmed for each grain. Table 10.1 lists 14 reflections of one selected grain of the cubic py-phase. This example has demonstrated that individual grains as small as 100 nm of a weak phase in a multiphase assemblage can be identified applying the multigrain method, because multigrain indexation can provide rigorous 3D orientation and geometrical relationship in addition to d spacing and thus allows for unambiguous identification of a weak phase. Secondly, powder diffraction patterns only contain d-spacing information, and successful indexing of several obvious d spacings could still be nonunique. For example, the unit cell parameters of the newly discovered hexagonal hydrous phase (HH-phase) in (Fe,Al)OOH was determined applying the multigrain method, with a = 10.5803(6) Å and c = 2.5897(3) Å at 110 GPa (Zhang et al., 2018). However, in the powder diffraction pattern, if the very weak (211) and (311) peaks were neglected, a small orthorhombic unit cell was obtained instead. It was only when the correct hexagonal unit cell was input for indexation that 27 individual grains were confirmed with up to 104 observed reflections for each grain. Through these examples, we can conclude that the combination of powder diffraction with multigrain indexation provides a powerful tool for phase identification and lattice determination at high pressures.

The main limitation for indexation lies in spot overlap on the diffraction images at megabar pressures. Use of quasihydrostatic pressure medium and/or performing stable heating at high temperatures can significantly promote crystal growth in an aggregate. As shown in the spotty XRD pattern of the (Mg,Fe)SiO_3-pPv phase (Figure 10.1) (Zhang et al., 2013), the spottiness is a combined result of using a microfocused X-ray beam and crystal growth in the assemblage upon heating. If the synchrotron X-ray beam is further reduced to submicron in size, grains smaller than 100 nm can possibly be indexed (Wang et al., 2010). In an ideal single-crystal diffraction measurement, a crystal must fully "bathe" within the X-ray beam. The X-ray beam size is commonly characterized by the full width at half maximum (FWHM) of peak intensity, with the X-ray beam outside the FWHM area regarded as the "tail." The measured crystals can be partly within the tail beam or rotating out of the beam if the rotation stage is not properly aligned, which will cause problems in multigrain indexation and produces unreliable intensity data. In practice, we have found

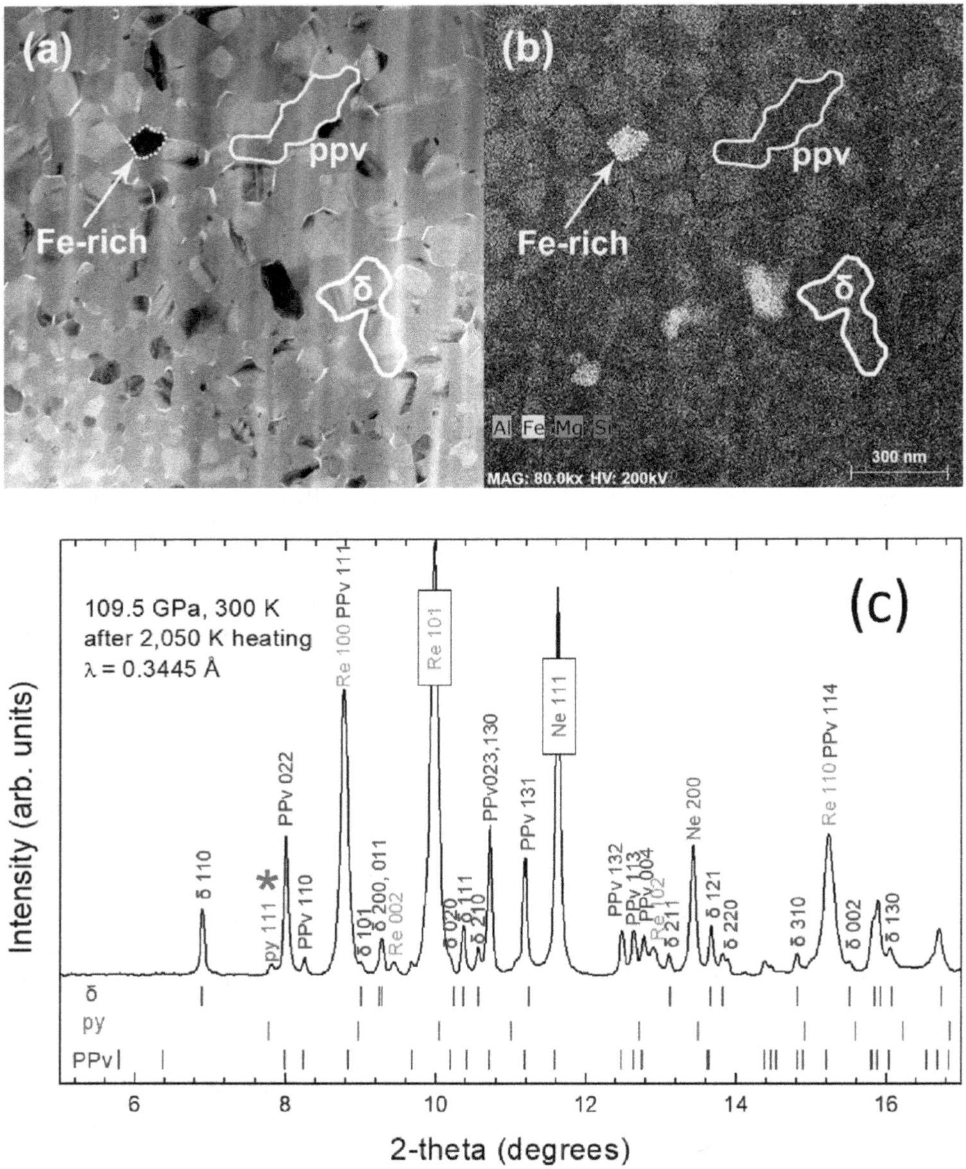

Figure 10.3 A multiphase assemblage contains pPv phase and hydrous δ-phase, coexisting with a minor pyrite-structured phase (py-phase) FeOOH at 110 GPa. (a and b) TEM mapping shows a few Fe-rich grains of about 100 nm in dimensions, which is consistent with the existence of a minor py-phase suggested by (c) in situ XRD where only the (111) peak is visible, marked by a red asterisk. A list of 14 reflections of one selected grain of the py-phase is shown in Table 10.1. For the color version, refer to the plate section.
Reproduced from Zhang et al. (2019).

Table 10.1. *A list of reflections from one selected py-phase $FeOOH_x$ grain with the lattice parameter of a=4.3820(4) Å applying the multigrain indexation. The X-ray wavelength was 0.34454 Å, and ω, η, and θ represent the rotation, azimuth, and Bragg angles, respectively. Only the peak (111) was visible in the integrated powder pattern, as shown in Figure 10.3. After Yuan et al. (2019).*

h	*k*	*l*	2*θ* (°)	ω (°)	η (°)	d_{obs} (Å)	d_{cal} (Å)	Δ*d*/*d* (%)
−1	−1	−1	7.807	12.9	271.74	2.531	2.530	−0.02
−1	1	1	7.806	−1.49	162.93	2.531	2.530	−0.03
1	1	1	7.803	5.08	91.93	2.532	2.530	−0.08
−2	0	0	9.012	15.38	216.78	2.193	2.191	−0.08
2	0	0	9.027	0.34	36.91	2.189	2.191	0.09
−2	0	−1	10.066	−7.5	238.82	1.964	1.960	−0.20
0	2	2	12.775	1.5	127.53	1.548	1.549	0.05
−3	−1	−1	14.985	16.97	241.95	1.321	1.321	0.01
3	1	1	14.984	0.07	62.14	1.321	1.321	0.00
−2	−2	−2	15.655	16.7	271.73	1.265	1.265	0.01
−2	2	2	15.662	−15.3	164.42	1.264	1.265	0.05
2	2	2	15.658	1.1	91.95	1.265	1.265	0.02
4	0	0	18.105	−7.3	36.26	1.095	1.096	0.05
−1	3	3	19.724	−5.7	141.62	1.006	1.005	−0.05

that using focused X-ray beam of 2–10 μm is suitable for multigrain indexation of grains 100–500 nm in size that are commonly present in mineral assemblages under high *P*–*T* conditions.

The indexing algorithms are included in the FABLE package (Sørensen et al., 2012) and the programs can be readily applied for DAC samples at high pressures. The indexing program *ImageD11* (Wright, 2005) has been used to index the orientations of individual grains of the $MgGeO_3$ pPv phase at 90 GPa, and then dislocations of the grains were analyzed (Nisr et al., 2012). The algorithm GrainSpotter efficiently determines individual crystallographic orientations through finding replicas of the set of theoretical scattering vectors in the measured data sets (Schmidt, 2014). The algorithm has been successfully applied for indexation of grains at megabar pressures, and that was followed by single-crystal structure determination for selected grains (Zhang et al., 2013, 2016b). Multigrain indexation assigns an individual orientation matrix to each of the grains in coarse-grained polystalline samples, and therefore each grain can be treated as a single crystal.

10.3 Single-Crystal Structure Determination at Megabar Pressures

One of the greatest motivations for applying the multigrain method to high-pressure data sets is to obtain structures of a quality exceeding that of structures determined by conventional single-crystal or powder X-ray diffraction methods. Each individual grain can be processed applying the existing single-crystal analysis software. However, in conventional high-pressure single-crystal diffraction measurements, the opening cones of DACs limit the possibility of performing the wide omega rotations, which is essential for obtaining high-quality structure data to solve a complicated high-pressure structure. Furthermore, when a single crystal is loaded in a DAC, its orientation has to be carefully arranged to ensure sufficient coverage of the structure (Miletich et al., 2000). Applying the multigrain method, we can select multiple grains with proper orientations to form a complete structure and thus overcome the major limitation in high-pressure single-crystal diffraction measurements. Compatibility between the HKL files from different grains can be checked by merging them using the XSCALE software available within the XDS package (Kabsch, 2010). A complete data set not only provides essential information for determination of a structure model but also can greatly improve the refinement results. Only with high-quality data set, one can distinguish the subtle differences between similar structure models and obtain the correct structure for a high-pressure phase.

10.3.1 Calibration and Powder Diffraction Data

Here we use an example of structure determination for the SiO_2 phase seifertite at 129 GPa to demonstrate the great potential of applying the multigrain method for structure determination at megabar pressures. Figure 10.4a and b shows seifertite as a minor phase coexisting with the dominant phase pPv. The pressure determined from the Ne pressure scale (Fei et al., 2007) was 129 GPa after T quench from 2,500 K. The sample was aligned to the ω-rotation center, and a set of XRD patterns were collected by rotating the DAC from –24 to 24° at steps

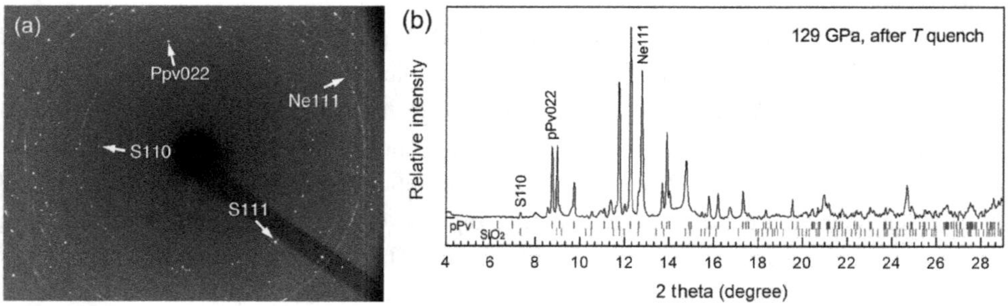

Figure 10.4 XRD data of seifertite at 129 GPa and after T quench (X-ray wavelength of 0.3738 Å). (a) A representative spotty XRD pattern of seifertite coexisting with the dominant phase pPv at a fixed rotation angle. S = silica, pPv = postperovskite, Ne = neon. (b) An integrated powder pattern at a fixed omega angle showing the intensity contrast between the major phase (Mg,Fe)SiO_3–pPv and seifertite. The dominant pPv phase (space group: *Cmcm*) has unit-cell parameters of a = 2.463(2) Å, b = 8.137(8) Å, and c = 6.119(2) Å. Only the characteristic peak (110) of the seifertite SiO_2 phase at 2.912 Å was visible in the powder pattern, and the unit-cell parameters of seifertite SiO_2 (space group: *Pbcn*) can be estimated as a = 3.724 Å, b = 4.667 Å, and c = 4.174 Å based on the observed d spacing of the 110 peak, using the a/c and b/c ratios from the first-principles calculations as a reference (Teter et al., 1998).
Reproduced from Zhang et al. (2016b).

of 0.2° (Zhang et al., 2016b). Knowing the pixel sizes of the detector, the experimental geometry (sample to detector distance, beam center, and detector tilt with respect to the omega rotation axis) can be calibrated against a reference single-crystal and/or a standard powder, e.g., CeO_2. The exposure time was 4 s/frame. The X-ray wavelength was 0.3738 Å and the sample-to-detector distance was 140 mm. In the diffraction patterns, because seifertite is a minor phase in the assemblage, most of the peaks are overshadowed by the strong pPv peaks and background noise (Figure 10.4). Only one characteristic peak, the 110 at 2.912 Å of the seifertite phase, was visible in the low *d*-spacing region (Zhang et al., 2016b). We can estimate the lattice parameters of seifertite as a = 3.724 Å, b = 4.667 Å, and c = 4.174 Å based on the observed d_{110} = 2.912 Å, using the *a/c* and *b/c* ratios of seifertite from first-principles total-energy calculations as a reference (Teter et al., 1998). In the powder diffraction of two-phase mixed assemblage, one characteristic peak in the high *d*-spacing region is indicative of the existence of seifertite. Confirmation of its existence will require additional information. Furthermore, in situ structure determination of a minor phase in an assemblage would be impossible using the conventional single-crystal or powder diffraction methods.

10.3.2 Multigrain Indexation and Grain Selection

To process the rotation data set applying the multigrain method, 2D image coordinates were translated to 3D diffraction scattering vectors using the code included in the program ImageD11 (Wright, 2005). A total of over 40,000 diffraction peaks were filtered. Using the estimated lattice parameters of seifertite as input, we can then index the grains for the phase using the algorithm GrainSpotter (Schmidt, 2014) available within the FABLE package. We set up a completeness as the threshold for the grain selection because a portion of reflections were blocked by the DAC body and weak reflections are not included at this stage for simplicity. A total of 153 grains were indexed for the seifertite phase and 91 grains for the pPv phase, respectively, indicating that the grain size of the dominant pPv phase is significantly larger than that of seifertite.

Through the multigrain indexation, each of the grains is assigned its own particular orientation matrix. Multigrain data sets allow selection of multiple grains with complementary orientations and merging them to form a complete structure. Out of 153 indexed grains for the seifertite phase, three grains with the maximum number of reflections were selected for the structure determination. The orientation matrix of a selected grain can be used to track the grain during data processing, which allows us to identify all the reflections belonging to this grain in the data set.

10.3.3 Single-Crystal Structure Determination from Multigrain Data

Several existing software packages can be used for data reduction of the rotation diffraction images collected in DACs. The CrysAlis software (Oxford Diffraction Ltd, 2006) has been widely used as it has included some useful routines designed to process high-pressure data sets (Merlini and Hanfland, 2013). The GSE_ADA/RSV software package has been developed to process high-pressure crystallographic data (Dera et al., 2013). The XDS package (Kabsch, 2010) is used for reduction of all types of single-crystal diffraction rotation data sets. All the aforementioned programs are performed assuming that reflections from only one crystal are present on the XRD images. That is why, among all the diffraction spots (>40,000), only those originated from a specific grain need to be selected. We select the XDS package for processing the high-pressure multigrain data sets because the XDS package (Kabsch, 2010) provides users the option to override the default parameters at each step during data processing.

The orientation matrix of a selected grain is calculated from reflections identified using the FABLE package (Busing and Levy, 1967), which in turn allowed us to identify all the reflections belonging to this grain in the data set. All further calculation and integration processes can be performed by XDS as normal, assuming there was only one crystal in the beam. Another program was developed to define the DAC cone shadow areas (Zhang et al., 2016b). Each single-crystal data set can be handled following the same procedures as in traditional single-crystal XRD methods. The structure can be solved and refined using the SHELX package (Sheldrick, 2015) or the algorithm *charge flipping* (Oszlanyi and Suto, 2004).

Here we used the OLEX2 program (Dolomanov et al., 2009), which can work seamlessly with other programs such as the SHELX package and provide a graphical user interface for structure solution, refinement, and report generation. The structures were solved by the direct method and refined by full matrix least-squares using the SHELX package (Sheldrick, 2015). However, attempts to obtain the crystallographic data from one crystal alone did not produce satisfactory results. We chose to merge selected grain to improve completeness of the structure. Compatibility between these three data sets was checked by merging them using the XSCALE software available within the XDS package. Eventually, high-quality crystallographic data were obtained from the data sets of two merged grains and three merged grains, respectively.

Details of the crystallographic data are shown in Table 10.2, where three sets of results refined from merged data sets are presented for comparison. Data completeness of 84–93% was achieved in the *d*-spacing range down to 0.73 Å, indicating a nearly full convergence of the structure. The merged data sets have provided a reasonable redundancy of 2.4–3.3. The merged multigrain data sets allow solving the structure of SiO_2 by the direct method at megabar pressures.

10.3.4 Advantages of Applying the Multigrain Method to High-Pressure Data Sets

Overall, multigrain crystallography shows great potential as a general solution for determination of new structures at megabar pressures. Applying multigrain crystallography has shown several obvious advantages for structure analysis at high pressures:

(1) Up to a thousand grains belonging to multiple phases are included in one data set of a typical multigrain sample at high pressures, from which specific grains can be selected for studies of different purposes. For instance, interactions of coexisting phases in a

Table 10.2. *Details of crystallographic data of the SiO_2 phase seifertite at 129 GPa applying the multigrain method. Three grains with maximum number of reflections were selected for structure analysis out of 153 indexed grains. Seifertite has an orthorhombic lattice with* a = *3.7350(7) Å,* b = *4.6650(9) Å,* c = *4.1650(8) Å, and* V = *72.57(2) $Å^3$ with the space group* Pbcn *(#50).*

Crystallographic data	Grains 1 + 2	Grains 1 + 3	Grains 1 + 2 + 3
*x*Si	0.5	0.5	0.5
*y*Si	0.8495(4)	0.8492(5)	0.8497(5)
*z*Si	0.75	0.75	0.75
*x*O	0.2425(10)	0.2405(18)	0.2421(10)
*y*O	0.6130(7)	0.6130(7)	0.6134(7)
*z*O	0.5798(17)	0.5280(15)	0.5809(17)
U_{eq}–Si	0.0049(8)	0.0052(9)	0.0063(10)
U_{eq}–O	0.0033(11)	0.0045(11)	0.0044(12)
Index ranges	$-4 \leq h \leq 5$	$-4 \leq h \leq 4$	$-4 \leq h \leq 5$
	$-6 \leq k \leq 5$	$-5 \leq k \leq 6$	$-6 \leq k \leq 6$
	$-4 \leq l \leq 5$	$-4 \leq l \leq 5$	$-4 \leq l \leq 5$
Completeness to $d = 0.8$ Å, %	85	84	93
Reflections	208	214	308
Independent reflections/Rint	79/0.0613	80/0.0921	87/0.0829
Data/restraints/parameters	79/0/7	80/0/7	87/0/7
GooF	1.194	1.222	1.316
Final *R* indices [$I > 2\sigma(I)$]			
R_1/wR_2	0.0539/ 0.1345	0.0909/0.1540	0.0655/0.1643
Final R indices (all data)			
R_1/wR_2	0.0640/0.1381	0.0659/0.1461	0.0822/0.1696

multiphase assemblage can be examined under high *P–T* conditions applying the multigrain analysis.

(2) One of the greatest advantages of using the multigrain method at high pressures is to obtain a nearly full convergence of the structure through the combination of several grains with different orientation into a merged data set. In conventional single-crystal X-ray diffraction measurements, the opening cones of DACs limit the possibility of performing the wide omega rotations. Applying the multigrain method overcomes this limitation, which in turn allows us to obtain structures of a quality exceeding that of structures determined by conventional single-crystal methods.

(3) To solve a previously unknown structure at high pressure, we are able to select the most suitable grains by applying the multigrain method. In addition, many grains, each with a particular orientation matrix, unambiguously confirm the same structure.

10.4 Online Multigrain Data Analysis during Synchrotron Sessions

In situ X-ray diffraction is used to track phase changes using under high *P–T* conditions during synchrotron beamtimes. Multigrain analysis may enable fast online data analysis and therefore guide ongoing experiments. First of all, we need to check if a sample is suitable for applying the multigrain method based on the level of overlap of the diffraction spots in powder diffraction patterns. Too many grains within the beam can produce severe spot overlap in the XRD patterns. Adjusting X-ray beam size relative to grain size in a polycrystalline sample can help to reduce the spot overlap (Zhang et al., 2019).

Multigrain indexation is not an automatic process in handling a multiphase sample, especially when unknown phases are included. The multigrain indexation for an unknown phase requires knowledge of the lattice dimensions. The lattice dimensions can be obtained from the powder XRD data and tested using the multigrain method. When a correct lattice is selected for indexation, multiple grains will be indexed each with its particular orientation matrix. All possible lattices consistent with powder diffraction data can be tested using the multigrain method repeatedly until the correct lattice is found and confirmed, that is, the multigrain method can be used to recognize an incorrect lattice. The combination of powder diffraction and multigrain indexation can provide unambiguous characterization of the lattice of a new phase.

To efficiently process multigrain analysis, we will need an integration of the existing tools with additional considerations for megabar samples in the DAC. Figure 10.5 shows a flow chart for multigrain indexation and structure determination. A DAC sample is aligned to the ω-rotation center, and the diffracted beam is acquired on a 2D detector (Figure 10.2). A monochromatic X-ray beam of 30–45 keV is commonly used to cover a sufficiently wide d-spacing range. Input parameters include calibration parameters and estimated lattice parameters of a phase. We will obtain (1) the number of indexed grains; (2) orientation matrices for each of the grains; and (3) refined lattice parameters of merged grains for each phase.

In situ monitoring of phase changes under high P–T conditions remains a top task in high-pressure synchrotron XRD experiments. A discontinuity in the P–V relation would suggest a phase transition or compositional change of a phase due to element partitioning in the system. Furthermore, a decreasing number of grains belonging to one phase would suggest disappearance of the phase and appearance of a new phase. The capability to perform fast multigrain analysis online during a synchrotron beamtime will allow in situ studies of phase transitions under high P–T conditions at the grain scale (Rosa et al., 2015) and reliable equation of state (EoS) measurements of individual phases up to extremely high pressures (Zhang et al., 2016a,b). In addition, the output orientation matrix for each grain can be used to track specific grains and perform further detailed crystallographic analysis. Taking into account the limitations of DAC designs and megabar samples, we will integrate existing analysis capabilities and prepare a wrapper program for fast online analysis of a multigrain sample to guide experiments during synchrotron beamtime.

10.5 Future Perspectives

The current capabilities of multigrain crystallography provide a general solution for indexation and structure determination of individual grains as small as 100 nm in a multiphase coarse-grained assemblage at megabar pressures. With the continuing

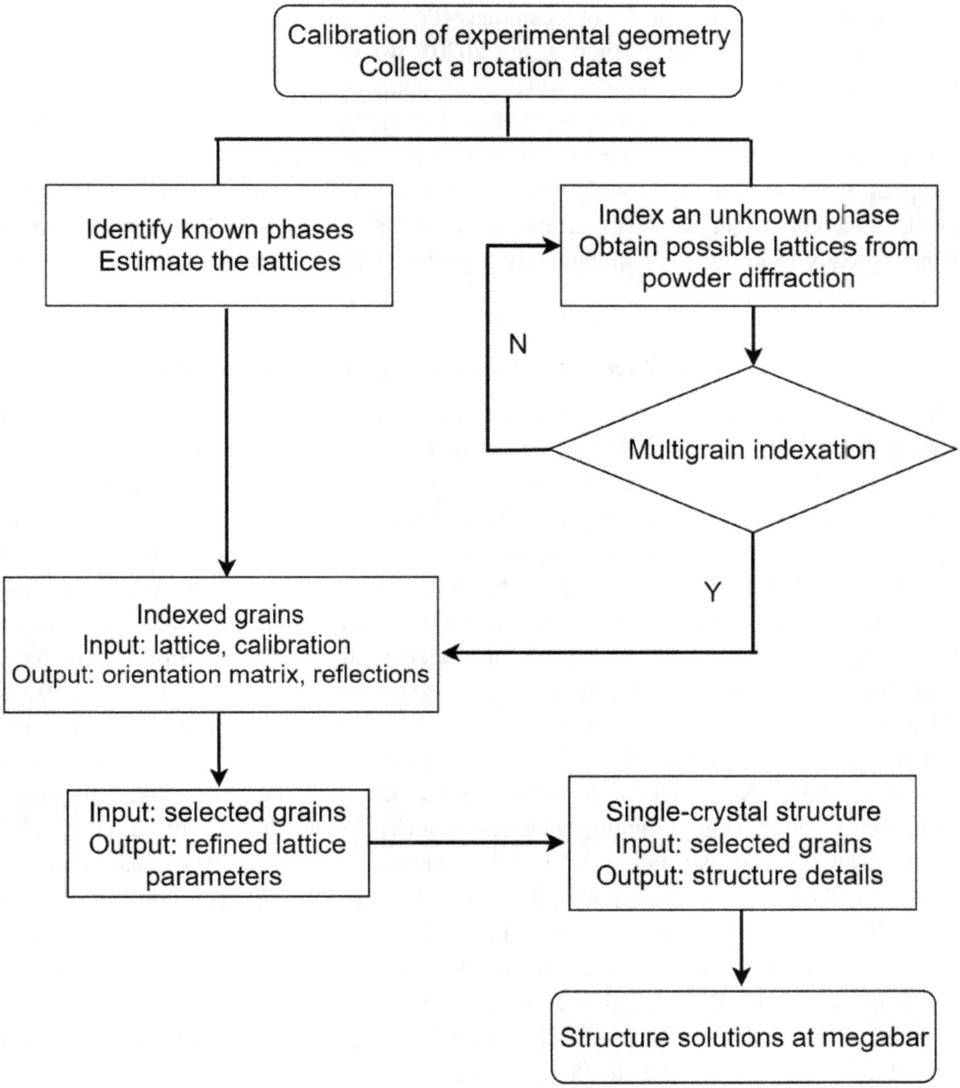

Figure 10.5 Flow chart for processing XRD data when applying the multigrain method. A DAC sample is aligned to the rotation center, and a set of XRD images is collected at small incremental steps by the rotation method. The diffraction spots on the spotty 2D diffraction images are translated to 3D diffraction scattering vectors, applying the program ImageD11 (Wright, 2005) that is included in the FABLE package (Sørensen et al., 2012). The algorithm GrainSpotter efficiently determines individual crystallographic orientations in the measured rotation data set (Schmidt, 2014), which is also included in the FABLE package. Indexation of an unknown phase requires a priori knowledge of its lattice dimensions. All possible lattices consistent with powder diffraction data can be tested using the multigrain method. When a correct lattice is found, multiple grains can be indexed each with its particular orientation matrix. The output orientation matrix can be used to track the grain in the following single-crystal structure analysis. Among all the software available for data reduction, the XDS package (Kabsch, 2010) provides users the option to override the default parameters and add routines for processing high-pressure multigrain data sets. A nearly full convergence of the structure can be achieved by combination of random crystallographic orientations from several grains in a merged data set. The structure can then be solved and refined applying the same procedures as in traditional single-crystal XRD methods.

advancement of synchrotron X-ray techniques (Prakapenka et al., 2008; Liermann et al., 2015; Shen and Mao, 2017; Hirao et al., 2020), we will be able to tune X-ray beam properties, improve the coupling between X-ray optics and laser-heating setup, and have more options for advanced detectors. Considering the unique features of the multigrain techniques, we are facing exciting opportunities to conduct in-depth research of materials at the grain scale under extremely high *P–T* conditions and use the multigrain approach to tackle some challenging problems. Here we propose a few research directions that might benefit greatly from future development of the multigrain techniques.

10.5.1 Pressure Determination in Ultrahigh-Pressure Experiments

The capability of structure determination at the grain scale implies great potential for applying the multigrain method to pressure determination in ultrahigh-pressure experiments, and the spatial resolution of an X-ray probe is the key factor. Recent reported developments show that routine experiments have been made possible in the multimegabar region up to 400 GPa using conventional beveled diamond anvils with a culet size of 20–30 μm (Li et al., 2018; Yagi et al., 2020). When a sample is compressed between two small culets to multimegabar pressures, the anvils are deformed significantly into a cup shape, and a large pressure gradient exists across the culet (Li et al., 2018). Various efforts have been made to generate static pressures beyond the conventional ~400 GPa limit of DACs (Dubrovinsky et al., 2012; Dewaele et al., 2018). The information will be useful to guide future designs of diamond anvils for multimegabar research. However, difficulties in obtaining reliable XRD patterns to determine the peak pressure have caused the major controversies in the field of multimegabar research (Yagi et al., 2020). Only when FWHM of the X-ray beam is as small as 1/10 of the culet size can sharp XRD signals from the sample be obtained at the peak pressure spot (see the details in the recent review by Yagi et al., 2020)). There are only a few beamlines where the submicron-sized X-ray beam is available for such tiny samples at multimegabar pressures. Furthermore, beam tail always brings unwanted signals to the diffraction patterns.

Multigrain analysis can be used to calculate the lattice parameters for each grain, and accordingly pressures can then be determined at the grain level. Three-dimensional orientation and geometrical relationships in the multigrain data set allow separation of reflections with close *d* spacings (Zhang et al., 2019) such that pressure gradient can be mapped at the grain scale. We would expect that the multigrain method will be readily applied for pressure determination with a spatial resolution of ~100 nm scale at multimegabar pressures.

10.5.2 Combination of In Situ X-Ray Diffraction and Ex Situ Chemical Analysis Techniques

Multigrain crystallographic tools are extremely useful in handling multiphase systems at the grain scale. Phase changes are often associated with element partitioning within a multiphase assemblage under high *P–T* conditions (Yuan et al., 2019; Ohira et al., 2014).

In high *P–T* experiments, combining ex situ chemical analysis on recovered samples with in situ powder diffraction and multigrain indexation can be used to determine phase changes and chemical compositions of individual grains in a multiphase assemblage. The focused ion beam (FIB) is used to cut and lift out a thin cross-section in a recovered DAC sample, and the thin section can then be analyzed in a TEM. The consistency between in situ XRD and ex situ TEM analysis should be carefully checked. In a high *P–T* experiment on the multicomponent system ~30 mol% Al_2O_3–10 mol% Fe_2O_3–60 mol% $MgSiO_3$ containing about 7 wt.% H_2O, the py-phase was unambiguously confirmed by sorting out 42 grains applying the multigrain indexation (Figure 10.3) (Zhang et al., 2019). Figure 10.3b shows compositional mapping of a thin cross section from the recovered sample, showing several grains belonging to a very Fe-rich phase. Energy-dispersive spectroscopy (EDS) analysis verified only the elements Fe and oxygen (O) in the grains, while hydrogen (H) is undetectable if present. The element mapping shows that the py-phase occupies less than 3% by volume in the assemblage. Therefore, the chemical composition of the grains and its volume proportion are consistent with in situ XRD observation of the py-phase under high *P–T* conditions.

Note that most high-pressure structures are unquenchable to ambient conditions, and therefore in situ structure determination is required for characterization of the high-pressure phases. Single-crystal structure determination requires input of a chemical formula of the phase. Chemical analysis on the recovered samples can be used to confirm elements for each phase and obtain a chemical formula in some cases. Combination of in situ X-ray powder diffraction and multigrain indexation with ex situ chemical analysis is a useful technique to sort out new phases in assemblages synthesized under high *P–T* conditions.

10.5.3 Limitations of the Multigrain Techniques

Overall, the multigrain method has shown great potential for phase identification and structure characterization at megabar pressures that are otherwise impossible using conventional single-crystal techniques. Still, this technique has limitations. First of all, it would be difficult to proceed with such multigrain analysis without a priori determination of the lattice dimensions of the newly formed phases under high *P–T* conditions. Powder diffraction patterns only contain *d*-spacing information, and indexing match of multiple peaks could suggest possible lattices. The multigrain indexation can be used to select the correct lattice with the stringent requirement that all reflections from each grain must satisfy the three-dimensional orientation and geometrical relationship predicted by its particular orientation matrix and the lattice parameters. It can be problematic if the correct lattice cannot be determined due to overlapping with peaks from coexisting phases in a multiphase assemblage. Another main limitation for applying the multigrain method is the overlap of diffraction spots on the 2D raw images. Spotty diffraction patterns are impossible to obtain for some materials under high *P–T* conditions.

Successful indexation and structure determination depends on availability of software that can be applied to high-pressure datasets. Through processing data sets obtained at megabar pressures, we have demonstrated that up to 1,000 grains can be indexed in a multiphase assemblage at megabar pressures applying the FABLE package (Sørensen et al., 2012). Among the software for data reduction, the XDS package (Kabsch, 2010) allows the users to include the routines that are essential for processing the high-pressure multigrain data sets. Further development of programs will promote wide application of the multigrain method in high-pressure research.

Crystal structure determination remains the dominant task in studies of physics, chemistry, Earth, and materials sciences at high pressures. Combination of in situ X-ray powder diffraction, multigrain indexation, and single-crystal structure determination on individual grains with ex situ chemical analysis can be readily applied to characterize new phases under high *P–T* conditions. In summary, multigrain crystallography enables high-quality crystal structure determination, processing multiphase assemblages at the grain scale, and fast online analysis of multiphase systems during synchrotron beamtime at megabar pressures. These capabilities can be further extended to multimegabar pressures.

Acknowledgments

The authors thank Dmitry Popov, Yue Meng, Guoyin Shen, Marco Merlini, and Junliang Sun for their helpful discussions and comments. This work was supported by the National Natural Science Foundation of China (42150103).

References

Boehler, R., De Hantsetters, K. (2004). New anvil designs in diamond-cells. *High Pressure Research*, **24**(3), 391–396.

Busing, W. R., Levy, H. A. (1967). Angle calculations for 3- and 4- circle X-ray and neutron diffractometers. *Acta Crystallographica*, **22**, 457–464.

Bykova, E., Dubrovinsky, L., Dubrovinskaia, N., et al. (2016). Structural complexity of simple Fe2O3 at high pressures and temperatures. *Nature Communications*, **7**(1), 10661.

Cerantola, V., Bykova, E., Kupenko, I., et al. (2017). Stability of iron-bearing carbonates in the deep Earth's interior. *Nature Communications*, **8**(1), 15960.

Dera, P., Zhuravlev, K., Prakapenka, V., et al. (2013). High pressure single-crystal micro X-ray diffraction analysis with GSE_ADA/RSV software. *High Pressure Research*, **33**(3), 466–484.

Dewaele, A., Loubeyre, P., Occelli, F., Marie, O., Mezouar, M. (2018). Toroidal diamond anvil cell for detailed measurements under extreme static pressures. *Nature Communications*, **9**(1), 2913.

Dolomanov, O. V., Bourhis, L. J., Gildea, R. J., Howard, J. A. K., Puschmann, H. (2009). OLEX2: a complete structure solution, refinement and analysis program. *Journal of Applied Crystallography*, **42**(2), 339–341.

Dubrovinsky, L., Boffa-Ballaran, T., Glazyrin, K., et al. (2010). Single-crystal X-ray diffraction at megabar pressures and temperatures of thousands of degrees. *High Pressure Research*, **30**(4), 620–633.

Dubrovinsky, L., Dubrovinskaia, N., Prakapenka, V. B., Abakumov, A. M. (2012). Implementation of micro-ball nanodiamond anvils for high-pressure studies above 6 Mbar. *Nature Communications*, **3**(1), 1163.

Fei, Y., Ricolleau, A., Frank, M., Mibe, K., Shen, G., Prakapenka, V. (2007). Toward an internally consistent pressure scale. *Proceedings of the National Academy of Sciences*, **104**(22), 9182.

Gregoryanz, E., Lundegaard, L. F., McMahon, M. I., Guillaume, C., Nelmes, R. J., Mezouar, M. (2008). Structural diversity of sodium. *Science*, **320**(5879), 1054.

Hirao, N., Kawaguchi, S. I., Hirose, K., Shimizu, K., Ohtani, E., Ohishi, Y. (2020). New developments in high-pressure X-ray diffraction beamline for diamond anvil cell at SPring-8. *Matter and Radiation at Extremes*, **5**(1), 018403.

Kabsch, W. (2010). XDS. *Acta Crystallographica Section D*, **66**(2), 125–132.

Lavina, B., Dera, P., Kim, E., et al. (2011). Discovery of the recoverable high-pressure iron oxide Fe₄O₅. *Proceedings of the National Academy of Sciences*, **108**(42), 17281.

Lavina, B., Meng, Y. (2015). Unraveling the complexity of iron oxides at high pressure and temperature: synthesis of Fe5O6. *Science Advances*, **1**(5), e1400260

Li, B., Ji, C., Yang, W., et al. (2018). Diamond anvil cell behavior up to 4 Mbar. *Proceedings of the National Academy of Sciences*, **115**(8), 1713.

Liermann, H.-P., Konopkova, Z., Morgenroth, W., et al. (2015). The extreme conditions beamline P02.2 and the extreme conditions science infrastructure at PETRA III. *Journal of Synchrotron Radiation*, **22**(4), 908–924.

Lundegaard, L. F., Weck, G., McMahon, M. I., Desgreniers, S., Loubeyre, P. (2006). Observation of an O8 molecular lattice in the ε phase of solid oxygen. *Nature*, **443** (7108), 201–204.

Mao, H. K., Bell, P. M. (1978). High-pressure physics: sustained static generation to 1.36–1.72 megabars. *Science*, 200, 1145–1147.

Meng, Y., Hrubiak, R., Rod, E., Boehler, R., Shen, G. (2015). New developments in laser-heated diamond anvil cell with in situ synchrotron X-ray diffraction at high pressure collaborative access team. *Review of Scientific Instruments*, **86**(7), 072201.

Merlini, M., Cerantola, V., Gatta, G. D., et al. (2017). Dolomite-IV: candidate structure for a carbonate in the Earth's lower mantle. *American Mineralogist*, **102**(8), 1763–1766.

Merlini, M., Hanfland, M. (2013). Single-crystal diffraction at megabar conditions by synchrotron radiation. *High Pressure Research*, **33**(3), 511–522.

Merlini, M., Hanfland, M., Salamat, A., Petitgirard, S., Müller, H. (2015). The crystal structures of Mg2Fe2C4O13, with tetrahedrally coordinated carbon, and Fe13O19, synthesized at deep mantle conditions. *American Mineralogist*, **100**(8-9), 2001–2004.

Miletich, R., Allan, D. R., Kuhs, W. F. (2000). High-pressure single-crystal techniques. *Reviews in Mineralogy and Geochemistry*, **41**(1), 445–519.

Murakami, M., Hirose, K., Kawamura, K., Sata, N., Ohishi, Y. (2004). Post-perovskite phase transition in MgSiO3. *Science*, **304**(5672), 855–858.

Nisr, C., Ribárik, G., Ungár, T., Vaughan, G. B. M., Cordier, P., Merkel, S. (2012). High resolution three-dimensional X-ray diffraction study of dislocations in grains of $MgGeO_3$ post-perovskite at 90 GPa. *Journal of Geophysical Research*, **117**(B3), B03201.

Ohira, I., Ohtani, E., Sakai, T., et al. (2014). Stability of a hydrous δ-phase, AlOOH–MgSiO2(OH)2, and a mechanism for water transport into the base of lower mantle. *Earth and Planetary Science Letters*, **401**, 12–17.

Oszlanyi, G., Suto, A. (2004). Ab initio structure solution by charge flipping. *Acta Crystallographica Section A*, **60**(2), 134–141.

Oxford Diffraction Ltd. (2006). CrysAlis Red.

Prakapenka, V. B., Kubo, A., Kuznetsov, A., et al. (2008). Advanced flat top laser heating system for high pressure research at GSECARS: application to the melting behavior of germanium. *High Pressure Research*, **28**(3), 225–235.

Rosa, A. D., Hilairet, N., Ghosh, S., et al. (2015). In situ monitoring of phase transformation microstructures at Earth's mantle pressure and temperature using multi-grain XRD. *Journal of Applied Crystallography*, **48**(5), 1346–1354.

Sørensen, H. O., Schmidt, S., Wright, J. P., et al. (2012). Multigrain crystallography. *Zeitschrift für Kristallographie*, **227**(1), 63–78.

Schmidt, S. (2014). GrainSpotter: a fast and robust polycrystalline indexing algorithm. *Journal of Applied Crystallography*, **47**(1), 276–284.

Sheldrick, G. M. (2015). SHELXT – integrated space-group and crystal-structure determination. *Acta Crystallographica. Section A, Foundations and Advances*, **71**(1), 3–8.

Shen, G., Mao, H. K. (2017). High-pressure studies with X-rays using diamond anvil cells. *Reports on Progress in Physics*, **80**(1), 016101.

Tateno, S., Hirose, K., Ohishi, Y., Tatsumi, Y. (2010). The structure of iron in Earth's inner core. *Science*, **330**(6002), 359–361.

Teter, D. M., Hemley, R. J., Kresse, G., Hafner, J. (1998). High pressure polymorphism in silica. *Physical Review Letters*, **80**(10), 2145–2148.

Wang, L., Ding, Y., Yang, W., et al. (2010). Nanoprobe measurements of materials at megabar pressures. *Proceedings of the National Academy of Sciences*, **107**(14), 6140.

Wright, J. P. (2005). ImageD11. https://sourceforge.net/p/fable/code/HEAD/tree/ImageD11/, ESRF, Grenoble.

Yagi, T., Sakai, T., Kadobayashi, H., Irifune, T. (2020). Review: high pressure generation techniques beyond the limit of conventional diamond anvils. *High Pressure Research*, **40**(1), 148–161.

Yuan, H., Zhang, L., Ohtani, E., Meng, Y., Greenberg, E., Prakapenka, V. B. (2019). Stability of Fe-bearing hydrous phases and element partitioning in the system MgO–Al2O3–Fe2O3–SiO2–H2O in Earth's lowermost mantle. *Earth and Planetary Science Letters*, **524**, 115714.

Zhang, L., Meng, Y., Dera, P., Yang, W., Mao, W. L., Mao, H. K. (2013). Single-crystal structure determination of (Mg,Fe)SiO3 postperovskite. *Proceedings of the National Academy of Sciences of the United States of America*, **110**(16), 6292–6295.

Zhang, L., Meng, Y., Mao, H.-k. (2016a). Unit cell determination of coexisting post-perovskite and H-phase in (Mg,Fe)SiO3 using multigrain XRD: compositional variation across a laser heating spot at 119 GPa. *Progress in Earth and Planetary Science*, **3**(1), 13.

Zhang, L., Popov, D., Meng, Y., et al. (2016b). In-situ crystal structure determination of seifertite SiO2at 129 GPa: studying a minor phase near Earth's core–mantle boundary. *American Mineralogist*, **101**(1), 231–234.

Zhang, L., Yuan, H., Meng, Y., Mao, H. K. (2018) Discovery of a hexagonal ultradense hydrous phase in (Fe,Al)OOH. *Proceedings of the National Academy of Sciences USA*, **115**(12), 2908–2911.

Zhang, L., Yuan, H., Meng, Y., Mao, H.-K. (2019) Development of high-pressure multigrain X-ray diffraction for exploring the Earth's interior. *Engineering*, **5**(3), 441–447.

11

Deformation and Plasticity of Materials under Extreme Conditions

SÉBASTIEN MERKEL

Understanding mechanical properties and their microscopic origins is fundamental for multiple fields in condensed matter research. They are controlled by defects, dislocations, diffusion, as well as microstructures, which are not trivial to study under extreme conditions. This chapter summarizes the last 25 years of advances in high-pressure devices, X-ray measurements, and data interpretation capabilities for addressing the deformation and plasticity of materials under extreme conditions, from experiments in large-volume presses or diamond anvil cells, texture and stress analysis in powder X-ray diffraction, multigrain crystallography, to self-consistent models of materials behavior. Examples of applications are then provided in the fields of geophysics and materials science along with perspectives for studies of plastic deformation under extreme conditions in the coming years.

11.1 Introduction

Mechanical properties are important for a range of applications, from the design of materials with advanced properties to the dynamics of planetary interiors. The plastic behavior of solids is controlled by defects, dislocations, diffusion, as well as microstructures, a generic term describing the arrangement of a material from the nanometer to the centimeter scale. Microstructures in minerals, for instance, are important for the dynamics of the deeper layers of the Earth (Karato et al., 2000). For such applications, materials of interest include the high-pressure phase of Fe, ε-Fe (Wenk et al., 2000; Lincot et al., 2016), bridgmanite and ferropericlase (Marquardt and Miyagi, 2015; Girard et al., 2016; Miyagi and Wenk, 2016; Nzogang et al., 2018), or postperovskite (Merkel et al., 2007; Miyagi et al., 2010; Dobson et al., 2013), most of which are not stable under ambient pressure and temperature and should be studied in situ. Hydrostatic pressure also induces phase transformations in metals. Titanium and zirconium, for instance, are hexagonal-close-packed metals under ambient conditions. Under high pressure, they transform to another hexagonal structure, the ω phase, which is not compact and for which mechanical properties are poorly constrained (Wenk et al., 2013; Yu et al., 2015; Kumar et al., 2020). In other cases, high pressure has been used as a fine-tuning parameter for optimizing strength and grain sizes (e.g., Zhou et al., 2020). As such, the study of mechanical properties in situ under extreme conditions is of general interest.

The last 25 years have seen tremendous advances in high-pressure instruments coupled with synchrotron radiation that allow studies of microstructures and mechanical properties under high pressure and high temperature (Raterron and Merkel, 2009). The group at the Geophysical Laboratory and H.-K. Mao provided key contributions to this field, through the design of new methods and groundbreaking publications (e.g., Hemley et al., 1997; Mao et al., 1998; Wenk et al., 2000; Merkel et al., 2002). Nowadays, millimeter size samples are deformed in large-volume presses (LVPs) up to Earth's uppermost lower-mantle pressures. High pressures and lower-dimension samples are deformed and studied in diamond anvil cells (DACs). In all cases, samples are submitted to a macroscopic deformation to induce plasticity and deformation microstructures. Their properties are then studied in situ using X-ray diffraction and imaging. X-ray radiography allows the measurement of macroscopic properties, such sample size and the applied macroscopic deformation. X-ray diffraction on polycrystals allows for extracting average sample properties by the study of lattice-preferred orientations (LPOs) and the average stress state. Multigrain X-ray diffraction (sometimes labeled as 3D-XRD) addresses single grains inside a polycrystalline matrix and, in some cases, allows for measuring densities of defects such as dislocations. As such, the properties of deforming materials can be studied in situ, in their stability field or conditions of application.

In this chapter, I will present the experimental techniques for the high-pressure study of materials plasticity. I will start with a description of the experimental devices, then present the characterization techniques using in situ X-rays, modeling using self-consistent calculations, sample results for Earth's and other materials, and perspectives for the years to come.

11.2 Experimental Techniques

11.2.1 Plasticity in the Large-Volume Press

The Deformation-DIA (D-DIA), Rotational Drickamer (RDA), and deformation T-Cup (D-Tcup) are the main LVPs for deformation experiments coupled with in situ characterization under high pressure. All allow the controlled deformation of millimeter-sized samples. In addition, new apparatuses such as the rotational tomographic Paris–Edinburgh cell (RoToPEc) and high-pressure X-ray tomography microscope (HPXTM) are devices under development (Wang et al., 2011; Philippe et al., 2016).

The D-DIA (Wang et al., 2003; Figure 11.1) is a hydraulic press in which a cubic assembly is deformed under the action of six anvils. Heating sleeves allow temperatures up to ≈2,000 K. Hydrostatic compression is obtained by advancing all six anvils (four horizontal and two vertical) at the same velocity. Axial compression deformation at constant pressure is obtained by advancing the upper and lower anvils while retracting the horizontal anvils. Lateral compression is obtained by retracting the vertical anvils and advancing the horizontal anvils. The D-DIA allows experiments at strain rates ranging from 10^{-7} to 10^{-2} s^{-1} and pressures and temperatures up to 18 GPa and 1,900 K (Kawazoe et al., 2013). It has also been adapted for deformation using a shear geometry up to 25 GPa and

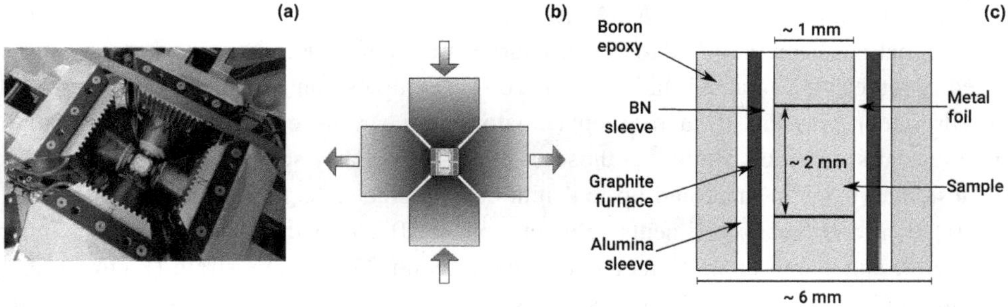

Figure 11.1 (a) Main module of the D-DIA deformation press at the ID06 beamline of the ESRF synchrotron. The main anvils, labeled as here E, W, N, and S, apply a load on secondary anvils to generate deformation. The lower (hidden) and top (opened) anvils are not visible in the image. (b) Principle of the D-DIA. Vertical and horizontal anvils can be moved independently. (c) Typical sample assembly placed between the six anvils for deformation experiments. For simplicity, the thermocouples measuring temperature are not shown.

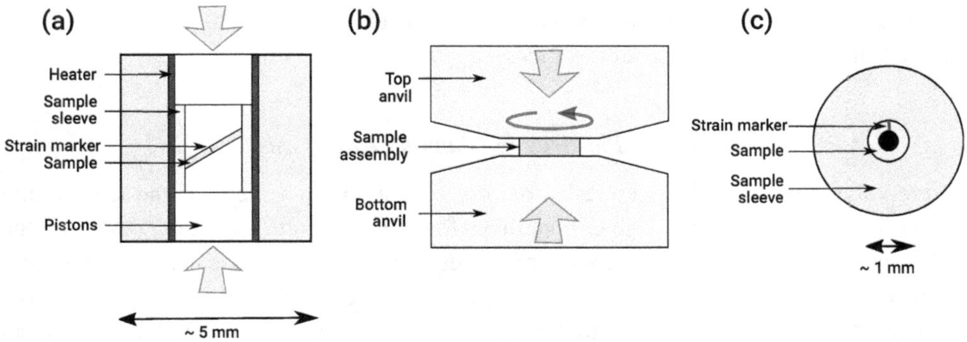

Figure 11.2 Typical sample assemblies for shear deformation in the LVP. (a) Shear deformation in the D-DIA (e.g., Tsujino et al., 2016). Axial strain (red arrows) is applied to the cell assembly, which translates to shear strain on the sample. (b) Torsion in the RDA. Pressure is applied by advancing the top and bottom anvils. The cell assembly is submitted to torsion by rotating one of the anvils. (c) Simplified top view of the cell assembly in the RDA in Girard et al. (2016). The sample is a ring of 0.2 mm thickness, 0.45 mm inner diameter, and 1 mm outer diameter.

1,873 K (e.g., Figure 11.2a; Tsujino et al., 2016; Nishihara et al., 2018). Work is under way to develop new generations of D-DIA-like presses such as the D-DIA-30, allowing reaching higher pressures and working on larger samples (e.g., Wang and Shen, 2014). The D-TCup (Hunt et al., 2014; Hunt and Dobson, 2017) is a similar setup with second-stage anvils being developed to allow deformation experiments at pressures and temperatures exceeding those of the D-DIA. In all cases, the imposed macroscopic strain of the cell assembly is axial and measured in situ using X-ray radiography (discussed later in this chapter). Lateral macroscopic strain is reconstructed based on the axial macroscopic strain and the sample's change of volume, deduced from X-ray diffraction.

The rotational Drickamer (RDA, Figure 11.2b) is an opposed anvil high-pressure device that has been modified to perform large strain deformation experiments in simple shear geometry (Yamazaki and Karato, 2001). Samples can be sheared between two anvils under pressure by a rotation actuator, with strains exceeding $\gamma \approx 6$ at high pressure at strain rates similar to those of the D-DIA. Early sample designs used disks of less than 0.8 mm thick and up to 4 mm outer diameter (e.g., Yamazaki and Karato, 2001), while recent experiments rely on rings of 0.2 mm thickness, 0.45 mm inner diameter, and 1 mm outer diameter for a better control of sample strain (Figure 11.2c; Girard et al., 2016). The RDA was successfully used to obtain quantitative creep results on wadsleyite (Nishihara et al., 2008; Kawazoe et al., 2010; Farla et al.. 2015), ringwoodite (Miyagi et al., 2014), and bridgmanite and magnesiowüstite aggregates (Girard et al., 2016) up to 27 GPa and 2,150 K. In the RDA, the stress applied to the sample is evaluated using X-ray diffraction and is a combination of pure shear and axial compression (Xu et al., 2005).

New advances in the field also include the coupling deformation in simple shear using torsion devices such as the RoToPEc and HPXTM with tomographic measurements. Proofs of principles have been published (Wang et al., 2011; Philippe et al., 2016) and the devices are currently in use at synchrotron sources.

11.2.2 Plasticity in Diamond Anvil Cells

Large-volume presses allow controlled deformation at constant pressure and temperature but do not cover the entire range of conditions found in the Earth's interior. DACs, on the other hand, allow static experiments at pressures exceeding that at the center of the Earth (Dewaele et al., 2018) and combined pressures and temperatures of the Earth's core (Tateno et al., 2010). In addition to confining pressure, diamonds in the DAC impose an axial compressive stress. In most experiments, the compressive stress is reduced by using pressure media around the sample. In plasticity studies, this comes at a benefit that allows macroscopic deformation of the sample. Unlike deformation in LVP, however, pressure and deformation cannot be decoupled. Deformation experiments in the DAC are not performed at constant pressure nor at controlled stress or strain rates. The total plastic strain in DAC experiments, estimated from the magnitude of texture evolution, is on the order of 10–30%. It is applied over a duration of minutes to a few hours, translating to a strain-rate estimate ranging from 10^{-5} to 10^{-3} s^{-1}.

In most DAC experiments, the sample properties are studied through the diamond anvils, which are transparent over a broad range of wavelength, including X-rays (Figure 11.3a). For plasticity studies, it can be useful to have the incoming X-ray beam perpendicular to the loading axis (Figure 11.3b). In the early radial diffraction experiments, the sample itself served as a gasket material (Kinsland and Bassett, 1977). The sample stress state, however, was complex, with large pressure gradients across the sample. The technique was then improved by the group at the Geophysical Laboratory with the use of beryllium gaskets (Hemley et al., 1997) for a better controlled sample pressure and stress state, and further improved with the use of amorphous boron–epoxy composite gaskets (Merkel and Yagi, 2005). Early radial X-ray diffraction experiments were performed at

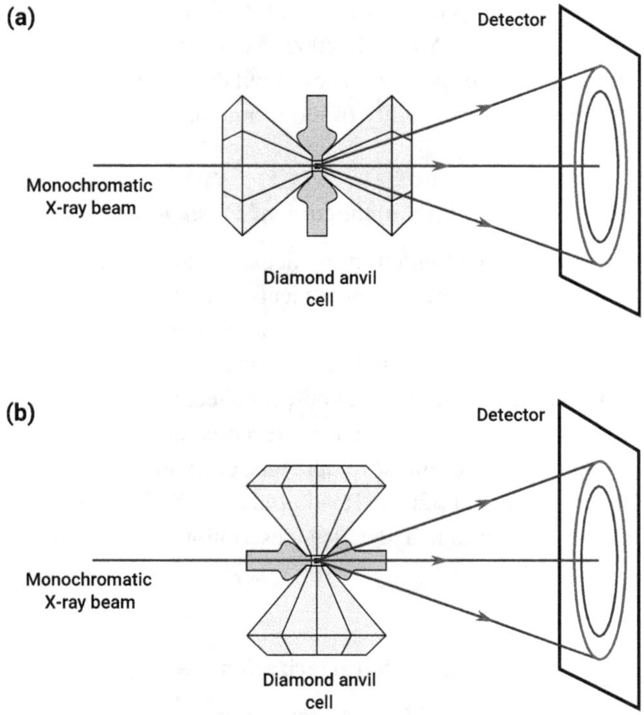

Figure 11.3 Geometries for X-ray diffraction DAC experiments. The axial geometry (a) is most common, with the incident X-ray beam aligned with the compression direction. The radial geometry (b) is commonly used in deformation experiments. The incoming and diffracted X-ray beams are perpendicular to compression and pass through an X-ray transparent gasket made of beryllium or composite materials.

ambient temperature (Hemley et al., 1997; Mao et al., 1998; Wenk et al., 2000; Merkel et al., 2002).

High-temperature deformation studies have been attempted by combining laser heating and diffraction in the radial geometry, with limited success due to large temperature gradients in the sample (Miyagi et al., 2008). In order to obtain homogeneous conditions of stress, pressure, and temperature, efforts were hence invested in developing dedicated DAC systems combined with resistive heating allowing for radial X-ray diffraction (Liermann et al., 2009; Immoor et al., 2020), sometimes also complemented with additional laser heating (Miyagi et al., 2013). Under ambient temperature, radial diffraction experiments reached pressures of nearly 300 GPa (Hemley et al., 1997). Combined conditions of 62 GPa at 1,400 K were obtained in the resistive heating setup (Immoor et al., 2018) and 69 GPa at 2,273 K for setups combining laser and resistive heating (Miyagi et al., 2013).

In addition, deformation experiments can be performed using a rotational diamond anvil cell (rDAC). In the rDAC, one of the diamond anvils is allowed to rotate, which produces a torsional deformation like in the RDA (Blank et al. 1984). The rDAC has the potential to achieve large strain under ultrahigh-pressure conditions while attaining a steady-state deformation (Levitas, 2004; Nomura et al., 2017). To this day, however, the use of the rDAC for true deformation experiments has been limited due to the complex stress and strain field in rDAC experiments (Ma et al., 2006, Zarechnyy et al., 2012). The reliability of rDAC experiments is being improved, with new anvil designs for slip-free experiments that could considerably improve the usability of the technique (Azuma et al., 2017).

11.2.3 Computational Plasticity

Experimental techniques can be limited. In particular, when addressing the plastic properties of minerals in planetary interiors, experiments are performed at strain rates that are orders of magnitude above those of nature. As such, numerical approaches can become useful to understand, model, and extrapolate materials behavior.

Plastic deformation is the result of the complex, collective behavior of a large collection of defects within a material microstructure. Until recently, this task was not within the reach of numerical techniques, but recent developments open the door to such studies (e.g., Cordier et al., 2012; Boioli et al., 2017; Reali et al., 2019). Such methods go beyond the scope of the current chapter, but a detailed description can be found in Cordier and Goryaeva (2018).

11.3 In Situ Characterization Techniques

11.3.1 Deformation

In LVPs, the sample's macroscopic deformation is measured using X-ray Radiography. Strong X-ray absorbing samples are visible directly while, for less absorbing samples, one places foils of gold, platinum, or molybdenum in the sample assembly (e.g., Figure 11.1c). Based on the displacement, shortening, or tilting of the metal makers, one can then reconstruct the macroscopic deformation applied to the sample (e.g., Raterron and Merkel, 2009; Farla et al., 2015). Strain versus time plots are then used to evaluate the sample strain rate.

In DACs, deformation is rarely controlled and occurs during pressure increases. Moreover, the sample thickness is on the order of 20 μm. Decent sample images can be obtained using changes in X-ray transmission contrast (Merkel and Yagi, 2005), but deformation is not very well resolved, and hence such images are rarely used. This could be improved using phase contrast imaging (e.g., Schropp et al., 2015), which allows for a much better resolved imaging of interfaces, and particularly so at upgraded synchrotron sources such as ESRF Extremely Brilliant Source (EBS), which offers improved performance by a factor of 100 in beam coherence. To my knowledge, this has not been attempted yet.

11.3.2 Polycrystal Properties

Powder X-ray diffraction is a standard technique for assessing average macroscopic sample properties and can be applied to both DAC and LVP experiments. The diameter of the

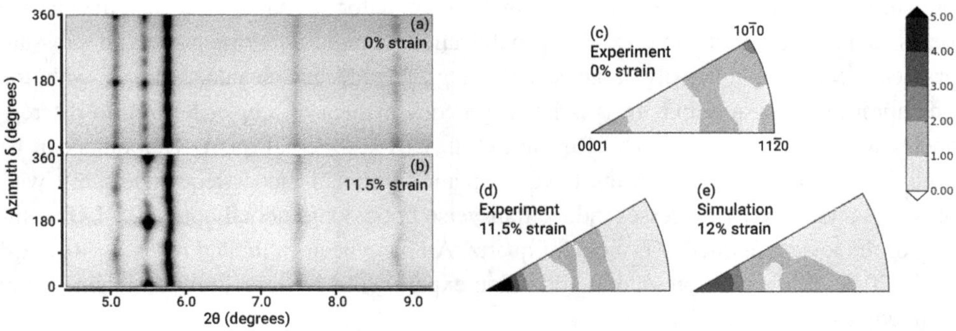

Figure 11.4 Diffraction images for ε-Fe at 17 GPa and 400 K (Merkel et al., 2012) measured at the start of the deformation (a) and after 11.5% axial strain in a D-DIA (b). Variations of peak intensities and positions with azimuth are indicative of the sample LPO and stress, respectively. Measured inverse pole figure of the compression direction at the start of deformation (c) and after 11.5% axial strain (d). (e) Results of a self-consistent model after 12% axial strain reproducing the measured sample strain and textures and constraining the strength of basal, prismastic, and pyramidal slip as well as tensile twinning in ε-Fe at those conditions. For the color version, refer to the plate section.

incident X-ray beam and sample microstructures, however, should be adjusted so that a statistically relevant number of grains contribute to the diffraction signal. Otherwise, statistical assumptions with the powder X-ray diffraction processing methods will fail and may lead to inconsistent results.

11.3.2.1 Lattice-Preferred Orientations

Often, crystallites in a polycrystal are not randomly distributed in orientation and materials develop LPOs. LPOs can arise from processes such as nucleation, phase transformation, or plastic deformation. LPOs have a strong influence on mechanical properties as physical properties become anisotropic (Kocks et al., 1998). They also affect the propagation of seismic waves and are hence of great interest in geosciences (Mainprice et al., 2000).

In high-pressure deformation experiments, LPO measurements are often used to constrain plastic deformation mechanisms (Wenk et al., 2006). In fact, plastic deformation mechanisms, and in particular dislocation glide, will give rise to specific LPOs in polycrystals. LPO measurements, combined with texture modeling (described later in this chapter), allow the identification of the active plastic deformation mode in a material.

Sample LPOs appear as variation of diffraction intensities with orientation (Figure 11.4) and can be extracted from the diffraction images using either Multifit (Merkel and Hilairet, 2015) and BEARTEX (Wenk et al., 1998) or MAUD (Lutterotti et al., 1999; Wenk et al., 2014).

11.3.2.2 Stress and Strains

Stress–strain and stress–pressure curves are a critical element to characterize the mechanical properties of materials. In situ determination of stress, however, is not trivial in high-pressure experiments. Unlike classical, low-pressure deformation experiments, the applied stress is not measured directly. Early publications estimated stress based on pressure gradients across a relatively large sample (e.g., Meade and Jeanloz, 1988), with the limitation that pressure and stress is heterogeneous in such a sample, leading to discrepancies with other measurements (Reynard et al., 2019). Internal quartz calibrants can be used for in situ piezometry in the DAC. Raman vibrational mode frequencies shift with pressure, and deviatoric stresses induce transverse optic–longitudinal optic (TO-LO) splitting of the lowest frequency E mode of quartz. All have been calibrated recently (Reynard et al., 2019). Most high-pressure deformation experiments, however, rely on in situ X-ray diffraction for stress evaluation.

Residual stress analysis using X-ray diffraction is a common technique in materials science (Noyan and Cohen, 1987) and was adapted to high-pressure research in the mid-1990s (e.g., Hemley et al., 1997; Singh et al. 1998). Hydrostatic pressure changes the unit cell parameters of a material and hence the average *d* spacings of all powder X-ray diffraction lines. Deviatoric stresses add an additional component with changes of the measured *d* spacings with orientation (Figure 11.4a). Theories relating deviatoric stress and the measured strains using X-ray diffraction are available in the literature, both for axial (Singh et al., 1998) and shear (Nishihara et al., 2008) deformation experiments. All rely on elasticity theory and assume continuity of stress or strain within the deforming aggregate.

The elastic assumption for stress inversion, however, fails when plastic deformation is at play (Li et al., 2004, Merkel et al., 2006). In fact, as materials deform, stresses will relax in grains in soft orientations (i.e., for which plastic deformation can be easily activated), while other grains, whose orientations do not favor plastic deformation, will remain highly stressed. For each orientation, the measured *d* spacing in powder X-ray diffraction is the average *d* spacing of all grains contributing to the diffraction, either soft or hard. This heterogeneity of stress within grains contributing to the same diffraction peak is not well captured by elasticity-based models. The average stress state and intergranular stress heterogeneities in a deforming material, however, can be modeled using self-consistent techniques (as discussed later in this section).

Another approach to circumvent the issue of stress evaluation using X-ray diffraction is to use an internal standard. Al_2O_3, for instance, is well known and was calibrated using self-consistent techniques (Raterron et al., 2013). Pyrope is another option as experiments have shown that it is plastically isotropic with less than 10% variation for stresses deduced between different diffraction lines (e.g., Girard et al., 2020).

11.3.2.3 Interpretation Using Self-Consistent Models

Deformation experiments are best interpreted using self-consistent (SC) methods. SC calculations treat each grain of the polycrystal as an inclusion in a homogeneous but anisotropic medium (Figure 11.5). The properties of the medium are determined by the average of all inclusions. At each deformation step, the inclusion and medium interact, and the macroscopic properties are updated iteratively until the average strain and stress of all inclusions equal the macroscopic strain and stress.

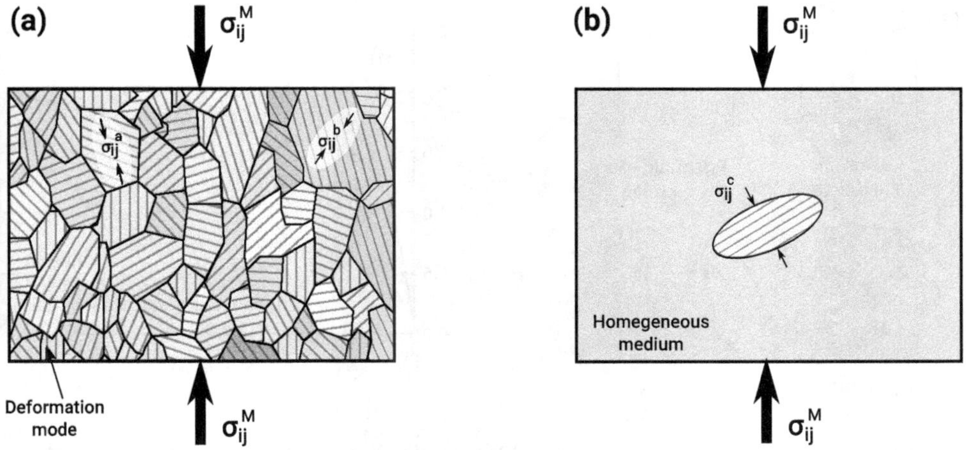

Figure 11.5 Polycrystalline simulation (after Amodeo et al., 2018). (a) In a polycrystal, stress and strain distributions are heterogeneous. The behavior of each grain depends on its local environment, elastic, and plastic properties. (b) In self-consistent models, each grain is treated as an elliptical inclusion inside a homogeneous medium. For the color version, refer to the plate section.

The elasto-plastic self-consistent (EPSC) approximation of Turner and Tomé (1994), later extended by Clausen et al. (2008) and Neil et al. (2010), accounts for elasticity of the material as well as stress relaxation and grain rotation due to twinning and dislocation slip. Plasticity in EPSC calculations, however, is not dependent on strain rates. Visco-plastic self-consistent (VPSC) calculations account for stress relaxation and grain rotation due to twinning and dislocation slip, accounting for a strain-rate-dependent plasticity, but do not account for elasticity (Lebensohn and Tomé, 1994). Finally, the elasto-visco-plastic self-consistent (EVPSC) formulation, such as in Wang et al. (2010), accounts for elastic stresses, grain rotation, and relaxation due to plasticity, as well as strain-rate dependence of the plastic behavior.

All SC calculations can be used to model the evolution of polycrystalline textures. Comparing the experimentally measured LPO and the results of SC calculations allows for the identification of dominant plastic deformation mechanisms and, in particular, dislocation slip or twinning. VPSC calculations, however, will not model macroscopic or intergranular stress, nor lattice strains measured in experiments. EPSC or EVPSC will allow interpreting both the experimental texture and lattice strains (Li et al., 2004; Merkel et al., 2009; Lin et al., 2017).

For ε-Fe at 17 GPa and 400 K, for instance, a comparison between experimental data and the results of EPSC models indicates that plastic deformation is dominated by the activity of basal slip and $\{10\bar{1}\bar{2}\}$ tensile twinning (Figure 11.6). The critical resolved shear stress (CRSS) of basal slip is low (0.05 GPa), but due to intergranular interactions and effects of grain orientations, the overall macroscopic stress reaches $t = 2.0$ GPa after 12% axial compression.

Figure 11.7 highlights one of the advantages on self-consistent models for analyzing the results of high-pressure deformation experiments. Stresses deduced from elastic models are

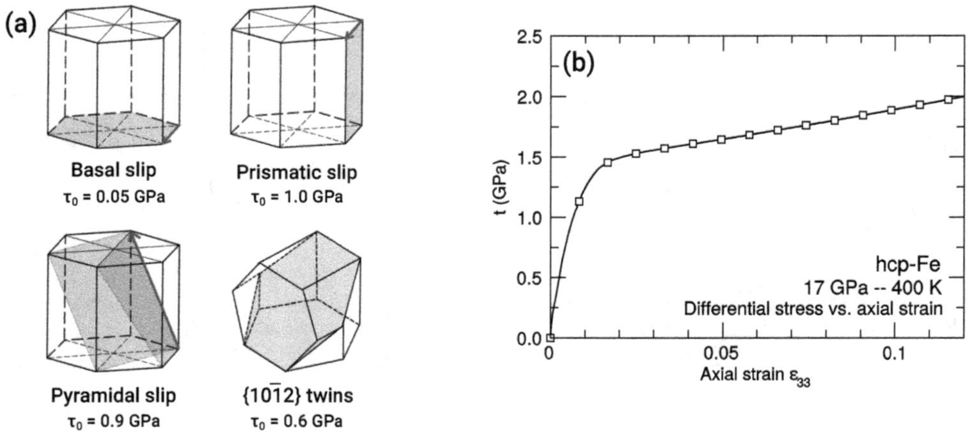

Figure 11.6 Plasticity of ε–Fe at 17 GPa and 400 K (after Merkel et al., 2012). (a) Dominant deformation mechanisms deduced from the data in Figure 11.4. τ_0 is the initial CRSS, before hardening later in the deformation. (b) Differential stress versus axial strain.

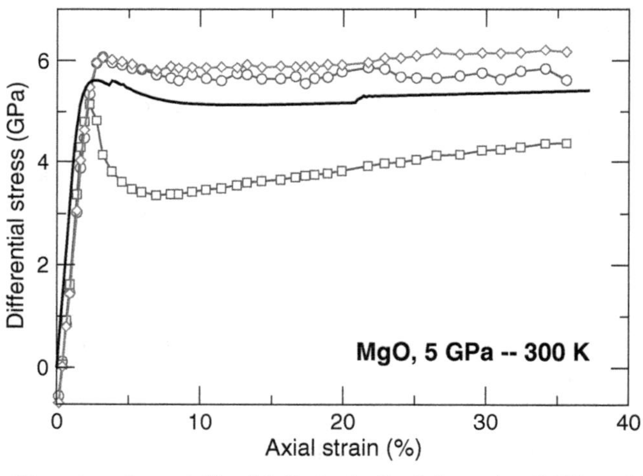

Figure 11.7 Stress in polycrystalline MgO plastically deformed at 5 GPa and 300 K in the D-DIA (Lin et al., 2017). The plot displays apparent stresses deduced from an elastic model and X-ray diffraction measurements on the 111 (circles), 200 (squares), and 220 (diamonds) diffraction lines. The solid black line is the result of EVPSC calculations. Plastic deformation results in different stress values in the elastic model, which is inconsistent: Average stress in the polycrystal should be unique. This can be solved by modeling the experiment using self-consistent calculations.

inconsistent due to stress heterogeneities between grains in soft orientations, for which stresses are relaxed through plastic deformation, and grains in hard orientations, which experience a higher stress. The results in apparent stresses can vary by up to 50% or more depending on the choice of measurement, which will later translate into inconsistent flow laws. One should keep in mind, however, that X-ray diffraction is sensitive to strains at the local scale, for grains contributing to diffraction and not stress. Self-consistent calculations

(EVPSC in Figure 11.7) allow determining the true stress value, consistent with all strain measurements using X-ray diffraction.

11.3.3 Plasticity at the Grain Scale

Recently, experimental studies of microstructures and plastic deformation reached a new milestone with techniques allowing the characterization of individual grains or subgrains within a polycrystalline material (Ludwig et al., 2009). Multigrain X-ray diffraction allows for a rapid, in situ, and nondestructive study of microstructural elements. These elements can also be followed in situ as a function of time or external parameters such as stress, pressure, or temperature. Moreover, samples in high-pressure experiments and, in particular, those of DAC are often small and do not contain enough grains for a statistically relevant powder diffraction analysis. They are, on the other hand, perfectly suited for multigrain X-ray diffraction.

11.3.3.1 Multigrain Crystallography

Multigrain X-ray diffraction allows for the extraction of the orientation and crystal structure of hundreds of elements inside a polycrystalline material (Poulsen, 2004). Extensions of the technique further allow the determination of the position and stress state for each of those elements (Oddershede et al., 2010).

The method has been applied to DAC experiments (Nisr et al., 2012; Zhang et al., 2014; Rosa et al., 2016; Langrand et al., 2017) including pioneering works involving H.-K. Mao (Ice et al., 2005). The method is not restricted to DAC experiments and could also be applied to LVP experiments, but this remains to be reported. Multigrain X-ray diffraction consists in a search for diffraction spots while exploring reciprocal space (Figure 11.8). A first analysis generates a database of experimental diffraction spots, along with their 2θ, ω, and η angles and intensities. Algorithms such as GrainSpotter (Schmidt, 2014) then use the experimental diffraction spots database and scan the grain orientation space to reconstruct the number and orientations of the individual diffracting sample grains.

In the work of Rosa et al. (2016), for instance, we used multigrain X-ray diffraction in order to study the effect of olivine–wadsleyite–ringwoodite chain of phase transformations on microstructures up to 22 GPa and 940 K. We follow the number of grains for each phase, their orientations, and a distribution of grain sizes at each step of the phase transformation (Figure 11.9). Such measurement allows the study of phase transformation mechanisms and associated microstructures, with important applications for constraining deep Earth processes. They can, as well, be extended to follow microstructures induced by plastic deformation.

11.3.3.2 Defects

X-ray line profile analysis (XLPA) is an effective technique for the study of grain-level microstructures using X-ray diffraction (Kerber et al., 2011). The shape of Bragg diffraction peaks can be quantitatively evaluated not only in terms of the crystallite size and its

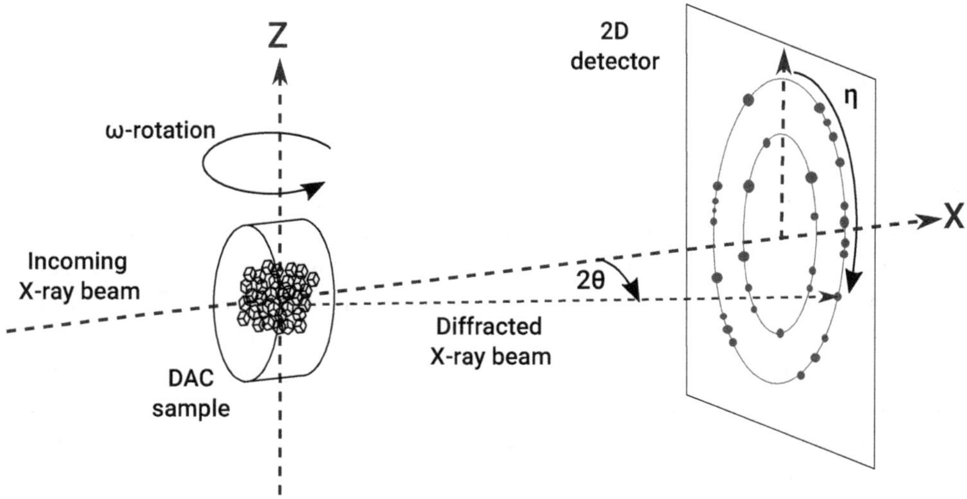

Figure 11.8 Setup for multigrain crystallography at high pressure. The sample is confined in a DAC in axial geometry with an ω rotation parallel to the *Z* direction. Diffraction patterns are collected on a flat panel detector orthogonal to the incoming X-ray beam over $\Delta\omega$ ranges of $\approx 50°$ in steps of $\delta\omega \approx 0.5°$. Individual grains inside the polycrystalline sample give rise to diffraction spots at specific 2θ, ω, and η angles.

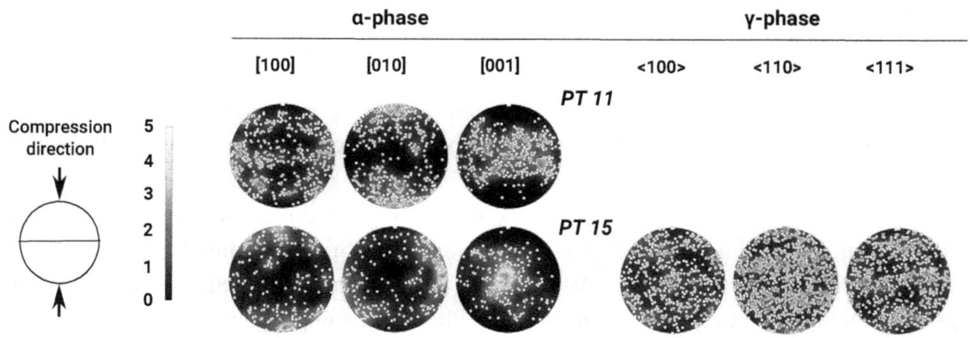

Figure 11.9 Sample results from multigrain crystallography in the DAC (Rosa et al., 2016). Pole figures presenting the orientations of individual olivine and ringwoodite grains during a phase transformation at ≈18 GPa and 880 K. White circles are individual grain orientations. Background color scale is recalculated based on an orientation distribution function fitted to the sample. Scale in multiples of a random distribution (m.r.d.). For the color version, refer to the plate section.

distribution, but also in terms of the density, type, and arrangement of dislocations, twins, and stacking faults (e.g., Figure 11.10).

The advantages of the XPLA method are that (i) the resolution increases with a decreasing grain size and increasing defect density, and (ii) it relies on X-ray diffraction and can hence be applied to high-pressure experiments. The effect of dislocations on peak broadening, for instance, can be measured when their density increases above 10^{12} m^{-2}. On the other hand, it requires a high angular resolution (below 0.01°) and relative accuracy of

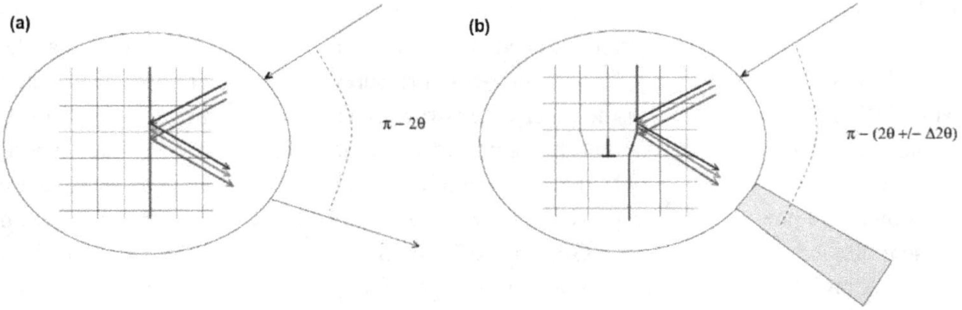

Figure 11.10 Principles of defect analysis using X-ray diffraction. Unlike a perfect crystal (a), defects such as dislocations induce a local distortion of the crystal structure (b). The presence of defects induces a broadening of the X-ray diffraction peak, which, unlike that of grain sizes, for instance, is anisotropic and depends on the hkl indices of the reflection and the type of dislocation.

10^{-4} on diffraction intensities at the edges of diffraction peaks. Also note that, unlike techniques such as high-resolution backscattered electron diffraction (HR-EBSD), XLPA is sensitive to the total dislocation density, including both geometrically necessary and statistically stored dislocations (Ribárik et al., 2019), and cannot distinguish between both.

Under high pressure, the XPLA technique, combined with multigrain crystallography, was used for the study of dislocations in deep Earth materials. Experiments on $MgGeO_3$ in the postperovskite structure, for instance, allowed the identification of dominant dislocation types in this material at 90 GPa (Nisr et al., 2012).

11.4 Sample Results

11.4.1 Deep Earth Materials

Iron is one of the most abundant metals, a widely used technological material, and the main constituent of the Earth's core. As such, the plastic properties of iron under pressure and temperature have received much attention during the past decades. Nothing was known on the plasticity of ε–Fe in the mid-1990s, and geophysical models of inner-core behavior often relied on simple assumptions such as the comparison of *c*/*a* ratios between ε–Fe and other hcp metals (e.g., Jeanloz and Wenk, 1988). Pioneering works were performed at the Geophysical Laboratory with radial X-ray diffraction on Fe up to 220 GPa (Mao et al., 1998) and paved the way to the determination of the plastic properties of ε–Fe under high pressure (Wenk et al., 2000; Merkel et al., 2004, 2012; Gleason and Mao, 2013; Nishihara et al., 2018).

Our current understanding is that basal slip and tensile twinning are the dominant plastic deformation mechanism of ε–Fe at high pressure and room temperature (Wenk et al., 2000; Merkel et al., 2012). The strength of each deformation was characterized with D-DIA measurements at 13–17 GPa and 400–700 K (Merkel et al., 2012; Nishihara et al., 2018). These new interpretations also highlight the important contribution of tensile twinning to understand the evolution of stress and texture at those conditions. Measurements up to 200 GPa at ambient temperature combined with a numerical model also indicate that the bulk shear strength of Fe is $\approx$1 GPa at inner-core pressures and temperatures (Gleason and Mao, 2013). This suggests that the inner core is rheologically weak and supports dislocation creep as the dominant creep mechanism influencing deformation. Deformation experiments on Fe, however, have only reached pressures of $\approx$200 GPa at ambient temperature and combined pressure and temperatures of $\approx$17 GPa and $\approx$700 K. These remain far from those in the inner core (above 300 GPa and 5,000 K). Future efforts will be hence invested in extending the pressure and temperature range of experiments analyzing the plastic behavior of Fe.

Ferropericlase, (Mg,Fe)O, is another interesting material regarding high-pressure plasticity studies. MgO is a very well-known crystalline ceramic whose mechanical properties are thoroughly studied in materials science (Amodeo et al., 2018), both experimentally (e.g., Tromas et al. 2000; Korte and Clegg, 2011) and using numerical plasticity techniques (e.g., Amodeo et al., 2012; Cordier et al., 2012). (Mg,Fe)O is also the second dominant phase in the Earth's lower mantle. For these reasons, numerous studies of the plastic properties of (Mg,Fe)O under high pressure and temperature can be found in the literature (e.g., Kinsland and Bassett, 1977; Stretton et al., 2001; Merkel et al., 2002; Yamazaki and Karato, 2002; Heidelbach et al., 2003; Uchida et al., 2004; Long et al., 2006; Tommaseo et al., 2006; Mei et al., 2008; Lin et al., 2009, 2017, 2019; Amodeo et al., 2012, 2016, Cordier et al., 2012; Marquardt and Miyagi, 2015; Immoor et al., 2018).

Under ambient temperature, pure MgO deforms through dislocation slip on $\{110\}\langle 1\bar{1}0\rangle$ (Merkel et al., 2002; Lin et al., 2017) with an increased activity of $\{110\}\langle 100\rangle$ with increasing temperature (Paterson and Weaver, 1970). Recent calculations also predict a transition between $\{110\}\langle 1\bar{1}0\rangle$ and $\{110\}\langle 100\rangle$ slip with increasing pressure (Amodeo et al., 2012), and that moreover, in the Earth's mantle, extremely low strain rates counteract the influence of pressure with MgO deforming in the athermal regime where dislocation motion is purely controlled by dislocation interactions rather than lattice friction (Cordier et al., 2012). Deformation experiments on single crystal do show a trend consistent with such a change of slip system with pressure (Girard et al., 2012) and deformation experiments on ferropericlase show evidence of both $\{110\}\langle 1\bar{1}0\rangle$ and $\{110\}\langle 100\rangle$ slip at $\approx$1,400 K in a range of 30–60 GPa (Immoor et al., 2018). Recent ambient temperature measurements and modeling on pure MgO up to 50 GPa, however, predict an increasing activity of $\{110\}\langle 1\bar{1}0\rangle$ slip with pressure (Lin et al., 2019). As such, the effect of pressure (and strain rate) on the plasticity of pure MgO remains an active field of research.

The addition of Fe in (Mg,Fe)O does not seem to drastically affect the type of dominant slip systems in (Mg,Fe)O (Stretton et al., 2001; Yamazaki and Karato, 2002; Long et al., 2006; Tommaseo et al., 2006; Immoor et al., 2018). Between 40–80 GPa, however, Fe in (Mg,Fe)O undergoes electronic spin-pairing transition from a high-spin to a low-spin state (Badro et al., 2003) with potential effects of on sample stress (Lin et al., 2009). Measurements up to 100 GPa at ambient temperature also seem to indicate that the strength of (Mg,Fe)O increases by a factor of three at pressures from 20–65 GPa (Marquardt and Miyagi, 2015), with important consequences regarding the stagnation of slabs sinking through the shallow lower mantle. These conclusions remain to be confirmed at mantle temperatures and strain rates.

11.4.2 Materials Science

Simple metals were quickly recognized as a field of study regarding high-pressure mechanical properties (e.g., Re, Mo, and Au; Duffy et al., 1999a, 1999b). These first publications were followed by studies on Pt (Kavner and Duffy, 2003), Os (Weinberger et al. 2008; Chen et al., 2010a), Gd (Xiong et al., 2014), W (He and Duffy, 2006; Xiong et al., 2018), Co (Merkel et al., 2009), or Al (Singh et al., 2007), slowly constituting a database of strength and deformation mechanisms in simple metals as a function of pressure. It turns out that the yield strength of Os is significantly larger than that of other stiff pure metals (≈12 GPa at 60 GPa), followed by Re and W. Other metals, such as Ti or Zr, undergo phase transitions at relatively low pressure. This motivated studies of texture and strength, both during the phase transformation and for the higher-pressure phase. Wenk et al. (2013), for instance, studied the $\alpha \leftrightarrow \omega$ transition in Zr and found (i) a martensitic mechanisms with $(0001)_\alpha // (11\bar{2}0)_\beta$, and (ii) remarkable orientation memory during the reverse transformation. This was then followed by further studies on the $\alpha \leftrightarrow \omega$ transition in Zr and the mechanical properties of both phases (Yu et al., 2015, Kumar et al., 2020), with the aim to design Zr microstructure and strengthen its mechanical properties for high-pressure applications.

The Hall–Petch relationship predicts that the strength of materials should increase with smaller grain sizes. It is verified down to grain sizes of about 30 nm for which an inverse Hall–Petch effect is often observed, with a decrease of strength with decreasing grain size. This inverse Hall–Petch effect is attributed to a transition from dislocation-based plasticity to grain boundary sliding, rotation, or diffusion, but this remains controversial (Naik and Walley, 2020). This controversy motivated studies on the effect of grain sizes on the plastic behavior of nanocrystalline materials under pressure (e.g., Singh et al., 2008). The application of hydrostatic pressure is found to have a strong effect on the transition from the Hall–Petch to inverse Hall–Petch effect. In fact, evidences for dislocation activity were observed in 3 nm Ni compressed to 18.5 GPa (Chen et al., 2012). Moreover, recent work on pure Ni report a continuous strengthening in samples with grain sizes from 200 nm down to 3 nm, with the strengthening enhanced (rather than reduced) at grain sizes smaller than 20 nm (Zhou et al., 2020). These recent works await confirmation in other materials and a better understanding of the combined effect of pressure and grain sizes on dislocation-based plasticity. They do, however, illustrate how studies under high pressure can help solve fundamental issues in materials science.

Other applications in materials science include the design of new, strong materials either as pure phases (He et al., 2004; Kiefer et al., 2005; Liermann et al., 2007; Dong et al., 2009; Chen et al., 2010b; Xiong et al., 2013) or with a controlled microstructure (Conil and

Kavner, 2006), as well as the peculiar effect of phase transitions on microstructures in oxides (Yue et al., 2016). All remain an active field of study.

11.5 Perspectives

11.5.1 Multiphase Aggregates

The vast majority of rocks that constitute the Earth are not made of a single material but rather an assemblage of multiple minerals. The microstructure, i.e., the arrangement of each phase, the orientation and size of each grain, has an influence on the overall mechanical properties of the rock. As such, the mechanical properties of multiphase aggregate are of great interest for the geosciences. They are also fundamental to materials science application, for which the design of composite materials with a predefined microstructure and physical properties are of great interest. Despite their relevance, however, the mechanical properties of polyphase aggregates under pressure remain poorly understood.

In the Earth's lower mantle, for instance, bridgmanite is substantially stronger than ferropericlase, and experiments indicate that ferropericlase accommodates most of the strain (Wang et al., 2013; Girard et al., 2016; Miyagi and Wenk, 2016). A simple approach indicates that, if the weaker ferropericlase is not interconnected, then rheology of the lower mantle will mostly depend on bridgmanite, but if ferropericlase is interconnected, then it will control deformation. The microstructural arrangement of bridgmanite and ferropericlase, however, is controlled by plastic deformation, which in turn is controlled by the properties of both phases. It is therefore necessary to understand this complex interplay between microstructure, the properties of each phase, and the material's macroscopic response. Moreover, stress percolates through polycrystalline materials that have heterogeneous elastic and plastic properties. The pattern of stress percolation is related to the degree of heterogeneity in and statistical distribution of the elastic and plastic properties of the constituent grains in the aggregate (Burnley, 2013). To this day, the understanding and modeling of the feedback between single-phase plasticity, microstructures, and macroscopic behavior of multiphase materials remain a challenge that has seldomly been attempted in high-pressure research (Kaercher et al., 2016; Kasemer et al., 2020).

In addition, pressure, temperature, and deformation can induce mineralogical reactions that will affect the mechanical properties of the aggregate. In olivine + serpentines aggregates, for instance, it has been shown that dehydration reactions of deforming samples containing only 5 vol% of antigorite suffices to trigger significant stress transfer and embrittlement (Ferrand et al., 2017). A strong weakening of cold subducting slab was also reported due to the olivine to ringwoodite phase transformation (Mohiuddin et al., 2020). The mechanical properties of aggregates undergoing phase transformations are relevant for mantle dynamics and the generation of deep earthquakes deep inside the Earth. Such approach could also find multiple applications for the understanding of mechanical properties of materials undergoing phase transformations at other conditions of extreme pressure and temperature.

11.5.2 Technical Developments

This chapter mostly has focused on coupling in situ deformation, X-ray diffraction, and radiography. In the coming years, high-pressure deformation studies could greatly benefit from further technological developments. For the study of deep earthquakes, for instance, the D-DIA deformation press has been coupled with devices recording the sample's acoustic emissions during deformation (Schubnel et al., 2013). The acoustic emission signals can even be further processed to understand the nature of the focal mechanisms and their distribution in both space and time during deformation, and further waveform analysis could be possible (Wang et al., 2017). The combination of high-pressure plastic deformation experiments with the tracking of the sample's acoustic emissions could be of great interest for high-pressure plasticity studies. In fact, plastic mechanisms such as twinning or dislocation slip are known to trigger acoustic emissions in ambient-pressure deformation experiments (Weiss et al., 2000; Vinogradov et al., 2016; Muránsky et al., 2010), and this field of study could open new doors to high-pressure research.

Presently, strain measurements in DAC deformation experiments are limited to average measurements in polycrystals. Multigrain crystallography allows strain mapping at the grain level (Oddershede et al., 2010) and can be used, for instance, to evaluate strain localization in deformation experiments (Sedmák et al., 2016). Strain mapping using multigrain crystallography has been tested in DAC experiments (Nisr et al., 2014), but the method remains to be strengthened, tested, and applied to topics such as stress percolation in heterogeneous aggregates.

Advanced imaging techniques, often relying on the high coherence of synchrotron X-ray beams, such as Bragg coherent diffractive imaging, ptychography, or dark-field X-ray microscopy, allow for revealing strains and heterogeneities within a single grain (Simons et al., 2015; Hruszkewycz et al., 2016; Yau et al., 2017), may be used under operando conditions, and some have been tested in DAC experiments (Yang et al., 2013). These proof-of-concept experiments show that three-dimensional strain distribution with a spatial resolution of 30 nm can be measured inside a 400 nm crystal within a DAC. At present, these techniques represent a steep technical challenge, but with the advent of new, powerful, and highly coherent sources at ERSF, PETRA, SPring-8, and APS, may become more routine measurements in the future. Local strain mapping techniques allowing, for instance, the determination and mapping of dislocation types and densities (Wallis et al., 2017) could then be performed in situ under high pressure.

Three-dimensional X-ray tomography also offers avenues for new experiments in LVP (e.g., Urakawa et al., 2010; Wang et al., 2011; Philippe et al., 2016; Yu et al., 2016) with the ability to resolve small heterogeneities under high pressure and temperature and various strain conditions. This allows in situ tracking of various components in complex materials, with implications for their mechanical properties, connectivity, or permeability. Wang et al. (2011) and Todd et al. (2016), for instance, studied the effect of deformation on microstructures of metal-silicate aggregates with implications for the mechanical properties of composite materials relevant to Earth's mantle and the formation of planetary cores. One could also refer to ongoing on work on serpentine–olivine aggregates, with implications for subduction zones (Mandolini et al., 2020) or the in-situ monitoring of the orientations and mobility of interfaces during phase transitions (e.g., Boulard et al., 2020).

Finally, the present chapter focused on measurements performed in "static" experiments. With the advent of gas-gun or laser-driven dynamic compression techniques coupled with either synchrotron or X-ray free electron lasers and faster detectors, time-resolved measurements at rapid strain rates are now achievable (e.g., Wehrenberg et al., 2017; Chen et al., 2019). Combined with deformation experiments at low strain rates, as described in this chapter, intermediate strain-rate experiments in a dynamic diamond anvil cell (Méndez et al., 2020), and dynamic compression experiments will allow investigating the effect of orders of magnitude in strain rate on deformation and plasticity.

11.6 Conclusion

This chapter summarized the last 25 years of development of high-pressure measurements of mechanical properties. The team around H.-K. Mao at the Carnegie Institution pioneered the development of radial X-ray diffraction in the DAC. The following years saw the development of LVP experiments, external-heating radial X-ray diffraction in the DAC, or multigrain crystallography. Conceptual advances, from the first theories of lattice strains in the 1990s to more advanced self-consistent calculations, have also been a key component of such research. Early results and interpretation of high-pressure deformation experiments lacked understanding of the fundamentals of plasticity and its effect on materials properties and X-ray diffraction measurements. The introduction of self-consistent calculations was hence an important contribution to the field.

Nowadays, the studies of mechanical properties under extreme conditions constitute a thriving field of research, with applications in geosciences, for which it was first intended, but also materials sciences, as highlighted in the previous section of this chapter. Recent works relying on multigrain crystallography now allow the study of microstructures and plasticity at the grain scale and will open new doors to this field of research, offering a quantitative and comprehensive description of materials microstructure at the micrometer length scale. Strain mapping within a grain, along with the identification of deformation defects and densities, is also within reach.

Plasticity depends on pressure, temperature, and microstructures but also strain rates. Further works should hence not only focus on extending the pressure and temperature conditions of the experiments, in order to reach those of the innermost sections of our planet, but also on exploring strain rates, from extremely fast processes at 10^{10} s^{-1} in a shock wave, to deep Earth conditions of 10^{-15} s^{-1}. To this day, this remains an experimental and numerical challenge.

Acknowledgments

The author would like to thank the two anonymous reviewers and N. Hilairet for constructive comments, as well as F. Lin for providing the raw data for Figure 11.7. S. M. received support from the MetalCore grant, which was supported by the French government through the Programme Investissement d'Avenir (I-SITE ULNE R-ERCGEN-19-006-MERKEL / ANR-16-IDEX-0004 ULNE) managed by the Agence Nationale de la Recherche..

References

Amodeo, J., Carrez, P., Cordier, P. (2012). Modelling the effect of pressure on the critical shear stress of MgO single crystals. *Philosophical Magazine*, **92**, 1523–1541.

Amodeo, J., Dancette, S., Delannay, L. (2016). Atomistically-informed crystal plasticity in MgO polycrystals under pressure. International Journal of Plasticity, **82**, 177–191.

Amodeo, J., Merkel, S., Tromas, C., et al. (2018). Dislocations and plastic deformation in MgO crystals: a review. *Crystals*, **8**, 240.

Azuma, S., Nomura, R., Uesugi, K., et al. (2017). Anvil design for slip-free high pressure deformation experiments in a rotational diamond anvil cell. *High Pressure Research*, **38**(1), 23–31.

Badro, J., Fiquet, G., Guyot, F., et al. (2003). Iron partitioning in Earth's mantle: toward a deep lower mantle discontinuity. *Science*, **300**, 789–791.

Blank, V., Konyaev, Y. S., Kuznetsov, A. I., Estrin, E. (1984). Diamond chamber for examining the effects of shear deformation on the structure and properties of solids at pressures up to 43 GPa. *Instruments and Experimental Techniques*, **27**, 1240–1242.

Boioli, F., Carrez, P., Cordier, P., et al. (2017). Pure climb creep mechanism drives flow in Earth's lower mantle. *Science Advances*, **3**(3), e1601958.

Boulard, E., Denoual, C., Dewaele, A., King, A., Godec, Y. L., Guignot, N. (2020) Following the phase transitions of iron in 3D with X-ray tomography and diffraction under extreme conditions. *Acta Materialia*, **192**, 30–39

Burnley, P. (2013). The importance of stress percolation patterns in rocks and other polycrystalline materials. *Nature Communications*, **4**(1), 2117.

Chen, B., Lutker, K., Raju, S. V., et al. (2012). Texture of nanocrystalline nickel: probing the lower size limit of dislocation activity. *Science*, **338**(6113), 1448–1451.

Chen, H., He, D., Liu, J., et al. (2010a). High-pressure radial X-ray diffraction study of osmium to 58 GPa. *European Physical Journal B*, **73**, 321–326.

Chen, H., Peng, F., Mao, H.-K., et al. (2010b). Strength and elastic moduli of TiN from radial X-ray diffraction under nonhydrostatic compression up to 45 GPa. *Journal of Applied Physics*, **107**(11), 113503.

Chen, S., Li, Y., Zhang, N., et al. (2019). Capture deformation twinning in Mg during shock compression with ultrafast synchrotron X-ray diffraction. *Physical Review Letters*, **123**, 255501

Clausen, B., Tomé, C. N., Brown, D. W., Agnew, S. R. (2008). Reorientation and stress relaxation due to twinning: modeling and experimental characterization for Mg. *Acta Materialia*, **56**, 2456–2468.

Conil, N., Kavner, A. (2006). Elastic behavior and strength of Al_2O_3 fiber/Al matrix composite and implications for equation of state measurements in the diamond anvil cell. *Journal of Applied Physics*, **100**, 043517.

Cordier, P., Amodeo, J., Carrez, P. (2012). Modelling the rheology of MgO under Earth's mantle pressure, temperature and strain rates. *Nature*, **481**, 177–180.

Cordier, P., Goryaeva, A. (eds.) (2018). Multiscale modeling of the mantle rheology. Lille, France. https://lilloa.univ-lille.fr/handle/20.500.12210/5756.

Dewaele, A., Loubeyre, P., Occelli, F., Marie, O., Mezouar, M. (2018). Toroidal diamond anvil cell for detailed measurements under extreme static pressures. *Nature Communications*, **9**, 2913.

Dobson, D. P., Miyajima, N., Nestola, F., et al. (2013). Strong inheritance of texture between perovskite and post-perovskite in the D" layer. *Nature Geoscience*, **6**, 575–578.

Dong, H., He, D., Duffy, T. S., Zhao, Y. (2009). Elastic moduli and strength of nanocrystalline cubic BC_2N from X-ray diffraction under nonhydrostatic compression. *Physical Review B*, **79**, 014105.

Duffy, T. S., Shen, G., Heinz, D. L., et al. (1999a). Lattice strains in gold and rhenium under non-hydrostatic compression to 37 GPa. *Physical Review B*, **60**, 15063–15073.

Duffy, T. S., Shen, G., Shu, J., Mao, H. K., Hemley, R. J., Singh, A. K. (1999b). Elasticity, shear strength and equation of state of molybdenum and gold from X-ray diffraction under nonhydrostatic compression to 24 GPa. *Journal of Applied Physics*, **86**, 6729–6736.

Farla, R., Amulele, G., Girard, J., Miyajima, N., Karato, S.-I. (2015). High-pressure and high-temperature deformation experiments on polycrystalline wadsleyite using the rotational drickamer apparatus. *Physics and Chemistry of Minerals*, **42**, 541–558.

Ferrand, T. P., Hilairet, N., Incel, S., et al. (2017). Dehydration-driven stress transfer triggers intermediate-depth earthquakes. *Nature Communications*, **8**, 15247.

Girard, J., Amulele, G., Farla, R., Mohiuddin, A., Karato, S.-I. (2016). Shear deformation of bridgmanite and magnesiowüstite aggregates at lower mantle conditions. *Science*, **351**, 144–147.

Girard, J., Chen, J., Raterron, P. (2012). Deformation of periclase single crystals at high pressure and temperature: quantification of the effect of pressure on slip-system activities. *Journal of Applied Physics*, **111**, 112607.

Girard, J., Silber, R. E., Mohiuddin, A., Chen, H., Karato, S. (2020) Development of a stress sensor for in-situ high-pressure deformation experiments using radial X-ray diffraction. *Minerals*, **10**, 166.

Gleason, A. E., Mao, W. L. (2013). Strength of iron at core pressures and evidence for a weak Earth's inner core. *Nature Geoscience*, **6**, 571–574.

He, D., Duffy, T. S. (2006). X-ray diffraction study of the static strength of tungsten to 69 GPa. *Physical Review B*, **73**(13) 134106.

He, D., Shieh, S., Duffy, T. (2004). Strength and equation of state of boron suboxide from radial X-ray diffraction in a diamond cell under nonhydrostatic compression. *Physical Review B*, **70**, 184121.

Heidelbach, F., Stretton, I., Langenhorst, F., Mackwell, S. (2003). Fabric evolution during high shear strain deformation of magnesiowüstite ($Mg_{0.8}Fe_{0.2}$ O). *Journal of Geophysical Research*, **108**, 2154.

Hemley, R. J., Mao, H. K., Shen, G., Badro, et al. (1997). X-ray imaging of stress and strain of diamond, iron & tungsten at megabar pressures. *Science*, **276**, 1242–1245.

Hruszkewycz, S. O., Allain, M., Holt, M. V., et al. (2016). High-resolution three-dimensional structural microscopy by single-angle Bragg ptychography. *Nature Materials*, **16**(2), 244–251.

Hunt, S. A., Dobson, D. P. (2017). Note: modified anvil design for improved reliability in DT-cup experiments. *Review of Scientific Instruments*, **88**, 126106.

Hunt, S. A., Weidner, D. J., McCormack, R. J., et al. (2014). Deformation T-cup: a new multi-anvil apparatus for controlled strain-rate deformation experiments at pressures above 18 GPa. *Review of Scientific Instruments*, **85**, 085103.

Ice, G. E., Dera, P., Liu, W., Mao, H.-K. (2005). Adapting polychromatic X-ray micro-diffraction techniques to high-pressure research: energy scan approach. *Journal of Synchrotron Radiation*, **12**, 608–617.

Immoor, J., Marquardt, H., Miyagi, L., et al. (2018). Evidence for {100}<011> slip in ferropericlase in Earth's lower mantle from high-pressure/high-temperature experiments. *Earth and Planetary Science Letters*, **489**, 251–257.

Immoor, J., Marquardt, H., Miyagi, L., et al. (2020). An improved setup for radial diffraction experiments at high pressures and high temperatures in a resistive graphite-heated diamond anvil cell. *Review of Scientific Instruments*, **91**(4), 045121.

Jeanloz, R., Wenk, H.-R. (1988). Convection and anisotropy of the inner core. *Geophysical Research Letters*, **15**, 72–75.

Kaercher, P., Miyagi, L., Kanitpanyacharoen, W., et al. (2016). Two-phase deformation of lower mantle mineral analogs. *Earth and Planetary Science Letters*, **456**, 134–145.

Karato, S.-I., Forte, A., Liebermann, R., Masters, G., Stixrude, L., eds. (2000). *Earth's Deep Interior: Mineral Physics and Tomography from the Atomic to the Global Scale*. Geophysical Monograph Series. American Geophysical Union.

Kasemer, M., Zepeda-Alarcon, E., Carson, R., Dawson, P., Wenk, H.-R. (2020). Deformation heterogeneity and intragrain lattice misorientation in high strength contrast, dual-phase bridgmanite/periclase. *Acta Materialia*, **189**, 284–298.

Kavner, A., Duffy, T. S. (2003). Elasticity and rheology of platinum under high pressure and nonhydrostatic stress. *Physical Review B*, **68**, 144101.

Kawazoe, T., Karato, S., Ando, J., Jing, Z., Otsuka, K., Hustoft, J. W. (2010). Shear deformation of polycrystalline wadsleyite up to 2100 K at 14–17 GPa using a rotational drickamer apparatus (RDA). *Journal of Geophysical Research*, **115**, B08208.

Kawazoe, T., Ohuchi, T., Nishihara, Y., Nishiyama, N., Fujino, K., Irifune, T. (2013). Seismic anisotropy in the mantle transition zone induced by shear deformation of wadsleyite. *Physics of Earth and Planetary Interiors*, **216**, 91–98.

Kerber, M. B., Zehetbauer, M. J., Schafler, E., Spieckermann, F. C., Bernstorff, S., Ungar, T. (2011). X-ray line profile analysis – an ideal tool to quantify structural parameters of nanomaterials. *JOM*, **63**, 61–70.

Kiefer, B., Shieh, S. R., Duffy, T. S., Sekine, T. (2005). Strength, elasticity, and equation of state of the nanocrystalline cubic silicon nitride γ-Si_3N_4 to 68 GPa. *Physical Review B*, **72**, 014102.

Kinsland, G. L., Bassett, W. A. (1977). Strength of MgO and NaCl polycrystals to confining pressures of 250 kbar at 25 C. *Journal of Applied Physics*, **48**, 978–984.

Kocks, U. F., Tomé, C. N., Wenk, H.-R. (1998). *Texture and Anisotropy: Preferred Orientations and their Effects on Material Properties*. Cambridge University Press.

Korte, S., Clegg, W. (2011). Discussion of the dependence of the effect of size on the yield stress in hard materials studied by microcompression of MgO. *Philosophical Magazine*, **91**(7-9), 1150–1162.

Kumar, M. A., Hilairet, N., McCabe, R., Yu, T., Wang, Y., Beyerlein, I., Tomé, C. (2020). Role of twinning on the omega-phase transformation and stability in zirconium. *Acta Materialia*, **185**, 211–217.

Langrand, C., Hilairet, N., Nisr, C., et al. (2017). Reliability of multigrain indexing for orthorhombic polycrystals above 1mbar: application to $MgSiO_3$ post-perovskite. Journal of Applied Crystallography, **50**, 120–130.

Lebensohn, R. A., Tomé, C. N. (1994). A self-consistent visco-plastic model: calculation of rolling textures of anisotropic materials. *Materials Science and Engineering A*, **175**, 71–82.

Levitas, V. I. (2004). High-pressure mechanochemistry: conceptual multiscale theory and interpretation of experiments. *Physical Review B*, **70**, 184118.

Li, L., Weidner, D. J., Chen, J., Vaughan, M. T., Davis, M., Durham, W. B. (2004). X-ray strain analysis at high pressure: effect of plastic deformation in MgO. *Journal of Applied Physics*, **95**, 8357–8365.

Liermann, H.-P., Merkel, S., Miyagi, L., et al. (2009). New experimental method for in situ determination of material textures at simultaneous high-pressure and temperature by means of radial diffraction in the diamond anvil cell. *Review of Scientific Instruments*, **80**, 104501.

Liermann, H. P., Singh, A. K., Somayazulu, M., Saxena, S. K. (2007). Compression behavior of NbC under nonhydrostatic conditions to 57 GPa. *International Journal of Refractory Metals and Hard Materials*, **25**, 386–391.

Lin, F., Couper, S., Jugle, M., Miyagi, L. (2019). Competing deformation mechanisms in periclase: implications for lower mantle anisotropy. *Minerals*, 9**,** 650.

Lin, F., Hilairet, N., Raterron, P., et al. (2017). Elasto-viscoplastic self consistent modeling of the ambient temperature plastic behavior of periclase deformed up to 5.4 GPa. *Journal of Applied Physics*, **122**, 205902.

Lin, J. F., Wenk, H.-R., Voltolini, M., Speziale, S., Shu, J., Duffy, T. S. (2009). Deformation of lower-mantle ferropericlase (Mg,Fe)O across the electronic spin transition. *Physics of Chemistry and Minerals*, **36**, 585–592.

Lincot, A., Cardin, P., Deguen, R., Merkel, S. (2016). Multiscale model of global inner-core anisotropy induced by hcp alloy plasticity. *Geophysical Research Letters*, **43**, 1084–1091.

Long, M. D., Xiao, X., Jiang, Z., Evansa, B., Karato, S. (2006). Lattice preferred orientation in deformed polycrystalline (Mg,Fe)O and implications for seismic anisotropy in D". *Physics of the Earth and Planetary Interiors*, **156**, 75–88.

Ludwig, W., King, A., Reischig, P., et al. (2009). New opportunities for 3D materials science of polycrystalline materials at the micrometre lengthscale by combined use of X-ray diffraction and X-ray imaging. *Materials Science and Engineering A*, **524**, 69–76.

Lutterotti, L., Matthies, S., Wenk, H.-R. (1999). MAUD: a friendly Java program for materials analysis using diffraction. *IUCr: Newsletter of the CPD*, **21**, 14–15.

Ma, Y., Selvi, E., Levitas, V. I., Hashemi, J. (2006). Effect of shear strain on the $\alpha-\varepsilon$ phase transition of iron: a new approach in the rotational diamond anvil cell. *Journal of Physics: Condensed Matter*, **18**, S1075–S1082.

Mainprice, D., Barruol, G., Ismaïl, W. B. (2000). The seismic anisotropy of the earth's mantle: from single crystal to polycrystal, in S. Karato, A. M. Forte, R. C. Liebermann, G. Master, L. Stixrude, eds. *Earth's Deep Interior. Mineral Physics and Tomography. From the Atomic to the Global Scale*, American Geophysical Union, pp. 237–264.

Mandolini, T., Hilairet, N., Chantel, J., et al. (2020). Microstructural evolution in deforming olivine-serpentine aggregates at subduction zones conditions using in-situ X-ray tomography. AGU Fall Meeting, MR028–06

Mao, H. K., Shu, J., Shen, G., Hemley, R. J., Li, B., Singh, A. K. (1998). Elasticity and rheology of iron above 220 GPa and the nature of the earth's inner core. *Nature*, 396, 741–743. Correction (1999), *Nature*, **399**, 280.

Marquardt, H., Miyagi, L. (2015). Slab stagnation in the shallow lower mantle linked to an increase in mantle viscosity. *Nature Geoscience*, **8**, 311–314.

Meade, C., Jeanloz, R. (1988) Yield strength of MgO to 40 GPa. *Journal of Geophysical Research*, **93**, 3261–3269.

Mei, S., Kohlstedt, D. L., Durham, W. B., Wang, L. (2008). Experimental investigation of the creep behavior of MgO at high pressures. *Physics of the Earth and Planetary Interiors*, **170**, 170–175.

Méndez, A. S. J., Marquardt, H., Husband, R. J., et al. (2020). A resistively-heated dynamic diamond anvil cell (RHdDAC) for fast compression X-ray diffraction experiments at high temperatures. *Review of Scientific Instruments*, **91**, 073906

Merkel, S., Gruson, M., Wang, Y., Nishiyama, N. & Tomé, C. N. (2012). Texture and elastic strains in hcp-iron plastically deformed up to 17.5 GPa and 600 K: experiment and model. *Modelling and Simululation in Materials Science and Engineering*, **20**, 024005.

Merkel, S., Hilairet, N. (2015). Multifit/polydefix: a framework for the analysis of polycrystal deformation using X-rays. *Journal of Applied Crystallography*, **48**, 1307–1313.

Merkel, S., McNamara, A. K., Kubo, A., et al. (2007). Deformation of (Mg,Fe)SiO_3 post-perovskite and D” anisotropy. *Science*, **316**, 1729–1732.

Merkel, S., Miyajima, N., Antonangeli, D., Fiquet, G., Yagi, T. (2006). Lattice preferred orientation and stress in polycrystalline hcp–Co plastically deformed under high pressure. *Journal of Applied Physics*, **100**, 023510.

Merkel, S., Tomé, C. N., Wenk, H.-R. (2009). A modeling analysis of the influence of plasticity on high pressure deformation of hcp–Co. *Physical Review B*, **79**, 064110.

Merkel, S., Wenk, H.-R., Gillet, P., Mao, H. K., Hemley, R. J. (2004). Deformation of polycrystalline iron up to 30 GPa and 1000 K. *Physics of the Earth and Planetary Interiors*, **145**, 239–251.

Merkel, S., Wenk, H.-R., Shu, J., et al. (2002). Deformation of polycrystalline MgO at pressures of the lower mantle. *Journal of Geophysical Research*, **107**, 2271.

Merkel, S., Yagi, T. (2005). X-ray transparent gasket for diamond anvil cell high pressure experiments. *Review of Scientific Instruments*, **76**, 046109.

Miyagi, L., Amulele, G., Otsuka, K., Du, Z., Farla, R., Karato, S. (2014). Plastic anisotropy and slip systems in ringwoodite deformed to high shear strain in the rotational Drickamer apparatus. *Physics of the Earth and Planetary Interiors*, **228**, 244–253.

Miyagi, L., Kanitpanyacharoen, W., Kaercher, P., Lee, K. K. M., Wenk, H.-R. (2010). Slip systems in $MgSiO_3$ post-perovskite: Implications for D” anisotropy. *Science*, **329**, 1639–1641.

Miyagi, L., Kanitpanyacharoen, W., Raju, S. V., et al. (2013). Combined resistive and laser heating technique for in situ radial X-ray diffraction in the diamond anvil cell at high pressure and temperature. *Review of Scientific Instruments*, **84**, 025118

Miyagi, L., Kunz, M., Knight, J., Nasiatka, J., Voltolini, M., Wenk, H.-R. (2008). In situ phase transformation and deformation of iron at high pressure and temperature. *Journal of Applied Physics*, **104**, 103510.

Miyagi, L., Wenk, H.-R. (2016). Texture development and slip systems in bridgmanite and bridgmanite + ferropericlase aggregates. *Physics and Chemistry of Minerals*, **43**(8), 597–613.

Mohiuddin, A., Karato, S., Girard, J. (2020). Slab weakening during the olivine to ringwoodite transition in the mantle. *Nature Geoscience*, **13**, 170–174

Muránsky, O., Barnett, M., Carr, D., Vogel, S., Oliver, E. (2010). Investigation of deformation twinning in a fine-grained and coarse-grained ZM20 mg alloy: combined in situ neutron diffraction and acoustic emission. *Acta Materialia*, **58**(5), 1503–1517.

Naik, S. N., Walley, S. M. (2020). The Hall–Petch and inverse Hall–Petch relations and the hardness of nanocrystalline metals. *Journal of Materials Science*, **55**(7), 2661–2681.

Neil, C. J., Wollmershauser, J. A., Clausen, B., Tomé, C. N., Agnew, S. R. (2010). Modeling lattice strain evolution at finite strains and experimental verification for copper and stainless steel using in situ neutron diffraction. *International Journal of Plasticity*, **26**, 1772–1791

Nishihara, Y., Ohuchi, T., Kawazoe, T., et al. (2018). Deformation-induced crystallographic-preferred orientation of hcp-iron: an experimental study using a deformation-DIA apparatus. *Earth and Planetary Science Letters*, **490**, 151–160.

Nishihara, Y., Tinker, D., Kawazoe, T., et al. (2008). Plastic deformation of wadsleyite and olivine at high-pressure and high-temperature using a rotational Drickamer apparatus (RDA). *Physics of the Earth and Planetary Interiors*, **170**, 156–169.

Nisr, C., Ribárik, G., Ungár, T., Vaughan, G. B., Merkel, S. (2014). Three-dimensional X-ray diffraction in the diamond anvil cell: application to stishovite. *High Pressure Research*, **34**, 158–166.

Nisr, C., Ribárik, G., Ungár, T., Vaughan, G. B. M., Cordier, P., Merkel, S. (2012). High resolution three-dimensional X-ray diffraction study of dislocations in grains of $MgGeO_3$ post-perovskite at 90 GPa. *Journal of Geophysical Research*, **117**, B03201.

Nomura, R., Azuma, S., Uesugi, K., et al. (2017). High-pressure rotational deformation apparatus to 135 GPa. *Review of Scientific Instruments*, **88**, 044501.

Noyan, I., Cohen, J. (1987). *Residual Stress: Measurements by Diffraction and Interpretation*. Springer-Verlag.

Nzogang, B. C., Bouquerel, J., Cordier, P., Mussi, A., Girard, J., Karato, S. (2018). Characterization by scanning precession electron diffraction of an aggregate of bridgmanite and ferropericlase deformed at HP-HT. *Geochemistry, Geophysics, Geosystems*, **19**(3), 582–594.

Oddershede, J., Schmidt, S., Poulsen, H. F., Sorensen, H. O., Wright, J., Reimers, W. (2010). Determining grain resolved stresses in polycrystalline materials using three-dimensional X-ray diffraction. *Journal of Applied Crystallography.*, **43**, 539–549.

Paterson, M. S., Weaver, C. W. (1970). Deformation of polycrystalline MgO under pressure. *Journal of the American Ceramic Society*, **53**, 463–471.

Philippe, J., Godec, Y. L., Mezouar, M., et al. (2016). Rotating tomography Paris–Edinburgh cell: a novel portable press for micro-tomographic 4-D imaging at extreme pressure/temperature/stress conditions. *High Pressure Research*, **36**, 512–532.

Poulsen, H. F. (2004). *Three-Dimensional X-Ray Diffraction Microscopy: Mapping Polycrystals and their Dynamics*, vol. 205 of Springer Tracts in Modern Physics. Springer.

Raterron, P., Merkel, S. (2009). In situ rheological measurements at extreme pressure and temperature using synchrotron X-ray diffraction and radiography. *Journal of Synchrotron Radiation*, **16**, 748–756.

Raterron, P., Merkel, S., Holyoke III, C. W. (2013). Axial temperature gradient and stress measurements in the deformation-DIA cell using alumina pistons. *Review of Scientific Instruments*, **84**, 043906.

Reali, R., Jackson, J., Orman, J. V., Bower, D., Carrez, P., Cordier, P. (2019). Modeling viscosity of (Mg,Fe)O at lowermost mantle conditions. *Physics of the Earth and Planetary Interiors*, **287**, 65–75.

Reynard, B., Caracas, R., Cardon, H., Montagnac, G., Merkel, S. (2019). High-pressure yield strength of rocksalt structures using quartz raman piezometry. *Comptes Rendus Geoscience.*, **351**(2–3), 71–79.

Ribárik, G., Jóni, B., Ungár, T. (2019). Global optimum of microstructure parameters in the CMWP line-profile-analysis method by combining Marquardt–Levenberg and

Monte-Carlo procedures. *Journal of Materials Science and Technology*, **35**, 1508–1514

Rosa, A. D., Hilairet, N., Ghosh, S., Perrillat, J.-P., Garbarino, G., Merkel, S. (2016). Evolution of grain sizes and orientations during phase transitions in hydrous Mg_2SiO_4. *Journal of Geophysical Research*, **121**, 7161–7176.

Schmidt, S. (2014). Grainspotter: a fast and robust polycrystalline indexing algorithm. *Journal of Applied Crystallography*, **47**, 276–284.

Schropp, A., Hoppe, R., Meier, V., et al. (2015). Imaging shock waves in diamond with both high temporal and spatial resolution at an XFEL. *Scientific Reports*, **5**, 11089.

Schubnel, A., Brunet, F., Hilairet, N., Gasc, J., Wang, Y., Green II, H. W. (2013). Deep-focus earthquake analogs recorded at high pressure and temperature in the laboratory. *Science*, **341**, 1377–1380.

Sedmák, P., Pilch, J., Heller, L., et al. (2016). Grain-resolved analysis of localized deformation in nickel–titanium wire under tensile load. **353**, 559–562.

Simons, H., King, A., Ludwig, W., et al. (2015). Dark-field X-ray microscopy for multi-scale structural characterization. *Nature Communications*, **6**(1), 6098.

Singh, A. K., Balasingh, C., Mao, H. K., Hemley, R. J., Shu, J. (1998). Analysis of lattice strains measured under non-hydrostatic pressure. *Journal of Applied Physics*, **83**, 7567–7575.

Singh, A. K., Liermann, H., Akahama, Y., Saxena, S. K., Menéndez-Proupin, E. (2008). Strength of polycrystalline coarse-grained platinum to 330 GPa and of nanocrystalline platinum to 70 GPa from high-pressure X-ray diffraction data. *Journal of Applied Physics*, **103**, 063524.

Singh, A. K., Liermann, H.-P., Akahama, Y., Kawamura, H. (2007). Aluminum as a pressure-transmitting medium cum pressure standard for X-ray diffraction experiments to 200 GPa with diamond anvil cells. *Journal of Applied Physics*, **101**, 123526.

Stretton, I., Heidelbach, F., Mackwell, S., Langenhorst, F. (2001). Dislocation creep of magnesiowüstite ($Mg_{0.8}Fe_{0.2}O$). *Earth and Planetary Science Letters*, **194**, 229–240.

Tateno, S., Hirose, K., Ohishi, Y., Tatsumi, Y. (2010). The structure of iron in Earth's inner core. *Science*, **330**, 359–361.

Todd, K. A., Watson, H. C., Yu, T., Wang, Y. (2016). The effects of shear deformation on planetesimal core segregation: results from in-situ X-ray micro-tomography. *American Mineralogist*, **101**, 1996–2004.

Tommaseo, C. E., Devine, J., Merkel, S., Speziale, S., Wenk, H.-R. (2006). Texture development and elastic stresses in magnesiowüstite at high pressure. *Physics and Chemistry of Minerals*, **33**, 84–97.

Tromas, C., Girard, J. C., Woirgard, J. (2000). Study by atomic force microscopy of elementary deformation mechanisms involved in low load indentations in MgO crystals. *Philosophy Magazine A*, **80**, 2325–2335.

Tsujino, N., Nishihara, Y., Yamazaki, D., Seto, Y., Higo, Y., Takahashi, E. (2016). Mantle dynamics inferred from the crystallographic preferred orientation of bridgmanite. *Nature*, **539**, 81–84.

Turner, P. A., Tomé, C. N. (1994). A study of residual stresses in zircaloy-2 with rod texture. *Acta Metallurgica et Materialia*, **42**, 4143–4153.

Uchida, T., Wang, Y., Rivers, M. L., Sutton, S. R. (2004). Yield strength and strain hardening of MgO up to 8 GPa measured in the deformation-DIA with monochromatic X-ray diffraction. *Earth and Planetary Science Letters*, **226**, 117–126.

Urakawa, S., Terasaki, H. P., Funakoshi, K., Uesugi, K., Yamamoto, S. (2010). Development of high pressure apparatus for X-ray microtomography at SPring-8. Journal of Physics: Conference Series, **215**, 012026.

Vinogradov, A., Vasilev, E., Seleznev, M., Máthis, K., Orlov, D., Merson, D. (2016). On the limits of acoustic emission detectability for twinning. *Materials Letters*, **183**, 417–419.

Wallis, D., Hansen, L. N., Britton, T. B., Wilkinson, A. J. (2017). Dislocation interactions in olivine revealed by HR-EBSD. *Journal of Geophysical Research. Solid Earth*, **122**, 7659–7678.

Wang, H., Wu, P. D., Tomé, C. N., Huang, Y. (2010). A finite strain elastic-viscoplastic self-consistent model for polycrystalline materials. *Journal of the Mechanics and Physics of Solids*, **58**, 594–612.

Wang, Y., Duhram, W. B., Getting, I. C., Weidner, D. J. (2003). The deformation-DIA: a new apparatus for high temperature triaxial deformation to pressures up to 15 GPa. *Review of Scientific Instruments*, **74**, 3002–3011.

Wang, Y. Lesher, C., Fiquet, G., et al. (2011). In situ high-pressure and high-temperature X-ray microtomographic imaging during large deformation: a new technique for studying mechanical behavior of multiphase composites. *Geosphere*, **7**, 40–53.

Wang, Y., Hilairet, N., Nishiyama, N., et al. (2013). High-pressure, high-temperature deformation of $CaGeO_3$ (perovskite)±MgO aggregates: Implications for multiphase rheology of the lower mantle. *Geochemistry, Geophysics, Geosystems*, **14**, 3389–3408.

Wang, Y., Shen, G. (2014). High-pressure experimental studies on geo-liquids using synchrotron radiation at the Advanced Photon Source. *Journal of Earth Science*, **25**, 939-958.

Wang, Y., Zhu, L., Shi, F., et al. (2017). A laboratory nanoseismological study on deep-focus earthquake micromechanics. *Science Advances*, **3**(7), e1601896.

Wehrenberg, C. E., McGonegle, D., Bolme, C., et al. (2017). In situ X-ray diffraction measurement of shock-wave-driven twinning and lattice dynamics. *Nature*, **550**, 496–499.

Weinberger, M. B., Tolbert, S. H., Kavner, A. (2008). Osmium metal studied under high pressure and nonhydrostatic stress. *Physical Review Letters*, **100**, 045506.

Weiss, J., Lahaie, F., Grasso, J. R. (2000). Statistical analysis of dislocation dynamics during viscoplastic deformation from acoustic emission. *Journal of Geophysical Research*, **105**, 433–442.

Wenk, H.-R., Kaercher, P., Kanitpanyacharoen, W., Zepeda-Alarcon, E., Wang, Y. (2013). Orientation relations during the α-ω phase transition of zirconium: in situ texture observations at high pressure and temperature. *Physical Review Letters*, **111**, 195701.

Wenk, H.-R., Lonardelli, I., Merkel, S., et al. (2006). Deformation textures produced in diamond anvil experiments, analysed in radial diffraction geometry. *Journal of Physics: Condensed Matter*, **18**, S933–S947.

Wenk, H.-R., Lutterotti, L., Kaercher, P., Kanitpanyacharoen, W., Miyagi, L.,Vasin, R. (2014). Rietveld texture analysis from synchrotron diffraction images. ii. Complex multiphase materials and diamond anvil cell experiments. *Powder Diffraction*, **29**, 220–232.

Wenk, H.-R., Matthies, S., Donovan, J., Chateigner, D. (1998). Beartex: a Windows-based program system for quantitative texture analysis. *Journal of Applied Crystallography*, **31**, 262–269.

Wenk, H.-R., Matthies, S., Hemley, R. J., Mao, H. K., Shu, J. (2000). The plastic deformation of iron at pressures of the Earth's inner core. *Nature*, **405**, 1044–1047.

Xiong, L., Li, B., Tang, Y., et al. (2018). Radial X-ray diffraction study of the static strength and texture of tungsten to 96 GPa. *Solid State Communications*, **269**, 83–89.

Xiong, L., Liu, J., Bai, L., et al. (2013). Radial X-ray diffraction of tungsten tetraboride to 86 GPa under nonhydrostatic compression. *Journal of Applied Physics*, **113**, 033507.

Xiong, L., Liu, J., Bai, L., Li, X., Lin, C., Lin, J.-F. (2014). Strength and structural phase transitions of gadolinium at high pressure from radial X-ray diffraction. *Journal of Applied Physics*, **116**, 243503.

Xu, Y., Nishihara, Y., Karato, S. (2005). Development of a rotational drickamer apparatus for large-strain deformation experiments at deep Earth conditions, in J. Chen, Y. Wang, T. S. Duffy, G. Shen, L. F. Dobrzhinetskaya, eds., *Advances in High-Pressure Technology for Geophysical Applications*, Elsevier, pp. 167–182.

Yamazaki, D., Karato, S. (2001). High-pressure rotational deformation apparatus to 15 GPa. *Review of Scientific Instruments*, **72**, 4207–4211.

Yamazaki, D., Karato, S. (2002). Fabric development in (Mg,Fe)O during large strain, shear deformation: implications for seismic anisotropy in Earth's lower mantle. Physics of the Earth and Planetary Interiors, **131**, 251–267.

Yang, W., Huang, X., Harder, R., Clark, J. N., Robinson, I. K., Mao, H. K. (2013). Coherent diffraction imaging of nanoscale strain evolution in a single crystal under high pressure. *Nature Communications*, **4**(1), 1680.

Yau, A., Cha, W., Kanan, M. W., Stephenson, G. B., Ulvestad, A. (2017). Bragg coherent diffractive imaging of single-grain defect dynamics in polycrystalline films. *Science*, **356**(6339), 739–742.

Yu, T., Wang, Y., Rivers, M. L. (2016). Imaging in 3D under pressure: a decade of high-pressure X-ray microtomography development at GSECARS. *Progress in Earth and Planetary Science*, **3**, 17

Yu, X., Zhang, R., Weldon, D., et al. (2015). High pressure phase-transformation induced texture evolution and strengthening in zirconium metal: experiment and modeling. *Scientific Reports*, **5**, 12552.

Yue, B., Hong, F., Merkel, S., et al. (2016). *Deformation behavior across the zircon–scheelite phase transition.* Physical Review Letters, **117**, 135701.

Zarechnyy, O. M., Levitas, V. I., Ma, Y. (2012). Coupled plastic flow and phase transformation under compression of materials in a diamond anvil cell: effects of transformation kinetics and yield strength. *Journal of Applied Physics*, **111**, 023518.

Zhang, L., Meng, Y., Yang, W., et al. (2014). Disproportionation of (Mg,Fe)SiO_3 perovskite in Earth's deep lower mantle. *Science*, **344**, 877–882.

Zhou, X., Feng, Z., Zhu, L., et al. (2020). High-pressure strengthening in ultrafine-grained metals. *Nature*, **579**(7797), 67–72.

12

Synthesis of High-Pressure Silicate Polymorphs Using Multi-Anvil Press

JIE LI

Multi-anvil press (MAP) is a major type of apparatus for phase equilibrium, material synthesis, and property investigations at up to 120 GPa of static compression. Here I summarize the fundamental aspects of the MAP techniques, highlighting uncertainties in pressure and temperature in experiments using tungsten carbide (WC) anvils with 3 mm and 5 mm truncation edge lengths (TEL). To illustrate the application of the MAP for synthesizing millimeter-sized single crystals of deep mantle silicates, I review the theory of crystal growth and compile relevant phase diagrams. The theory informs synthesis methods such as slow cooling of a fluid solution and growing crystals in a thermal gradient. The updated phase diagrams involve Mg_2SiO_4, $MgSiO_3$, and H_2O at pressures up to 27 GPa and touch upon iron-bearing compositions. Also reviewed briefly are techniques to characterize the compositions and structures of synthesis products.

12.1 Introduction

Multi-anvil press (MAP) or multi-anvil apparatus, along with piston-cylinder apparatus and diamond anvil cell (DAC), are major types of experimental apparatus for generating static high pressure. The MAP and DAC both came into existence about half a century ago and have evolved along parallel pathways (Bassett, 2009; Lieberman, 2011; Mao et al., 2016; Yagi, Chapter 2). In comparison, the DAC achieves higher pressures at small sample size (e.g., nanograms), and it is accessible to a much wider range of in situ observations, including optical access to compressed samples, whereas the MAP can accommodate larger samples (e.g., milligrams). They play complementary roles in high-pressure research and are sometimes combined to validate phase equilibria in overlapping pressure range (Hirose et al., 1999; Kulka et al., 2020). Crystals of high-pressure phases may be synthesized in the MAP and then investigated for structure and properties using the DAC (e.g., Jacobsen et al., 2004).

Typical MAP experiments are conducted at 2–27 GPa and up to ~2,000°C, using tungsten carbide (WC) anvils (Leinenweber et al., 2012). The newly developed "binder-less" WC (TJS01) with <0.2% Ni is ~40% harder than conventional WC and has been used to reach pressure greater than 60 GPa at room temperature and ~50 GPa at ~1,700°C

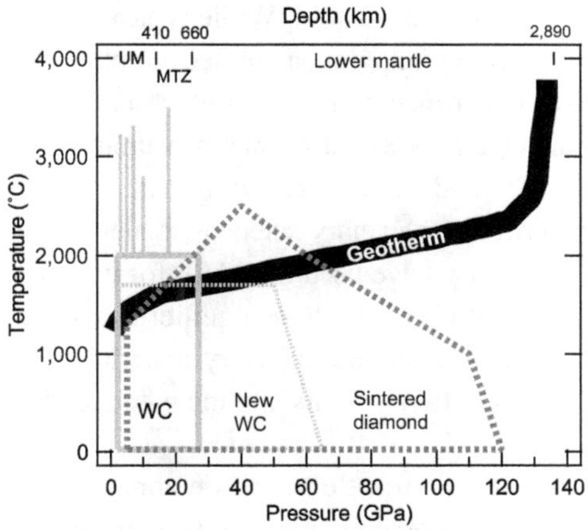

Figure 12.1 Pressure and temperature ranges of multi-anvil press. With tungsten carbide (WC) anvils (box with thick yellow border and vertical bars), routine operations cover 2–27 GPa and 25–2,000°C (Leinenweber et al., 2012), and 2,800°C using a doped-$LaCrO_3$ heater at 10 GPa (Walker, 1993), ~3,200°C using a graphite heater at 3–7 GPa (Zhang and Fei, 2008) and ~3,500°C using a boron-doped diamond heater at 17.5±0.5 GPa (Shatskiy et al., 2009a). With binderless WC anvils (nWC, box with thin yellow border), experiments have reached 65 GPa (Ishii et al., 2017). With sintered diamond (SD) anvils (domain with blue dotted border), a record pressure of 120 GPa was achieved at room temperature (Yamazaki and Ito, 2020). The operation temperature of the SD anvil is limited by conversion to graphite at low pressures and by heat loss and mechanical instability at high pressures (Stewart et al., 2007; Tange et al., 2009; Yamazaki et al., 2014, 2019; Arimoto et al., 2019; Yamazaki and Ito, 2020). The pressure–temperature range of the MAP overlaps with the geotherm (Katsura et al., 2010) in the top 1,000 km of the solid Earth, including the entire upper mantle (UM), mantle transition zone (MTZ), and the top 300 km of the lower mantle.

(Ishii et al., 2019). The MAP limits extend to ~120 GPa with sintered diamond (SD) anvils (Yamazaki and Ito, 2020) and to 3,500°C using boron-doped diamond heaters (Shatskiy et al., 2009a), thus encompassing most of the pressure and temperature conditions of the Earth's mantle (Figure 12.1).

Ito et al. (2007) and Liebermann (2011) gave comprehensive accounts of the invention and development of the large-volume press since the late 1950s, and Yagi (Chapter 2) provides an historical perspective. Shortly after the 2011 review, a record pressure of 120 GPa was reached for the first time using sintered diamond anvils (Yamazaki and Ito, 2020). Moreover, the standardized COMPRES assemblies became available, which not only improved experimental precision but also significantly enhanced the productivity of many MAP laboratories worldwide (Leinenweber et al., 2012). The MAP has been combined with synchrotron radiation for in situ X-ray diffraction, radiography, and tomography investigations, using modified assemblies and WC anvils (e.g., Irifune et al., 1998, Fei et al., 2004) or X-ray transparent SD anvils (e.g., Tange et al., 2009). With the enormous weight of a typical hydraulic press (>5,000 kg), a MAP is usually placed on the lowest level of a building and rarely moved, with the exception of portable MAPs that combine T-cup multi-anvil with the Paris–Edinburgh press for in situ experiments at synchrotron facilities (Le Godec et al., 2009). While typical phase equilibrium studies rely on ex situ analyses of recovered samples, a limited number of in situ observations are available to inhouse MAP experiments, including electrical conductivity (e.g., Secco and Schloessin, 1989; Yoshino et al., 2008), ultrasonic measurements (e.g., Li et al., 2004), and melting detection from ionic conduction (e.g., Zhou et al., 2020).

The communities of Earth and planetary sciences, materials science, condensed matter physics, and solid-state chemistry have used the MAP for material synthesis, structure and property investigations, and studies of phase equilibria and chemical reactions. MAP experiments on a wide range of Earth and planetary materials with simplified and realistic compositions have yielded important insights into the nature and dynamics of rocky bodies in the solar system.

Through MAP experiments, the mantle transition zone (MTZ) has been recognized as a potential underground water reservoir, with more than one ocean mass of structural water possibly stored in its principal minerals wadsleyite and ringwoodite (Inoue et al., 1995). Various dense-hydrous magnesium silicates, including phase A, B, D, and H, have been identified as potential carriers to transport water from Earth's surface to its interior through subduction (Ohtani et al., 2000; Poli and Schmidt, 2002). Moreover, MAP experiments revealed that ferrous iron in silicates would disproportionate in the deep mantle to produce metal and ferric iron (Frost et al., 2004; Rohrbach et al., 2007), potentially influencing the long-term evolution of the Earth's oxidation state (Wood et al., 2006). In addition, MAP experiments have led to the discoveries of redox freezing and melting (Rohrbach and Schmidt, 2011) and a melting ledge in carbonated rocks (Thomson et al., 2016), with interesting implications for the genesis of superdeep diamonds (Zhu et al., 2019a), the water consumption capacity of the MTZ (Zhu et al., 2019b), and the time scale and pathways of the Earth's deep carbon cycle (Stagno et al., 2013).

Much has been learned from MAP experiments about the differentiation and evolution processes of Earth-like planets. The models of core formation in a deep magma ocean received initial support from MAP experiments and have been refined through further studies using the MAP and DAC (e.g., Li and Agee, 1996; Wood et al., 2006; Siebert et al., 2012). The melting behaviors of iron alloys observed in MAP experiments suggest that Mars may restart its core dynamo through "snowing" iron or floating iron sulfide (Stewart et al., 2007), and that Mercury's weak magnetic field may be explained by double layers of iron "snow" in its core (Chen et al., 2008). These exotic scenarios of core solidification bear on the sustainability of planetary dynamo and habitable worlds.

Equally fruitful are the applications of the MAP for thermodynamics, elasticity, and rheological studies, which have provided the necessary material parameters for interpreting seismic observations and modeling mantle dynamics (e.g., Karato and Wu, 1993; Dobson et al., 1996; Chen et al., 1998; Li et al., 2004; Schmandt et al., 2014). The negative Clapeyron slope of the postspinel transition in olivine, determined in MAP experiments, implies that the lower boundary of the MTZ in colder regions shifts downward, leading to stagnation of slabs at 660 km (Irifune and Ringwood, 1993; Christensen, 1995). The electric conductivities of hydrous wadsleyite and ringwoodite measured in MAP experiments were compared magnetotelluric observations to infer a relatively dry MTZ (Yoshino et al., 2008; Karato, 2011).

The MAP is considered a type of large-volume apparatus, which was invented to synthesize diamonds (Hall, 1958, 1962). A major area of its application remains in material synthesis. High-pressure polymorphs of oxides, silicates, and iron alloys have been grown in the MAP for subsequent chemical and structural analyses (e.g., Inoue et al., 1998; Smyth et al., 2003) and for further investigations using the DAC and shock compression (e.g., Jacobsen et al., 2004; Millot et al., 2015). The MAP is widely used for synthesizing novel and functional materials, including polymerized C_{60}, nanopolycrystalline diamonds (NPD) and other superhard materials (Iwasa et al., 1994; Haines et al., 2001; Irifune et al., 2003; Huppertz, 2004).

In this chapter, I will review basic aspects of the MAP technique in terms of the generation and characterization of high pressures and high temperatures, with an attempt to quantify the pressure and temperature uncertainties of MAP experiments using tungsten carbide anvils and the COMPRES 8/3 and 10/5 standard assemblies. The rest of the chapter will focus on strategies to synthesize millimeter-sized single crystals of principal silicate phases in the Earth's MTZ and lower mantle.

12.2 Multi-Anvil Press

Various aspects of the MAP techniques have been discussed extensively in the literature (Hall, 1958; Walker et al., 1990; Walter et al., 1995; Bertka and Fei, 1997; Rubie, 1999; Frost et al., 2004; Katsura et al., 2004a; Keppler et al., 2005; Ito et al., 2007; Shatskiy et al., 2011; Leinenweber et al., 2012; Knibbe et al., 2018; Liu et al., 2020; Walker and Li, 2020). This chapter will only discuss the commonly used 6-8 Kawaii-type, where the load or the hydraulic pressure is transmitted through six wedge-shaped anvils to eight cubic anvils, which in turn compress the sample. The six first-stage anvils are typically made of tungsten-rich tool steel with Rockwell hardness in the low 60s, and they are commonly housed in a Walker-style cylindrical chamber (Walker et al., 1990). The eight second-stage anvils are usually made of WC or SD, as mentioned earlier.

12.2.1 Pressure Generation and Measurement

12.2.1.1 Pressure Generation and Limits on Capacity

The MAP generates static high pressure by focusing a large load on a small volume. With the 6-8 type MAP, the final pressure is supported by the triangular-shaped corners of eight cubic anvils, and it inversely scales with the truncation edge length (TEL). The actual pressure generated by the MAP falls well below the theoretical value (Figure 12.2). At small loads, the loss in pressure is mainly caused by plastic deformation of the gaskets. From the difference between the theoretical and actual sample pressures at a load of 400 tons, we estimate that the gaskets bear 50% and 90% of the load in 12 mm and 5 mm TEL experiments, respectively. With the Walker-style apparatus, additional loss due

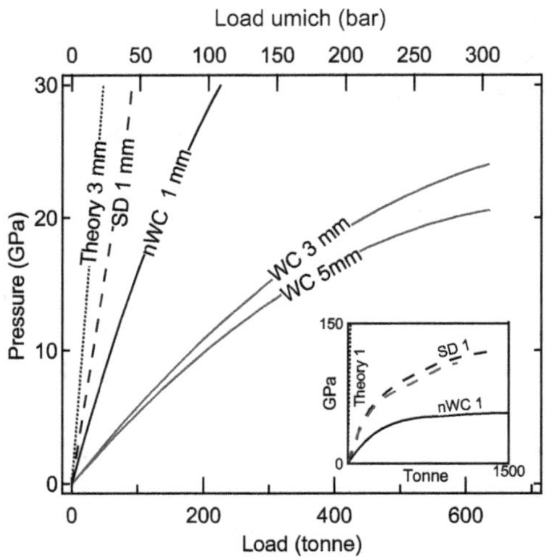

Figure 12.2 Pressure–load relation. At a given load, the pressure at high temperature (red curves) may be higher or lower than that at room temperature (black curves). The main figure shows that sample pressures fall well below the theoretical values (dotted lines) for WC anvils of Tungaloy-Toshiba grade-F (solid curves) with 3 mm TEL (this chapter). The inset shows similar deviations from the theoretical relation for SD anvils (dashed curves) with 1 mm TEL at room temperature (Yamazaki et al., 2019) or at high temperature (Yamazaki et al., 2014). All curves flatten at sufficiently large loads. 1 MN = 10^6 newton in force and corresponds to ~102 tons in mass. Tonnage is given in tonnes (1 US ton ~0.9 metric tonnes).

to friction between first-stage wedges and their container can be reduced to a negligible extent through lubrication (Walker, 1991).

The achievable pressure of a MAP is usually limited by the strength of load-bearing anvils rather than the maximum load of the hydraulic press. The strongest WC anvils with a Knoop hardness of 2,400 kg/mm^2 have been used to achieve pressures up to 31 GPa, and SD anvils with a Knoop hardness of 5,000 kg/mm^2 can support 150 GPa (Zhai and Ito, 2011; Yamazaki and Ito, 2020). Tapering of the truncated corners has been shown to extend the maximum pressure (Ishii et al., 2016). The bevels help to stop the propagation of fracture, typically originating at the centers of truncated corners.

At large loads, anvil deformation reduces the pressurization efficiency, as indicated by the flattening of the pressure–load curve (Figure 12.2). Further loading may lead to anvil and gasket failure known as a "blowout." It is generally assumed that fast compression and decompression increase the chance of anvil failure, especially if they have been weakened or stressed in previous exposure to high pressures and high temperatures for long durations. Leinenweber et al (2012) recommended not to exceed 150 tons per hour for compression. This rate may be treated as a general guidance rather than a hard limit or safe recipe, as anvil performance may vary significantly from batch to batch, and their use histories are difficult to track. One set of WC anvils with 3 mm TEL survived multiple experiments within just a few days, where they were first compressed to 10 GPa in 30 minutes and then compressed further to 25 GPa in another 30 minutes, followed by decompression at the same rates. On the other hand, anvils have often failed while resting on a bench after experiments.

The pressurization efficiency of the MAP also depends on the compositions and dimensions of the gasket material and pressure medium, as shown by a comparison between the 584 castable, 646 castable, and the COMPRES assembly (Liu et al. 2020; Walker and Li, 2020) and by the use of semisintered gaskets with WC anvils to achieve 30 GPa pressure (Shatskiy et al., 2010a). Incompatibility between the Toshiba-brand WC anvils and the 584 cast gasket consistently led to blowout in dozens of attempts to reach the bridgmanite stability field at pressures above 23 GPa. In addition, hardened alumina components in a cell assembly tend to produce local stress, as indicated by deformation in the heater and sample.

12.2.1.2 Pressure Calibration and Uncertainties

Conventional methods for pressure calibration use fixed-point references of known phase transitions. Pressure–load relation at room temperature can be calibrated by monitoring resistance changes associated with phase transitions in semiconducting materials such as Bi, Tl, Pb, ZnS, ZnTe, GaP, and GaAs (Schwarz, 2010). The pressure–load relation at high temperature can be constrained by quenchable high-temperature phase transitions in oxides and silicates such as SiO_2, $CaGeO_3$, and Mg_2SiO_4 (Bertka and Fei, 1997; Rubie, 1999; Walker and Li, 2020). For ex situ calibration of high-temperature transitions, demonstrating approach to equilibrium is crucial, because kinetics of phase transition may cause as much as ±2 GPa uncertainty (Zhang et al., 1996; Rubie, 1999). Using an inhouse ionic conduction method, pressure at 3–10 GPa along the melting curves of NaCl and KCl can be calibrated (Walker and Li, 2020; Zhou et al., 2020).

Pressure calibration using in situ X-ray diffraction (XRD) revealed considerable uncertainties in press–load relations at high temperatures. The effect of temperature on sample pressure was recognized in ex situ experiments and quantified by in situ XRD measurements. At a fixed load, the sample pressure may increase or decrease with temperature, depending on the competing effects of thermal pressure and thermal relaxation (Fei and Bertka, 1999). With the COMPRES 10/5 assembly at 15–20 GPa, the sample pressure at a fixed press load varied by as much as 1.5 GPa upon heating from room temperature to 1,500°C, reaching a peak pressure at 600–1,000°C (Leinenweber et al., 2012). Similarly, with the COMPRES 8/3 assembly at 21–27 GPa, the sample pressure at a fixed load decreased by 1–2 GPa upon heating from 1,200°C to 1,600°C (Fei et al., 2004). Interestingly, with the COMPRES 8/3 assembly at 21–27 GPa, after a previous temperature cycle at a smaller load, the sample pressure increased by 2 GPa upon heating from 1,200°C to 2,000 °C (Fei et al. 2004). More significantly, with the COMPRES 8/3 assembly at 20–30 GPa, the sample pressure at a fixed load decreased by 5 to 10 GPa during heating from room temperature to 2,000–2,300°C (Leinenweber et al., 2012). In an experiment using the binderless WC anvils, a pressure loss of nearly 20 GPa at 65 GPa was observed upon heating from room temperature to ~1,700°C (Ishii et al., 2017). Furthermore, sample pressure at a constant load and temperature may drift with time. With the COMPRES 8/3 assembly, relaxation at a constant load and 1,600°C caused downward drift in sample pressure by ~0.5 GPa within the first 30–60 minutes (Fei et al., 2004).

Because of irreversible deformation of gasket and pressure medium, the pressure–load relation only holds for the first round of compression, and not for decompression or recompression. In situ observations during decompression revealed that sample pressure remains high or falls more slowly than the initial release of press load, a hysteresis caused by the friction between the wedges and container ring in the Walker-style apparatus (Walker et al., 1990). To avoid blowout during decompression, the load must be released at a sufficiently slow rate. Considering the hysteresis, the load may be released quickly to an intermediate value and then slowly to 1 bar. With tungsten carbide anvils and the COMPRES 10/5 and 8/3 assemblies in this study, the load is released to 200–250 tonnes in 8–12 hours, and then to 1 bar in another 8–12 hours, or to 450 tonnes in 1 hour, and then to 1 bar in 16–20 hours.

Considering various sources of errors, the overall uncertainty in pressure measurement is estimated to be ±10%, including ±5% variations between duplicate experiments, ±5% systematic errors arising from the thermal effects on pressure calibration, and additional variations associated with temperature cycle and drift over time. Uncertainties in temperature measurements contribute additional errors to pressure calibration, as discussed in the following section.

12.2.2 Temperature Generation and Measurement

The majority of MAP experiments are conducted at high temperatures. The material and geometry of the heater, as well as the type and placement of the thermocouple, influence the distribution and characterization of temperature across the sample.

12.2.2.1 Heater

In MAP experiments, high temperature is generated through resistive heating. Typical heaters are made of rhenium (Re), graphite (C), or doped lanthanum chromite, $LaCrO_3$. Other heater materials include metal foils such as molybdenum Mo and tantalum Ta and semiconducting materials such as boron-doped diamond (Shatskiy et al., 2009a) and TiB_2-BN, TiC–diamond, or TiC–MgO composite (e.g., Xu et al., 2020). At 1 bar, Re, Ta, and Mo melt at 3,185°C, 3,017°C, and 2,623°C, respectively. Although considerably less expensive, Ta is more vulnerable to oxidation, and Mo is less refractory than Re. The melting points of TiB_2, TiC, and graphite at 1 bar are 3,230°C, 3,160°C, 3,550°C, respectively. As heater materials, doped TiB_2, TiC, and diamond stand out in their transparency to X-ray and higher temperature limits, in addition to the low cost of the raw ingredients (Shatskiy et al., 2009a). However, they are not yet commercially available in reproducible compositions and convenient forms.

Designing a MAP experiment involves selecting suitable heater material and geometry to generate desired temperature distribution in the sample. The resistance of a metal increases with temperature, whereas that of a semiconductor decreases with temperature (Figure 12.3a). With the commonly used shape of straight-walled cylinder, the temperature-resistance relations imply a steeper temperature gradient in a metal heater than a semiconducting heater. The idealized temperature profiles of some standard assemblies were predicted by finite-element simulations (Hernlund, 2006; Leinenweber et al., 2012), and the general features were confirmed by measurements using multiple thermocouples (Walker and Li, 2020) and in situ X-ray diffraction (Nishihara et al., 2020). With the 8/3 COMPRES assembly, the middle quarter of the Re heater length is isothermal for a peak temperature of 1,050°C or 1,200°C (Figure 12.3b). With the 10/5 COMPRES assembly, the center isothermal zone extends to one third of the Re heater length through the use of a more insulating zirconia end cap. With an 8 mm TEL assembly, the temperatures in the middle half of the $LaCrO_3$ heater length fall within ±50°C of the peak value, and most of

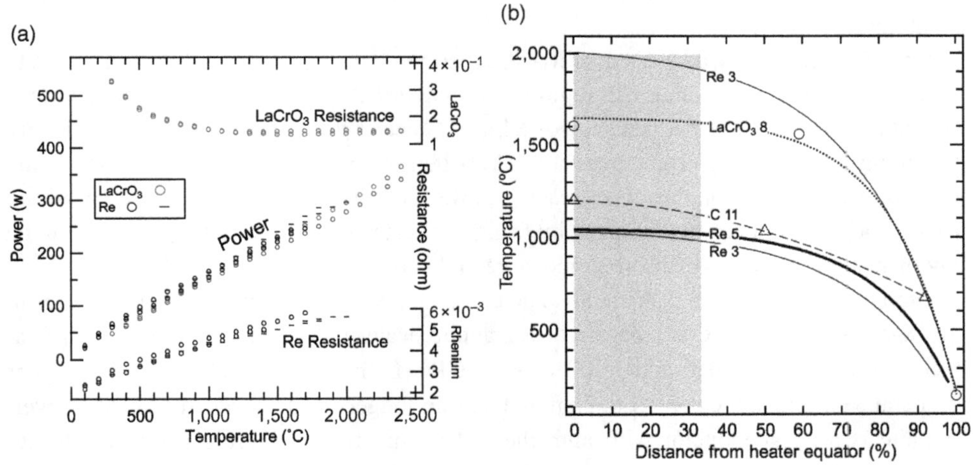

Figure 12.3 Heating in a multi-anvil experiment. (a) Power and resistance of heater as a function of temperature. The data are from multi-anvil experiments conducted at the University of Michigan, using Toshiba–Tungaloy grade-F anvils with 5 mm TEL. Experiments with Re heaters used the COMPRES 10/5 assembly (M062319, black open circles; M100319, black bars). Experiments with $LaCrO_3$ heaters had radial thermocouples through the heater equator and 1 mm thick $LaCrO_3$ discs at both ends (M080719, M080819, red open circles). The power and resistance are calculated from measured voltage and current. (b) Temperature distribution. The thermal profiles inside Re heaters of the COMPRES 8/3 and 10/5 assemblies (thin and thick solid curve, respectively) are fitted to the results of finite-element simulations for a peak temperature of 1,050°C (Leinenweber et al., 2012), which are scaled up to a peak temperature of 2,000°C. Those in a $LaCrO_3$ heater of an 8 mm TEL assembly (dotted curve) and a graphite heater of a 11 mm TEL assembly with (dashed curve) are fitted to measured temperatures (open circles, Walker and Li, 2020, and open triangles, Nishihara et al., 2020, respectively). All the curves take the form $T = a_0 - a_1\exp(-a_2x)$, where T is temperature in °C, x is the relative distance to the equator of a cylindrical heater (0 at the equator, and 100 at the end), and a_0, a_1, a_2, are fitting parameters. The heater lengths of the COMPRES 8/3 and 10/5 assemblies at pressures above 5 GPa are about 4.7 mm and 6.1 mm respectively.

the thermal gradient concentrates near the cold ends. In comparison, a thermal gradient of – 200°C/mm near the heater equator at 1,500°C was measured using the two-pyroxene solvus thermometry in a smaller (10/4) assembly with a straight $LaCrO_3$ heater (Walter et al., 1995). The isothermal zone is further extended in a stepped cylindrical heater of graphite or $LaCrO_3$, where the resistance and hence the heating power in the middle section is reduced (e.g., Walter et al., 1995; Shatskiy et al., 2009b). This is verified by the shape of sintering halos outside the heater: The inward concave halos correspond to isotherms, and they appear straighter in stepped heaters

A metal heater usually occupies much less space in an MA cell assembly than a semiconducting heater. In the cylindrical form, the wall thickness of a Re heater is typically 0.025–0.075 mm, compared to ~1 mm for a $LaCrO_3$ heater. However, because the resistance increases with temperature, a metal heater not only has a relatively small quasi-isothermal region but is also more vulnerable to the formation of hot spot and meltdown failure. Moreover, some metal foils are expensive and may require additional cost to fabricate.

Doped $LaCrO_3$ is widely used in MAP experiments because of its ability to generate relatively large quasi-isothermal regions in cylindrical heaters, even at temperatures exceeding 2,000°C. The performance of a $LaCrO_3$ heater depends critically on the quality of doping and sintering. Sometimes a $LaCrO_3$ heater fails at 1,200–1,400°C, where the temperature stops responding to increasing power and then crashes to room temperature. This "droop" failure can be mitigated by firing the heater at >500–600°C for 1 hour on the day of loading, to remove absorbed moisture and carbon dioxide.

Graphite heaters have a distinct advantage of being X-ray transparent and are used for in situ X-ray studies. However, the use of a graphite heater is limited to roughly 1,600°C at 5 GPa, 1,400°C at 8 GPa, and 1,200°C at 10 GPa (Leinenweber et al., 2012). At higher temperatures, graphite converts to diamond and no longer conducts electricity. Moreover, graphite is a semiconductor with high thermal conductivity, therefore a graphite heater consumes more power than a metal or $LaCrO_3$ heater for the same peak temperature, and it may overheat fusible parts in the experimental assembly, such as G10 squares used to hold the eight anvils together.

12.2.2.2 Thermocouple

In MAP experiments, the sample temperature is usually monitored with a thermocouple. Common types include the Pt–Rh-based type S, R, and B thermocouples, the Re–W-based type C and D thermocouples, and the Ni–Cr-based type K thermocouple. In comparison, Pt–Rh alloys are more chemically inert than Re–W alloys, which are vulnerable to oxidation and therefore sometimes embedded in boron–nitride (BN) for long-duration experiments (Müller et al., 2017). The temperature ranges of type S and R thermocouples are limited by the melting point of Pt, which is 1,768°C at 1 bar, whereas the Re–W-based thermocouples are more refractory and have been used up to 2,800°C at 10 GPa (Walker et al., 1993) and up to 3,500°C at higher pressures (Shatskiy et al., 2009a), although the thermocouples are not calibrated at these temperatures. Mechanical deformation of thermocouple wires may introduce random errors to temperature measurements, especially at high temperatures (Mao and Bell, 1971). At 5–15 GPa, the random error is estimated at 1.6% for the type S thermocouple at temperatures up to 1,700°C, and 0.8% for the type C thermocouple at temperatures up to 2,100°C (Li et al., 2003).

There are a number of options to place a thermocouple in an MA cell assembly. The thermocouple wires are often placed along the axis of rotation of a cylinder-shaped heater and exit the heater near the cold ends to minimize perturbation to the hot middle section of the heater and maximize its thermal stability (e.g., Zhou et al., 2020). The thermocouple junction and sample may be positioned symmetrically with respect to the heater equator, or the thermocouple junction may be placed adjacent to the sample container at the heater equator. With the axial placement, the pointed junction of strong W–Re thermocouple wires may be supported by a sintered alumina (Al_2O_3) disc or placed in a cavity to prevent it from piercing through the sample container during compression (Leinenweber et al., 2012). It is also common to place the thermocouple wires along the radius of a cylindrical heater, perpendicular to its axis of rotation (e.g., Zhou et al., 2020). With the radial placement, the thermocouple may be placed on one side of the heater, with the junction outside or inside the heater, or the two wires may be placed on opposite sides of the heater, connected directly or through a metal heater. If placed at the heater equator to measure the peak temperature, a radial thermocouple may cause irregular deformation and local cold or hot spots in the heater. To prevent a breaking failure caused by large shear stress, wires or coils of matching Seebeck coefficients may be added to mechanically stabilize thermocouple wires in shear zones. Note that including copper coils in the thermocouple circuit may cause considerable errors in the temperature measurements (e.g., Liebske and Frost, 2012).

Reliable temperature measurements typically require that both the sample and thermocouple junction are placed in the isothermal zone near the equator of a cylindrical heater. Using thinner wires reduces the junction size and improves the spatial resolution of temperature measurement. In a cylindrical $LaCrO_3$ heater, the isothermal zone at ~1,700°C covers nearly half of the distance between the equator and end of the heater, whereas in a cylindrical Re heater, the isothermal zone at 1,200°C only extends to one quarter of the distance (Figure 12.3b). With the COMPRES 8/3 assembly, the sample and thermocouple junction are placed within the middle third of the heater length. The uncertainties in temperature measurements due to thermal gradient are estimated to be negligible at temperatures below 1,050°C, ±3% at 1,200°C, and by extrapolation, ±6% at 2,000°C.

12.2.2.3 Pressure Effect on Thermocouple's emf

A thermocouple monitors the temperature at the hot junction of two wires by measuring the electromotive force (emf) at the separate cold ends, usually at ambient temperature. The emf of the type C thermocouple increases by roughly 2 mV per 100°C to reach ~20 mV at 1,000°C, and that of the type S thermocouple increases by ~1 mV per 100°C to reach ~10 mV at 1,000°C. The emf–temperature relations are calibrated at 1 bar, to 2,320°C for the type C and 1,760°C for the type S.

Pressure is known to influence the emf–temperature relation. The effect depends on the type of thermocouple and generally increases with pressure and temperature. At 4 GPa and

1,500°C, the type D thermocouple measures below the true temperature by 25°C (Mao and Bell, 1971). At 15 GPa and 900°C, the type S thermocouple measures below the true temperature by 30°C (Nishihara et al., 2020). By extrapolation to 15 GPa and 1,500°C, the type S thermocouple measures below the true temperature by as much as 80°C. When placed at the same location in a multi-anvil cell, the temperature reading of the type D thermocouple is higher than that of the type R thermocouple by 12–27°C at 600–1,600°C and 6 GPa (Walter et al., 1995), and the readings of the type S and type C thermocouples differ by –35 to 25°C at 15 GPa and temperatures up to 1,800°C (Li et al., 2003).

In earlier studies, the effect of pressure on thermocouple emf was measured by connecting two identical thermocouple wires at the hot end, compressing one wire and keeping the other wire at 1 bar inside a hollow and hard alumina tubing (Mao and Bell, 1971). Recently, the effect was studied by comparing the temperature readings, pressures, and deformation of two identical wires in a crushable MgO tubing and a hard Al_2O_3 tubing, using synchrotron X-ray radiography and diffraction (Nishihara et al., 2020). This method worked up to 15 GPa and ~1,000°C, upon which the Al_2O_3 tubing collapses or gets filled by MgO. Due to limited knowledge, the pressure effect on the emf–temperature relation is ignored in most MAP studies. In fact, correcting for the pressure effect would be a daunting task considering its complex dependence on pressure, temperature, and composition. Moreover, no universal correction is available to account for the effects of pressure on thermocouple emf because the correction depends on the detailed and variable temperature distribution in the gasket regions surrounding the pressure cell, where pressure is reduced to ambient conditions (Walter et al., 1995). As a practical approach, we can continue to conduct experiments and report temperature readings on the basis of the 1 bar calibrations and apply corrections afterward when necessary and possible.

12.2.2.4 Power Curve

Using a power curve to estimate sample temperature is not ideal because irregular heater deformation may cause large deviations from a calibrated power–temperature relation. Nevertheless, a power curve provides a rough measure of the sample temperature when the thermocouple fails and when the target temperature is beyond the calibrated emf–temperature range. Sometimes the thermocouple is left out to free up space, to avoid contamination, or simply for convenience, where precise measurement of temperature is unnecessary. Even when thermocouple readings are available, a power curve can be used to detect erroneous measurements due to a misplacement of the junction or an unintended shorting of the thermocouple. At a constant temperature reading, fluctuations in power consumption usually indicate abnormal mechanical or chemical changes.

With properly assembled experiments, power curves at a given pressure can be reproducible within ±100°C at temperatures up to 1,200°C (Figure 12.3a). At higher temperatures, the temperature at a given power becomes less certain and may scatter by as much ±300°C at 2,000°C. As pressure increases, more power is needed to generate the same temperature, because the assembly gets more compacted and the lattice thermal conductivities of its components increase with pressure, both facilitating heat loss. With the 10/5 COMPRES assembly at 15–20 GPa and 1,200–1,800°C, power consumption in this study drifted over time, usually decreasing by 1% in the first hour, and then remaining constant within 5% for up to 48 hours. The time dependence can be attributed to initial heat expenditure to warm up the surrounding materials until a steady state is reached.

12.3 Theoretical Basis for High-Pressure Synthesis

Synthesizing single crystals at high pressure is important for understanding the origin and evolution of large planetary bodies. For instance, to reconstruct the process of Moon formation during the Earth's early history, we need to model the Earth-Moon system undergoing melting or vaporization during giant impacts (Davies et al., 2020). Essential model inputs include equation-of-state data from dynamically compressed samples of deep mantle phases. Clear single crystals of at least 300 μm, and preferably >1 mm in size, are required for in situ temperature measurements in laser-shock experiments. Here I review the synthesis strategies and phase diagrams relevant to growing large and transparent crystals of deep mantle phases wadsleyite, ringwoodite, and bridgmanite using a MAP.

12.3.1 Nucleation and Growth from a Melt

Crystal growth may be viewed as a two-step process: The formation of nuclei from source is followed by the growth of the nuclei into the final product. In the case of growing crystals from a liquid, the rates of nucleation and growth chiefly depend on the temperature, as illustrated in the following subsection.

12.3.1.1 Nucleation as a Function of Temperature

The growth of a crystal begins with nucleation, unless a seed crystal is already present. A very low but nonzero nucleation rate is key to synthesizing large crystals. As explained in a theoretical account of igneous rock formation (Brandeis et al., 1984), nucleation is either homogeneous, resulting from thermal, chemical, or stress fluctuation in the source material, or heterogeneous, facilitated by an existing solid phase such as the container. At temperatures near the liquidus, where the amount of undercooling is small, nuclei do not survive because of excess surface energy. Therefore, the nucleation rate is negligible within a temperature interval just below the liquidus. This interval is smaller for heterogeneous nucleation than homogeneous nucleation. For crystallization of magma near Earth's surface, the critical temperature interval for heterogeneous nucleation, δ, is typically a few degrees, and that for homogeneous nucleation, Δ, is tens of degrees (Figure 12.4).

Following the general theory outlined in Brandeis et al. (1984), the nucleation rate is governed by the thermodynamic driving force arising from the Gibbs energy difference between the source and a nucleus of the product, and by the suppressing force associated with the kinetic barrier to nucleation. As a system continues to cool below the liquidus, the driving force increases, but the probability to overcome the kinetic barrier decreases.

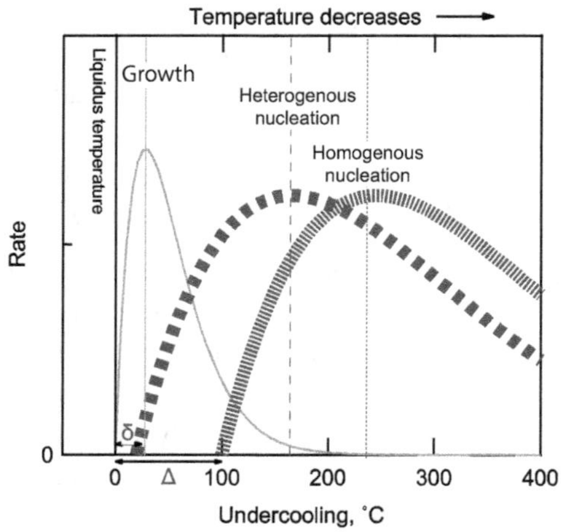

Figure 12.4 Nucleation and growth of crystals from a liquid. Upon cooling below the liquidus temperature, the growth curve (thin blue) peaks first, followed by the heterogeneous nucleation curve (thick red dashed curve) and then the homogeneous nucleation curve (thick red dotted curve). While the growth of existing crystals may start as soon as the temperature falls below the liquidus, heterogeneous nucleation requires a minimum amount of undercooling, δ, and homogeneous nucleation requires a minimum amount of undercooling, Δ.

Here the nucleation rate is formulated as a function of temperature, T, and three parameters, the liquidus temperature, T_L, the Gibbs energy difference between the nucleus and liquid, ΔG, and the activation energy for nucleation, ΔG_a:

$$\text{Nucleation rate} = k_1 \bullet \{1 - \exp[(T - T_L) \bullet \Delta S/RT)\} \bullet \exp(\Delta G_a/RT),$$

where k_1 is a preexponential constant; T_L determines the amount of undercooling at a given T; ΔG provides a measure of the thermodynamic driving force for nucleation and is assumed to take the form $(T_L - T) \bullet \Delta S$, where ΔS, the entropy difference between liquid and crystal, is independent of temperature; and ΔG_a provides a measure of the kinetic barrier to nucleation and is also assumed to be independent of temperature. According to this formulation, the peak rate of either homogeneous or heterogenous nucleation occurs at an intermediate amount of undercooling (Figure 12.4).

12.3.1.2 Growth as a Function of Temperature

A crystal grows through interface reactions that convert the source to the product. The rates of interface reactions are governed by the driving force arising from the difference in Gibbs free energy between the source and product, and by the opposing force associated with kinetic barrier to the conversion. After initial nucleation, chemical diffusion may become a limiting process for further growth.

According to Brandeis et al. (1984), the growth rate is a function of temperature T and three parameters: the liquidus temperature, T_L, the Gibbs energy difference between the crystal and liquid, $\Delta G'$, and the activation energy for interface reaction that converts liquid to solid, $\Delta G_a'$:

$$\text{Growth rate} = k_2 T \bullet \{1 - \exp[(T_L - T) \bullet \Delta S'/RT]\} \bullet \exp(\Delta G_a'/RT),$$

where k_2 is a preexponential constant; T_L determines the amount of undercooling at a given T, $\Delta G'$ provides a measure of the thermodynamic driving force for crystal growth and is assumed to take the form $(T_L - T) \bullet \Delta S'$, where $\Delta S'$ is independent of temperature; and $\Delta G_a'$ provides a measure of the kinetic barrier to crystal growth and is also assumed to be independent of temperature. In general, $\Delta G' \neq \Delta G$ and $\Delta G_a' \neq \Delta G_a$, because a nucleus has a higher surface energy than a crystal. In this formulation, the additional T term accounts for faster transportation of the source to the interface to facilitate crystal growth at higher temperatures. There is no growth at the phase boundary, where the driving force is zero. The growth rate peaks at an intermediate amount of undercooling (Figure 12.4). For crystallization of magma near Earth's surface, peak growth likely occurs at a few tens of degrees below the liquidus (Brandeis et al., 1984).

The presence of a seed crystal allows for fast crystal growth just below the liquidus temperature, where homogeneous nucleation is energetically unfavorable (Figure 12.4). This is the basis of some classical methods for growing large crystals. For example, the Czochralski method involves pulling single crystal from a melt using a seed crystal. The method was invented in 1915 and is used for producing bulk single crystals of a wide range of electronic and optical materials, including semiconductors, metals, salts, and synthetic gemstones (Kitamura et al., 1992). Similarly, the Bridgman–Stockbarger method used for growing single crystal ingots (boules) involves heating polycrystalline material above its melting point and then slowly cooling the melt from one end of its container, where a seed crystal is located (Bridgman, 1925).

12.3.1.3 Crystal Growth with Time

At a constant temperature, the transformation from source to product follows a sigmoidal curve, where the rate is low at the beginning and the end of the transformation, but rapid in between, as described by the Avrami equation, also known as the Johnson–Mehl–Avrami-Kolmogorov (JMAK) equation (Avrami 1939). The initial rate is low because it takes time to form nuclei that are sufficiently large to survive. Toward the end of the transformation, the rate diminishes because the source is nearly exhausted. In the intermediate stage, the source is readily available and the kinetic barrier to growth of existing nuclei is relatively small, and therefore rapid transformation can be sustained.

Crystal size tends to increase with time through Oswald ripening, a term that describes the growth of large droplets or crystals at the expense of small ones. Coarsening happens because excess surface energy makes smaller particles less stable and more readily dissolve during thermal, chemical, or stress fluctuations. Temperature cycling can enhance the ripening process to produce large crystals. During a heating cycle, small crystals disappear and large crystals shrink, by converting into another phase or dissolving into a melt or solution. In the subsequent cooling cycle, large crystals recover the lost portion and grow further from the newly released source. By oscillating between high and low temperature in a suitable range at appropriate rates, the average crystal size may increase rapidly with time.

12.3.2 Growing Large Crystals from a Fluid Solution

Shatskiy et al. (2010b) clearly explained the underlying mechanisms for growing large crystals from a fluid solution, by cooling slowly or setting up a thermal gradient. In the case of crystal growth from a fluid solution, the amount of undercooling translates into supersaturation of the solution as the controlling parameter for the rates of nucleation and growth. The stability field of coexisting solid and liquid may be schematically divided into regions of growth, heterogeneous nucleation, and homogeneous nucleation (Figure 12.5). At small degrees of supersaturation, existing crystals can grow, but nucleation is difficult because excess surface energy exceeds the driving force for partial crystallization of the fluid. At intermediate degrees of supersaturation, only a small number of viable nuclei can form through heterogeneous nucleation on the container wall, and they may grow into large crystals. At large degrees of supersaturation, this large driving force for crystallization enables homogeneous nucleation when the rate of heterogeneous nucleation is already high. As a result, many viable nuclei may form and compete for the source, likely producing small crystals.

With the slow cooling approach, viable nuclei form through heterogeneous nucleation at a small degree of supersaturation. The remaining fluid becomes diluted and shifts into the growth region. The temperature is further reduced slowly to keep the system in the growth region, so that the existing crystals continue to grow until the final temperature T_L is reached (Figure 12.5a). The slow cooling method has been successfully implemented in MAP experiments (e.g., Shatskiy et al., 2010b; Okuchi et al., 2015; Fu et al., 2019).

With the thermal gradient approach, a solid source is initially placed at the hot end, and it dissolves into the fluid at the interface (Figure 12.5b). If the fluid is well mixed, heterogeneous nucleation would begin near the cold end. The formation of nuclei shifts the fluid composition into the growth region. Further growth is sustained as the fluid is replenished by the dissolution of the solid source at the interface, which moves up along the thermal gradient with time (Figure 12.5b). Shatskiy et al. (2010b) showed that using metastable solid source promotes homogeneous nucleation, and therefore is not recommended for growing large crystals. The thermal gradient method has been successfully implemented in MAP synthesis experiments by shifting the sample away from the heater equator (Shatskiy et al., 2007) or using asymmetric insulating end caps (Millot et al., 2015).

12.3.3 Nucleation and Growth through Solid-State Transformation

Polymorphic transformation from a solid source is a straightforward approach to synthesize a crystal at high pressure. Transformation of one solid structure into another resembles the

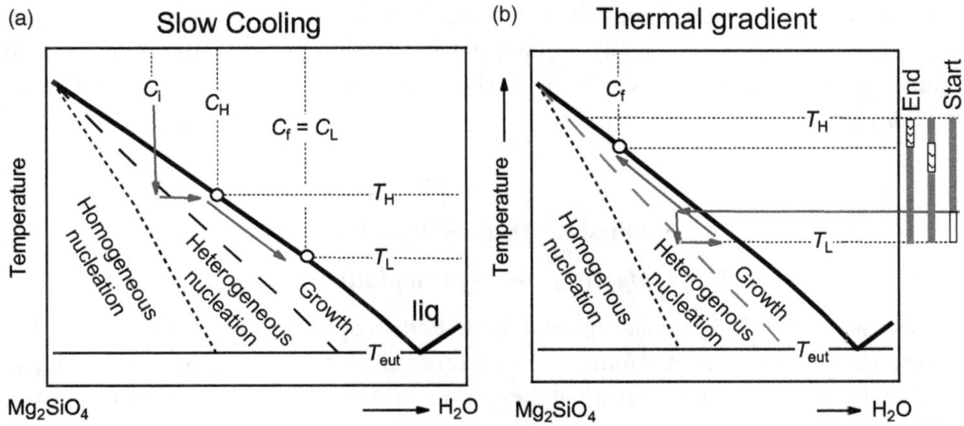

Figure 12.5 Crystal growth from a fluid solution. In this schematic binary phase diagram of Mg_2SiO_4–H_2O at 18 GPa, the two-phase region on the Mg_2SiO_4-rich side of the eutectic point is divided into a growth region with a small amount of supersaturation of Mg_2SiO_4; a heterogeneous nucleation region with an intermediate amount of supersaturation, where nucleation may take place on the walls of sample container; and a homogeneous nucleation region with a large amount of supersaturation, where spontaneous nucleation may occur within the solution. (a) Slow cooling method. Red arrows denote the evolution path of the fluid composition as the system cools. With sufficient cooling, heterogeneous nucleation begins and shifts the composition of the solution from the initial value C_i into the growth region. Slow cooling keeps the fluid composition within the growth region and allows existing crystals to grow without further nucleation. As temperature decreases, the fluid composition evolves from C_H at high temperature T_H, to C_L at low temperature, T_L, which is also the final composition C_f. (b) Thermal gradient method. Red arrows denote the evolution path of the fluid composition with time, assuming instantaneous mixing in the fluid. Initially, solid Mg_2SiO_4 is placed at the hot end (solid bar) and H_2O is placed at the cold end (open bar). At the start, solid Mg_2SiO_4 dissolves in H_2O at the interface. Through convection and diffusion in the fluid, solute propagates down the temperate gradient. Heterogeneous nucleation occurs where the coldest fluid composition crosses the long-dashed curve. As a result of nucleation, the system shifts into the growth region. In the intermediate stage, solid Mg_2SiO_4 continues to dissolve at the interface. The solute diffuses toward the cold end to replenish the source fluid for crystal growth. Finally, the solid at the hot end is replaced by a residual fluid of the composition C_f (patterned bar), which depends on the temperature distribution and bulk mass ratio of Mg_2SiO_4 and H_2O. In practice, H_2O may permeate through grain boundaries and disperse in Mg_2SiO_4 powder, but the fluid in the end product will have the same final composition, C_f.

congruent melting of a solid in terms of nucleation and growth, as applied to the olivine–wadsleyite–ringwoodite system (Rubie and Ross, 1994; Perrillat et al., 2016). For solid growth at high pressures, undercooling in the general theory of crystal growth from a liquid translates into overpressure (Kubo et al., 2004), the magnitude of which is likely larger for reconstructive transformation than for displacive transformation.

Compared with the liquid growth method, solid growth can take place at lower temperatures and hence is less vulnerable to chemical contamination. Moreover, there is no need to weld the sample container. However, using the solid-state synthesis approach requires a high-purity source to avoid inclusions and minor phases in the product, which would impede crystal growth. The purity requirement is less stringent for growth from a fluid solution or binary melt, where deviations in chemical composition may be absorbed by the fluid or melt.

12.4 Synthesis of Dense Silicate Polymorphs

12.4.1 Mg_2SiO_4 Wadsleyite and Ringwoodite

Olivine $(Mg,Fe)_2SiO_4$ is a major mineral in the Earth's upper mantle. Upon compression along a geotherm, olivine transforms into wadsleyite at 13–15 GPa and into ringwoodite at 17–21 GPa, before breaking down into bridgmanite $(Mg,Fe)SiO_3$ and ferropericlase (Mg, Fe)O near 23 GPa (Figure 12.6).

The magnesium endmember of wadsleyite and ringwoodite may be synthesized from their own melts, from a fluid solution, or through solid-state transformation. Relevant phase diagrams provide guidance for the synthesis strategy.

12.4.1.1 Growth of Wadsleyite and Ringwoodite from Anhydrous Melt

Congruent melting is a prerequisite for growing crystals from their own melt. The Mg_2SiO_4 composition melts congruently between 1 bar and 10 GPa, at temperatures above 1,900°C. At 10–14 GPa, forsterite starts to melt incongruently to form periclase and liquid. At higher pressure, incongruent melting of wadsleyite produces anhydrous phase B, an Mg-rich silicate with the nominal composition of $Mg_{14}Si_5O_{24}$, where 1/6 of the Si in Mg_2SiO_4 is replaced by Mg. Wadsleyite always melts incongruently, into anhydrous B and liquid at pressures up to 22 GPa, and into periclase and liquid at 22–24 GPa. Ringwoodite does not melt directly. It transforms into wadsleyite upon heating and breaks down to form $MgSiO_3$ bridgmanite and MgO periclase upon compression. As a result, wadsleyite or ringwoodite with Mg_2SiO_4 composition cannot be grown from its own melt.

The phase diagram of Mg_2SiO_4 presented here resolves a number of discrepancies in the literature. The incongruent melting features included here are not shown in some simplified versions (e.g., Fei and Bertka, 1999). The pressures in the original experimental data (Ito and Takahashi, 1989; Katsura and Ito, 1989) are revised to be consistent with updated high-temperature pressure scales (Fei et al., 2004). Some of the phase boundaries in Shatskiy et al. (2009b) are corrected upward by 80–300°C to account for the use of copper coils to mechanically stabilize thermocouple wires in shear zones (Liebske and Frost, 2012).

The phase diagram of the $MgO–SiO_2$ binary system (Figure 12.7) suggests that it is possible to grow wadsleyite from an anhydrous melt with slightly more Mg-rich composition than the $Mg_2SiO_4–MgSiO_3$ eutectic at 16 GPa. This is challenging, though, because wadsleyite only coexists with silica-rich liquid in a narrow temperature range of ~100°C above 2,200°C. Cooling from a liquid, the relative mass fraction of wadsleyite is small according to the lever rule. With increasing pressure, the wadsleyite-liquid field shrinks further and shifts to even higher temperatures (Liebske and Frost, 2012).

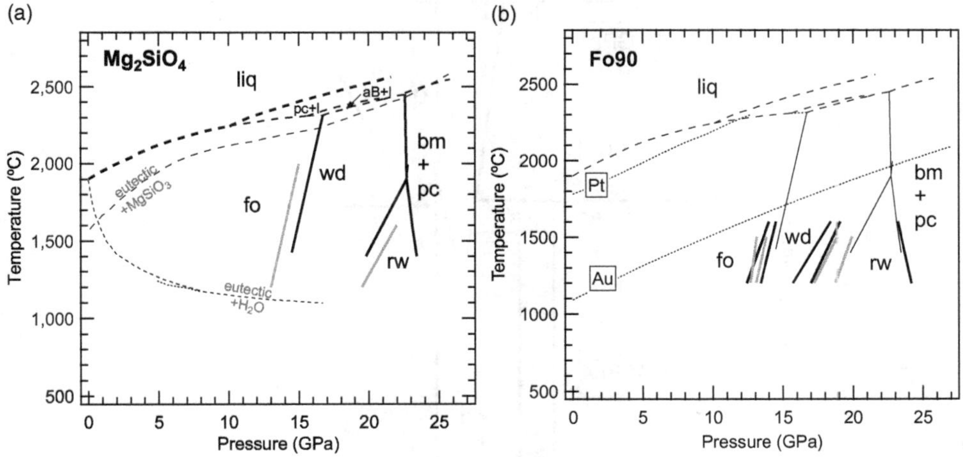

Figure 12.6 Phase diagram of Mg_2SiO_4 and $(Mg_{0.9}Fe_{0.1})_2SiO_4$. (a) Mg_2SiO_4. Phase boundaries are shown as thick black traces. Thick black dashed curves represent melting boundaries involving forsterite (fo), wadsleyite (wd), bridgmanite (bm), periclase (pc), and anhydrous-B (aB) and liquid (liq or l) (Presnall et al., 1998). Thick black solid lines represent solid phase boundaries: fo/wd (Presnall et al., 1998), wd/bm+pc (Presnall et al., 1998; Fei et al., 2004), wd/rw and rw/bm+pc (Fei et al., 2004). A small amount of H_2O (<0.2 wt.%) expands the stability field of wadsleyite (thick gray lines; Mrosko et al., 2015; data at 1,200°C, assuming the same slope as the anhydrous composition). Also shown are the binary eutectic melting curve of Mg_2SiO_4–$MgSiO_3$ (thin long-dashed curve; Presnall et al., 1998; Liebske and Frost, 2012), the binary eutectic melting curve of Mg_2SiO_4+H_2O (thin short-dashed curve; Inoue et al., 1994), and the ternary eutectic melting curve of Mg_2SiO_4–$MgSiO_3$+H_2O (thin dotted curve; Inoue et al., 1994). (b) $(Mg_{0.9}Fe_{0.1})_2SiO_4$. Phase boundaries are shown as thick black traces. With the addition of a small amount of iron, the univariant boundaries between fo and wd and that between wd and ringwoodite (rw) split into two-phase fields, and they shift to lower pressures (thick black solid lines). Data sources: fo/wd at 1,200°C and 1,600°C from Fei et al., 1991; fo/wd at 1,400°C from Fei and Bertka, 1999, fo/wd and rw/bm+pc at 1,400°C from Ito and Takahashi, 1989; wd/bm+pc from Fei et al., 2004. The presence of a small amount of water (<0.2 wt.%) shifts the wd+rw field up by more than 2 GPa but has little effect on the fo+wd field (thick gray solid lines, the widths of which indicate uncertainties; Chen et al., 2002; Mrosko et al., 2015). Also shown are (a) the melting curve and solid phase boundary of anhydrous Mg_2SiO_4 as thin curves, and (b) the melting curves of Au (Akella and Kennedy, 1971) and Pt (Mitra et al., 1967) as thin dotted curves.

12.4.1.2 Growth of Wadsleyite and Ringwoodite from Hydrous Melt

Water is commonly used as a solvent for growing crystals from a fluid solution. Growing wadsleyite or ringwoodite crystals from a hydrous melt proves to be successful, as reported in the literature and here (Figure 12.8). Shatskiy et al. (2007) used 6 mm TEL anvils to generate 24 GPa at a load of 1,900 tonnes. With a stepped heater, they reduced the thermal gradient in the sample region from 100 °C/mm to 20°C/mm and applied the thermal gradient method to grow crystals of wadsleyite and ringwoodite larger than 1 mm in at least two dimensions. In this study, we grew millimeter-sized crystals using WC carbide anvils

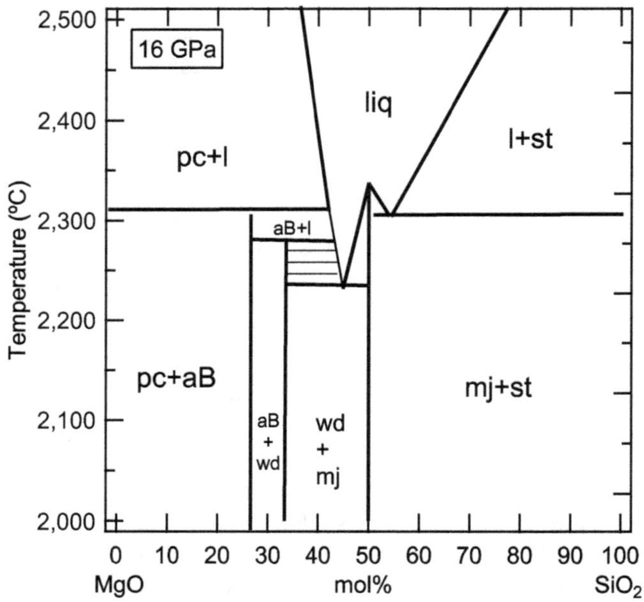

Figure 12.7 MgO–SiO_2 binary phase diagram at 16 GPa. The diagram is drawn on the basis of experimentally determined triple points involving periclase (pc), anhydrous B (aB), wadsleyite (wd), majorite (mj), stishovite (st), and liquid (l or liq) in the binary system (Liebske and Frost, 2012), and the melting points of MgO (Du et al., 2014) and SiO_2 (Shen and Lazor, 1995). Tie-lines in the stability field of wd+liq are shown. Note that the temperatures in Liebske and Frost (2012) are corrected upward by 80–300°C from those of Shatskiy (2009b) to account for the use of copper coils to mechanically stabilize thermocouple wires in shear zones.

and the COMPRES 10/5 assembly at ~540 tonnes. All the crystals larger than 0.3 mm were grown in hydrous compositions, in as little as 7 hours. The largest crystals exceeding 1 mm in at least two dimensions were synthesized with water in 24–48 hours. In contrast, anhydrous experiments produced crystals smaller than 100 μm and down to submicron. Some wet experiments produced small crystals despite long duration and/or high temperature, possibly due to water loss or homogeneous nucleation.

Water can be added directly as a liquid or in the form of a hydroxide, such as brucite $Mg(OH)_2$ and $H_2Pt(PH)_6$ (Myhill et al., 2017), and sealed in a gold or platinum capsule by arc welding or pressure welding. With the Lampert arc welder, the energy and duration of electric pulse can be precisely controlled to weld millimeter-sized capsules. For pressure welding, the sample may be packed into a matching pair of open-end capsules (Inoue et al., 1994) or a capsule formed from thin foil. In this study, we were able to load ~3 mg and ~10 mg forsterite into COMPRES 8/3 and 10/5 cells, respectively, and add 10 wt.% water by soaking a closed capsule in deionized water for at least 30 minutes. The amount of sample can be doubled by excluding the thermocouple. Although the container size in this study is only 2.0 × 2.0 × 1.8 mm, compared with 2.0 × 2.0 × 2.5 mm in Shatskiy et al. (2007) and 4 × 4 × 5 mm in Okuchi et al. (2015), millimeter-sized wadsleyite crystals were produced using the fluid solution method.

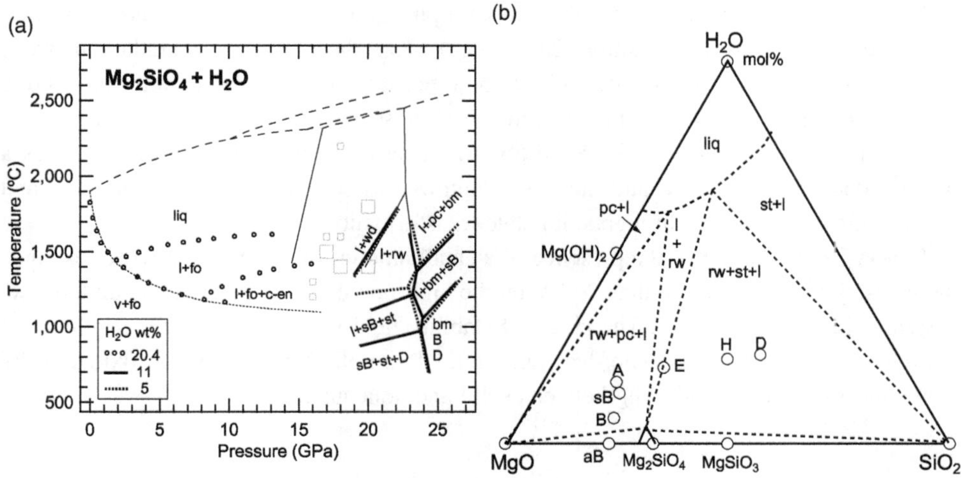

Figure 12.8 Phase diagrams of the Mg_2SiO_4–H_2O binary system and MgO–SiO_2–H_2O ternary system. (a) Mg_2SiO_4–H_2O binary system. Phase boundaries are shown as thick traces. At pressures up to 15 GPa, forsterite (fo) and high-pressure clinoenstatite (c-en) may coexist with aqueous liquid (liq or l) in a system containing 20.4 wt.% H_2O (open circles; Inoue et al., 1994). Near 15 GPa and at least up to 1,300°C, phase E (not shown) is stable (Kanzaki, 1991; Purevjav et al., 2020). Between 15–27 GPa, wadsleyite (wd), ringwoodite (rw), superhydrous B (sB), stishovite (st), periclase (c), and bridgmanite (bm) may coexist with liquid, whereas phase B (B), phase D (D), and sB, bm occur as subsolidus phases in systems containing 5 or 11 wt.% water (thick dotted and solid lines; Ohtani et al., 2000). Also shown are the melting curve and solid phase boundary of anhydrous Mg_2SiO_4 (Figure 12.6a) as thin traces. Large and small green squares denote conditions where large and small wadsleyite crystals, respectively, were grown in this study. (b) MgO–SiO_2–H_2O ternary system. At 20 GPa and 1,400°C, the system contains phase A, B, E, H, D, superhydrous B (sB), ringwoodite (rw), periclase (pc), stishovite (st), and liquid (Ohtani et al., 2000; Ohtani, 2015).

12.4.1.3 Growth of Wadsleyite and Ringwoodite through Solid-State Transformation

Through solid-state transformation, single crystals of iron-bearing wadsleyite with dimensions up to ~1 mm were synthesized at 16 GPa and ~1,600°C in 1–3 hours (Kawazoe et al., 2015), and ringwoodite crystals of hundreds of microns were produced at 22 GPa and 1,523°C in 6 hours (Yingwei Fei, personal communication). In this study, iron-free wadsleyite or ringwoodite crystals synthesized from anhydrous forsterite are no more than 10 μ in average size, even after equilibrating at 1,400°C for 48 hours. Attempts were made to cross the forsterite–wadsleyite boundary by compressing slowly or oscillating in pressure at a constant high temperature, to no avail. Our failure to synthesize large crystals through solid-state transformation is consistent with kinetic studies suggesting large nucleation rates at <0.5 GPa overpressure (Kubo et al., 2004). Given uncertainties in pressure calibration and pressure drifts with temperature and time, it is difficult to control nucleation and growth rates in solid state.

Natural olivine and its high-pressure polymorphs contain a substantial amount of iron. With the iron-bearing composition $(Mg_{0.9}Fe_{0.1})_2SiO_4$, the univariant boundary between iron-free forsterite and wadsleyite and that between iron-free wadsleyite and ringwoodite split into divariant loops and shift toward lower pressures, whereas the postspinel phase transition is barely affected (Figure 12.6b). As a result, iron-bearing ringwoodite has a much broader stability field and therefore requires less stringent control of pressure and temperature for synthesis through solid-state transformation.

In iron-bearing systems, the presence of a small amount of water (<0.2 wt.%) expands the wadsleyite field by more than 2 GPa on the high-pressure side and slightly on the low-pressure side (Figure 12.6b). The enlarged stability field would facilitate the synthesis of hydrous iron-bearing wadsleyite, barring potential complications from iron oxidation by water and from iron partitioning between solid and aqueous melt.

12.4.2 $MgSiO_3$ Bridgmanite

Pyroxene $(Mg,Fe)SiO_3$ is a major mineral in the Earth's upper mantle, and its high-pressure polymorph bridgmanite is the predominant constituent of the lower mantle and the most abundant mineral in the planet. Bridgmanite was known as silicate perovskite before being found in nature and given the mineral name (Tschauner et al., 2014).

12.4.2.1 Grow $MgSiO_3$ Crystals from Anhydrous Melt

The incongruent melting behavior of $MgSiO_3$ at 1 bar is well known, but it ends at ~0.13 GPa. The high-temperature clinoenstatite, orthoenstatite, high-pressure clinoenstatite, majorite, and bridgmanite all melt congruently. Like ringwoodite, akimotoite does not melt directly. It transforms to majorite upon heating and to bridgmanite upon compression.

Here several updates are made to the low-temperature region of the $MgSiO_3$ phase diagram (Figure 12.9a). First, the invariant point involving akimotoite, majorite, and bridgmanite is modified from 21.9 GPa and 2,060°C (Fei and Bertka, 1999) to 21.4 GPa and 1,900°C, to be consistent with the updated high-temperature pressure scales (Fei et al., 2004). The location of this triple point remains uncertain, however, as a recent study placed it at 19.9 ± 0.4 GPa and 1,723 ± 50°C on the basis of their MAP data, and at 23.8 ± 0.6 GPa and 1,713 ± 100°C according to their DAC results (Kulka et al., 2020). Second, downward revisions in pressures and temperatures are made to several phase boundaries at 15–20 GPa in Fei and Bertka (1999) to ensure conformation with the metastable extension rule (Figure 12.9a). Finally, the boundary between orthoenstatite and high-pressure clinoenstatite is shifted downward by 1 GPa, following the latest in situ X-ray diffraction study (Akashi et al., 2009).

12.4.2.2 Growth of $MgSiO_3$ Crystals from Hydrous Melt

In the $MgSiO_3$+H_2O binary system (Figure 12.10), orthoenstatite and bridgmanite crystals can be grown from hydrous melts. Under water-saturated conditions, millimeter-sized single crystals of iron-free enstatite were grown at relatively low pressures and moderate

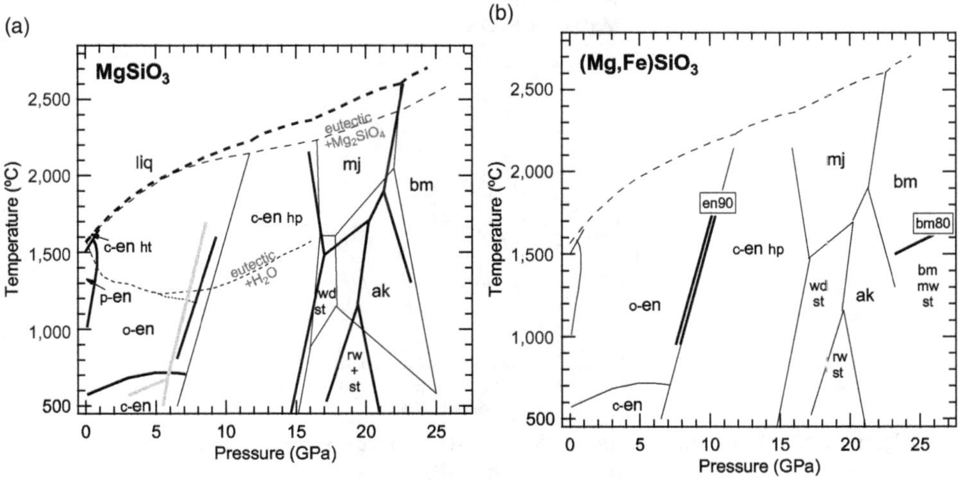

Figure 12.9 Phase diagram of $MgSiO_3$ and (Mg,Fe)SiO_3. (a) $MgSiO_3$. Phase boundaries are shown as thick black traces. The thick black dashed curve represents the melting curve of the high temperature clinoenstatite(c-en ht), orthoenstatite (o-en), high-pressure clinoenstatite(c-en hp), majorite (mj), and bridgmanite (bm) (Presnall et al., 1998). Not shown is a singular point at ~0.13 GPa, below which enstatite melts incongruently to produce forsterite and liquid. The thick black solid curves and lines represent boundaries involving the low-pressure and low-temperature clinoenstatite (c-en), protoenstatite (p-en), and c-en ht (Presnall, 1995), o-en and c-en hp (Akashi et al., 2009), and wadsleyite (wd), stishovite (st), ringwoodite (rw), akimotoite (ak), mj, and bm (Fei and Bertka, 1999; Fei et al., 2004). The thin curves and lines represent the same boundaries in Presnall (1995). Note considerable discrepancies in the stability fields of akimotoite, wadsleyite plus stishovite, and ringwoodite plus stishovite (Fei and Bertka, 1999; Presnall, 1995), even though both refer to Presnall and Gasparik (1990). Adding water shifts the transition pressure down by 1.5 GPa per 1,000 ppm water (thick gray lines; Jacobsen et al., 2010). Also shown are the binary eutectic melting curve of $MgSiO_3$+Mg_2SiO_4, the binary eutectic melting curve of $MgSiO_3$+H_2O, and the ternary eutectic melting curve of $MgSiO_3$+Mg_2SiO_4+H_2O (Figure 12.6a) as thin curves. (b) (Mg,Fe)SiO_3. Phase boundaries are shown as thick lines. For a composition containing 90 mol.% Mg in the divalent cation site, the univariant boundary between o-en and c-en hp splits into a two-phase field and shifts to lower pressures (Presnall, 1995; Woodland, 1998). For a composition containing 80 mol.% Mg in the divalent cation site, a three-phase field exists at low temperatures near 25 GPa (Arimoto et al., 2019). Also shown are the melting curve and solid phase boundary of anhydrous $MgSiO_3$ (Figure 12.9a) as thin traces.

temperatures (Stalder, 2002). Using H_2O as a solvent and 85$MgSiO_3$–15Mg_2SiO_4 as a silicate source, Shatskiy et al. (2007) synthesized high-quality single crystals of bridgmanite that were larger than 1 mm. The success was attributed to the low viscosity and high silicate solubility in water. A $KHCO_3$–$Mg(OH)_2$ solvent was also used to grow bridgmanite crystals, but the carbonate solvent produced melt inclusions (Shatskiy et al., 2007). Where a hydrous melt coexists with more than one solid phase, growing very large crystals may be more difficult because one solid phase may impede the growth of another.

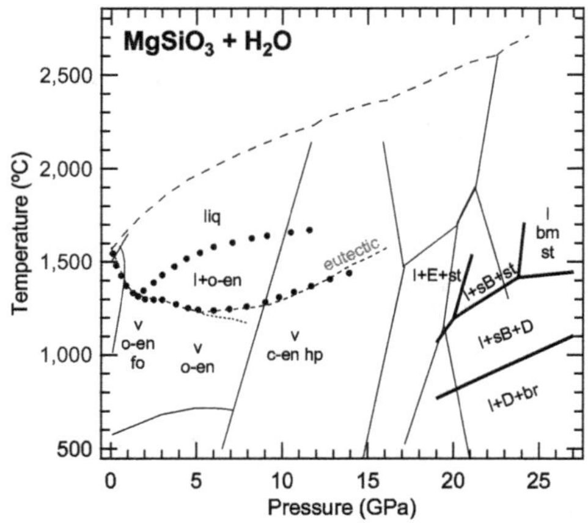

Figure 12.10 Phase diagram of the $MgSiO_3$–H_2O binary system. Phase boundaries are shown as thick traces. At pressures up to ~15 GPa, orthoenstatite (o-en), forsterite (fo), or high-pressure clinoenstatite (c-en hp) may coexist with liquid (liq or l) in a system containing 15 wt.% H_2O (Inoue et al., 1994). Between 15–27 GPa, superhydrous B (sB), stishovite (st), phase E (E), phase D (D), bridgmanite (bm), or brucite (br) may coexist with liquid in a system containing 10 wt.% H_2O (Ohtani et al., 2000). Also shown are the melting curve and solid phase boundary of anhydrous $MgSiO_3$, the eutectic melting curve of $MgSiO_3$+H_2O binary, and the eutectic melting curve of $MgSiO_3$+Mg_2SiO_4+H_2O ternary (Figure 12.9a) as thin traces.

12.4.2.3 Growth of $MgSiO_3$ Crystals through Solid- State Transformation

Natural enstatite usually contains a substantial amount of iron. With the iron-bearing composition $(Mg_{0.9}Fe_{0.1})SiO_3$, the univariant boundary between iron-free orthoenstatite and clinoenstatite splits into a narrow divariant loop ((Figure 12.9b). The iron endmember of pyroxene, known as ferrosilite ($FeSiO_3$), does not occur in nature, but a monoclinic phase of this composition was synthesized at a few GPa and moderate temperatures using a multi-anvil press (Lindsley et al., 1964). The monoclinic ferrosilite can be obtained free of twinning, while the orthorhombic ferrosilite turns into twinned monoclinic polymorphs in a few hours at room conditions.

Growing large crystals of bridgmanite through solid-state transformation is challenging for various reasons, including nonstoichiometry and back transformation to low-pressure polymorphs. A silicate source with $MgSiO_3$ composition produced stishovite inclusions, which caused splitting of bridgmanite crystals (Shatskiy et al., 2007). Bridgmanite may contain variable amounts of ferrous iron (Fe^{2+}), ferric iron (Fe^{3+}), and aluminum (Al^{3+}). The stability field of bridgmanite shifts to lower pressures with the addition of iron and to higher pressures with the presence of aluminum (e.g., Hirose et al., 2001). Ferric-iron-only bridgmanite with a small amount of iron was recovered at the ambient condition (Hummus and Fei, 2012), but the more iron-rich $(MgFe^{3+})(Fe^{3+}Si)O_6$ bridgmanite and $(MgFe^{3+})(AlSi)O_6$ bridgmanite transform to the akimotoite structure and $LiNbO_3$ structure, respectively, upon decompression to 1 bar at room temperature (Liu et al., 2018; Zhu et al., 2020).

12.5 Characterization of Synthesis Products

Synthesis is incomplete without characterization of the phase and chemical compositions of the product. In addition to achieving sufficient size, the crystals must also be free of chemical zoning and defects such as inclusions, vacancies, or stacking faults (Shatskiy et al., 2009b; Okuchi et al., 2015). An optical microscope can be used for an initial assessment of the crystal quality and phase composition (Figure 12.11). In this study, a Leica stereoscope was used to check large fragments for color, transparency, and grain size and to identify domains of quenched melt, inclusions, and bubbles. Small fragments were examined using a Zeiss polarizing microscope to detect minor phases and inclusions, estimate refractive index, and assist phase identification.

Using a Raman spectrometer, forsterite, wadsleyite, and ringwoodite can be distinguished quickly with little sample preparation and no sample damage (Walker and Li, 2020). In this study, wadsleyite crystals grown from a hydrous melt show clear peaks of OH vibration near 3,400/cm in their Raman and Fourier transform infrared (FTIR) spectra (Figure 12.11), whereas ringwoodite crystals show a large background between 3,000–4,000/cm in the Raman spectra. With suitable standards, the water contents can be quantified through FTIR measurements on doubly polished samples (e.g., Zhu et al., 2019a).

A scanning electron microscope (SEM) with an energy-dispersive spectrometer (EDS) is an effective instrument for detecting minor phases and examining compositional homogeneity. In some synthesized ferric-iron-rich bridgmanite samples, a minor oxide phase was not detected in the optical and Raman analyses, but it appeared as bright domains in the backscattered-electron (BSE) image (Figure 12.12). The chemical composition of the oxide was determined using an electron probe microanalyzer (EPMA). Furthermore, Mössbauer spectroscopy and X-ray emission spectroscopy (XES) were applied to determine the valence and spin states of iron in bridgmanite samples (e.g., Liu et al., 2018; Zhu et al., 2020).

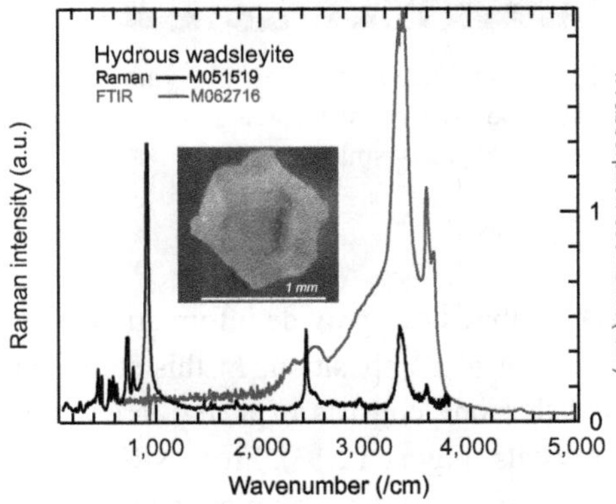

Figure 12.11 Raman and Fourier transform infrared (FTIR) spectra of synthetic hydrous wadsleyite crystals. Optical image (yellow traces are relict fragments of a gold capsule) and Raman spectrum of a hydrous wadsleyite crystal from experiment M051519 (black) and FTIR spectrum of a hydrous wadsleyite crystal from experiment M062716 (red). The peaks near 3,400/cm correspond to OH vibration modes.

Figure 12.12 Backscattered-electron image of coexisting akimotoite and oxide. The sample consists of $(MgFe^{3+}{}_4)O_7$ (bright minority grains) and $(MgFe^{3+})(Fe^{3+}Si)O_6$ akimotoite (gray majority grains), which presumably adopted the bridgmanite structure at 24 GPa and 1,600°C in a high-pressure synthesis experiment (Liu et al., 2018).

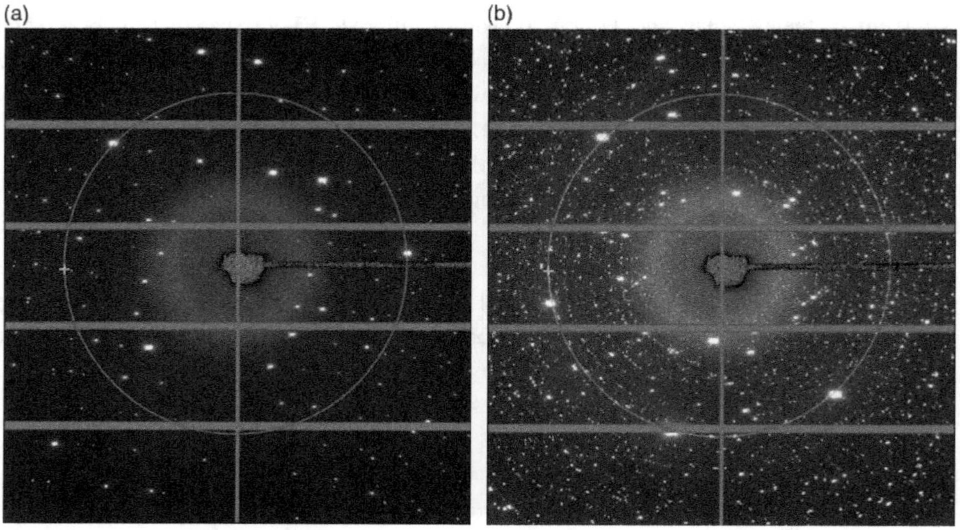

Figure 12.13 X-ray diffraction (XRD) patterns of synthesized wadsleyite grains. (a) A grain from experiment M061419 consists of a few large crystals. (b) A grain from experiment M062319 consists of many small crystals.

X-ray diffraction (XRD) analyses can provide information on crystal structure and size, as well as constraints on chemical composition. In this study, the XRD patterns of two transparent grains revealed that one consists of a few large crystals and the other is an aggregate of many small crystals (Figure 12.13). From XRD patterns, we were also able to extract the lattice parameters of wadsleyite samples, which allowed us to estimate their water contents (Chang et al., 2015). This method works for micron-sized crystals, is nondestructive, and requires less sample preparation than FTIR. In addition, neutron diffraction has been applied to locate hydrogen in the single crystal of phase E synthesized using a multi-anvil press (Purevjav et al., 2020).

12.6 Conclusions

The multi-anvil press can generate static pressures exceeding 1 Mbar at hundreds of °C, and temperatures exceeding 3,500°C at tens of GPa. The applications of the MAP in Earth science have led to important discoveries of material behavior inside rock planets, with profound implications for their nature, evolution, and dynamics. For crystal synthesis using the multi-anvil press, uncertainties in the control and characterization of pressure and temperature must be considered. With tungsten carbide anvils and the COMPRES standard 10/5 and 8/3 assemblies, thermal expansion and relaxation may produce 0.5–1.5 GPa and as much as 5–10 GPa uncertainties in sample pressure at high temperatures, whereas thermal gradient in a cylinder-shaped rhenium heater may introduce ±100°C uncertainties in temperature measurement. The effect of pressure on the electromotive force (emf) of thermocouples gives rise to systematic errors in temperature measurements, which likely exceed 100°C at temperatures above ~1,000°C and pressures above 15 GPa, by extrapolation. Synthesis methods involving slowly cooling a fluid solution or a thermal gradient have been successfully applied to grow 0.1–1 mm-sized crystals using the MAP at pressures up to ~30 GPa. These strategies can be understood from the basic theory of crystal growth. The phase diagrams of dominant constituents of the Earth's mantle compiled here resolve some of the inconsistencies in the literature, including those caused by revisions in the high-temperature pressure scale (Fei et al., 2004) and errors associated with the use of copper coils in temperature measurement circuit (Lieske and Frost, 2012). They provide the basis for selecting synthesis methods and conditions.

In 2020, the US National Science Foundation announced a major initiative to construct a very large multi-anvil press, as described in the Decadal Survey Results "Earth in Time." Such apparatus have existed outside the USA for more than one decade, including the 6,000 tonne BOTCHAN press at Ehime University in Japan, and the 5,000 tonne presses at Bayreuth Geoinstitut in Germany and Okayama University in Japan. Larger presses allow for more efficient synthesis of large quantities of samples, which are needed for neutron diffraction, calorimetry measurements, and classic shock wave experiments. More importantly, they increase the maximum size of crystals that can be synthesized for structure and property investigations. In material science, giant presses exceeding 60,000 tonnes have been constructed for synthesis of superhard materials, including polycrystalline nanodiamonds, carbides, and nitrides (Liebermann, 2011).

The MAP and DAC techniques have evolved in parallel for many decades, but the gap between them is gradually closing up. Through the use of sintered polycrystalline diamond anvils, a multidiamond-anvil press is about to break the megabar pressure barrier, and it is more accessible to in situ investigations using synchrotron X-rays. On the other hand, the sample size of the DAC is approaching millimeters through the use of large single-crystal diamonds grown by the chemical vapor deposition (CVD) technique. With new techniques such as the focused ion beam technique (FIB) and multigrain diffraction, it is now possible to study complex systems containing multiple phases and chemical reactions using the DAC (e.g., Baron et al., 2017, and other chapters in this volume). While technical innovations are still important in high-pressure research, exploring the compositional dimension with existing capabilities will likely be fruitful in the near future.

Acknowledgments

Ho-kwang (Dave) Mao has mentored and supported many MAP users including myself, and participated in projects involving the MAP. Dave's own large-volume press work used the Boyd–England apparatus at pressures up to 5 GPa. His vision that the DAC would enter every major university, which I first heard in 1999, has become a reality. The number of MAP laboratories in the world has also increased significantly, albeit at a slower pace, partly because of the large spatial footprint and maintenance requirements.

I thank many fellow MAP experimentalists for providing guidance and kind help over the years. My Ph.D. advisor Carl Agee introduced me to the MAP and showed me an elegant style to apply cutting-edge capabilities to tackle important scientific problems. Bob Liebermann welcomed me as a student visitor to the Mineral Physics Institute at Stony Brook, where Yanbin Wang showed me how to reach the bridgmanite stability field using Toshiba WC anvils and pyrophyllite gaskets. With Yingwei Fei, I had the unforgettable experience of conducting my first in situ MAP experiments at SPring-8 in Japan. At the former Geophysical Laboratory, I also learned to make parts using a lathe and then run MAP experiments using a highly automated control system made by Chris Hadidiacos. Dave Walker hosted my sabbatical visit as a Tharp Fellow at the Lamont Doherty Earth Observatory. He showed me lots of tricks, and we had numerous discussions on almost all aspects of the MAP techniques. Dave also helped me replace the gaskets of the 1,000 ton press at the University of Michigan. Eiji Ohtani hosted my sabbatical visit as a Global Excellence Scholar at the Tohoku University and provided opportunities for me to know the MAP community in Japan. Kurt Leinenweber generously shared his keen insights, invaluable vendor contacts, and nifty tools. I thank Bin Chen for helping me revive the MAPs at the University of Illinois and the University of Michigan, and for sharing his invention of 3D printed octahedra casting device. I have benefited from discussions with many colleagues, including Toru Inoue, Michael Walter, Shuangmeng Zhai, Jiuhua Chen, Don Weidner, Baosheng Li, Konstantin Litasov, Youmo Zhou, Hidenori Terasaki, Yu Nishihara, Kenji Mibe, Tetsu Irifune, Max Schmidt, Christian Liebske, Dan Frost, Monica Koch-Müller, and Lin Wang. I thank David Payne and Youxue Zhang for kindly offering their MAPs to me. Former members of my group Liang Li, Ryan Meredith, Heather Kirkpatrick, Jiachao Liu, Feng Zhu, Junjie Dong, and Yanhan Si contributed significantly to the calibration and synthesis projects described in this manuscript. Dongyuan Zhou helped me with data processing. The synchrotron XRD data were collected by Dongzhou Zhang at the PX^2 of the Advanced Photo Source, Argonne National Laboratory.

I thank Dave Walker, Michael Walter, and two anonymous reviewers for their meticulous reviews, insightful comments, and valuable suggestions. This work is partially supported by NASA grant NNX15AG54G through a collaboration with Sarah Stewart, Dylan Spaulding, and Bethany Chidester.

References

Akashi, A., Nishihara, Y., Takahashi, E., Nakajima, Y., Tange, Y., Funakoshi, K.i. (2009). Orthoenstatite/clinoenstatite phase transformation in $MgSiO_3$ at high-pressure and high-temperature determined by in situ X-ray diffraction: implications for nature of the X discontinuity. *Journal of Geophysical Research: Solid Earth*, **114**, B04206.

Akella, J., Kennedy, G. C. (1971). Melting of gold, silver, and copper – proposal for a new high-pressure calibration scale. *Journal of Geophysical Research*, **76**, 4969.

Arimoto, T., Irifune, T., Nishi, M., Tange, Y., Kunimoto, T., Liu, Z. (2019). Phase relations of $MgSiO_3$–$FeSiO_3$ system up to 64 GPa and 2300 K using multianvil apparatus with sintered diamond anvils. *Physics of the Earth and Planetary Interiors*, **295**, 106297.

Avrami, M. (1939). Kinetics of phase change. I General theory. *Journal of Chemical Physics*, **7**, 1103.

Baron, M. A., Lord, O. T., Myhill, R., et al. (2017). Experimental constraints on melting temperatures in the MgO–SiO_2 system at lower mantle pressures. *Earth and Planetary Science Letters*, **472**, 186.

Bassett, W. A. (2009). Diamond anvil cell, 50th birthday. *High Pressure Research* **29**, 163–186.

Bertka, C. M., Fei, Y. (1997). Mineralogy of the Martian interior up to core–mantle boundary pressures. *Journal of Geophysical Research: Solid Earth*, **102**, 5251–5264.

Brandeis, G., Jaupart, C., Allègre, C. J. (1984). Nucleation, crystal growth and the thermal regime of cooling magmas. *Journal of Geophysical Research*, **89**, 10161.

Bridgman, P. W. (1925). Certain physical properties of single crystals of tungsten, antimony, bismuth, tellurium, cadmium, zinc, and tin. *Proceedings of the American Academy of Arts and Sciences*. American Academy of Arts and Sciences, pp. 305–383.

Chang, Y.-Y., Jacobsen, S. D., Bina, C. R., et al. (2015). Comparative compressibility of hydrous wadsleyite and ringwoodite: effect of H_2O and implications for detecting water in the transition zone. *Journal of Geophysical Research: Solid Earth*, **120**, 8259–8280.

Chen, B., Li, J., Hauck, I. S. A. (2008). Non-ideal liquidus curve in the Fe-S system and Mercury's snowing core. *Geophysical Research Letters*, **35**, L07201, doi:07210.01029/02008GL033311.

Chen, J., Inoue, T., Weidner, D. J., Wu, Y., Vaughan, M. T. (1998). Strength and water weakening of mantle minerals, olivine, wadsleyite and ringwoodite. *Geophysical Research Letters*, **25**, 575–578.

Chen, J., Inoue, T., Yurimoto, H., Weidner, D. J. (2002). Effect of water on olivine–wadsleyite phase boundary in the $(Mg, Fe)_2SiO_4$ system. *Geophysical Research Letters*, **29**, 22-21–22-24.

Christensen, U. (1995). Effects of phase transitions on mantle convection. *Annual Review of Earth and Planetary Sciences*, **23**, 65–87.

Davies, E. J., Carter, P. J., Root, S., et al. (2020). Silicate melting and vaporization during rocky planet formation. *Journal of Geophysical Research (Planets)*, **125**, e06227.

Dobson, D. P., Jones, A. P., Rabe, R., et al. (1996). In-situ measurement of viscosity and density of carbonate melts at high pressure. *Earth and Planetary Science Letters*, **143**, 207–215.

Du, Z., Lee, K. K. M. (2014). High-pressure melting of MgO from (Mg,Fe)O solid solutions. *Geophysical Research Letters*, **41**, 8061.

Fei, H., Yamazaki, D., Sakurai, M., et al. (2017). A nearly water-saturated mantle transition zone inferred from mineral viscosity. *Science Advances*, **3**, e1603024.

Fei, Y., Bertka, C. M. (1999). Phase transitions in the Earth's mantle and mantle mineralogy. *Mantle Petrology: Field Observations and High Pressure Experimentation*, **6**, 189–207.

Fei, Y., Mao, H. K., Mysen, B. O. (1991). Experimental determination of element partitioning and calculation of phase relations in the MgO–FeO–SiO_2 system at high pressure and high temperature. *Journal of Geophysical Research: Solid Earth*, **96**, 2157–2169.

Fei, Y., Van Orman, J., Li, J., et al. (2004). Experimentally determined postspinel transformation boundary in Mg_2SiO_4 using MgO as an internal pressure standard and its geophysical implications. *Journal of Geophysical Research: Solid Earth*, **109**, B02305.

Frost, D., Poe, B., Trønnes, R., Liebske, C., Duba, A., Rubie, D. (2004). A new large-volume multianvil system. *Physics of the Earth and Planetary Interiors*, **143**, 507–514.

Fu, S., Yang, J., Karato, S.-i., et al. (2019). Water concentration in single-crystal (Al,Fe)-bearing bridgmanite grown from the hydrous melt: implications for dehydration melting at the topmost lower mantle. *Geophysical Research Letters*, **46**, 10,346–10,357.

Haines, J., Leger, J., Bocquillon, G. (2001). Synthesis and design of superhard materials. *Annual Review of Materials Science*, **31**, 1.

Hall, H. T. (1958). Some high-pressure, high-temperature apparatus design considerations: equipment for use at 100000 atmospheres and 3000°C. *Review of Scientific Instruments*, **29**, 267–275.

Hall, H. T. (1962). Anvil guide device for multiple-anvil high pressure apparatus. *Review of Scientific Instruments*, **33**, 1278.

Hernlund, J. (2006). A numerical model for steady-state temperature distributions in solid-medium high-pressure cell assemblies. *American Mineralogist*, **91**, 295.

Hirose, K., Fei, Y., Ma, Y., Mao, H.-K. (1999). The fate of subducted basaltic crust in the Earth's lower mantle. *Nature*, **397**, 53–56.

Hirose, K., Komabayashi, T., Murakami, M., Funakoshi, K.i. (2001). In situ measurements of the majorite–akimotoite–perovskite phase transition boundaries in $MgSiO_3$. *Geophysical Research Letters*, **28**, 4351–4354.

Hummer, D. R., Fei, Y. (2012). Synthesis and crystal chemistry of Fe^{3+}-bearing (Mg, Fe^{3+})(Si, Fe^{3+})O_3 perovskite. *American Mineralogist*, **97**, 1915–1921.

Huppertz, H. (2004). Multianvil high-pressure/high-temperature synthesis in solid state chemistry. *Zeitschrift für Kristallographie – Crystalline Materials*, **219**, 330–338.

Inoue, T. (1994). Effect of water on melting phase relations and melt composition in the system Mg_2SiO_4–$MgSiO_3$–H_2O up to 15 GPa. *Physics of the Earth and Planetary Interiors*, **85**, 237–263.

Inoue, T., Weidner, D. J., Northrup, P. A., Parise, J. B. (1998). Elastic properties of hydrous ringwoodite (γ-phase) in Mg_2SiO_4. *Earth and Planetary Science Letters*, **160**, 107–113.

Inoue, T., Yurimoto, H., Kudoh, Y. (1995). Hydrous modified spinel, $Mg_{1.75}SiH_{0.5}O_4$: a new water reservoir in the mantle transition region. *Geophysical Research Letters*, **22**, 117–120.

Irifune, T., Kurio, A., Sakamoto, S., Inoue, T., Sumiya, H. (2003). Materials: ultrahard polycrystalline diamond from graphite. *Nature*, **421**, 599.

Irifune, T., Nishiyama, N., Kuroda, K., et al. (1998). The postspinel phase boundary in Mg_2SiO_4 determined by in situ X-ray diffraction. *Science*, **279**, 1698–1700.

Irifune, T., Ringwood, A. (1993). Phase transformations in subducted oceanic crust and buoyancy relationships at depths of 600–800 km in the mantle. *Earth and Planetary Science Letters*, **117**, 101–110.

Ishii, T., Liu, Z., Katsura, T. (2019). A breakthrough in pressure generation by a Kawai-type multi-anvil apparatus with tungsten carbide anvils. *Engineering*, **5**, 434–440.

Ishii, T., Shi, L., Huang, R., et al. (2016). Generation of pressures over 40 GPa using Kawai-type multi-anvil press with tungsten carbide anvils. *Review of Scientific Instruments*, **87**, 024501.

Ishii, T., Yamazaki, D., Tsujino, N., et al. (2017). Pressure generation to 65 GPa in a Kawai-type multi-anvil apparatus with tungsten carbide anvils. *High Pressure Research*, **37**, 507.

Ito, E. (2012). Development of the Kawai-type multi-anvil apparatus (KMA) and its application to high pressure Earth science. *Journal of Physics: Conference Series*, **377**, 012001.

Ito, E., Schubert, G., Romanowicz, B., Dziewonski, A. (2007). Theory and practice-multianvil cells and high-pressure experimental methods. *Treatise on Geophysics*, **2**, 197–230.

Ito, E., Takahashi, E. (1989). Postspinel transformations in the system Mg_2SiO_4–Fe_2SiO_4 and some geophysical implications. *Journal of Geophysical Research: Solid Earth*, **94**, 10637–10646.

Iwasa, Y., Arima, T., Fleming, R. M., et al. (1994). New phases of C_{60} synthesized at high pressure. *Science*, **264**, 1570.

Jacobsen, S. D., Liu, Z., Ballaran, T. B., Littlefield, E. F., Ehm, L., Hemley, R. J. (2010). Effect of H_2O on upper mantle phase transitions in $MgSiO_3$: Is the depth of the seismic X-discontinuity an indicator of mantle water content? *Physics of the Earth and Planetary Interiors*, **183**, 234–244.

Jacobsen, S. D., Smyth, J. R., Spetzler, H., Holl, C. M., Frost, D. J. (2004). Sound velocities and elastic constants of iron-bearing hydrous ringwoodite. *Physics of the Earth and Planetary Interiors*, **143**, 47–56.

Kanzaki, M. (1991). Stability of hydrous magnesium silicates in the mantle transition zone. *Physics of the Earth and Planetary Interiors*, **66**, 307.

Karato, S.-i. (2011). Water distribution across the mantle transition zone and its implications for global material circulation. *Earth and Planetary Science Letters*, **301**, 413–423.

Karato, S.-i., Wu, P. (1993). Rheology of the upper mantle: a synthesis. *Science*, **260**, 771–778.

Katsura, T., Funakoshi, K.-i., Kubo, A., et al. (2004a). A large-volume high-pressure and high-temperature apparatus for in situ X-ray observation, "SPEED-Mk. II." *Physics of the Earth and Planetary Interiors*, **143**, 497–506.

Katsura, T., Ito, E. (1989). The system Mg_2SiO_4–Fe_2SiO_4 at high pressures and temperatures: precise determination of stabilities of olivine, modified spinel, and spinel. *Journal of Geophysical Research: Solid Earth*, **94**, 15663–15670.

Katsura, T., Shatskiy, A., Manthilake, M. G. M., et al. (2009). P–V–T relations of wadsleyite determined by in situ X-ray diffraction in a large-volume high-pressure apparatus. *Geophysical Research Letters*, **36**, L11307.

Katsura, T., Yamada, H., Nishikawa, O., et al. (2004b). Olivine–wadsleyite transition in the system $(Mg, Fe)_2SiO_4$. *Journal of Geophysical Research: Solid Earth*, **109**, B20209.

Katsura, T., Yoneda, A., Yamazaki, D., Yoshino, T., Ito, E. (2010). Adiabatic temperature profile in the mantle. *Physics of the Earth and Planetary Interiors*, **183**, 212–218.

Kawazoe, T., Buchen, J., Marquardt, H. (2015). Synthesis of large wadsleyite single crystals by solid-state recrystallization. *American Mineralogist*, **100**, 2336–2339.

Keppler, H., Frost, D. J., Miletich, R. (2005). Introduction to minerals under extreme conditions. *EMU Notes in Mineralogy*, **7**, 1–30.

Kitamura, K., Yamamoto, J. K., Iyi, N., Kirnura, S., Hayashi, T. (1992). Stoichiometric $LiNbO_3$ single crystal growth by double crucible Czochralski method using automatic powder supply system. *Journal of Crystal Growth*, **116**, 327.

Knibbe, J., Luginbühl, S., Stoevelaar, R., et al. (2018). Calibration of a multi-anvil high-pressure apparatus to simulate planetary interior conditions. *EPJ Techniques and Instrumentation*, **5**, 1–14.

Kubo, T., Ohtani, E., Funakoshi, K.-i. (2004). Nucleation and growth kinetics of the α-β transformation in Mg_2SiO_4determined by in situ synchrotron powder X-ray diffraction. *American Mineralogist*, **89**, 285.

Kulka, B. L., Dolinschi, J. D., Leinenweber, K. D., Prakapenka, V. B., Shim, S.-H. (2020). The Bridgmanite–Akimotoite–Majorite triple point determined in large volume press and laser-heated diamond anvil cell. *Minerals*, **10**, 67.

Le Godec, Y., Hamel, G., Solozhenko, V. L., et al. (2009). Portable multi-anvil device for in situ angle-dispersive synchrotron diffraction measurements at high pressure and temperature. *Journal of Synchrotron Radiation*, **16**, 513–523.

Leinenweber, K. D., Tyburczy, J. A., Sharp, T. G., et al. (2012). Cell assemblies for reproducible multi-anvil experiments (the COMPRES assemblies). *American Mineralogist*, **97**, 353–368.

Li, B., Kung, J., Liebermann, R. C. (2004). Modern techniques in measuring elasticity of Earth materials at high pressure and high temperature using ultrasonic interferometry in conjunction with synchrotron X-radiation in multi-anvil apparatus. *Physics of the Earth and Planetary Interiors*, **143**, 559–574.

Li, J., Agee, C. B. (1996). Geochemistry of mantle–core differentiation at high pressure. *Nature*, **381**, 686–689.

Li, J., Hadidiacos, C., Mao, H. K., Fei, Y., Hemley, R. J. (2003). Effect of pressure on thermocouples in a multi-anvil apparatus. *High Pressure Research*, **23**, 389–401.

Liebermann, R. C. (2011). Multi-anvil, high pressure apparatus: a half-century of development and progress. *High Pressure Research*, **31**, 493–532.

Liebske, C., Frost, D. J. (2012). Melting phase relations in the MgO–$MgSiO_3$ system between 16 and 26 GPa: implications for melting in Earth's deep interior. *Earth and Planetary Science Letters*, **345**, 159–170.

Lindsley, D. H., Davis, B. T. C., Macgregor, D. (1964). Ferrosilite ($FeSiO_3$): synthesis at high pressures and temperatures. *Science*, **144**, 73–74.

Liu, J., Dorfman, S. M., Zhu, F., et al. (2018). Valence and spin states of iron are invisible in Earth's lower mantle. *Nature Communications*, **9**, 1284.

Liu, Y., Li, H., Lai, X., Zhu, F., Rapp, R. P., Chen, B. (2020). Casting octahedra for reproducible multi-anvil experiments by 3D-printed molds. *Minerals*, **10**, 4.

Mao, H., Bell, P. (1971). Behavior of thermocouples in the single-stage piston-cylinder apparatus. *Carnegie Institution of Washington Yearbook*, **69**, 207–216.

Mao, H.-K., Chen, B., Chen, J., et al. (2016). Recent advances in high-pressure science and technology. *Matter and Radiation at Extremes*, **1**, 59–75.

Millot, M., Dubrovinskaia, N., Černok, A., et al. (2015). Shock compression of stishovite and melting of silica at planetary interior conditions. *Science*, **347**, 418–420.

Mitra, N. R., Decker, D. L., Vanfleet, H. B. (1967). Melting curves of copper, silver, gold, and platinum to 70 kbar. *Physical Review*, **161**, 613.

Mrosko, M., Koch-Müller, M., McCammon, C., Rhede, D., Smyth, J. R., Wirth, R. (2015). Water, iron, redox environment: effects on the wadsleyite–ringwoodite phase transition. *Contributions to Mineralogy and Petrology*, **170**, 9.

Müller, J., Koch-Müller, M., Rhede, D., Wilke, F. D., Wirth, R. (2017). Melting relations in the system $CaCO_3$–$MgCO_3$ at 6 GPa. *American Mineralogist*, **102**, 2440–2449.

Myhill, R., Frost, D. J., Novella, D. (2017). Hydrous melting and partitioning in and above the mantle transition zone: insights from water-rich $MgO–SiO_2–H_2O$ experiments. *Geochimica et Cosmochimica Acta*, **200**, 408.

Nishihara, Y., Doi, S., Kakizawa, S., Higo, Y., Tange, Y. (2020). Effect of pressure on temperature measurements using WRe thermocouple and its geophysical impact. *Physics of the Earth and Planetary Interiors*, **298**, 106348.

Ohtani, E. (2015). Hydrous minerals and the storage of water in the deep mantle. *Chemical Geology*, **418**, 5–15.

Ohtani, E., Mizobata, H., Yurimoto, H. (2000). Stability of dense hydrous magnesium silicate phases in the systems $Mg_2SiO_4–H_2O$ and $MgSiO_3–H_2O$ at pressures up to 27 GPa. *Physics and Chemistry of Minerals*, **27**, 533–544.

Okuchi, T., Purevjav, N., Tomioka, N., et al. (2015). Synthesis of large and homogeneous single crystals of water-bearing minerals by slow cooling at deep-mantle pressures. *American Mineralogist*, **100**, 1483–1492.

Perrillat, J. P., Chollet, M., Durand, S., et al. (2016). Kinetics of the olivine–ringwoodite transformation and seismic attenuation in the Earth's mantle transition zone. *Earth and Planetary Science Letters*, **433**, 360.

Poli, S., Schmidt, M. W. (2002). Petrology of subducted slabs. *Annual Review of Earth and Planetary Sciences*, **30**, 207–235.

Presnall, D., Gasparik, T. (1990). Melting of enstatite ($MgSiO_3$ from 10 to 16.5 GPa and the forsterite (Mg_2SiO_4)-majorite ($MgSiO_3$) eutectic at 16.5 GPa: implications for the origin of the mantle. *Journal of Geophysical Research: Solid Earth*, **95**, 15771–15777.

Presnall, D. C. (1995). Phase diagrams of Earth-forming minerals. *Mineral Physics and Crystallography: A Handbook of Physical Constants*, **2**, 248–268.

Presnall, D. C., Weng, Y.-H., Milholland, C. S., Walter, M. J. (1998). Liquidus phase relations in the system $MgO–MgSiO_3$ at pressures up to 25 GPa—constraints on crystallization of a molten Hadean mantle. *Physics of the Earth and Planetary Interiors*, **107**, 83–95.

Purevjav, N., Okuchi, T., Hoffmann, C. (2020). Strong hydrogen bonding in a dense hydrous magnesium silicate discovered by neutron Laue diffraction. *IUCrJ*, **7**, 370–374.

Rohrbach, A., Ballhaus, C., Golla-Schindler, U., Ulmer, P., Kamenetsky, V. S., Kuzmin, D. V. (2007). Metal saturation in the upper mantle. *Nature*, **449**, 456.

Rohrbach, A., Schmidt, M. W. (2011). Redox freezing and melting in the Earth's deep mantle resulting from carbon–iron redox coupling. *Nature* **472**, 209–212.

Rubie, D. C. (1999). Characterising the sample environment in multianvil high-pressure experiments. *Phase Transitions*, **68**, 431–451.

Rubie, D. C., Ross, C. R., II (1994). Kinetics of the olivine–spinel transformation in subducting lithosphere: experimental constraints and implications for deep slab processes. *Physics of the Earth and Planetary Interiors*, **86**, 223.

Schmandt, B., Jacobsen, S. D., Becker, T. W., Liu, Z., Dueker, K. G. (2014). Dehydration melting at the top of the lower mantle. *Science*, **344**, 1265–1268.

Schwarz, M. R. (2010). Multianvil calibration and education: a four probe method to measure the entire force-versus-pressure curve in a single run – performed as an interdisciplinary lab-course for students. *Journal of Physics: Conference Series*, **215**, 012193.

Secco, R., Schloessin, H. (1989). The electrical resistivity of solid and liquid Fe at pressures up to 7 GPa. *Journal of Geophysical Research: Solid Earth*, **94**, 5887–5894.

Shatskiy, A., Fukui, H., Matsuzaki, T., et al. (2007). Growth of large (1 mm) $MgSiO_3$ perovskite single crystals: a thermal gradient method at ultrahigh pressure. *American Mineralogist*, **92**, 1744–1749.

Shatskiy, A., Katsura, T., Litasov, K., et al. (2011). High pressure generation using scaled-up Kawai-cell. *Physics of the Earth and Planetary Interiors*, **189**, 92–108.

Shatskiy, A., Litasov, K. D., Matsuzaki, T., et al. (2009b). Single crystal growth of wadsleyite. *American Mineralogist*, **94**, 1130–1136.

Shatskiy, A., Litasov, K. D., Terasaki, H., Katsura, T., Ohtani, E. (2010a). Performance of semi-sintered ceramics as pressure-transmitting media up to 30 GPa. *High Pressure Research*, **30**, 443–450.

Shatskiy, A., Yamazaki, D., Borzdov, Y. M., et al. (2010b). Stishovite single-crystal growth and application to silicon self-diffusion measurements. *American Mineralogist*, **95**, 135–143.

Shatskiy, A., Yamazaki, D., Morard, G., et al. (2009a). Boron-doped diamond heater and its application to large-volume, high-pressure, and high-temperature experiments. *Review of Scientific Instruments*, **80**, 023907.

Shen, G., Lazor, P. (1995). Measurement of melting temperatures of some minerals under lower mantle pressures. *JGR*, **100**, 17,699–617,713.

Siebert, J., Badro, J., Antonangeli, D., Ryerson, F. J. (2012). Metal–silicate partitioning of Ni and Co in a deep magma ocean. *Earth and Planetary Science Letters*, **321**, 189.

Smyth, J. R., Holl, C. M., Frost, D. J., Jacobsen, S. D., Langenhorst, F., McCammon, C. A. (2003). Structural systematics of hydrous ringwoodite and water in Earth's interior. *American Mineralogist*, **88**, 1402–1407.

Sokol, A. G., Borzdov, Y. M., Palyanov, Y. N., Khokhryakov, A. F. (2015). High-temperature calibration of a multi-anvil high pressure apparatus. *High Pressure Research*, **35**, 139–147.

Stagno, V., Ojwang, D. O., McCammon, C. A., Frost, D. J. (2013). The oxidation state of the mantle and the extraction of carbon from Earth's interior. *Nature*, **493**, 84.

Stalder, R. (2002). Synthesis of enstatite single crystals at high pressure. *European Journal of Mineralogy*, **14**, 637–640.

Stewart, A., Van Westrenen, W., Schmidt, M., Melekhova, E. (2006). Effect of gasketing and assembly design: a novel 10/3.5 mm multi-anvil assembly reaching perovskite pressures. *High Pressure Research*, **26**, 293–299.

Stewart, A. J., Schmidt, M. W., van Westrenen, W., Liebske, C. (2007). Mars: a new core-crystallization regime. *Science*, **316**, 1323–1325.

Tange, Y., Takahashi, E., Nishihara, Y., Funakoshi, K.-I., Sata, N. (2009) Phase relations in the system MgO–FeO-SiO_2 to 50 GPa and 2000°C: an application of experimental techniques using multianvil apparatus with sintered diamond anvils. *Journal of Geophysical Research (Solid Earth)*, **114**, B02214.

Thomson, A. R., Walter, M. J., Kohn, S. C., Brooker, R. A. (2016). Slab melting as a barrier to deep carbon subduction. *Nature*, **529**, 76–79.

Tschauner, O., Ma, C., Beckett, J. R., Prescher, C., Prakapenka, V. B., Rossman, G. R. (2014). Discovery of bridgmanite, the most abundant mineral in Earth, in a shocked meteorite. *Science*, **346**, 1100.

Walker, D. (1991). Lubrication, gasketing, and precision in multi-anvil experiments. *American Mineralogist*, **76**, 1092–1100.

Walker, D., Carpenter, M. A., Hitch, C. M. (1990). Some simplifications to multianvil devices for high-pressure experiments. *American Mineralogist*, **75**, 1020–1028.

Walker, D., Li, J. (2020). Castable solid pressure media for multi-anvil devices. *Matter and Radiation at Extremes*, **5**(1) 018402.

Walker, D., Norby, L., Jones, J. H. (1993). Superheating effects on metal–silicate partitioning of siderophile elements. *Science*, **262**, 1858–1860.
Walter, M. J., Thibault, Y., Wei, K., Luth, R. W. (1995). Characterizing experimental pressure and temperature conditions in multi-anvil apparatus. *Canadian Journal of Physics*, **73**, 273.
Wood, B. J., Walter, M., Wade, J. (2006). Accretion of the Earth and segregation of its core. *Nature*, **442**, 825–833. doi:810.1038/nature04763.
Woodland, A. B. (1998). The orthorhombic to high-P monoclinic phase transition in Mg–Fe Pyroxenes: Can it produce a seismic discontinuity? *Geophysical Research Letters*, **25**, 1241.
Xu, F., Xie, L., Yoneda, A., et al. (2020). TiC–MgO composite: an X-ray transparent and machinable heating element in a multi-anvil high pressure apparatus. *High Pressure Research*, **40**, 257.
Yamazaki, D., Ito, E., Yoshino, T., et al. (2014). Over 1 Mbar generation in the Kawai-type multianvil apparatus and its application to compression of $(Mg_{0.92}Fe_{0.08})SiO_3$ perovskite and stishovite. *Physics of the Earth and Planetary Interiors*, **228**, 262.
Yamazaki, D., Ito, E., Yoshino, T., et al. (2019). High-pressure generation in the Kawai-type multianvil apparatus equipped with tungsten–carbide anvils and sintered-diamond anvils, and X-ray observation on $CaSnO_3$ and $(Mg,Fe)SiO_3$ perovskite and stishovite. *Comptes Rendus Geoscience*, **351**, 253–259.
Yamazaki, D., Ito, E. (2020) High pressure generation in the Kawai-type multianvil apparatus equipped with sintered diamond anvils. *High Pressure Research*, **40**, 3.
Yoshino, T., Manthilake, G., Matsuzaki, T., Katsura, T. (2008). Dry mantle transition zone inferred from the conductivity of wadsleyite and ringwoodite. *Nature*, **451**, 326–329.
Zhai, S., Ito, E. (2011). Recent advances of high-pressure generation in a multianvil apparatus using sintered diamond anvils. *Geoscience Frontiers*, **2**, 101–106.
Zhang, J., Li, B., Utsumi, W., Liebermann, R. C. (1996). In situ X-ray observations of the coesite–stishovite transition: reversed phase boundary and kinetics. *Physics and Chemistry of Minerals*, **23**, 1.
Zhang, L., Fei, Y. (2008). Melting behavior of (Mg,Fe)O solid solutions at high pressure. *Geophysical Research Letters*, **35**, L13302.
Zhou, D., Dong, J., Si, Y., Zhu, F., Li, J. (2020). Melting curve of potassium chloride from in situ ionic conduction measurements. *Minerals*, **10**(3), 250.
Zhu, F., Li, J., Liu, J., Dong, J., Liu, Z. (2019a). Metallic iron limits silicate hydration in Earth's transition zone. *Proceedings of the National Academy of Science*, **116**, 22526.
Zhu, F., Li, J., Liu, J., Lai, X., Chen, B., Meng, Y. (2019b). Kinetic control on the depth distribution of superdeep diamonds. *Geophysical Research Letters*, **46**, 1984.
Zhu, F., Liu, J., Lai, X., et al. (2020). Synthesis, elasticity and spin state of an intermediate $MgSiO_3$–$FeAlO_3$ bridgmanite: implications for iron in Earth's lower mantle. *Journal of Geophysical Research: Solid Earth*, **127**(7), e2020JB019964.

13

Investigation of Chemical Interaction and Melting Using Laser-Heated Diamond Anvil Cell

YINGWEI FEI, MICHAEL J. WALTER, JAMES BADRO, KEI HIROSE, OLIVER T. LORD, ANDREW J. CAMPBELL, AND EIJI OHTANI

The combination of double-sided laser heating in the diamond anvil cell and detailed chemical analysis of the recovered samples is a promising approach to explore the chemistry of the Earth's deep interior from the lower mantle to the core. Routine recovery of laser-heated samples coupled with chemical and textural characterization at the submicron scale is the key to expand knowledge of chemical interactions and melting at extreme conditions, particularly in complex systems. Recent technical developments have allowed us to investigate element partitioning and melting relations at pressures approaching the Earth's inner-core boundary. In this chapter, we review the techniques used for recovering tiny laser-heated samples and analyzing their chemical compositions and quenched textures, while highlighting key experiments that address silicate–metal element partitioning during mantle–core differentiation, silicate melting relations with applications to early magma ocean crystallization and deep mantle melting, and melting relations in iron-alloy systems relevant to the core. The results have drastically expanded our understanding of element redistribution at deep chemical boundaries and the chemical evolution of the deep mantle and the inner core. Finally, we emphasize the need for standardized protocols to obtain consistent, reproducible results and streamlined procedures to promote good practice and increase productivity. A broad collaboration with a systematic approach would further advance the field of high-pressure geochemistry.

13.1 Introduction

Development of piston-cylinder and multi-anvil devices revolutionized experimental petrology, allowing researchers to simulate conditions corresponding to the Earth's upper mantle (to ~100 km) and the shallow lower mantle (~750 km), respectively. A range of experiments conducted in these high-pressure devices, focused on phase transformations, melting relations, and element partitioning, have led to a better understanding of the mineralogy, composition, and geochemical processes of the upper mantle, transition zone, and upper part of the lower mantle. These studies have been fundamental for understanding deep mantle mineralogy, particularly the 410 km and 660 km mantle discontinuities, as well as the chemistry associated with metal–silicate differentiation during core formation.

The conditions of the deep mantle and core in the Earth lie beyond the reach of the piston cylinder and multi-anvil devices. The development of the laser-heated diamond anvil cell (LHDAC) technique has provided additional capability to study phase transitions, equations of state, and melting under extreme pressure–temperature conditions. Experiments on silicate minerals in the LHDAC led to the discovery of the $MgSiO_3$–perovskite phase (now called bridgmanite) stable at lower-mantle conditions (Liu, 1975, 1976) and the determination of phase relations in the MgO–FeO–SiO_2 system (e.g., Yagi et al., 1979). These early studies focused on the structure of quenched phases based on conventional X-ray diffraction (XRD). The use of the LHDAC as a reaction cell to study chemical equilibrium reactions faced many challenges because of the small, heated volume and large thermal gradients, and the difficulty of achieving stable heating and recovery of the tiny sample for chemical analysis at high-spatial resolution. Early attempts were made to study possible reactions between iron and silicate and Mg–Fe partitioning by both X-ray diffraction and electron microprobe (e.g., Knittle and Jeanloz, 1989; Fei et al., 1991).

The development of the double-sided laser-heating technique (Shen et al., 1996) addressed some of the challenges associated with temperature gradients within the laser-heated spot. The use of the double-sided laser-heating technique at different high-pressure laboratories and at synchrotron user facilities such as the Advanced Photon Sources (APS) (USA), European Synchrotron Radiation Facility (ESRF) (France), and Super Photon ring-8 GeV (SPring-8) (Japan) (Shen et al., 2001; Shultz et al., 2005; Meng et al., 2006; Ohishi et al., 2008; Prakapenka et al., 2008; Fukui et al., 2013; Hirao et al., 2020) have offered new opportunities for studying geophysical and geochemical problems related to the Earth's deep interior. In particular, the combination of laser heating with tightly focused X-ray beams (<5 μm) and fast data acquisition with area detectors has expanded our ability to obtain high-quality XRD data from materials at high pressure and temperature. This improved capability has played an important role in studying the phase transitions and physical properties of materials under extreme high pressures and temperatures, including the discovery of postperovskite at core–mantle boundary conditions (Murakami et al., 2004). However, investigating the chemistry of the deep interior in the LHDAC has been limited primarily due to the difficulty of sample recovery and chemical analysis at small spatial scales and ambiguity regarding the attainment of chemical equilibrium.

Examination of the texture and chemical composition of recovered high-pressure samples using analytical tools is a key procedure in experimental petrology and geochemistry. The combination of the LHDAC with analytical transmission electron microscopy (TEM) opened new research directions for studying element partitioning between coexisting solid solutions (e.g., Kesson et al., 2002; Kobayashi et al., 2005; Murakami et al., 2005; Sakai et al., 2006) and examining element partitioning between metal and silicate (Tschauner et al., 1999; Bouhifd and Jephcoat, 2003; Hirao et al., 2006). Those technically challenging experiments typically had low success rates because it is extremely difficult to reliably recover the very small samples involved and to prepare them for ex situ textual and chemical analyses.

Focused ion beam (FIB) technology has revolutionized preparation of small specimens (Heaney et al., 2001; Benzerara et al., 2005; Fei, 2013), and it provides unprecedented ability to quantitatively analyze the recovered high-pressure LHDAC samples (e.g., Auzende et al., 2008; Miyahara et al., 2008; Ricolleau et al., 2008; Frost et al., 2010; Piet et al., 2016; Dorfman et al., 2018; Lv et al., 2021). The use of FIB technology has opened a new avenue to reliably obtain cross sections of the laser-heated spot in DAC samples for textual and chemical composition analyses. Here, we describe FIB applications in experimental petrology, particularly involving recovery of LHDAC samples, and highlight results from liquid silicate–metal equilibrium experiments (e.g., Siebert et al., 2012, 2013; Fischer et al., 2015; Badro et al., 2016, 2018; Chidester et al., 2017; Suer et al., 2017; Du et al. 2017, 2019; Jackson et al., 2018), melting relations in the Fe–FeS system (e.g., Kamada et al., 2012; Mori et al., 2017), in the Fe–S–Si and Fe–Si–O ternary systems (e.g., Hirose et al., 2017; Sakairi et al., 2017; Tateno et al., 2018), and in the Fe–C and Fe–C–H systems (e.g., Hirose et al., 2019; Mashino et al., 2019; Takahashi et al., 2019), and melting relations in silicate systems (e.g., Fiquet et al., 2010; Nomura et al., 2014; Ohira et al., 2014; Tateno et al., 2014; Thomson et al., 2014; Pradhan et al., 2015; Walter et al., 2015; Ohtani et al., 2016; Baron et al., 2017; Caracas et al., 2019; Yuan et al., 2019). The results represent significant advances in our understanding of the chemistry of the deep interior.

13.2 Experimental Techniques and Procedures

13.2.1 Temperature Measurement in the Laser-Heated DAC

Measuring temperature precisely and accurately and understanding how it varies in samples laser heated in diamond anvil cell experiments is critical for evaluating phase relationships, melting curves, and element partitioning. Large uncertainties (e.g., >200 K) related to both inaccurate temperature measurement and poorly characterized temperature gradients can plague laser-heated DAC experiments and depend on the details of the heating geometry and approach to temperature measurement. Since the first pioneering temperature measurements in the laser-heated DAC (Ming and Bassett, 1974; Jeanloz and Heinz, 1984; Bassett and Weathers, 1986; Heinz and Jeanloz, 1987), much attention has been given to understanding and mitigating these sources of error and in developing approaches to reducing temperature gradients (Jeanloz and Kavner, 1996; Jephcoat and Besedin, 1996; Manga and Jeanloz 1996; Dewaele et al., 1998; Panero and Jeanloz, 2001; Kavner and Panero, 2004; Walter and Koga, 2004; Campbell, 2008; Kavner and Nugent, 2008; Prakapenka et al., 2008; Du et al., 2013; Rainey and Kavner, 2014; Giampaoli et al., 2018; Kunz et al., 2018; Zinin et al., 2019).

Laser-heated spots in the DAC are generally of the order 10–40 μm in diameter, typically reducing in size with pressure as a more tightly focused beam is required to generate high temperatures as sample thickness decreases. Beam-shaping optics that aim for “flat-top” laser power profiles help reduce large radial gradients (Prakapenka et al., 2008), but because heating occurs by sample absorption of the laser, large (e.g., >500 K) radial temperature gradients occur especially near the edges of the focused laser beam. The highly conductive diamond anvils rapidly remove heat from the sample such that axial temperature gradients can also be large (>100 K), depending on sample and heating configuration. Double-sided laser heating mitigates axial gradients and is now the standard approach (Shen et al., 1996, 2001; Meng et al., 2006; Prakapenka et al., 2008), but gradients are present even in this geometry. Many factors determine the temperature distribution throughout the sample. These include the absorption and thermal properties of the sample material, the sample size and thickness, the thermal properties and thickness of the insulation between the sample and anvils and the nature of the laser absorber (e.g., bulk metallic samples, dispersed metallic particles, metal coatings or layers, or transition metals dissolved in silicates), and the energy density distribution of the impinging laser(s). Careful choice of sample geometry, sample and insulation thicknesses, and laser beam shape and size can reduce radial gradients (e.g., <50 K) in regions of the sample that are comparatively large (e.g., 5–20 μm) relative to X-ray beam sizes (2–10 μm) and chemical probes (e.g., electron probe microanalyzer [EPMA] or TEM) so that accurate measurements of crystal structure, material properties, and chemistry are possible.

Temperature measurement in the LHDAC is typically achieved using spectroradiometry, wherein incandescent light from the heated sample spot is collected and dispersed with appropriate optical components, and its wavelength-dependent intensity measured with an array detector (Bassett and Weathers, 1986; Heinz and Jeanloz, 1987; Boehler and Chopelas, 1991; Boehler, 1993; Jeanloz and Kavner, 1996; Jephcoat and Besedin, 1996). The resulting spectra are calibrated relative to a source with a known spectral radiance, and the normalized spectra are then fitted with either a Wien or Planck function to retrieve sample temperature. Early systems employed photodiode array detectors (Bassett and Weathers, 1986), whereas modern systems typically employ an imaging spectrograph with 2D multichannel CCD detectors (Jeanloz and Kavner, 1996; Jephcoat and Besedin, 1996; Andrault and Fiquet, 2001; Shen et al., 2001; Watanuki et al., 2001; Walter and Koga, 2004;Meng et al., 2006; Ohishi et al., 2008; Prakapenka et al., 2008).

The largest sources of potential uncertainty in reported temperatures in LHDAC experiments include chromatic and spherical aberrations in refracting and reflecting optical components, errors in indexing the location of measured temperatures to the location analyzed, and the unknown effect of pressure and temperature on sample emissivity. The first source of uncertainty relates to the optical components and their fidelity in accurately preserving the wavelength-dependent intensity distribution from a given region at the sample to the reciprocal region on the detector. When a temperature gradient exists at the sample, chromatic aberration from optical components, including the diamond anvils, can lead to large uncertainties in precision and accuracy (Boehler and Chopelas, 1991; Jeanloz and Kavner, 1996; Panero and Jeanloz, 2001; Benedetti and Loubeyre, 2004; Walter and Koga, 2004). The amount of aberration depends on many factors (e.g., quality of collecting optics, maximum temperature, size of the heated spot, temperature gradient) but can be detected through errors in the precision of spectral fitting to black-body equations. Numerous studies have quantified the potential magnitude of errors and presented methods for recognizing and mitigating against chromatic effects (Boehler and Chopelas, 1991; Jeanloz and Kavner, 1996; Panero and Jeanloz, 2001; Benedetti and Loubeyre, 2004; Walter and Koga, 2004; Benedetti et al., 2007; Campbell, 2008; Rainey and Kavner, 2014; Deng et al., 2017; Kunz et al., 2018). Fit precision to individual black-body spectra of <5 K are routinely achieved and indicate that uncertainties in temperatures related to chromatic effects are likely <50 K. However, at very high temperatures (e.g., >4,000 K) and small spot sizes (e.g., <20 μm), where temperature gradients can be large, chromatic effects can still lead to large (>500 K) temperature errors if extreme care is not taken in analyzing the spectral data. Reflective optics that utilize mirrors for light collection do not impart chromatic dispersion, although the optical pathway may still have some minor chromatic dispersion due to the diamond anvil (Boehler, 2000) and can have poorer image quality that can potentially affect measured temperature (Jeanloz and Kavner, 1996), but may be preferred, especially at higher temperatures where temperature gradients are large in small heated spots (Giampaoli et al., 2018). However, modern near-infrared apochromatic refractive optics coupled with elimination of nonparaxial light through a reduced aperture (f-stop) largely eliminate chromatic effects even at extreme temperatures.

The second source of uncertainty that is especially critical for understanding phase relations and chemical variations within samples is quantifying the temperature distribution across the heated sample area. There are many approaches to measuring the spatial temperature distribution depending on the setup and aim of the experiment. For example, a common approach at synchrotron beamlines is to focus the light from the sample onto a pinhole mounted at the entrance slit to a spectrometer (Shen et al., 2001; Meng et al., 2006; Mezouar et al., 2017). In this geometry, a single temperature is measured from a small region of the sample (e.g., <10 μm diameter) that ideally matches the diameter of the X-ray beam. This approach has the advantage that if the X-ray beam and temperature measurement pinhole are coaligned to the same location on the sample, then the measured temperature will represent the temperature of the analyzed volume. The disadvantage of this approach is that much of the information regarding the spatial and temporal changes in temperature across the sample surface is lost, which can make it challenging to relate postquench observations of the sample to temperature. Another common approach is to focus the light from the sample directly onto the entrance slit of the spectrometer, which results in a one-dimensional line profile of temperature across the heated spot (Figure 13.1). This mode of operation takes full advantage of the 2D nature of the CCD detector, with one dimension being spatial and the other wavelength. This allows temperature gradients to be measured and can provide a better correlation between temperature and post-quench features of the sample, as illustrated in Figure 13.1.

While 1D temperature profiles can be related to sample features it is clearly advantageous to have a full 2D rendering of temperature across the laser-heated spot when attempting to relate specific observations to temperature, for example in deducing melting points or in post-quench analysis of sample textures and phase relations (Du and Lee, 2014). A recent advance to obtain full 2D temperature maps is to split the light from the sample into multiple wavelengths (typically four) at a narrow bandpass of 10 nm or less, and then focus each of the resulting monochromatic images onto a large-format, high-resolution CCD camera (Campbell, 2008). The 2D images can then be calibrated for normalized intensity and spatially correlated in software (e.g., the open source MIRRORS graphical user interface; Lord and Wang, 2018) so that each pixel has intensity information for multiple colors over a specified wavelength range (e.g., ~400 nm).

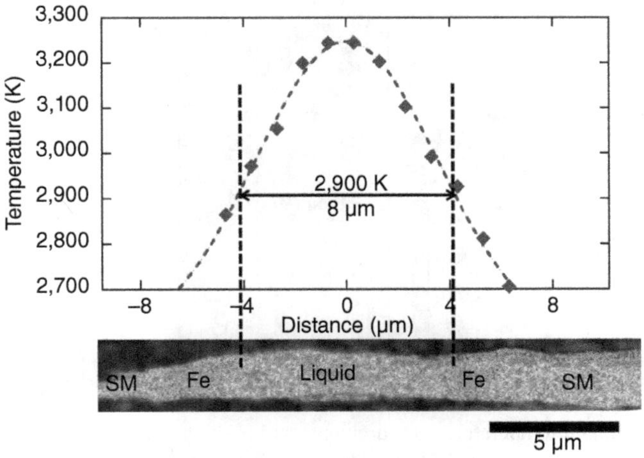

Figure. 13.1 A representative one-dimensional temperature profile across a laser-heated hot spot and its relation to composite X-ray maps for an Fe-C sample recovered from 138 GPa (Mashino et al., 2019). In melting experiments such as this one, by combining such a temperature profile with a sample cross section, one can obtain the temperature at the liquid–solid boundary, which indicates the liquidus temperature. This procedure is important for precisely determining melting temperatures because some temperature variations always exist in laser-heated DAC samples, even with a liquid portion, and therefore the temperature at the center of a hot spot may be higher than the melting temperature. For the color version, refer to the plate section.

A temperature can then be fitted for each pixel and a full 2D temperature map of the sample surface obtained (Campbell, 2008; Du et al., 2013; Lord et al., 2014). Temperatures measured using a limited number of wavelengths have lower precision (e.g., >20 K) than in standard spectroradiometry but can have equivalent or even higher accuracy as chromatic aberrations are virtually eliminated as each wavelength is independently focused onto the detector (Campbell, 2008).

An example illustrating the advantages of two-dimensional multispectral temperature mapping is shown in Figure 13.2. In the experiment, a sample of Ni was laser heated from both sides with a flat-top laser profile at ~28 GPa until melting. Temperatures were measured using both standard spectroradiometry from both sides using a slit geometry and with 2D multispectral imaging radiometry (four-color) from one side (Lord et al., 2014). During the temperature ramp as laser power increases (Figure 13.2a), the maximum spectroradiometric temperatures are close on both sides and the multispectral peak temperatures are similar but somewhat higher, suggesting that the highest temperatures may have been missed by the slit geometry. The sample temperature plateaus just at the power where a large change occurs in the 2D temperature distribution (Figure 13.2c), with the highest temperatures corresponding to sample melting occurring first in a rind at the edge of the sample. Note how the 2D temperature maps give a clear indication of melting that can easily be missed if a limited aperture geometry is used, especially a pinhole geometry, where the highest temperatures may not occur at the center of the sample. Further melting leads to changes in the hot regions of the sample with time that may represent movement of

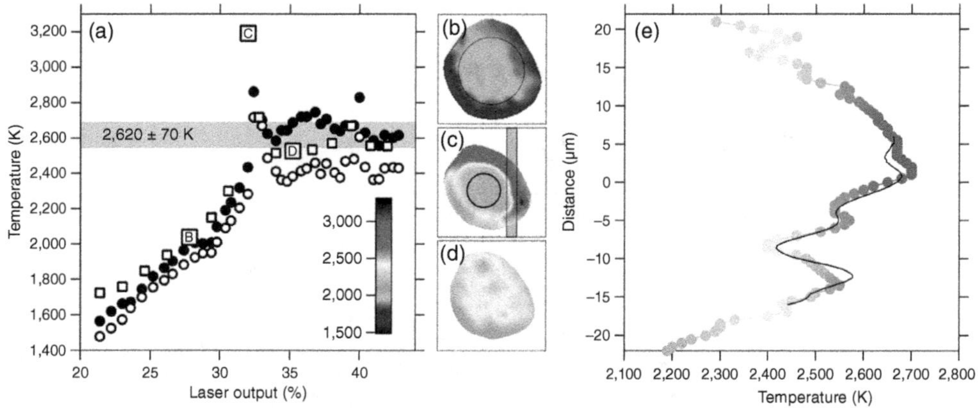

Figure 13.2 In situ temperature measurements and melting determination of Ni metal in a KCl pressure medium for run 77A at ~28 GPa (after Lord et al., 2014). (a) Temperature versus laser power plot showing spectroradiometric measurements from two sides of the sample (circles) together with multispectral imaging radiometry measurements (squares, made on the same side as the open circles). (b–d) Two-dimensional temperature maps determined by multispectral (four-color) imaging radiometry, color coded as a function of temperature. The black circle in (b) represents the approximate location of the ~20μm diameter incident laser beam. The gray bar in (c) represents the approximate location of the slit aperture (~3μm wide at the sample) used for 1D spectroradiometry, whereas the circle shows an example of the area interrogated using a 10μm diameter pinhole aperture. (e) A spectroradiometric line profile (circles) collected nearly simultaneously with the image in (c). The black line is an averaged line profile taken from the region in (c) denoted by the gray bar. For the color version, refer to the plate section.

a melt phase in the laser beam. Such spatial information is lost without the benefit of 2D temperature mapping. In fact, the shape of a line profile taken across the 2D temperature map can be matched with a spectroradiometric line profile collected simultaneously (Figure 13.2e) to determine the location of the spectroradiometric slit relative to the hot spot. In this case, the spectroradiometric slit was not perfectly aligned with the heated spot (Figure 13.2c). This error would not be detectable without 2D mapping; the potential to miss key details of sample dynamics when using a pinhole aperture is potentially even more significant. In Figure 13.2c, a hypothetical pinhole aperture, even well aligned to the center of the laser-heated spot, could easily miss the transient development of the rind and subsequent dynamic melt motion. Such 2D temperature maps are also hugely advantageous for relating temperatures to features in the quenched sample and for relating temperature to phase relations and measured chemical composition of phases (e.g., element partitioning).

13.2.2 Pressure Determination

Pressure at high temperature may be determined from XRD data of an internal pressure standard loaded with the sample of interest, by using the measured volume and its established *P–V–T* equation of state (EoS) when experiments are conducted at synchrotron facilities. In situ XRD measurements during laser heating in the DAC are standard for several dedicated high-pressure beamlines, such as GSECARS and HPCAT at the APS, ID27 at ESRF, and BL10XU at SPring-8, where high-resolution large-area detectors with wide dynamic range are used (e.g., Meng et al., 2006; Prakapenka et al., 2008; Hirao et al., 2020). Chapter 8 provides a comprehensive comparison of pressure determination using common pressure standards, such as Au, Pt, MgO, Ne, and NaCl. Experiments designed to obtain chemical information must consider possible contamination from the chosen pressure standard, especially for melting experiments. Hence, the choice of internal standards may be limited. MgO or NaCl is commonly used as both a thermal insulator and pressure medium in LHDAC experiments. It is useful to use the diffraction data of the pressure medium to monitor the pressure change during heating and estimate the pressure at high temperature based on the pressure response of the "cold" pressure medium during heating (Campbell et al., 2009).

The pressure during the heating cycle deviates from the initial pressure prior to heating due to relaxation and thermal pressure (Andrault et al., 1998; Morard et al., 2011; Jackson et al., 2018). The combination of these effects determines the pressure at high temperature prior to quenching. For purely isochoric heating, the thermal pressure is written as $\Delta P = \alpha K_T \times T$, in which α is the thermal expansivity and K_T is the isothermal bulk modulus. This represents the maximum thermal pressure to be added to the pressure measured at 300 K after heating. From the studies of the melting of Fe-bearing systems (see Ichikawa et al., 2014 for the αK_T value for pure Fe), the true thermal pressure component of the total pressure at high temperature is 60–90% of this maximum value (Andrault et al., 1998; Morard et al., 2011).

It is recognized that the estimated pressure at high temperature based on such an empirical method may strongly depend on the heating history. For example, the prerelaxation of the heating spot due to previous heating cycles completed in the same cell often results in a large thermal-pressure contribution (Jackson et al., 2018). The best practice for consistent pressure estimation at high temperature is to establish an empirical thermal-pressure contribution based on a set of in situ X-ray diffraction measurements during heating, and then run inhouse quenching experiments using the same heating protocols. It has been demonstrated that consistent results can be obtained with different methods of estimation if proper calibration is established (Lord et al., 2014; Hirose et al., 2019). The overall errors in liquidus temperature and pressure for melting experiments in the Fe–Fe_3S system, for example, may be ±5% and ±10%, respectively (Mori et al., 2017).

At very high temperatures (>3,000 K), one may observe a second set of MgO, NaCl, or Ne peaks within the XRD patterns when MgO, NaCl, or Ne is used as the pressure medium. This second set of peaks is shifted toward larger volume compared to the peaks of the "cold" pressure medium, resulting from extreme heating of the medium at the sample interface. These peaks may be used to directly estimate thermal pressure. The challenge of this approach is to have the right estimate of temperature gradient across the interface and also to account for potential composition change of the calibrant at such high temperatures. This approach has the advantage of being able to monitor thermal pressure over the duration of the experiment even beyond the melting temperature (Rainey et al., 2013; Jackson et al., 2018).

13.2.3 Preparation of Starting Material

It is crucial for every DAC experiment to have a homogeneous starting material at very fine scale (much smaller than the size of a laser-heated spot, ~20–40 μm). While naturally occurring samples (single-crystal minerals or glassy rock compositions such as mid-ocean ridge basalt, or MORB) can sometimes fulfill this requirement, the coarse-grained textures of most lithologies, such as peridotite, preclude their use even with extended milling, and therefore it is necessary to develop other synthesis routes. Additionally, geochemical investigation requires us to sometimes dope the starting materials with minor amounts of other elements of interest, such as siderophile or lithophile trace elements. These also need to be homogeneously distributed in the sample.

There are several ways to prepare homogeneous starting materials. Silicate starting materials can be prepared as a gel, which is obtained from nitric acid solutions and organic silicon compounds. For example, the starting materials for studies of Fe^{3+} bearing bridgmanite and the bridgmanite + aluminium hydrate system can be synthesized by the sol–gel method using a mixture of TEOS [$(C_2H_5O)_4Si$, tetraethyl orthosilicate] and other reagents (Hamilton and Henderson, 1968; Kawamura, 1994). A mixture of reagent-grade magnesium nitrate $Mg(NO_3)_2{\cdot}6H_2O$, powdered ^{57}Fe metal (96.63% enriched in ^{57}Fe), and TEOS is used for Fe^{3+}-bearing bridgmanite, whereas a mixture of $Mg(NO_3)^{\cdot}26H_2O$, Al $(NO_3)^{\cdot}39H_2O$ and TEOS($(C_2H_5O)_4Si$) is used for hydrous starting materials. The water contents of the starting gel materials can be characterized by ignition loss and by the FTIR spectra of the gel materials. Since dehydrated gel powder easily adsorbs water (Nomura et al., 2014), it is important to close the DAC in an argon atmosphere right after drying in a vacuum oven if the experiments are intended to study a nominally dry system.

It is desirable to start with a glass starting material. Some materials can be easily quenched to glass. For example, basaltic glasses can be prepared by melting and quenching mixtures of reagent oxides in a gas-mixing furnace under controlled H_2–CO_2 or CO–CO_2 atmospheres. On the other hand, other starting materials may not be possible to quench to glass using conventional quenching methods. The latest technique to prepare glass starting materials is to use an aerodynamic levitation laser furnace, which is a contactless furnace where the sample levitates in a gas flow and is heated by an impinging CO_2 laser beam (Badro et al., 2021). Glasses are usually synthesized from high-purity oxide or carbonate powders (MgO, FeO, SiO_2, Al_2O_3, $CaCO_3$, Na_2CO_3, K_2CO_3), thoroughly mixed in the proportion required to attain the desired starting composition. The powder mix is then pressed into pellets of roughly 300 mg. Chunks of the pellet are broken off and put in the nozzle of the levitator. The gas flux is then initiated, and the sample is laser heated and fused above the liquidus temperature of the mix to ensure full melting. In the case of a peridotitic composition (pyrolite), it should reach a temperature of 2,000°C. Once the sample is molten, it turns into a sphere and levitates stably in the gas stream. The sample is then quenched by switching off the laser. Because of the rapid quench rates and the absence of contact (nucleation sites), glasses can easily be made for a range of terrestrial compositions ranging from basalts, to pyrolite, to olivine or forsterite. The levitation method also has the advantage of controlling the sample environment by using different gas flux, allowing the preparation of glass starting materials under a range of oxygen fugacity conditions (Sossi et al., 2020).

For characterization and preparation of DAC samples, the recovered glass balls are cut in half using a diamond disk or wire saw. The two halves are embedded in epoxy resin with the flat surface exposed and polished to mirror-finish using a suite of diamond lapping films. The samples are carbon coated and imaged (backscattered electrons) and analyzed with an SEM equipped with an energy-dispersive X-ray (EDX) detector to check for (1) chemical homogeneity and (2) the absence of crystals. Each is then analyzed for major and trace elements using EDX or EPMA to obtain the bulk composition. The chemical variability is typically of the order of 0.5% across the sample, which is sufficiently homogeneous to be used as a starting material. The characterized glasses are then double polished to the desired thicknesses for DAC loading (as described later).

It is difficult to break down metallic powder to micron-sized grains with mechanical grinding. Compositional variations between heating spots within the same DAC sample are commonly observed for starting materials with relatively large grain size. This affects not only the ability to control the starting composition but also prevents uniform laser coupling with the sample. Homogeneous metal alloy starting materials can be prepared by the ball-mill technique. Figure 13.3 shows homogeneously mixed Fe–FeS starting material after ball milling. The duration and speed of the ball milling should be closely monitored, and the mixed sample should be characterized with SEM and XRD because long duration and high-speed milling could cause substantial heating, resulting in oxidation of the starting materials. It is often observed that ball-milled samples contain a small amount of oxygen.

For iron alloy samples, chemically homogeneous starting materials can be formed from liquids by ultrarapid quenching. Figure 13.3 shows typical dendritic quenched texture of Fe–FeS melt that has a homogeneous composition at the 2–3 μm scale. After grinding the quenched melt, the composition of the starting mixture is expected to be uniform at the micron scale. This method can also easily be used to dope trace elements into the starting material. Magnetron sputtering of Fe in an inductively coupled radio frequency (RF) reactive oxygen-rich argon plasma is another method to produce extremely homogeneous metal alloy starting materials with a range of compositions (Morard et al., 2017), particularly those containing large amounts of light elements beyond their solubility in liquid Fe. Homogeneity at a fine spatial scale is imperative for achieving uniform heating and tightly controlling the starting composition.

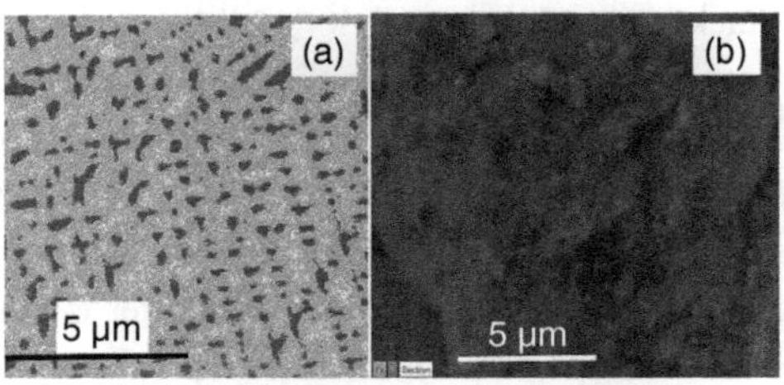

Figure 13.3 Element maps of the Fe–FeS starting materials (a) quenched Fe–FeS melt showing homogeneous distribution of S (green) and Fe (blue), and (b) made by ball milling showing the distribution of S (brown) and Fe (green).

13.2.4 Sample Loading Configuration

Sample loading is key for a successful experiment in the laser-heating diamond anvil cell. To achieve high temperature, excellent thermal insulation between the sample and diamond anvils is essential. The choice of insulators is largely dictated by possible chemical reactions between the sample and insulator. Typically, a thin layer of sample is loaded between insulation layers. Simple oxides such as SiO_2 glass, Al_2O_3, and MgO are commonly used as thermal insulators in experiments involving metallic samples, which also serve as pressure media and sometimes as pressure standards. Soft NaCl and KCl are also used as insulation, but they make dry loading and recovery more challenging. Noble gases such as Ne and Ar may be ideal thermal insulators in terms of nonreactivity and hydrostaticity, although sample recovery is tricky, and loading must be done with specialist equipment at high pressure (Ne) or at cryogenic temperatures (Ar). Symmetric loading in a fluid medium is also challenging as the sample is more likely to stick to one of the anvils during loading. One solution is to place small ruby spheres between the sample and lower anvil to keep the gaps between the sample anvils equal.

Samples may be pre-pressed to a thin plate from which a piece with the right size is selected to be loaded between the insulation layers. A more controlled loading is possible with quenched spherical glass by the levitation quenching method. A prepared half-ball can be thinned from the backside (the spherical part) by slow polishing parallel to the flat side, using a series of diamond lapping films. At this stage, each half-ball has turned into a wafer with thickness in the range of 20–40 μm that is ready for subsequent loading. Thinner wafers are chosen for higher-pressure experiments. These wafers are machined to the desired diameter using a laser micromachining system, such as the one constructed at IPGP (Badro et al., 2021). The system uses a picosecond 532 nm laser and a nanoresolution (linear motors and piezo) XYZ computer-controlled stage. Hundreds of disks can be cut in this way from the wafer, matching the diameter of the DAC sample chamber (usually in the range 50–120 μm diameter). Each disk needs to be individually cleaned in acetone to remove all traces of glue, epoxy, or deposits before DAC loading.

For single phase studies, a single disk is loaded with or without a pressure medium, and then pressurized in the DAC. More importantly, however, such a preparation technique allows multiple pieces to be stacked in a well-designed assemblage, with a different phase in the middle. For example, two disks of silicate glass with a piece of metal in the middle may be stacked to study metal–silicate interactions, or two disks of different silicate composition could be stacked to study reactivity at the interface. One can also consider drilling out a smaller inner disk on one such sample, and stack "full disk + drilled disk + full disk" to make a closed assemblage that is similar to sample chamber designs in piston-cylinder or multi-anvil experiments. Figure 13.4 shows an example of such a loading before and after laser heating. A metal foil is loaded between two silicate disks of identical composition, diameter, and thickness. The center is heated to melting to study metal–silicate interaction. Two important advantages of preparing such loadings are that (1) the sample is perfectly symmetrical both radially and axially, and (2) there is no possibility for the metal to react with the diamonds as the surrounding monolithic slabs of silicate have no cracks or grain boundaries.

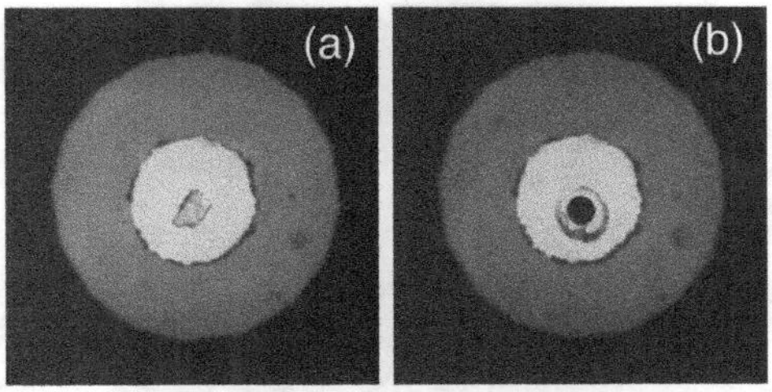

Figure 13.4 A metal–silicate LHDAC experiment before (a) and after (b) melting by laser heating. The sample assembly consists here of a stack of two silicate glass disks (transparent) with a small chunk of metal (shiny rectangle) in between. Using a short-pulse laser-machining instrument, the silicate disks are cut to match the size of the hole in the rhenium gasket, and the metal rectangle is cut similarly from a metal foil. After melting (a), the metal has turned into a spherical blob at the center, and the surrounding molten silicate can be seen optically as a rim around the opaque metal.

In other experiments, single-crystal sapphire is employed to minimize contamination by carbon from diamonds (Oka et al., 2019). Al_2O_3 is a desirable pressure medium in experiments on iron alloys, because the solubility of Al in solid and liquid Fe is known to be minimal except at very high temperatures (>5,000 K). To avoid adsorption of moisture, it is a good practice to place a loaded, but not pressurized, DAC in a vacuum oven at 150°C for at least 12 hours. For moisture sensitive samples, the DAC should be quickly moved from the oven to a glovebox or a plastic bag filled with argon gas, and then sealed by applying pressure. Alternatively, a bespoke oven can be employed that allows the DAC to be sealed under an Ar atmosphere while being held at high temperature, ensuring the sample is completely anhydrous (Lord et al., 2009).

It is now possible for more sophisticated loading using FIB fabrication. For example, one can develop a sample chamber configuration with isolated small samples (15 μm in diameter and 10 μm in thickness) using FIB microfabrication (Figure 13.5). The sample configuration design involves multiple steps for preparation. Figure 13.5 illustrates multiple sample containers for Fe–S mixtures with 5 wt.%S in a NaCl medium. The homogenous Fe–FeS mixture was first pressed into a thin plate (10 μm thickness) and loaded onto a 10 μm thick NaCl plate in the predrilled sample chamber. Several cylindrical islands (15 μm × 10 μm) were milled with FIB (Figure 13.5a). After completing the milling, a top layer of NaCl was added and then pressed by the opposite anvil. Figure 13.5b shows a top view of the sample chamber after being compressed to 50 GPa. The transparent area consists of NaCl pressure medium, whereas the dark islands are the sample surrounded by NaCl. Each sample was then heated to a target temperature at the APS GESCARS beamline and a series of X-ray diffraction patterns collected at different temperatures during heating. The sample was recovered and prepared for chemical composition analysis by FIB/SEM.

It is not possible to identify individual sample islands in the SEM image of the recovered DAC gasket. One can use area milling to remove the top NaCl layer to expose the metallic sample islands. Figure 13.5c shows the SEM image of the recovered sample after the upper NaCl layer has been removed and each laser-heated sample has been exposed using FIB milling. Each sample in the chamber can be heated to well-controlled target temperatures to

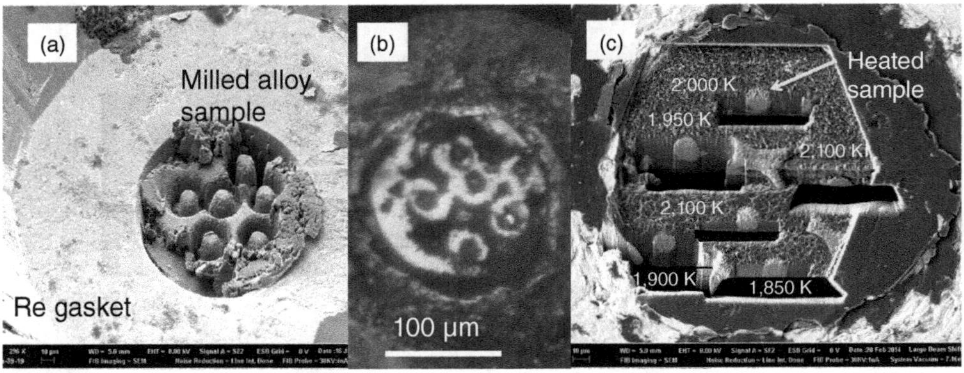

Figure 13.5 (a) A design of six small cylindrical samples (15 μm × 10 μm) milled on a NaCl plate in a prepressed, predrilled Re gasket. (b) A view of the pressurized sample through the diamond anvil. (c) The recovered sample after each heated sample was exposed and polished by FIB. Each sample was nicely confined in the NaCl pressure medium, which is critical to prevent melt migration during heating.

obtain the effect of temperature on the partitioning coefficient in one experiment, significantly increasing the efficiency of the experiments. Such a design also reduces thermal gradients, since the laser heating spot is larger than the sample size and provides stable temperature even during melting because the sample is well confined and its melt cannot migrate.

13.2.5 FIB Sample Recovery

Very small sample sizes (a few microns in thickness and several tens of microns in diameter) are characteristic of conventional ultrahigh pressure experiments that simulate conditions of the deep mantle and core. Laser heating is limited to heated spots at high pressure of about 20–40 μm in diameter. In order to obtain high-quality compositional data, it is critical to “polish” the heated spot for chemical analysis. FIB technology provides the capability to explore cross sections of the heated spot in a recovered DAC sample with high precision with a high success rate. The cross section can be prepared by either the lift-out method or site-specific milling.

Miyahara et al. (2008) described the procedure for the lift-out method to prepare thin samples for STEM-EDS analyses using a FIB system. This procedure is commonly adopted for recovering LHDAC samples, particularly for TEM analysis. Once the laser-heated spot is located in an SEM image, a narrow strip of platinum (2 × 30 μm) is deposited across the heated spot to protect the sample. To prepare for lift-out, two trapezoids are milled through on both sides if it is a thin sample (<10 μm), leaving a cross section with a width of 1 μm (Figure 13.6a). After cutting one side of the cross section, a tungsten or platinum needle may approach the top corner of the cross section to be welded by platinum deposition (Figure 13.6a and 13.6b). Once the cross section is firmly attached to the needle, it can be freed from the rest of the sample by cutting the other side and

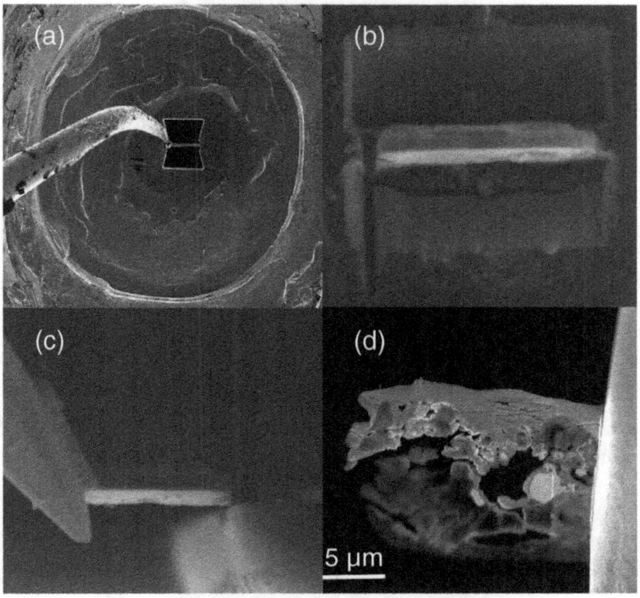

Figure 13.6 Representative lift-out preparation of the LHDAC sample using a FIB/SEM dual-beam instrument with an Omniprobe lift-out system. (a) A top view of the recovered DAC sample with milled trapezoids and the tip of the nanomanipulator. (b) A close-up view of the milled slice ready for lift-out. (c) Transferring the lifted slice from the probe tip to a TEM grid. (c) SEM image of the thinned slice mounted on the TEM grid. (d) SEM image of a recovered laser-heating spot containing a lower-mantle silicate assemblage after thinning to 100 nm thickness.

transporting the slice to a TEM grid (Figure 13.6c). Platinum deposition is used again to attach the lifted section to the TEM grid. Once the rectangular section is securely welded to the grid, it can be cut free from the needle and then thinned to a desirable thickness for STEM or TEM analysis (Figure 13.6d).

Making thin TEM sections can be challenging depending on the nature of the recovered samples. For a recovered lower-mantle silicate assemblage produced in the LHDAC (Figure 13.6d), the cross section becomes very fragile when it is thinned to 100 nm. For a recovered core iron alloy sample, the cross section may bend when it is too thin. Mounting on a V-grid with two-point welding can prevent the problem. Often extensive practice is needed to perfect the lift-out and thinning techniques.

It is not necessary to perform a lift-out for every laser-heated sample, and it is a time-consuming process with some risk of losing the sample in the process. The alternative is to obtain a "polished" cross section of the heated spot perpendicular to the compression axis of the DAC for chemical and texture analyses by milling down into the sample. Figure 13.7 shows images of heated spots that subsequently were milled to expose the interior of the heated spot. The milled surface can be rotated to horizontal for SEM image and quantitative EDX analysis.

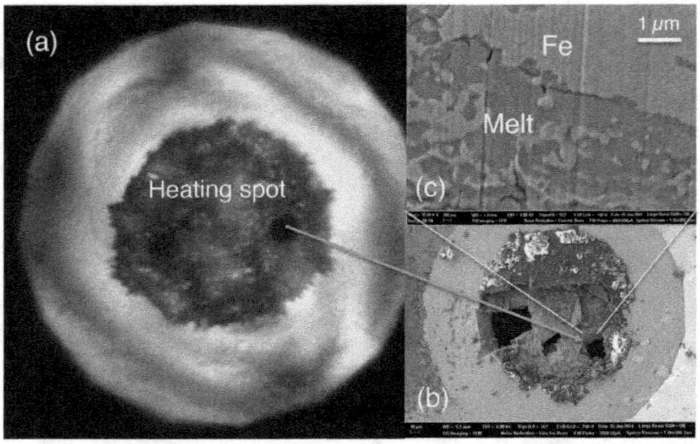

Figure 13.7 Image of a temperature-quenched sample in the diamond anvil cell (300 μm culet) with multiple laser-heating spots visible under transmitting and reflecting light (a) and an SEM image of the same recovered sample with FIB-milled spots (b). Three milled areas are shown to expose the laser-heated spots. High-resolution SEM image of the partially molten area in the Fe–FeO system, heated to 2,300 K at 53 GPa (c), showing a melting texture similar to that of the quenched multi-anvil sample, but at much smaller scale.

13.3 Sample Characterization

13.3.1 Imaging and Element Mapping of the Recovered Samples

Assessing quenched coexisting phases and textures of the recovered samples by imaging is the first step toward understanding of melting and element partitioning. The sample loading configuration may influence the observed quenching texture. Samples recovered from melting experiments in an LHDAC exhibit a quenched liquid pool at the center, surrounded by solid phases on both sides, as shown by scanning ion microscope (SIM) images (Figure 13.8). SEM images of the exposed heated spots are often acquired immediately after milling in the FIB/SEM dual-beam instrument (Figure 13.7). With the EDX silicon drift detector (SDD), element mapping can be very effective for identifying quenched mineral phases and melts (Figure 13.9).

Various imaging techniques provide direct visual examinations of the quench texture. For example, in accordance with the temperature gradient in a laser-heated sample, solid phases segregate from each other and form layers of a single phase (Figure 13.8). Such textures are similar to those obtained in melting experiments in a multi-anvil press (e.g., Tronnes and Frost, 2002). The solid phase next to a liquid pool may be interpreted to be a liquidus phase. Figure 13.8 shows the quenched phases in a melting experiment in the Fe–C system (Mashino et al., 2019). The liquidus phase Fe layer was thin (~1–2 μm) and thus the temperature difference between the liquid–Fe and Fe–Fe_3C boundaries should be very small (~100 K), suggesting it is close to the Fe–Fe_3C eutectic point.

Sample chamber design with precision loading, as shown in Figure 13.5, can improve heating stability by confining the melt pool. The small sample relative to the laser-heating spot size would also reduce thermal gradients. Figure 13.9 shows element mapping of the quenched sample recovered from an Fe–FeS melting experiment at 50 GPa and 2,200 K,

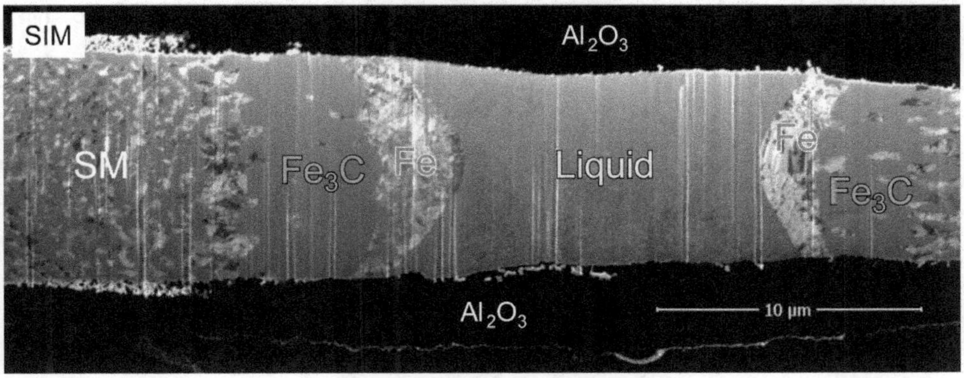

Figure 13.8 Representative scanning ion microscope (SIM) image of a laser-heated spot recovered from an Fe–C melting experiment at 40 GPa (Mashino et al., 2019). The Fe–C starting material was sandwiched between Al_2O_3 insulator layers.

Figure 13.9 Element mapping of the recovered sample at 50 GPa and 2,200 K with field-emission SEM EDX, showing existence of solid Fe with Fe–S melt.

showing Fe–S melt pools at both sides of the heated spot coexisting with solid Fe in the middle of the sample. Compared to the texture of the starting material, the Fe–S melt is well separated from the solid iron phase, providing a large enough area for quantitative chemical composition measurements.

STEM imaging is another way to obtain information about the quenched sample at the nanoscale. In particular, STEM-EDS analyses of FIB-milled slices of the samples can provide elemental maps. Ohira et al. (2014) demonstrated that bright-field STEM images overlaid with elemental maps could provide definitive phase identification for the coexisting hydrous δ-phase AlOOH and $MgSiO_3$ bridgmanite at very fine scale.

13.3.2 Quantitative Chemical Analyses of the Recovered Samples

Quantitative chemical analysis of quenched samples is conventionally done with wavelength-dispersive X-ray (WDX) spectroscopy in the electron microprobe because this

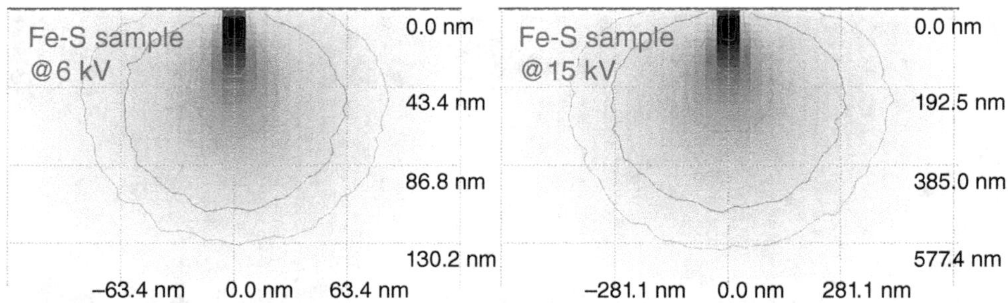

Figure 13.10 Energy transfer in XZ plane from Monte Carlo simulations for an Fe–S sample at 6 kV (left) and 15 kV (right). The excited volume is substantially smaller at 6 kV than at 15 kV, leading to high spatial resolution.

technique provides reliable elemental compositions, including concentrations of minor and trace elements, and light elements such as oxygen and carbon in metallic iron alloys. However, the relatively large excitation volume can be problematic for small LHDAC samples. Conventional WDX measurements at 15 kV have an excitation volume of ~3 μm in diameter. With the small grain size in the recovered sample, measurements with much higher spatial resolution are required. The development of large-area SDDs and thermal field-emission electron microprobes can considerably improve quantitative chemical analysis at the submicron scale.

The large-area SDD has much lower electronic noise and higher count rate over the traditional Si(Li) and Ge solid-state detectors. The use of the SDD allows accurate determination of chemical compositions with EDX for most of the major elements. Using a low accelerating voltage can significantly increase spatial resolution. Monte Carlo simulations (Drouin et al., 2007) for Fe–S samples at 15 kV show that 95% of X-rays come from a sphere with a diameter of 600 nm (Figure 13.10). Even higher spatial resolution can be achieved by reducing the accelerating voltage further. According to the simulations, 150 nm resolution is achievable at 6 kV for major element measurements (Figure 13.10). We have performed line analysis across a sharp boundary between Fe and melt and confirmed the correlation between the spatial resolution and acceleration voltage predicted by the simulations.

With the SDD EDX measurements in the field-emission SEM, it is possible to achieve higher spatial resolution than the WDX analysis with the electron microprobe. In comparison, the analytical SDD EDX measurements can produce equally good quantitative results for the common major elements in Earth materials. Table 13.1 compares the measured compositions of the quenched Fe–FeS melting sample from 21 GPa and 2,023 K, using WDX on the electron microprobe and analytical SDD-EDX on the FIB/SEM instrument. Using the same beam condition and standards, the measured compositions are in excellent agreement. The significant improvement of spatial resolution in composition analysis will allow us to precisely determine element partitioning between solid and melt from small melting areas within the laser-heated spot.

Use of the thermal field-emission electron microprobe for quantitative chemical analysis has improved spatial resolution for WDX measurements too. Typical beam operational

Table 13.1 *Comparison of chemical compositions of the 21 GPa quenched samples by WDX analysis with the JEOL JXA-8900 electron microprobe and by EDX analysis with the Zeiss Auriga FIB/SEM dual beam instrument equipped with an SDD. All analyses were done at 15 kV and using FeS_2 and pure Fe as standards.*

	WDX analysis			SDD	EDX analysis	
Sample	S, wt.%	Fe, wt.%	Total	S, wt.%	Fe, wt.%	Total
340a_Fe	0.16(0.01)	99.14(0.32)	99.31(0.32)	0.18(0.01)	99.05(0.12)	99.23(0.18)
340a_L	4.29(0.07)	94.40(0.07)	98.68(0.09)	4.42(0.24)	94.46(0.26)	98.87(0.40)
340b_L	7.66(0.21)	90.71(0.20)	98.37(0.33)	7.77(0.46)	91.09(0.44)	98.86(0.15)

conditions (15 kV and 30 nA) can produce good quantitative measurements for a grain size of 1 μm. At a low accelerating voltage of 7 kV, the estimated excited volume is much smaller, resulting in spatial resolution at about 300 nm. The sizes of the quenched liquid metal and molten silicate (e.g., Jackson et al., 2018) are sufficiently large for good quantitative chemical analysis.

The WDX measurements show advantages of quantification of elements such as O, N, and C that are important for composition analysis in LHDAC samples. For example, the Carnegie JEOL 6500F thermal field-emission electron microprobe is equipped with five wavelength dispersive spectrometers (WDX), each containing at least one lithium fluoride (LiF), one polyethylene terephthalate (PET), one triallyl phosphate (TAP), and one layered dispersion element (LDE) crystal, enabling analysis of C, N, and O. It also includes a fully integrated SDD-EDX detector that can be used in parallel with the WDX detectors for simultaneous quantitative elemental analysis and X-ray mapping. With these features, it is possible to compare the O measurements using both WDS and SDD-EDX by direct measurement of O content rather than calculated "by stoichiometry." This capability greatly benefits O measurements in partitioning experiments, particularly for the quenched liquid phases.

Analysis of carbon is challenging, particularly for recovered LHDAC samples, because of their small size and limited ability for replicate measurements. Carbon contamination in LHDAC samples under extreme conditions is unavoidable, and it is important to establish protocols for C measurements (Yang et al., 2019). As with previous studies (e.g., Dasgupta and Walker, 2008; Zhang et al., 2018), the LDE6 crystal is usually used to first determine the optimum conditions for C analysis. The optimized condition can be obtained by monitoring peak/background count per second (cps) values of the Fe_7C_3 standard, with variable peak/background count times (10/5, 20/10, and 30/15s), beam currents (20–80 nA), and accelerating voltages (12 and 15 kV). The correction is made by the measured C cps values as a function of C concentrations in C standards (Fe_7C_3 and Fe_3C) and pure Fe metal. The C concentration in the samples is determined based on the measured C cps values for each analytical session, and it is essential to perform calibration for each analytical session.

13.4 Results from Representative Experiments

The Earth's core–mantle boundary (CMB) and inner-core boundary (ICB) are the most important deep thermochemical boundaries, as they control the dynamics of the interior and the generation of Earth's magnetic field. These boundaries are defined by metal–silicate differentiation and inner-core crystallization, respectively. In order to understand the nature of chemical interactions and evolution of the CMB at 136 GPa and the ICB at 330 GPa, we need to determine liquid metal–silicate element partitioning at conditions relevant to the CMB and light element partitioning between solid iron and melt at the even more extreme conditions relevant to the ICB. The combination of the double-sided LHDAC and ex situ chemical analyses of recovered samples is the only experimental path toward these goals.

In addition to interactions at deep thermochemical boundaries, understanding solidification of the early deep magma ocean requires knowledge of the melting relations of model mantle compositions at high pressure. Using the techniques described in this chapter, we highlight results from metal–silicate reaction experiments in simulating core–mantle differentiation, silicate melting experiments relevant to crystallization of the deep magma ocean, and melting experiments of core materials to decipher the distribution of light elements in the core and its thermal structure.

13.4.1 Metal–Silicate Interactions

The LHDAC technique and chemical analysis with TEM-EDX was applied to examine the solubilities of O and Si in liquid iron in equilibrium with (Mg,Fe)SiO_3–perovskite to 97 GPa and 3,150 K (Takafuji et al., 2005). While the simultaneous solubilities of O and Si are known to be limited at 1 bar in steelmaking (O'Neill et al., 1998), they are enhanced with increasing temperature. Ozawa et al. (2008) performed similar chemical reaction experiments between molten Fe and ferropericlase and found that the pressure dependence of the solubility of oxygen is negative below 40 GPa but becomes positive at higher pressures. Combining these two experiments, Ozawa et al. (2009) argued that a pyrolitic mantle material cannot be in chemical equilibrium with the liquid outer core at the CMB.

Sakai et al. (2006) conducted experiments on the interaction between metallic iron and the postperovskite (PPv) phase at the CMB condition of 139 GPa and 3,000 K by combination of FIB sample recovery and analytical transmission electron microscope (ATEM) analysis. These experiments revealed that significant amounts of oxygen, up to 6.3 wt.%, and silicon, up to 4.0 wt.%, can be dissolved into molten iron, with an Fe/(Fe +Mg) ratio of ~0.02 in postperovskite. The use of FIB-ATEM analysis for LHDAC samples demonstrated the power of this combination of techniques for not only quantitative analysis, but also the imaging capability at very fine scale. The analysis provided measurements of dihedral angles between PPv and molten iron, which indicated a small amount of core metal could be trapped without separation in the PPv region at the CMB.

Frost et al. (2010) also examined oxygen partitioning between magnesiowustute and molten iron at 30–70 GPa and 2,800–3,500 K by analyzing FIB-recovered LHDAC samples. The partitioning of oxygen determined experimentally together with the thermodynamic extrapolation of the experiments to CMB conditions revealed that an outer core containing 6–9 wt.% oxygen can coexist with magnesiowustite with an Fe/(Fe+Mg) of 0.02–0.08, which is close to that of postperovskite and is below the value of ~0.2 for magniesiowustite in the normal lower mantle. Thus, an oxygen rich outer core and a normal mantle composed of magnesuowustite with Fe/(Fe+Mg)~0.12 may not be in equilibrium at the CMB.

These experiments (Takafuji et al., 2005; Sakai et al., 2006; Ozawa et al., 2008; Frost et al., 2010) provided the first look at the partitioning behavior between liquid iron and silicate mantle minerals to address equilibration at the CMB. Recent experiments in the LHDAC (Siebert et al., 2012; 2013; Fischer et al., 2015; Badro et al., 2016; Du et al., 2017; Jackson et al., 2018) produced liquid–liquid metal–silicate equilibrium at temperatures above the silicate liquidus (Figure 13.11). Compilation of Si and O partitioning data shows a general trend between the partition coefficient and temperature, with a stronger effect of temperature on Si than O (Fischer et al., 2015; Du et al., 2017). These data provide quantitative constraints on the Si and O contents in the Earth's core through core formation models (Fischer et al., 2017).

Modeling of the metal–silicate partitioning of siderophile elements as a function of pressure, temperature, oxygen fugacity, and composition has pointed to the existence of a

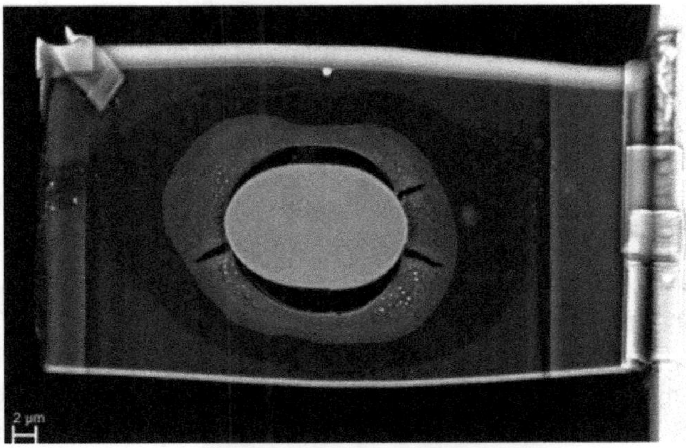

Figure 13.11 FIB thinned section removed from the center of the molten sample in Figure 13.4 showing the molten metal that has coalesced to the center (yellow) surrounded by molten silicate around it (light blue), itself surrounded by solidified silicate (dark blue) and with the untransformed starting silicate occupying the far field (turquoise). Such samples can be measured by traditional chemical analytical techniques (electron microprobe, scanning electron microscopy with EDX analysis) or more advanced techniques such as analytical transmission electron microscopy (Auzende et al., 2008; Piet et al., 2016) or ion probe analysis (Badro et al., 2007; Suer et al., 2017). It is noteworthy that loading samples by stacking monolithic disks of silicate of identical thickness and composition produces perfectly symmetrical samples at the end of the run, and one can note that the metal is far from the diamonds and isolated from them by the starting material, which bears no cracks or grain boundaries, hence efficiently blocking chemical reactions between the melt and the diamonds. For the color version, refer to the plate section.
(Huang and Badro, 2018).

deep magma ocean (35–55 GPa and 3,200–3,600 K) during the early differentiation of Earth. For example, Wade and Wood (2005) concluded that the metal–silicate partitioning of V, Ni, Co, Mn, and Si was compatible with single-stage core formation at 40 GPa and 3,773 K. Mann et al. (2009) showed that pressures of metal–silicate partitioning at 30–60 GPa would be required to reproduce the chondritic element ratios in the mantle. Germanium (Ge) partitioning data, combined with other data, suggested metal–silicate equilibrium in a magma ocean at 40 GPa and 3,473 K (Righter et al., 2011). Re-examination of Co and Ni partitioning also placed equilibrium at 35–50 GPa (Kegler et al., 2008). These conclusions were largely based on the pressure dependence derived from partitioning data below 25 GPa. Recent advances in analysis of the recovered samples from LHDAC element partitioning experiments performed at pressures above 25 GPa allow us to further assess the conditions of the deep magma ocean. For example, the mantle abundances of several siderophile elements (Ni, Co, V, Cr, Mn, Mo, W, Zn) could be reproduced by core–mantle equilibration at pressures from 32–42 GPa along the solidus of a magma ocean (~3,000 K) (Siebert et al., 2011) or 40–60 GPa under oxidizing conditions (Siebert et al., 2012, 2013). Similarly, Fischer et al. (2015) suggested single-stage core–mantle equilibration at 54 GPa and 3,300–3,400 K based on Ni, Co, V, and Cr partitioning data in the LHDAC.

Recent debate on the thermal conductivity of the Earth's core (e.g., Pozzo et al., 2012; Gomi et al., 2013; Seagle et al., 2013; Gomi and Hirose, 2015; Konopkova et al., 2016; Ohta et al., 2016; Zhang et al., 2020) has prompted discussion on energy sources that powered an early geodynamo prior to inner-core crystallization. Exsolution of light components such as MgO and SiO_2 in a liquid core could be a buoyancy source to drive the early geodynamo (O'Rourke and Stevenson, 2016). Subsequent experimental studies on the solubility of Mg and Si in liquid Fe at extreme conditions by analysis of LHDAC samples (Badro et al., 2016, 2018; Hirose et al., 2017; Du et al., 2017, 2019) provide necessary data needed to test the hypothesis. Experimental results have showed that significant MgO may be dissolved into the core during late-stage high-energy impacts (Badro et al., 2016; Du et al., 2017). Yet questions remain as to how much MgO could exsolve to drive the early geodynamo during core cooling. On the other hand, a thermodynamic model developed on the basis of melting in the Fe–Si–O system and metal–silicate partitioning data (Hirose et al., 2017) demonstrated that the Si+O solubility in molten Fe is limited at the present-day CMB pressure and temperature, suggesting that the liquid core has evolved to its composition due to SiO_2 crystallization and sustained convection and dynamo action.

13.4.2 Melting Relations in Mantle Phases and Solidification of the Deep Magma Ocean

Understanding deep mantle melting and the solidification of early deep magma oceans requires knowledge of melting relations in mantle compositions at lower-mantle pressures. Combination of in situ X-ray diffraction measurements and ex situ chemical analysis of recovered LHDAC samples allows the exploration of melting relations in the silicate systems beyond the P–T range of conventional multi-anvil devices. For example, the liquidus phase relations in the $MgO–SiO_2$ system have been experimentally examined to lowermost mantle pressures (Baron et al., 2017; Ozawa et al., 2018). On the basis of analyses of recovered LHDAC samples, the liquidus field of $MgSiO_3$ bridgmanite expands with increasing pressure, leading to more MgO-rich and SiO_2-rich eutectic liquid compositions in the $MgO–MgSiO_3$ and $MgSiO_3–SiO_2$ systems respectively at higher pressures.

The melting of peridotite has been extended to CMB pressure (Fiquet et al., 2010; Nomura et al., 2014) by textural and chemical characterizations of quenched LHDAC samples. The measured low solidus temperature for a natural primitive mantle (peridotite) composition, with about 400 ppm H_2O, would provide tight constraints on the temperature and H content at the CMB (Nomura et al., 2014). Tateno et al. (2014) reported that the liquidus phase of a peridotitic lower mantle changes from ferropericlase to bridgmanite above 34 GPa. Stepwise crystallization experiments performed in Caracas et al. (2019) demonstrated that bridgmanite crystallizes as a single phase until 50% solidification. Figure 13.12 shows images acquired with a field emission electron probe microanalyzer (JEOL JXA-8500F/8530F), providing quantitative chemical mapping of the melting spot at 59 GPa.

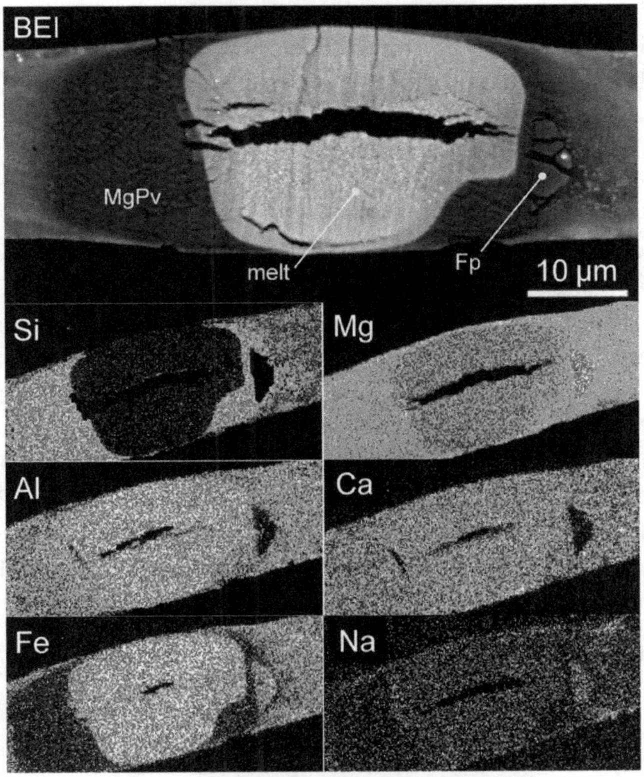

Figure 13.12 Melting experiment on the evolution of the magma ocean from Caracas et al. (2019). Backscattered electron images (top) and the X-ray maps for Si, Mg, Al, Ca, Fe, and Na obtained at 59 GPa. The Fe-rich melt crystallizes $MgSiO_3$-rich bridgmanite (Bdg) and a trace amount of (Mg,Fe)O ferropericlase (Fp) at 63% solidification from a fully molten pyrolite. For the color version, refer to the plate section.

Measurements of the melting of natural MORB have also been extended up to CMB pressures using LHDAC experiments combined with textural and chemical characterizations of quenched samples with ATEM and in situ X-ray diffraction measurements (Pradhan et al., 2015). The measured solidus temperature of MORB at the CMB (~3,970 K) is about 400 K higher than that of "wet" peridotite (~3,570 K) (Nomura et al., 2014), whereas the solidus temperature of MORB at the top of the lower mantle (at 25 GPa) is 2,600 K (Hirose and Fei, 2002), similar to that of anhydrous peridotite (Zhang and Herzberg, 1994). The complexity in melting determinations for various mantle compositions demands detailed characterization of the chemical compositions of the quenched phases and the melting texture. These experimental results also stress the importance of understanding the role of H_2O and other volatiles in melting experiments.

Thomson et al. (2014) took a systematic approach to define the solidus along the $MgSiO_3$–$MgCO_3$ join and the $MgCO_3$–$CaCO_3$ join from 15–80 GPa using in situ melting criteria coupled with ex situ textural and chemical analysis of recovered samples. Using the same approach, LHDAC experiments were performed on bulk compositions in the systems MgO–SiO_2–H_2O (MSH) and MgO–Al_2O_3–SiO_2–H_2O (MASH) up to 85 GPa to constrain the stability of hydrous phases in Earth's lower mantle (Walter et al., 2015).

In the $MgSiO_3$–Al_2O_3–H_2O system, Ohira et al. (2014) reported the phase relations and stability of a hydrous δ-phase, AlOOH–$MgSiO_2(OH)_2$ solid solution, at pressures from 43–128 GPa and temperatures from 1,580–2,430 K. The hydrous phase is stable to the bottom of the lower mantle and can coexist with bridgmanite and postperovskite phases. The detailed elemental mapping of the recovered samples by STEM-EDS analyses provided definitive phase identification and chemical compositions of the coexisting phases. The major lower-mantle minerals contain very low alumina contents when they coexist with the hydrous δ-phase. Additionally, the stability field of hydrous phase H, $(MgSi)O_4H_2$, has been investigated up to 60 GPa (Ohtani et al., 2016). Recently, Yuan et al. (2019) investigated the stability of Fe-bearing hydrous phases and element partitioning in the system MgO–Al_2O_3–Fe_2O_3–SiO_2–H_2O under Earth's deep mantle conditions. The results show that the hydrous δ-phase AlOOH–FeOOH–$MgSiO_2(OH)_2$–SiO_2 coexists with bridgmanite, PPv, or both in a broad P–T range of 104–126 GPa and 1,750–2,500 K. It is also observed that the hydrous pyrite-type FeOOHx phase coexists with the PPv phase over a P-T range of 112-123 GPa and 1,750–2,300 K. These studies demonstrate the power of combining in situ X-ray diffraction measurements with ex situ chemical characterizations using analytical tools at the sub-micron scale.

13.4.3 Melting of Core Materials

13.4.3.1 Melting Relations in the Fe–FeS System

Sulfur preferentially partitions into liquid iron in the Fe–FeS eutectic system. This system has been used as a model to explain the basic observations of the Earth's core system, including the liquid outer core and solid inner–core configuration and the large density jump at the ICB. To definitively evaluate the role of S during core formation and evolution, we must have full knowledge of the phase relations in the Fe–FeS system as a function of pressure and temperature up to ICB conditions. High-pressure experiments on Fe–FeS melting relations using piston-cylinder apparatus and multi-anvil devices have provided fundamental knowledge of the phase relations in the system up to 25 GPa (e.g., Fei et al., 1997, 2000; Li et al., 2001; Chen et al., 2008; Buono and Walker, 2011; Tao and Fei, 2021). The eutectic melting curve was further extended to 40 GPa using sintered diamond anvils in the multi-anvil device (Stewart et al., 2007). The eutectic melting measurements have been extended to higher pressure using the LHDAC (Campbell et al. 2007; Chudinovskikh and Boehler, 2007; Morard et al., 2008, 2011; Kamada et al., 2010, 2012; Mori et al., 2017).

The liquidus curves in the Fe-rich region are critical for determining the temperature at the ICB and the S distribution between the solid inner and liquid outer core. Kamada et al. (2012) determined the melting relations in the Fe–Fe_3S system up to 175 GPa and 3,500 K. Partial melting in the Fe–Fe_3S system was observed based on the reproducible disappearance/appearance of X-ray diffraction peaks of solid Fe_3S and quench texture observations of the recovered samples. More recently, the melting experiments by Mori et al. (2017) extended the eutectic melting curve up to 254 GPa. The reported reduction in sulfur concentration in the Fe–Fe_3S eutectic liquid down to ~5 wt.% at 254 GPa (Figure 13.13)

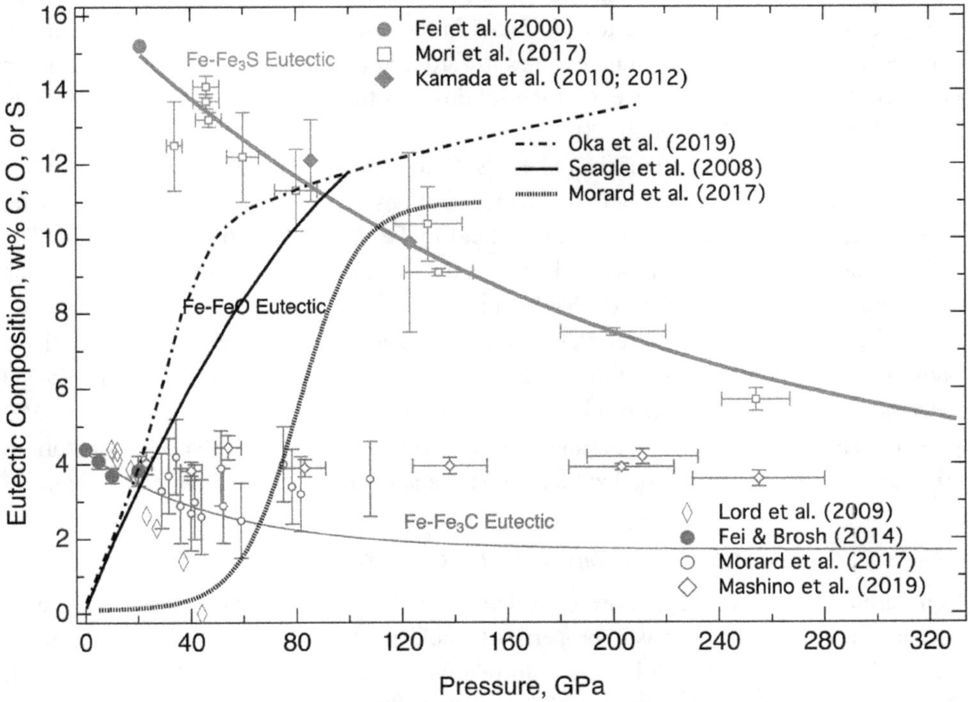

Figure 13.13 Compilation of eutectic compositions as a function of pressure in the Fe–Fe_3C system (Lord et al., 2009; Fei and Brosh, 2014; Morard et al., 2017; Mashino et al., 2019), the Fe–FeO system (Seagle et al., 2008; Morard et al., 2017; Oka et al., 2019), and the Fe–Fe_3S system (Fei et al., 2000; Kamada et al., 2010, 2012; Mori et al., 2017).

has implications for S distribution between solid and liquid iron. The effect of pressure on the S partitioning likely changes the melting relations at core conditions, as Fe_3S decomposes into Fe + Fe_2S above ~250 GPa (Ozawa et al., 2013; Tateno et al., 2019). The characterization of these critical changes has been demonstrated by a combination of in situ structure determination and ex situ chemical analysis.

13.4.3.2 Melting Relations in the Fe–S–Si, Fe–S–O, and Fe–Si–O Systems

The phase and melting relations in the Fe–S–Si system were determined up to 60 GPa in LHDAC experiments combined with in situ X-ray diffraction and ex situ characterization (Sakairi et al., 2017). The disappearance of the diffraction peaks of Fe_3S observed at 27, 44, 54, and 60 GPa was used as the criteria to define the solidus. The determination of the solidus curve was also confirmed by texture observations of the recovered sample in a separate run. The measured solidus and liquidus temperatures are significantly lower than the melting temperature of pure Fe. Examination of the inferred melting relation as a function of pressure in the Fe–S–Si system in relation to the adiabatic temperature gradients in the liquid cores of Mercury and Mars would allow crystallization of the cores to start at the core–mantle boundary region even in the cores containing both silicon and sulfur as light elements (Sakairi et al., 2017). On the other hand, Tateno et al. (2018) investigated the system up to 146 GPa. The melting curve was obtained through the appearance of diffuse X-ray scattering and/or melting texture found in the cross section of a recovered sample. The result indicates that the liquid evolves toward a Si-poor and S-rich composition upon crystallization. The measured slope of the solidus as a function of pressure is significantly higher than that of Sakairi et al. (2017).

The liquidus phase relations in the Fe–S–O ternary system have been examined to 208 GPa on the basis of microprobe characterizations of the recovered samples (Yokoo et al., 2019). They estimated the liquidus field of Fe at 330 GPa by extrapolation. The analyses indicate that the outer-core density and velocity and the density jump across the ICB cannot be explained by an Fe–S–O core.

On the basis of melting experiments on Fe–Si–O and literature data on metal–silicate partitioning experiments, a thermodynamic model developed by Hirose et al. (2017) demonstrated that the Si + O solubility in molten Fe is limited at the present-day CMB pressure and temperature, suggesting that the liquid core has evolved to its current composition due to SiO_2 crystallization and sustained convection and dynamo action.

13.4.3.3 Melting Relations in the Fe–C, Fe–O, and Fe–C–H Systems

Iron carbides (Fe_3C or Fe_7C_3) are considered possible phases in the Earth's inner core because of their favorable elastic properties (Mookherjee, 2011; Mookherjee et al., 2011; Lai et al., 2018) and high Poisson's ratio relative to that of pure Fe (Chen et al., 2014; Prescher et al., 2015) that may explain seismic observations. Knowledge of melting relations in the Fe–C system and the bulk C content in the core is required to determine whether iron carbide could be the crystalline phase in the inner core. Recent melting experiments in the Fe–C binary system have provided information on the change in the eutectic liquid composition up to 255 GPa (Mashino et al., 2019), and melting relations of Fe_3C to 300 GPa (Takahashi et al., 2019). It appears to be a consensus that Fe_7C_3 is a liquidus phase at pressures above ~200 GPa. However, Takahashi et al. (2019) concluded that Fe_3C melts incongruently to form Fe_7C_3 and liquid, whereas Mashino et al. (2019) suggested that Fe_7C_3 coexists with liquid at the eutectic point above 200 GPa. Thermodynamic calculations (Fei and Brosh, 2014) predicted that Fe_3C melts incongruently to form Fe_7C_3 and liquid at 136 GPa, but then the system transitions to Fe–Fe_7C_3 eutectic melting at ~270 GPa, in general agreement with the observation by Mashino et al. (2019). The calculated transition pressure is sensitive to the parameters of the thermal equations of state of iron carbides.

In the pressure range of the multi-anvil press, melting experiments showed that the eutectic C composition in the Fe–Fe_3C system shifts from ~4.4 wt.% at ambient pressure to about 3.5 wt.% at 25 GPa (Fei and Brosh, 2014). LHDAC experiments extended the measurements of the eutectic C composition up to 40 GPa (Lord et al., 2009) and 130 GPa (Morard et al., 2017), but the reported effect of pressure on the eutectic composition showed large uncertainties (Figure 13.13). Recently, Mashino et al. (2019) reported that the Fe–C binary eutectic liquids maintain a constant C concentration of ~3.5 wt.% with increasing pressure up to 255 GPa (Figure 13.13). This carbon content is slightly higher than the predicted value of ~ 2 wt.% C (Fei and Brosh, 2014), but is in general agreement. The precise determination of the eutectic C content is important because it determines if iron carbide Fe_7C_3 could be an inner-core phase for a given bulk C content in the core.

The Fe–FeO eutectic system is less studied in the LHDAC. Seagle et al. (2008) reported a steep shift of the eutectic composition toward the oxygen-rich side with pressure, reaching 11 wt.% at 91 GPa. A recent study shows high oxygen contents above 100 GPa, but it differs significantly from previous studies in the pressure range of 40–80 GPa (Morard et al., 2017). More recent experiments have extended the determination up to 204 GPa (Oka et al., 2019) (Figure 13.13). The influence of FeO metallization on the melting relations needs further experimental confirmation.

It is challenging to study melting relations in ternary Fe alloy systems, particularly involving hydrogen. There have been relatively few studies using the LHDAC at very high pressure. Nevertheless, they are important systems to study because of their relevance to the composition of the Earth's core. It is also known that strong interactions in multi-element systems are not always predictable based on the binary systems. Recently, Hirose et al. (2019) examined the effect of hydrogen on the solubility of carbon in liquid iron. The results show that hydrogen limits carbon concentration in molten iron, suggesting the incorporation of limited amount of carbon into core-forming metals when they are enriched in hydrogen.

13.5 Perspectives

Significant progress has been made in expanding our ability to experimentally probe the interior of the Earth by combining the LHDAC with various analytical techniques, including in situ synchrotron X-ray diffraction and ex situ characterization tools at a submicron scale. Notably, quantitative chemical analysis and high-resolution imaging of quenched LHDAC samples have revealed complex chemical interactions and melting relations of Earth-forming materials under extreme conditions in the last 15 years. A sustained effort is needed to advance our understanding of deep Earth chemistry. Currently, there are relatively few groups that have been actively pursuing this research direction, hindering fast-paced progress. In addition to increasing throughput by broad collaborations and training, establishment of protocols for sample recovery and analysis is an important step for achieving high-quality, reproducible results. Ingenious sample chamber design with FIB technology would further enhance sample environment control and produce repeatable, consistent results under extreme conditions.

Availability of homogeneous starting materials in various systems of interest with a range of chemical compositions is crucial for systematic investigation of mantle and core chemistry in multicomponent systems. We advocate to share well-characterized starting materials for coordinated collaborative projects that would have different experimental goals or cross-check different experimental techniques. These collaborations could drastically expand the scope of research and help to resolve discrepancies. Aided by thermodynamic calculations and theoretical predictions, well-chosen starting compositions in ternary or higher-order multicomponent systems could go a long way to improve our understanding of melting relations and element partitioning behavior at extreme pressures with a minimum number of required experiments.

Direct determination of melting relations at the ICB pressure (330 GPa) and recovery of samples at these conditions for quantitative chemical analysis should be one of the top priorities in high-pressure geochemistry. Any proposed core compositions could be tested against the observed density and velocity jumps at the ICB if the partitioning of light elements between solid and liquid iron is determined by the approach outlined in this chapter. Such experimental data would ultimately lead to a solution to the longstanding question of the composition of Earth's core.

Acknowledgments

The review represents a collective work from several laboratories. We acknowledge the contributions by many students and postdocs at each laboratory. Y.F. also acknowledges the supports by NSF grant EAR-2022492 and the Carnegie Institution for Science. O.T.L. would like to acknowledge support from the Royal Research Fellowship (UF150057).

References

Andrault, D., Bolfan-Casanova, N., Lo Nigro, G., Bouhifd, M.A., Garbarino, G., Mezouar, M. (2001). Melting curve of the deep mantle applied to properties of early magma ocean and actual core–mantle boundary. *Earth and Planetary Science Letters*, **304**, 251–259.

Andrault, D., Fiquet, G. (2001). Synchrotron radiation and laser heating in a diamond anvil cell. *Review of Scientific Instruments*, **72**, 1283–1288.

Andrault, D., Fiquet, G., Itie, J. P., et al. (1998). Thermal pressure in the laser-heated diamond-anvil cell: an X-ray diffraction study. *European Journal of Mineralogy*, **10**, 931–940.

Andrault, D., Pesce, D., Bouhifd, M. A., Bolfan-Casanova, N., Hénot, J-M., Mezouar, M. (2014). Melting of subducted basalt at the core–mantle boundary. *Science*, **344**, 892–895.

Andrault, D., Petitgirard, S., Lo Nigro, G., et al. (2012). Solid–liquid iron partitioning in Earth's deep mantle. *Nature*, **487**, 354–357.

Auzende, A., Badro, J., Ryerson, F. J., et al. (2008). Element partitioning between magnesium silicate perovskite and ferropericlase: new insights into bulk lower-mantle geochemistry. *Earth and Planetary Science Letters*, **269**, 164–174.

Badro, J., Aubert, K., Hirose, R., Nomura, I., Blanchard, S. B., Siebert, J. (2018). Magnesium partitioning between Earth's mantle and core and its potential to drive an early exsolution geodynamo. *Geophysical Research Letters*, **45**, 13,240–13,248. https://doi.org/10.1029/2018GL080405.

Badro, J., Ryerson, F. J., Weber, P. K., Ricolleau, A., Fallon, S. J., Hutcheon, I. D. (2007). Chemical imaging with NanoSIMS: a window into deep-Earth geochemistry. *Earth and Planetary Science Letters*, **262**, 543–551.

Badro, J., Siebert, J., Nimmo, F. (2016). An early geodynamo driven by exsolution of mantle components from Earth's core. *Nature*, **536**(7616), 326–328. https://doi.org/10.1038/nature18594.

Badro, J, Sossi, P. A., Deng, Z., Borensztajn, S., Wehr, N., Ryerson, F. J. (2021). Experimental investigation of elemental and isotopic evaporation processes by laser heating in an aerodynamic levitation furnace. *Comptes Rendus Geoscience*, **353**, 101–114. https://doi.org.10.5802/crgeos.56.

Baron, A. Q. R., Ishikawa, D., Fukui, H., Nakajima, Y. (2019). Auxiliary optics for meV-resolved inelastic X-ray scattering at SPring-8: Microfocus, analyzer masks, soller slit, soller screen, and beam position monitor. *AIP Conference Proceedings*, **2054**, 020002. https://doi.org/10.1063/1.5084562.

Baron, M. A., Lord, O. T., Myhill, R., et al. (2017). Experimental constraints on melting temperatures in the MgO–SiO_2 system at lower mantle pressures. *Earth and Planetary Science Letters*, **472**, 186–196.

Bassett, W. A., Weathers, M. S. (1986). Temperature-measurement in laser heated diamond anvil cells. *Physica B & C*, **139**, 900–902.

Benedetti, L. R., Guignot, N., Farber, D. L. (2007). Achieving accuracy in spectroradiometric measurements of temperature in the laser-heated diamond anvil cell: diamond is an optical component. *Journal of Applied Physics*, **101**, 013109.

Benedetti, L. R., Loubeyre, P. (2004). Temperature gradients, wavelength-dependent emissivity, and accuracy of high and very-high temperatures measured in the laser-heated diamond cell. *High Pressure Research*, **24**, 423–445.

Benzerara, K., Menguy, N., Guyot, F., Vanni, C., Gillet, P. (2005). TEM study of a silicate–carbonate–microbe interface prepared by focused ion beam milling. *Geochimica et Cosmochimica Acta*, **69**, 1413–1422.

Blanchard, I., Siebert, J., Borensztajn, S., Badro, J. (2017). The solubility of heat-producing elements in Earth's core. *Geochemical Perspectives Letters*, **5**, 1–5.

Bouhifd, M. A., Jephcoat, A. P. (2003). The effect of pressure on partitioning of Ni and Co between silicate and iron-rich metal liquids: a diamond-anvil cell study. *Earth and Planetary Science Letters*, **209**(1–2), 245–255.

Boehler, R. (1993). Temperatures in the Earth's core from melting-point measurements of iron at high static pressures. *Nature*, **363**, 534–536.

Boehler, R. (2000). High-pressure experiments and the phase diagram of lower mantle materials. *Reviews of Geophysics*, **38**, 221–245.

Boehler, R., Chopelas, A. (1991). A new approach to laser heating in high pressure mineral physics. *Geophysical Research Letters*, **18**, 1147–1150.

Buono, A. S., Walker, D. (2011). The Fe-rich liquidus in the Fe–FeS system from 1 bar to 10 GPa. *Geochimica et Cosmochimica Acta*, **75**, 2072–2087.

Campbell, A. J. (2008). Measurement of temperature distributions across laser-heated samples by multispectral imaging radiometry. *Review of Scientific Instruments*, **79**, 015108.

Campbell, A. J., Danielson, L., Righter, K., Seagle, C. T., Wang, Y., Prakapenka, V. B. (2009). High pressure effects on the iron–iron oxide and nickel–nickel oxide oxygen fugacity buffers. *Earth and Planetary Science Letters*, **286**(3–4), 556–564. https://doi.org/10.1016/j.epsl.2009.07.022.

Campbell, A. J., Seagle, C. T., Heinz, D. L., Shen, G., Prakapenka, V. B. (2007). Partial melting in the iron–sulfur system at high pressure: a synchrotron X-ray diffraction study. *Physics of the Earth and Planetary Interiors*, **162**, 119–128.

Caracas, R., Hirose, K., Nomura, R., Ballmer, M. D. (2019). Melt–crystal density crossover in a deep magma ocean. *Earth and Planetary Science Letters*, **516**, 202–211. https://doi.org/10.1016/j.epsl.2019.03.031.

Chen, B., Li, J., Hauck, S. A. II (2008). Non-ideal liquidus curve in the Fe–S system and Mercury's snowing core. *Geophysical Research Letters*, **35**, L07201.

Chen, B., Li, Z., Zhang, D., et al. (2014). Hidden carbon in Earth's inner core revealed by shear softening in dense Fe_7C_3. *Proceedings of the National Academy of Sciences USA*, **111**, 17755–17758.

Chidester, B. A., Rahman, Z., Righter, K., Campbell, A. J. (2017). Metal–silicate partitioning of U: implications for the heat budget of the core and evidence for reduced U in the mantle. *Geochimica et Cosmochimica Acta*, **199**, 1–12, 2017.

Chudinovskikh, L., Boehler, R. (2007). Eutectic melting in the system Fe–S to 44 GPa. *Earth and Planetary Science Letters*, **257**(1-2), 97–103. https://doi.org/10.1016/j.epsl.2007.02.024.

Dasgupta, R., Walker, D. (2008). Carbon solubility in core melts in a shallow magma ocean environment and distribution of carbon between the Earth's core and the mantle. *Geochimica et Cosmochimica Acta*, 72, 4627–4641.

Deng, J., Du, Z. X., Benedetti, L. R., Lee, K. K. M. (2017). The influence of wavelength-dependent absorption and temperature gradients on temperature determination in laser-heated diamond-anvil cells. *Journal of Applied Physics*, **121**, 025901.

Dewaele, A., Fiquet, G., Gillet, P. (1998). Temperature and pressure distribution in the laser-heated diamond-anvil cell. *Review of Scientific Instruments*, **69**, 2421–2426.

Dorfman, S. M., Badro, J., Nabiei, F., Prakapenka, V. B., Cantoni, M., Gillet, P. (2018). Carbonate stability in the reduced lower mantle. *Earth and Planetary Science Letters*, **489**, 84–91.

Drouin, D., Couture, A. R., Joly, D., Tastet, X., Aimez, V., Gauvin, R. (2007). CASINO V2.42 – a fast and easy-to-use modeling tool for scanning electron microscopy and microanalysis users. *Scanning*, **29**(3), 92–101.

Du, Z., Jackson, C., Bennett, N., et al. (2017). Insufficient energy from MgO exsolution to power early geodynamo. *Geophysical Research Letters*, **44**, L077283.

Du, Z. X., Amulele, G., Benedetti, L. R., Lee, K. K. M. (2013). Mapping temperatures and temperature gradients during flash heating in a diamond-anvil cell. *Review of Scientific Instruments*, **84**, 075111.

Du, Z., Boujibar, A., Driscoll, P., Fei, Y. (2019) Experimental constraints on an MgO exsolution-driven geodynamo. *Geophysical Research Letters*, **46**, L083017.

Du, Z. X., Lee, K. K. M. (2014). High-pressure melting of MgO from (Mg,Fe)O solid solutions. *Geophysical Research Letters*, **41**, 8061–8066.

Fei, Y. (2013). Simulation of the planetary interior differentiation processes in the laboratory. *Journal of Visualized Experiments*. **81**, e50778. doi:10.3791/50778.

Fei, Y., Bertka, C. M., Finger, L. W. (1997). High-pressure iron-sulfur compound, Fe_3S_2, and melting relations in the system Fe–FeS at high pressure. *Science*, **275**, 1621–1623.

Fei, Y., Brosh, E. (2014). Experimental study and thermodynamic calculations of phase relations in the Fe-C system at high pressure. *Earth and Planetary Science Letters*, **408**, 155–162.

Fei, Y., Li, J., Bertka, C. M., Prewitt, C. T. (2000). Structure type and bulk modulus of Fe_3S, a new iron–sulfur compound. *American Mineralogist*, **85**(11-12), 1830–1833. https://doi.org/10.2138/am-2000-11-1229.

Fei, Y., Mao, H. K., Mysen, B. O. (1991). Experimental determination of element partitioning and calculation of phase equilibrium relations in the MgO–FeO–SiO_2 system at high pressure and high temperature. *Journal of Geophysical Research*, **96**, 2157–2169.

Fiquet, G., Auzende, A. L., Siebert, J., et al. (2010). Melting of peridotite to 140 gigapascals. *Science*, **329**, 1516–1518.

Fischer, R. A., Campbell, A. J., Ciesla, F. J. (2017). Sensitivities of Earth's core and mantle compositions to accretion and differentiation processes. *Earth and Planetary Science Letters*, **458**, 252–262.

Fischer, R. A., Nakajima, Y., Campbell, A. J., et al. (2015). High pressure metal–silicate partitioning of Ni, Co, V, Cr, Si, and O. *Geochimica et Cosmochimica Acta*, **167**, 177–194.

Frost, D. J., Asahara, Y., Rubie, D. C., et al. (2010). Partitioning of oxygen between the Earth's mantle and core. *Journal of Geophysical Research*, **115**, B02202. doi:10.1029/2009JB006302.

Fukui, H., Sakai, T., Sakamaki, T., et al. (2013). A compact system for generating extreme pressures and temperatures: an application of laser-heated diamond anvil cell to inelastic X-ray scattering. *Review of Scientific Instruments*, **84**, 113902. doi: 10.1063/1.4826497.

Giampaoli, R., Kantor, I., Mezouar, M., et al. (2018). Measurement of temperature in the laser heated diamond anvil cell: comparison between reflective and refractive optics. *High Pressure Research*, **38**, 250–269.

Gomi, H., Hirose, K. (2015). Electrical resistivity and thermal conductivity of hcp Fe–Ni alloys under high pressure: implications for thermal convection in the Earth's core. *Physics of the Earth and Planetary Interiors*, **247**, 2–10.

Gomi, H., Ohta, K., Hirose, K., et al. (2013). The high conductivity of iron and thermal evolution of the Earth's core. *Physics of the Earth and Planetary Interiors*, **224**, 88–103.

Hamilton, D. L., Henderson, C. M. B. (1968). The preparation of silicate compositions by a gelling method. *Mineralogical Magazine and Journal of the Mineralogical Society*, **36**(282), 832–838.

Heaney, P. J., Vicenzi, E. P., Giannuzzi, L. A., Livi, K. J. T. (2001). Focused ion beam milling: a method of site-specific sample extraction for microanalysis of Earth and planetary materials. *American Mineralogist*, 86, 1094-1099.

Heinz, D. L., Jeanloz, R. (1987). Temperature measurement in the laser-heated diamond anvil cell, in M. Manghnani, Y. Syono, eds., *High Pressure Research in Mineral Physics*. American Geophysical Union, pp. 113–127.

Hirao, N., Ohtani, E., Kondo, T., et al. (2006). Partitioning of potassium between iron and silicate at the core–mantle boundary. *Geophysical Research Letters*, **33**, L08303. doi:10.1029/2005GL025324.

Hirao, N., Kawaguchi, S. I., Hirose, K., Shimizu, K., Ohtani, E., Ohishi, Y. (2020). New developments in high-pressure X-ray diffraction beamline for diamond anvil cell at SPring-8. *Matter and Radiation at Extremes*, **5**, 018403. https://doi.org/10.1063/1.5126038.

Hirose, K., Fei, Y. (2002). Subsolidus and melting phase relations of basaltic composition in the uppermost lower mantle. *Geochimica et Cosmochimica Acta*, **66**, 2099–2108.

Hirose, K., Morard, G., Sinmyo, R., et al. (2017). Crystallization of silicon dioxide and compositional evolution of the Earth's core. *Nature*, **543**(7643), 99–102. https://doi.org/10.1038/nature21367.

Hirose, K., Tagawa, S., Sinmyo, R., Morard, G., Ohishi, Y., Genda, H. (2019). Hydrogen limits carbon in liquid iron. *Geophysical Research Letters*, **46**, 5190–5197.

Huang, D., Badro, J. (2018). Fe–Ni ideality during core formation on Earth. *American Mineralogist*, **103**, 1707–1710.

Ichikawa, H., Tsuchiya, T., Tange, Y. (2014). The P–V–T equation of state and thermodynamic properties of liquid iron. *Journal of Geophysical Research: Solid Earth*, **119**, 240–252.

Jackson, C. R. M., Bennett, N. R., Du, Z., Cottrell, E., Fei, Y. (2018). Early episodes of high-pressure core formation preserved in plume mantle. *Nature*, **553**, 491–495.

Jeanloz, R., Heinz, D. L. (1984). Experiments at high-temperature and pressure - laser-heating through the diamond cell. *Journal de Physique*, **45**, 83–92.

Jeanloz, R., Kavner, A. (1996). Melting criteria and imaging spectroradiometry in laser-heated diamond-cell experiments. *Philosophical Transactions of the Royal Society A – Mathematical Physical and Engineering Sciences*, **354**, 1279–1305.

Jephcoat, A. P., Besedin, S. P. (1996). Temperature measurement and melting determination in the laser-heated diamond-anvil cell. *Philosophical Transactions of the Royal Society A – Mathematical Physical and Engineering Sciences*, **354**, 1333–1360.

Kamada, S., Ohtani, E., Fukui, H., et al. (2014). The sound velocity measurements of Fe_3S. *American Mineralogist*, **99**, 98–101.

Kamada, S., Ohtani, E., Terasaki, H., et al. (2012). Melting relationships in the Fe–Fe_3S system up to the outer core conditions. *Earth and Planetary Science Letters*, **359–360**, 26–33.

Kamada, S., Terasaki, H., Ohtani, E., et al. (2010). Phase relationships of the Fe–FeS system in conditions up to the Earth's outer core. *Earth and Planetary Science Letters*, **294**(1-2), 94–100. https://doi.org/10.1016/j.epsl.2010.03.011.

Kavner, A., Nugent, C. (2008). Precise measurements of radial temperature gradients in the laser-heated diamond anvil cell. *Review of Scientific Instruments*, **79**, 024902.

Kavner, A., Panero, W. R. (2004). Temperature gradients and evaluation of thermoelastic properties in the synchrotron-based laser-heated diamond cell. *Physics of the Earth and Planetary Interiors*, **143**, 527–539.

Kawamura, K. (1994). Preparation of starting materials of multi-component silicates using sol-gel method, in T. Iyama, K. Kawamura, S. Nakashima, eds., *Experiments of Geochemistry*, Tokyo University Press, pp. 47–52.

Kegler, Ph., Holzheid, A., Frost, D. J., Rubie, D. C., Dohmen, R., Palme, H. (2008). New Ni and Co metal–silicate partitioning data and their relevance for an early terrestrial magma ocean. *Earth and Planetary Science Letters*, **268**, 28–40.

Kesson, S. E., Fitz Gerald, J. D., O'Neil S. C., Shelley, J. M. G. (2002). Partitioning of iron between magnesian silicate perovskite and magnesiowüstite at about 1 Mbar. *Physics of the Earth and Planetary Interiors*, **131**, 295–310.

Knittle, E., Jeanloz, R. (1989). Simulating the core–mantle boundary – an experimental study of high-pressure reactions between silicates and liquid iron. *Geophysical Research Letters*, **16**, 609–612.

Kobayashi, Y., Kondo, T., Ohtani, E., et al. (2005). Fe–Mg partitioning between (Mg, Fe) SiO_3 post-perovskite, perovskite, and magnesiowüstite in the Earth's lower mantle. *Geophysical Research Letters*, **32**, L19301.

Konopkova, Z., McWilliams, R. S., Gomez-Perez, N., Goncharov, A. F. (2016). Direct measurement of thermal conductivity in solid iron at planetary core conditions. *Nature*, **534**(7605), 99–101.

Kunz, M., Yan, J. Y., Cornell, E., et al. (2018). Implementation and application of the peak scaling method for temperature measurement in the laser heated diamond anvil cell. *Review of Scientific Instruments*, **89**, 083903.

Lai, X., Zhur, F., Liu, J. et al. (2018). The high-pressure anisotropic thermoelastic properties of a potential inner core carbon-bearing phase, Fe_7C_3, by single-crystal X-ray diffraction. *American Mineralogist*, **103**, 1568–1574. https://doi.org/10.2138/am-2018-6527.

Li, J., Fei, Y., Mao, H. K., Hirose, K., Shieh, S. R. (2001). Sulfur in the Earth's inner core. *Earth and Planetary Interiors*, **193**(3-4), 509–514. https://doi.org/10.1016/S0012-821X(01)00521-0.

Li, Y., Vocadlo, L., Brodholt, J., Wood, I. G. (2016). Thermoelasticity of Fe_7C_3 under inner core conditions. *Journal of Geophysical Research Solid Earth*, **121**, 5828–5837.

Lin, J.-F., Struzhkin, V. V., Sturhahn, W., et al. (2003). Sound velocities of iron–nickel and iron–silicon alloys at high pressures. *Geophysical Research Letters*, **30**(21), 2112. doi:10.1029/2003GL018405.

Liu, L. G. (1975). Post-oxide phases of olivine and pyroxene and mineralogy of the mantle. *Nature*, **258**, 510–512.

Liu, L. G. (1976). The post-spinel phase of forsterite. *Nature*, **262**, 770–772.

Lord, O. T., Walter, M. J., Dasgupta, R., Walker, D., Clark, S. M. (2009). Melting in the Fe–C system to 70 GPa. *Earth and Planetary Science Letters*, **284**, 157–167.

Lord, O. T., Wang, W. (2018). MIRRORS: A MATLAB ®GUI for temperature measurement by multispectral imaging radiometry. *Review of Scientific Instruments*, **89**(10), 104903. https://doi.org/10.1063/1.5041360.

Lord, O. T., Wood, I. G., Dobson, D. P., et al. (2014). The melting curve of Ni to 1 Mbar. *Earth and Planetary Science Letters*, **408**, 226–236.

Lv, M., Dorfman, S. M., Badro, J., Borensztajn, S., Greenberg, E., Prakapenka, V. B. (2021). Reversal of carbonate–silicate cation exchange in cold slabs in Earth's lower mantle. *Nature Communications*, **12**, 1712.

Maeda, F., Ohtani, E., Kamada, S., Sakamaki, T., Hirao, N., Ohishi, Y. (2007). Diamond formation in the deep lower mantle: a high-pressure reaction of $MgCO_3$ and SiO_2. *Scientific Reports*, **7**, 40602. doi:10.1038/srep40602.

Manga, M., Jeanloz, R. (1996). Axial temperature gradients in dielectric samples in the laser-heated diamond cell. *Geophysical Research Letters*, **23**, 1845–1848.

Mann, U., Frost, D. J., Rubie, D. C. (2009). Evidence for high-pressure core–mantle differentiation from the metal–silicate partitioning of lithophile and weakly-siderophile elements. *Geochimica et Cosmochimica Acta*, **73**, 7360–7386.

Mashino, I., Miozzi, F., Hirose, K., Morard, G., Sinmyo, R. (2019). Melting experiments on the Fe–C binary system up to 255 GPa: constraints on the carbon content in the Earth's core. *Earth and Planetary Science Letters*, **515**, 135–144.

Meng, Y., Shen, G., Mao, H. K. (2006). Double-sided laser heating system at HPCAT for in situ X-ray diffraction at high pressures and high temperatures. *Journal of Physics: Condensed Matter*, **18**, S1097–S1103.

Mezouar, M., Giampaoli, R., Garbarino, G., et al. (2017). Methodology for in situ synchrotron X-ray studies in the laser-heated diamond anvil cell. *High Pressure Research*, **37**, 170–180.

Ming, L., Bassett, W. A. (1974). Laser-heating in diamond anvil press up to 2000 degrees C sustained and 3000 degrees C pulsed at pressures up to 260 kilobars. *Review of Scientific Instruments*, **45**, 1115–1118.

Miyahara, M., Sakai, T., Kobayashi, Y., et al. (2008). Application of FIB system to ultra-high-pressure Earth science. *Journal of Mineralogical and Petrological Sciences*, **103**, 88–93.

Mookherjee, M. (2011). Elasticity and anisotropy of Fe_3C at high pressures. *American Mineralogist*, **96**, 1530–1536.

Mookherjee, M., Nakajima, Y., Steinle-Neumann, G., et al. (2011). High pressure behaviour of iron carbide (Fe_7C_3) at inner core conditions. *Journal of Geophysical Research*, **116**, B04201.

Morard, G., Andrault, D., Antonangeli, D., et al. (2017). Fe–FeO and Fe–Fe_3C melting relations at Earth's core–mantle boundary conditions: implications for a volatile-rich or oxygen-rich core. *Earth and Planetary Science Letters*, **473**, 94–103.

Morard, G., Andrault, D., Guignot, N., et al. (2008). In situ determination of Fe–Fe_3S phase diagram and liquid structural properties up to 65 GPa. *Earth and Planetary Science Letters*, **272**(3-4), 620–626. https://doi.org/10.1016/j.epsl.2008.05.028.

Morard, G., Andrault, D., Guignot, N., Siebert, J., Garbarino, G., Antonangeli, D. (2011). Melting of Fe–Ni–Si and Fe–Ni–S alloys at megabar pressures: implications for the core–mantle boundary temperature. *Physics and Chemistry of Minerals*, **38**, 767–776.

Mori, Y., Ozawa, H., Hirose, K., et al. (2017). Melting experiments on Fe–Fe_3S system to 254 GPa. *Earth and Planetary Science Letters*, **464**, 135–141. https://doi.org/10.1016/j.epsl.2017.02.021.

Murakami, M., Hirose, K., Kawamura, K., Sata, N., Ohishi, Y. (2004). Post-perovskite phase transition in MgSiO3. *Science*, **304**(5672), 855–858.

Murakami, M., Hirose, K., Sata, N., Ohishi, Y. (2005). Post-perovskite phase transition and mineral chemistry in the pyrolitic lowermost mantle. *Geophysical Research Letters*, **32**, L03304.

Nomura, R., Hirose, K., Uesugi, K., Ohishi, Y., Tsuchiyama, A., Miyake, A. (2014). Low core–mantle boundary temperature inferred from the solidus of pyrolite. *Science*, **343**, 522–525.

Ohira, I., Ohtani, E., Sakai, T., et al. (2014). Stability of a hydrous δ-phase, AlOOH-$MgSiO_2(OH)_2$, and a mechanism for water transport into the base of lower mantle. *Earth and Planetary Science Letters*, **401**, 12–17.

Ohishi, Y., Hirao, N., Sata, N., Hirose, K., Takata, M. (2008). Highly intense monochromatic X-ray diffraction facility for high-pressure research at SPring-8. *High Pressure Research*, **28**, 163–173. https://doi.org/10.1080/08957950802208910.

Ohta, K., Kuwayama, Y., Hirose, K., Shimizu, K., Ohishi, Y. (2016). Experimental determination of the electrical resistivity of iron at Earth's core conditions. *Nature*, **534**(7605), 95–98. https://doi.org/10.1038/nature17957.

Ohtani E., Amaike, Y., Kamada, S., Ohira, I., Mashino, I. (2016). Stability of hydrous minerals and water reservoirs in the deep Earth's interior, in H. Terasaki, R. A. Fischer, eds., *Deep Earth: Physics and Chemistry of the Lower Mantle and Core*. Geophysical Monograph 217, 1st ed. John Wiley & Sons, Inc., pp. 265–275.

Ohtani, E., Yuan, L., Ohira, I., Shatskiy, A., Litasov, K. (2018). Fate of water transported into the deep mantle by slab subduction. *Journal of Asian Earth Sciences*, **167**, 2–10. https://doi.org/10.1016/j.jseaes.2018.04.024.

Oka, K., Hirose, K., Tagawa, S., et al. (2019). Melting in Fe–FeO to 204 GPa: implications for oxygen in Earth's core. *American Mineralogist*, **104**, 1603–1607.

O'Neill, H. S. C., Canil, D., Rubie, D. C. (1998). Oxide–metal equilibria at 2500°C and 25 GPa: implications for core formation and the light component in the Earth's core. *Journal of Geophysical Research: Solid Earth*, **103**, 12239–12260.

O'Rourke, J. G., Stevenson, D. J. (2016). Powering Earth's dynamo with magnesium precipitation from the core. *Nature*, **529**(7586), 387–389. https://doi.org/10.1038/nature16495.

Ozawa, H., Hirose, K., Mitome, M., Bando, Y., Sata, N., Ohishi, Y. (2008). Chemical equilibrium between ferropericlase and molten iron to 134 GPa and implications for iron content at the bottom of the mantle. *Geophysical Research Letters*, **35**, L05308.

Ozawa, H., Hirose, K., Mitome, M., Bando, Y., Sata, N., Ohishi, Y. (2009). Experimental study of reaction between perovskite and molten iron to 146 GPa and implications for chemical equilibrium at the core–mantle boundary. *Physics and Chemistry of Minerals*, **36**, 355–363.

Ozawa, H., Hirose, K., Ohishi, Y., Hirao, N. (2013). Decomposition of Fe_3S above 250 GPa. *Geophysical Research Letters*, **40**, 4845–4849. https://doi.org/10.1002/grl.50946.

Ozawa, K., Anzai, M., Hirose, K., Sinmyo, R., Tateno, S. (2018). Experimental determination of eutectic liquid compositions in the MgO–SiO_2 system to the lowermost mantle pressures. *Geophysical Research Letters*, **45**, https://doi.org/10.1029/2018GL079313.

Panero, W. R., Jeanloz, R. (2001). Temperature gradients in the laser-heated diamond anvil cell. *Journal of Geophysical Research: Solid Earth*, **106**, 6493–6498.

Piet, H., Badro, J., Nabiei, F., et al. (2016). Spin and valence dependence of iron partitioning in Earth's deep mantle. *Proceedings of the National Academy of Sciences USA*, **113**, 11127–11130.

Pozzo, M., Davies, C., Gubbins, D., Alfe, D. (2012). Thermal and electrical conductivity of iron at Earth's core conditions. *Nature*, **485**(7398), 355–358. https://doi.org/10.1038/nature11031.

Pradhan, G. K., Fiquet, G., Siebert, J., et al. (2015). Melting of MORB at core–mantle boundary. *Earth and Planetary Science Letters*, **431**, 247–255.

Prakapenka, V. B., Kubo, A., Kuznetsov, A., et al. (2008). Advanced flat top laser heating system for high pressure research at GSECARS: application to the melting behavior of germanium. *High Pressure Research*, **28**, 225–235.

Prescher, C., Dubrovinsky, L., Bykova, E., et al. (2015). High Poisson's ratio of Earth's inner core explained by carbon alloying. *Nature Geoscience*, **8**, 220–223.

Rainey, E. S. G., Hernlund, J. W., Kavner, A. (2013). Temperature distributions in the laser-heated diamond anvil cell from 3-D numerical modeling. *Journal Applied Physics*, **114**, 204905.

Rainey, E. S. G., Kavner, A. (2014). Peak scaling method to measure temperatures in the laser-heated diamond anvil cell and application to the thermal conductivity of MgO. *Journal of Geophysical Research: Solid Earth*, **119**, 8154–8170.

Ricolleau, A., Fiquet, G., Addad, A., et al. (2008). Analytical transmission electron microscopy study of a natural MORB sample assemblage transformed at high pressure and high temperature. *American Mineralogist*, **93**, 144–153.

Righter, K., King, C., Danielson, L., Pando, K., Lee, C. T. (2011). Experimental determination of the metal/silicate partition coefficient of germanium: implications for core and mantle differentiation. *Earth and Planetary Science Letters*, **304**, 379–388.

Sakai, T., Kondo, T., Ohtani, E., et al. (2006). Interaction between iron and post-perovskite at core–mantle boundary and core signature in plume source region. *Geophysical Research Letters*, **33**, L15317. doi:10.1029/2006GL026868.

Sakairi, T., Ohtani, E., Kamada, S., Sakai, T., Sakamaki, T., Hirao, N. (2017). Melting relations in the Fe–S–Si system at high pressure and temperature: implications for the planetary core. *Progress in Earth and Planetary Science*, **4**, 10. doi 10.1186/s40645–017-0125-x.

Sakairi, T., Sakamaki, T., Ohtani, E., et al. (2018). Sound velocity measurements of hcp Fe–Si alloy at high pressure and high temperature by inelastic X-ray scattering, *American Mineralogist*, **103**, 85–90.

Sakamaki, T., Ohtani, E., Fukui, H., et al. (2016). Constraints on Earth's inner core composition inferred from measurements of the sound velocity of hcp–iron in extreme conditions. *Science Advances*, **2**, e1500802. doi:10.1126/sciadv.1500802.

Seagle, C., Cottrell, E., Fei, Y., Hummer, D. R., Prakapenka, V. B. (2013). Electrical and thermal transport properties of iron and iron–silicon alloy at high pressure. *Geophysical Research Letters*, **40**, 1–5, doi:10.1002/2013GL057930.

Seagle, C. T., Heinz, D. L., Campbell, A. J., Prakapenka, V. B., Wanless, S. T. (2008). Melting and thermal expansion in the Fe–FeO system at high pressure. *Earth and Planetary Science Letters*, **265**, 655–665.

Shen, G., Mao, H. K., Hemley, R. J. (1996). Laser-heating diamond- anvil cell technique: double-sided heating with multimode Nd:Y AG laser, in *Advanced Materials '96 – New Trends in High Pressure Research*, National Instrument Research, Inorganic Materials, pp. 149–152.

Shen, G., Rivers, M. L., Wang, Y., Sutton, S. R. (2001). A laser heated diamond cell system at the Advanced Photon Source for in situ X-ray measurements at high pressure and temperature. *Review of Scientific Instruments*, **72**, 1273–1282.

Shibazaki Y., Ohtani E., Fukui H., et al. (2012). Sound velocity measurements in dhcp–FeH up to 70 GPa with inelastic X-ray scattering: implications for the composition of the Earth's core. *Earth and Planetary Science Letters*, **313–314**, 79–85.

Shultz, E., Mezouar, M., Crichton, W., et al. (2005). Double-sided laser heating system for in situ high pressure–high temperature monochromatic X-ray diffraction at the ESRF. *High Pressure Research*, **25**, 71–83.

Siebert, J., Badro, J., Antonangeli, D., Ryerson, F. J. (2012). Metal–silicate partitioning of Ni and Co in a deep magma ocean. *Earth and Planetary Science Letters*, **321–322**, 189–197.

Siebert, J., Badro, J., Antonangeli, D., Ryerson, F. J. (2013). Terrestrial accretion under oxidizing conditions. *Science*, **339**, 1194–1197.

Siebert, J., Corgne, A., Ryerson, F. J. (2011). Systematics of metal–silicate partitioning for many siderophile elements applied to Earth's core formation. *Geochimica et Cosmochimica Acta*, **75**, 1451–1489.

Sossi, P. A., Burnham, A. D., Badro, J., Lanzirotti, A., Newville, M., O'Neill, H. S. C. (2020). Redox state of Earth's magma ocean and its Venus-like early atmosphere. *Science Advances*, **6**, eabd1387.

Stewart, A. J., Schmidt, M. W., Westrenen, W. V., Liebske, C. (2007). Mars: a new core crystallization regime. *Science*, **316**, 1323–1325.

Suer, T.-A., Siebert, J., Remusat, L., Menguy, N., Fiquet, G. (2017). A sulfur-poor terrestrial core inferred from metal–silicate partitioning experiments. *Earth and Planetary Science Letters*, **469**, 84–97. https://doi.org/10.1016/j.epsl.2017.04.016.

Takafuji, N., Hirose, K., Mitome, M., Bando, Y. (2005). Solubilities of O and Si in liquid iron in equilibrium with (Mg,Fe)SiO_3 perovskite and the light elements in the core. *Geophysical Research Letters*, **32**, L06313.

Takahashi, S., Ohtani, E., Sakai, T., et al. (2019). Phase and melting relations of Fe3C to 300 GPa and carbon in the core, in C. E. Manning, J-F. Lin, W. Mao, eds., Carbon in Earth's interior. AGU monograph volume, pp. 25–36.

Tao, R., Fei, Y. (2021). High-pressure experimental constraints of partitioning behavior of Si and S at the Mercury's inner core boundary. *Earth and Planetary Science Letters*, **562**, 116849.

Tateno, S., Hirose, K., Ohishi, Y. (2014). Melting experiments on peridotite to lowermost mantle conditions, *Journal of Geophysical Research: Solid Earth*, **119**, 4684–4694. doi:10.1002/2013JB010616.

Tateno, S., Hirose, K., Sinmyo, R., Morard, G., Hirao, N., Ohishi, Y. (2018). Melting experiments on Fe–Si–S alloys to core pressures: silicon in the core? *American Mineralogist*, **103**(5), 742–748. https://doi.org/10.2138/am-2018-6299.

Tateno, S., Ozawa, H., Hirose, K., Suzuki, T., Kawaguchi, S., Hirao, N. (2019). Fe_2S: the most Fe-rich iron–sulfide at the Earth's inner core pressures. *Geophysical Research Letters*, **46**, 11,944–11,949. https://doi.org/10.1029/2019GL085248.

Thomson, A. R., Walter, M. J., Lord, O. T., Kohn, S. C. (2014). Experimental determination of melting in the systems enstatite–magnesite and magnesite–calcite from 15 to 80 GPa. *American Mineralogist*, **99**, 1544–1554.

Tronnes, R. G., Frost, D. J. (2002). Peridotite melting and mineral-melt partitioning of major and minor elements at 22–24.5 GPa. *Earth and Planetary Science Letters*, **197**, 117–131.

Tschauner, O., Zerr, A., Specht, S., Rocholl, A., Boehler, R., Palme, H. (1999). Partitioning of nickel and cobalt between silicate perovskite and metal at pressures up to 80 GPa. *Nature*, **398**, 604–607.

Umemoto, K., Hirose, K. (2020). Chemical compositions of the outer core examined by first principles calculations. *Earth and Planetary Science Letters*, **531**, 116009.

Wade, J., Wood, B. J. (2005). Core formation and the oxidation state of the Earth. *Earth and Planetary Science Letters*, **236**, 78–95.

Walter, M. J., Koga, K. T. (2004). The effects of chromatic dispersion on temperature measurement in the laser-heated diamond anvil cell. *Physics of the Earth and Planetary Interiors*, **143**, 541–558.

Walter, M. J., Thomson, A. R., Wang, W., et al. (2015). The stability of hydrous silicates in Earth's lower mantle: experimental constraints from the systems MgO–SiO_2_H_2O and MgO–Al_2O_3_SiO_2–H_2O. *Chemical Geology*, **418**, 16–29.

Watanabe, K., Ohtani, E., Kamada, S., Sakamaki, T., Miyahara, M., Ito, Y. (2014). The abundance of potassium in the Earth's core. *Physics of the Earth and Planetary Interiors*, **237**, 65–72.

Watanuki, T., Shimomura, O., Yagi, T., Kondo, T., Isshiki, M. (2001). Construction of laser-heated diamond anvil cell system for in situ X-ray diffraction study at SPring8. *Review of Scientific Instruments*, **72**, 1289–1292.

Yagi, T., Bell, P. M., Mao, H. K. (1979). Phase relations in the system MgO–FeO–SiO_2 between150 and 700 kbar at 1000°C. *Year Book Carnegie Institution of Washington*, **78**, 614–618.

Yang, J., Fei, Y., Hu, X., Greenberg, E., Prakapenka, V. B. (2019). Effect of carbon on the volume of solid iron at high pressure: implications for carbon substitution in iron structures and carbon content in the Earth's inner core. *Minerals*, **9**, 720

Yokoo, S., Hirose, K., Sinmyo, R., Tagawa, S. (2019). Melting experiments on liquidus phase relations in the Fe–S–O ternary system under core pressures, *Geophysical Research Letters*, **46**, https://doi.org/10.1029/2019GL082277.

Yuan, L., Ohtani, E., Ikuta, D., et al. (2018). Chemical reactions between Fe and H_2O up to megabar pressures and implications for water storage in the Earth's mantle and core. *Geophysical Research Letters*, **45**(3), 1330–1338.

Yuan, H., Zhang, L., Ohtani, E., Meng, Y., Greenberg, E., Prakapenka, V. B. (2019). Stability of Fe-bearing hydrous phases and element partitioning in the system MgO–Al_2O_3–Fe_2O_3–SiO_2–H_2O in Earth's lowermost mantle. *Earth and Planetary Science Letters*, **524**, 115714.

Zhang, J., Herzberg, C. (1994). Melting experiments on anhydrous peridotite KLB-1 from 5.0 to 22.5 GPa. *Journal of Geophysical Research: Solid Earth*, **99**, 17729–17742.

Zhang, Z., Hastings, P., Von der Handt, A., Hirschmann, M. M. (2018). Experimental determination of carbon solubility in Fe–Ni–S melts. *Geochimica et Cosmochimica Acta*, **225**, 66–79.

Zhang, Y., Hou, M., Liu, G., et al. (2020). Reconciliation of experiments and theory on transport properties of iron and the geodynamo. *Physical Review Letters*, **125**, 078501.

Zinin, P. V., Bykov, A. A., Machikhin, A. S., et al. (2019). Measurement of the temperature distribution on the surface of the laser heated specimen in a diamond anvil cell system by the tandem imaging acousto-optical filter. *High Pressure Research*, **39**, 131–149.

14

Molecular Compounds under Extreme Conditions

ALEXANDER F. GONCHAROV

This chapter reviews the tremendous progress over the past several decades in experimental research of molecular solids at high pressures. The interatomic interactions in these materials are greatly modified under pressure and generally strengthen intermolecular and weaken intramolecular bonds. This leads to the formation of structurally complex crystals and inclusion compounds at moderate pressures, where a variety of intermolecular bonds can exist. Pressing on, a great majority of molecular solids demonstrate transformations to extended (e.g., polymeric) states, which vary drastically in bonding and electronic properties. The most prominent example of such behavior is the symmetrization of hydrogen bonds in ionic ice X and metallization of hydrogen in monatomic solid. Dave Mao's legacy in this research has been remarkable ranging from discovering and establishing the structure and properties of hydrogen clathrate hydrates at 200 MPa to investigating the structure of a mixed atomic-molecular phase IV of hydrogen at 260 GPa. New generations of scientists continue to use and build upon his technical developments, which have enabled multimegabar investigations of molecular solids, including diamond anvil cell (DAC) design, the DAC gas-loading system, and a variety of optical, electric, magnetic, and X-ray DAC probes.

14.1 Introduction

Application of high pressure dramatically changes materials' properties compared to ambient conditions. These changes are especially remarkable in the case of molecular materials, which are usually gases or liquids at ambient pressure, depending on the temperature. They do not typically crystallize because the molecules in these materials are normally light (composed from the elements in the first two rows in the periodic table) and are bound by weak (e.g., van der Waals) bonds, which results in their low boiling and melting points. At high pressures, all molecular materials solidify, forming crystals held together by a variety of intermolecular bonds, including hydrogen and van der Waals types. Upon compression, the balance between these intramolecular and intramolecular (covalent) bonds changes until the system finds a new stable chemical state. It is worth emphasizing the importance of the pressure variable in the investigation of these new chemical states.

Application of pressures in the multimegabar range (hundreds of gigapascals) changes the free energy of the system via pressure–volume work by several electron volts, which exceeds the strength of even the strongest chemical bonds. The physical and chemical properties of these altered states are drastically different compared to the starting molecular system. The pressure–temperature (*P–T*) conditions at which these fundamental changes occur strongly depend on the system, and simple scaling rules do not apply. Addressing the overarching questions about the pathways that molecular systems undergo under extremes is of great interest for fundamental physics and chemistry as well as materials and planetary sciences.

Molecular compounds that include light elements are fundamental to understanding the physics and chemistry of the Earth and planets. These materials are among the most abundant in the Solar System, and they constitute a very sizable portion of planetary interiors (Hubbard, 1981; Guillot, 2005), such as layers of deep planetary ices and metallic hydrogen in the gas and ice giants. Thus, knowledge of the material properties under extreme *P–T* conditions is critical for modeling planetary structure, composition, dynamics, and evolution (Nettelmann et al., 2012). Moreover, knowledge of electrical and thermal conductivity provides valuable insight into planetary observations such as internally generated magnetic field (Ness, 1979; Stanley and Bloxham, 2004; Connerney et al., 2017).

Theoretical calculations can provide direction for static and dynamic compression experiments probing molecular systems. The work of Wigner and Huntington (1935), which predicted the existence of metallic hydrogen at high pressure, inspired many subsequent theoretical and experimental research. This includes the prediction that metallic hydrogen can be metastable at $P = 0$ (Brovman et al., 1972), which can be of great technological and practical importance. These works are ongoing research, and I will briefly review these works later in this chapter (see also recent review articles: Goncharov, 2020; Gregoryanz et al., 2020). In another fundamental system, water, it has been predicted that at sufficiently high pressure, the hydrogen bonds become symmetric, making all the O–H bond lengths equal in the crystal called ice X (Holzapfel, 1972). At these conditions, water will be ionic, somewhat similar to rock salt. However, determining hydrogen positions in water and other hydrogen-bearing systems at megabar pressures remains a great experimental challenge up to now; the ongoing efforts using the existing techniques are presented in this chapter. Nitrogen, revealing an astonishing number of molecular phases below 150 GPa, has been predicted to polymerize into a single bonded polymeric crystal (Mailhiot et al., 1992), which could potentially be an excellent environmentally benign energy source. This prediction resulted in the experimental discovery of three polymeric nitrogen phases, thus extending the remarkable polymorphism of nitrogen, which I will review in this chapter. I will also discuss the available data on formation of various mixed molecular crystals (or cocrystals at low pressures), which may be the building blocks of planetary interiors.

In this chapter, I will mainly discuss the results of static compression experiments; dynamic compression experiments and theoretical calculations represent other dimensions of materials research under extreme conditions, which are closely connected to this main topic. Dynamic experiments are complementary to static methods in many aspects, including the *P–T* coverage, ability to use the most informative probes, ability to investigate the thermodynamically stable state, and others. It is a combination of these techniques that often results in groundbreaking results such as a static high-pressure calibration (Mao et al., 1978), which has continued to be utilized in almost the same form for more than 40 years. Theoretical methods to investigate complex molecular systems under extreme conditions have greatly improved over the last few years, especially concerning the predictive power for stable phases and composition (Oganov et al., 2019).

This chapter is organized as follows. Section 14.2 describes the most recent technical advances, which were crucial for the development of the field. In Section 14.3, I present the research progress in archetypal molecular systems (Van der Waals compounds, nitrogen, ice, and hydrogen) that advanced our understanding of molecular systems at high pressures. Finally, Section 14.4 gives a brief perspective of future research in the field.

14.2 Technical Developments

Molecular systems are very compressible; for example, hydrogen increases its density by a factor of 10 at 150 GPa compared to the liquid state at 1 atmosphere (Mao and Hemley, 1994). Thus, reaching high pressures relevant for the topics of interest (see Section 14.1) with molecular systems is challenging. Diamond anvil cells (DACs) are compact devices for generating very high pressures in excess of 100 GPa. The essential features of their design, such as the use of beveled anvils and gasket, the alignment, and operational procedures, were designed by Dave Mao (Mao and Bell, 1977), and these are still being exploited without substantial modifications. DACs that are now capable of generating pressures in excess of 400 GPa (3.95 million of atmospheres) (Li et al., 2018) have been adapted to study simple molecular materials by Mao's group. In 1979, Mao and Bell reported cryogenic loading of liquid hydrogen in the DAC, which they solidified at 5.7 GPa at 25°C reaching the top pressure of 50 GPa (Mao and Bell, 1979). With the development of a gas-loading technique at room temperature (Mills et al., 1980), it becomes a routine task to use a variety of different gases (including noble gases such as He, Ne, and Ar) as a pressure medium to provide hydrostatic conditions in the sample cavity (Bell and Mao, 1981). However, pressing on low-Z molecules, such as He and especially H_2, continues to be an exceptionally challenging task that requires a careful choice of diamonds for the anvils, gasket material, and experimental strategy. Pioneering works led by Dave Mao on He (Loubeyre et al., 1993) and H_2 (Mao and Bell, 1979; Loubeyre et al., 1996) opened a new era in high-pressure research.

Probing molecular materials requires extremely refined experimental tools. The most critical is the structural information, which presents a great challenge because the majority of materials described here are low-Z ones. Thus, traditional X-ray diffraction (XRD) techniques are very difficult, which was at least partially circumvented by using single-crystal diffraction techniques (Mao et al., 1988; Loubeyre et al., 1993; Loubeyre et al., 1996; Ji et al., 2020; Ji et al., 2019). Probing the electronic properties of molecular materials at extreme conditions presents another great challenge. Optical spectroscopy techniques (discussed later), which are commonly used in this case, have significant limitations in terms of the spectral range that can be accessed, because diamonds are not transparent in the ultraviolet (UV) spectral range. X-ray spectroscopy techniques in a DAC have been shown to be very useful in probing the electronic properties of the materials under extremes (Shen and Mao, 2016). While probing the electronic structure of hydrogen continues to be a challenge (Mao et al., 2018), the electronic structure of compressed helium (He) has been measured up to 17 GPa in the spectral range of 20–40 eV using inelastic X-ray scattering (Mao et al., 2010).

Optical spectroscopy microprobe techniques developed for DACs by Dave Mao often remain indispensable for studies of molecular crystals. Raman and Brillouin spectroscopy and laser heating in DAC (Hemley et al., 1987) opened several new directions in research under extremes. Raman spectroscopy (Goncharov, 2012) is extremely informative as a probe of a particular phase, especially in the case of molecular compounds, where the intramolecular fundamental vibrations (often called vibrons for diatomic molecules) serve as the fingerprints of stable molecules (chemical state), while intermolecular vibrations (including free and restricted rotational motions) reflect the crystal lattice structure via Raman selection rules. The spectrum of these optical phonon modes provides direct information about the vibrational dynamics of the materials, shedding light upon the proximity to the phase and chemical transformations and changes in electronic structure, including ionization. Infrared (IR) spectroscopy is similar to Raman in that it also probes the vibrational excitations; however, the information is complementary, as this technique is sensitive to the dipole moment oscillations, unlike Raman, which probes the polarizability. Moreover, IR spectroscopy probes the electronic properties of materials by virtue of coupling of radiation with the electronic excitations of materials. Brillouin spectroscopy probes low-frequency vibrational excitations, corresponding to acoustic modes, thus providing access to the elastic properties of these materials.

Adapting these three optical spectroscopy techniques for the DAC, where the samples are extremely small and the diamond anvil often creates a strong parasitic background, represented a big step toward making a DAC the widely used instrument as we know it now. That includes the development of specialized optics interfacing the DAC and finding the most suitable DAC experimental conditions (Hemley et al., 1987, 1998; Goncharov et al., 2000). IR spectroscopy presents an example of the experimental method, where technical developments were particularly extensive involving the use of synchrotron radiation (Goncharov, et al., 2019; Hemley et al., 1998; Hemley et al., 2005). The development tremendously improves sensitivity, spatial resolution, and spectral coverage (extending to the far-IR range) of the technique (Goncharov et al., 2000; Chen et al., 2004; Koch-Müller et al., 2016), thus enabling the experiments to very high pressures exceeding 400 GPa (Loubeyre et al., 2020). Combining synchrotron IR spectroscopy with laser visible/near IR (supercontinuum) optical and Raman spectroscopy in one instrument (Goncharov et al., 2019) (Figure 14.1) makes it possible for concomitant investigations of electronic transitions such as pressure-induced metallization, and vibrational properties at ultrahigh pressure. This instrument enables investigation of very small samples with dimensions down to 5 μm, which can be studied over the entire IR-visible spectral range both in transmission and reflectivity modes. This is achieved via the use of easily

(a)

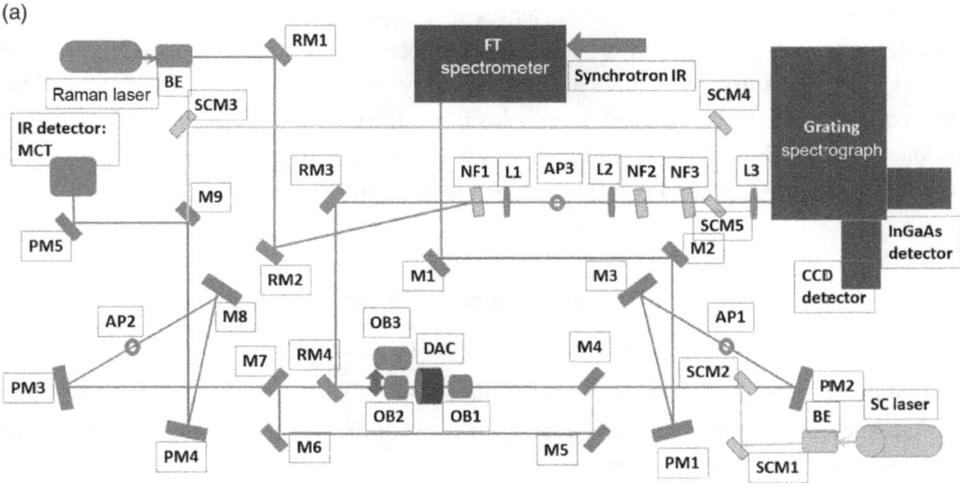

(b)

Figure 14.1 Combined synchrotron IR and Raman/optical spectroscopy system at the beamline BL01B of the SSRF (Shanghai). Left panel: Optical layout. Optical elements shown in blue correspond to synchrotron IR, while green and yellow indicate the optical elements that belong to Raman and visible/nearly IR parts of the system, respectively (Goncharov et al., 2019). Right panel: Photograph of the system with a continuous flow He cryostat. For the color version, refer to the plate section.

interchangeable specialized objective lenses and separate optical pathways for IR and visible light. Yet another and arguably more effective approach is to combine Raman and visible spectroscopy with synchrotron X-ray spectroscopy and diffraction. This approach has yet to be realized in one instrument, even though the use of an offline Raman system (Holtgrewe et al., 2019) was found to be extremely effective. The results of employing these systems will be discussed later in this chapter.

14.3 Experimental Research

14.3.1 Van der Waals Compounds

Solubility of molecules and noble gases at high pressures and formation of weakly bound stoichiometric compounds remained an open question before DAC experiments became routinely available. Understanding such phenomena, especially under very high pressures, where chemical reactivity between the components may take place, is of great interest for materials and planetary sciences.

In 1992, Vos et al. (1992) discovered the high-pressure $He(N_2)_{11}$ compound at 9 GPa. The stability of this compound is explained by a higher density of the compound compared to the randomly mixed constitutes, which is common for the size ratios between 0.57–0.85 (e.g., Murray and Sanders, 1980). The lattice symmetry and composition of this compound have been confirmed by Ninet et al. (2011), where the structure was directly solved; the material is stable up to 135 GPa, where it amorphizes very much like pure nitrogen (Goncharov et al., 2000) (details are provided later). It is interesting that molecular nitrogen forms a compound with Ne (Plisson et al., 2014), which remains stable up to at least 20 GPa, and likely to much higher pressures (perhaps up to 130 GPa). However, no compounds have been reported experimentally for the He–N system, despite theoretical calculations predicting the formation of a number of stable phases including clathrates, where the host structure is formed of the single-bonded nitrogen cages (Li et al., 2018).

Molecular hydrogen, which is a very small molecule (the bond length is 0.74 Å), can form Van der Waals compounds with many other molecules and noble gases as described in this chapter. However, it does not form a compound with He up to 250 GPa (Turnbull et al., 2018; Wang et al., 2019; Ramsey et al., 2020), perhaps because of the similarity in the sizes of He and H_2.

One of the most intriguing systems is the H_2–H_2O binary compound (see a recent review article by Bazarkina et al., 2020). Water tends to form many solid compounds with other molecules called clathrate hydrates (Sloan and Koh, 2007). These compounds consist of cages of hydrogen-bonded water molecules that host guest molecules via van der Waals bonds. Two major cubic clathrate structures are well known: They have water molecules forming cages that are different from any known ice structure. The guest molecular occupancy is limited to 1/6 (Villard's rule). As hydrogen molecules are normally too small for these cages, the H_2–H_2O system must be stabilized by pressure. In 2002, W. Mao et al. (2002) reported that the classical sII clathrate can have up to 1/2 H_2 occupancy. This result

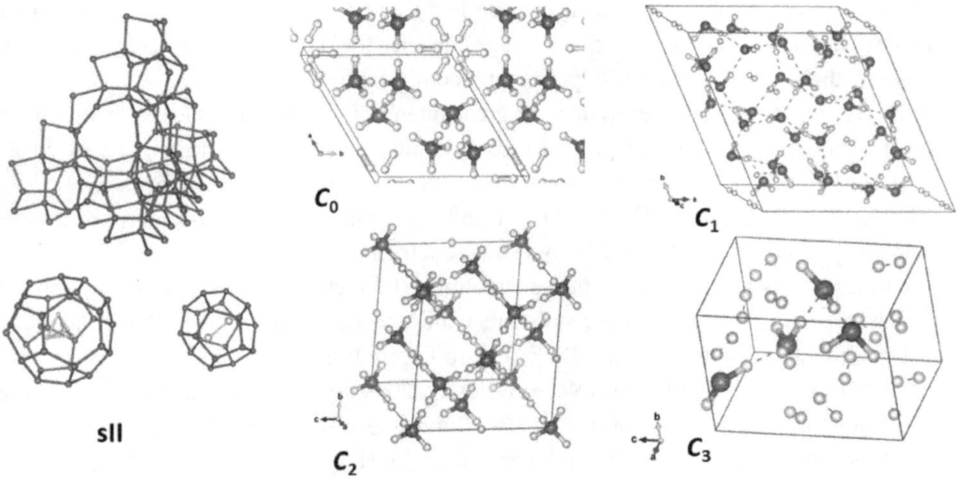

Figure 14.2 Structures of hydrogen hydrates
(Mak and McMullan, 1965; Londono et al., 1988; Vos et al., 1993; Qian et al., 2014; Strobel et al., 2016).

was later largely supported by direct neutron diffraction measurements that showed up to 0.35 occupancy (Lokshin et al., 2004) (Figure 14.2).

In 1993, Vos et al. (1993) discovered two new clathrate structures C_1 and C_2 with different compositions, which are stable between 0.75 and 3.1 GPa (C_1) and above 2.3 GPa (C_2), respectively. In C_1 clathrate, the H_2O sublattice is similar to the ice II structure, while H_2 molecules are positioned in a channel-like arrangement; the overall structure is similar to helium clathrate (Londono et al., 1988); it has a classic 1/6 H_2 occupancy. Recently, a new orientationally disordered C_1' clathrate has been discovered (Wang et al., 2020), which can be considered a realization of disordered ice II – the only ice for which a disordered counterpart was not known. In contrast, the C_2 structure is similar to ice VII in that there are two interpenetrating diamond-like sublattices, one of which is populated by H_2 molecules, yielding 1:1 H_2:H_2O occupancy. The C_2 structure is reported to be stable to at least 80 GPa with only minor structural modifications (Machida et al., 2008).

More recently, a C_0 clathrate has been discovered between the pressure ranges of stability of the sII and C_1 compounds (Efimchenko et al., 2011). According to a synchrotron diffraction structural study (Strobel et al., 2016), C_0 clathrate possesses a hexagonal chiral structure consisting of interpenetrating spiral chains of hydrogen-bonded H_2O molecules and rotationally disordered H_2 molecules bearing a topological similarity to the mineral quartz. This structure has 1:2 H_2:H_2O occupancy.

Since it is challenging to determine the structure and composition of the H_2–H_2O system at high pressure, our understanding of this system at the extreme conditions of giant planets is often based on theoretical calculations. Recent first-principles calculations have been successful in finding the structures and compositions of sII, C_0, C_1, and C_2 clathrates in good agreement with experiments. Moreover, these calculations made a prediction of a new C_3 clathrate (Qian et al., 2014) (see also Saleh and Oganov, 2016) with a large hydrogen–water ratio (H_2:H_2O = 2:1) (Figure 14.2). The C_3 structure becomes stable above 38 GPa. This structure could represent a huge capacity for H_2 storage in planetary interiors. However, their existence awaits experimental confirmation.

CH_4–H_2 represents another system of great interest for planetary science. While there are no compounds at ambient pressure, experiments up to 10 GPa (Somayazulu et al., 1996) revealed four compounds with a variety of compositions, such as $CH_4(H_2)_2$, $(CH_4)_2H_2$, $CH_4(H_2)_4$, and CH_4H_2. The 1:1 and 1:2 compounds have been reported to be stable up to 30 GPa. CH_4H_2 has a body-centered tetragonal (*bct*) structure, while $CH_4(H_2)_2$ crystallizes in a hexagonal Laves phase structure. The methane-rich $(CH_4)_2H_2$ compound also has been found to form a *bct* structure with a space group *I4/mcm*; this structure is similar to the Al_2Cu-type intermetallic structure family (Grin et al., 2006).

Recent theoretical calculations (Naumova et al., 2019) support the stability of this phase and suggest it is stable up to 90 GPa. In addition, these first-principles calculations also predict two more stable hydrogen-rich compounds $(CH_4)_2(H_2)_7$ and $CH_4(H_2)_9$. However, other theoretical calculations (Liu et al., 2014; Conway and Hermann, 2019) found that different compositions are more stable, calling for more extensive experimental and theoretical research. Elucidating this issue would help in the understanding of stability of hydrocarbons at high pressures, which remains controversial.

The study of inclusion compounds (or cocrystals) is paving the way for future studies of the chemistry in these materials; however, many of these compounds appear to be very resilient and survive at very high pressure. For example, $Ar(H_2)_2$ remains stable in the Laves phase up to 358 GPa (Ji et al., 2017). The mechanochemistry of these compounds is not well understood. For example, in the case of the N_2–H_2 system, where molecules (which form the inclusion compounds above 7 GPa; Spaulding et al., 2014; Laniel et al., 2018) are dissociated, and oligomeric highly disordered material forms at nearly 50 GPa at room temperature (Goncharov et al., 2015; Wang et al., 2015). Subjecting molecular compounds to extensive X-ray (Mao et al., 2006; Meng et al., 2006) and ultraviolet (UV) radiation (Ceppatelli et al., 2009; Goncharov et al., 2015; Holtgrewe et al., 2016) represents another possibility for initiating chemical reactions, which can be of great practical importance as it can be performed at lower pressures. Water was reported to react in the ice VII phase region when exposed to X-ray radiation above 2.5 GPa, forming an alloy of H_2 and O_2 – a somewhat puzzling result that still awaits further experimental confirmation and comprehensive theoretical explanation; the tentative mechanism of this reaction is presented in the work of Liang and Tse (2009). The Raman spectra and XRD of the reaction products and of the H_2–O_2 compound crystallized from molecular precursors (Loubeyre and LeToullec, 1995) differ substantially. Better understanding of high-pressure behavior of these and other inclusion compounds formed of simple molecules (for example, N_2–O_2; Baer and Nicol, 1990; Sihachakr and Loubeyre, 2004) would serve greatly for modeling of planetary interiors and synthesis of new energetic compounds.

Finally, it is worth noting that the recent discovery of a family of high-temperature superconductors is intriguingly related to the behavior of simple molecular materials at high pressure. The 200 K H_3S superconductor (at 150 GPa) (Drozdov et al., 2015) was synthesized from molecular H_2S. Later, the same cubic H_3S was synthesized from molecular H_2 and S (Goncharov et al., 2017). Interestingly, a molecular system $(H_2S)_2H_2$ (and a Se substituted system, $(H_2Se)_2H_2$; Pace et al., 2017; Zhang et al., 2018) synthesized at low pressures from H_2 and H_2S (Strobel et al., 2011) and from the elements (Goncharov et al., 2017; Pace et al., 2020) has the same composition as H_3S. The high-pressure behavior of this system was outlined theoretically by (Duan et al., 2014) including the emergence of high-temperature superconductivity. Recently, (Snider et al., 2020) reported room temperature superconductivity in a similar system but doped with carbon. Moreover, superhydrides of metals revealing very high superconducting temperatures (e.g., of La [Drozdov et al., 2019; Somayazulu et al., 2019] and Y [Kong et al., 2019]) have clathrate structures, where "guest" metals are hosted by atomic hydrogen cages (Liu et al., 2017; Peng et al., 2017; Struzhkin et al., 2020).

14.3.2 Rich Nitrogen Polymorphism

Molecular nitrogen is an archetypal diatomic molecule. The intramolecular covalent bonds generally destabilize under pressure because the electrons tend to find the states with the minimum kinetic energy, which increases very steeply for the bonds with high electron density (e.g., Hemley, 2000). This is exemplified in the triple-bonded N_2 molecule, which results in stabilization of a single-bonded polymeric nitrogen at high pressure (McMahan and LeSar, 1985; Martin and Needs, 1986; Mailhiot et al., 1992). It is interesting that N is the only group V element forming a diatomic molecule; phosphorous (P) is normally polymeric at ambient conditions except metastable P_4 modification (Seitz et al., 2018). Polymerization of molecular nitrogen has been predicted at rather moderate pressures (e.g., 50 GPa; Mailhiot et al., 1992), which seem to corroborate well with the shock experiments that detected such transition at around 30 GPa and 6,000 K (Nellis et al., 1984). However, it was realized that the transition pressure strongly depends on the structure of molecular and nonmolecular nitrogen allotropes (e.g., Pickard and Needs, 2009). The phase diagram of nitrogen is extremely rich for the molecular and nonmolecular parts (Figure 14.3), as there are numerous *P–T*-induced phase transformations, and many of them are kinetically controlled.

The phase diagram of nitrogen is especially complicated in the region where the molecular phases are stable. There are 12 molecular phases that have been reported and numerous theoretically predicted phases, many of which have not been confirmed (see Turnbull et al., 2018, and references therein). At room temperature, fluid N_2 crystallizes in the β-phase at 2.5 GPa, which is also stable between 36–63 K at ambient pressure. β-N_2 forms a hexagonal closed packed (hcp) lattice of molecular centers, and the molecules are rotationally disordered (e.g., Schuch and Mills, 1970). Other low-temperature N_2 phases include ordered cubic α-N_2 (Donohue, 1961) (below 36 K at $P \leq 0.6$ GPa) and ordered tetragonal γ-N_2 (Schuch and Mills, 1970) (at low T and 0.6 GPa $<P<$2.3 GPa). At 5.4 GPa (300 K), β-N_2 transforms to δ-N_2, which becomes contiguous with the fluid phase above 10 GPa. δ-N_2 is also orientationally disordered, but it has a cubic lattice with molecules occupying two distinct crystallographic positions with different patterns of rotational

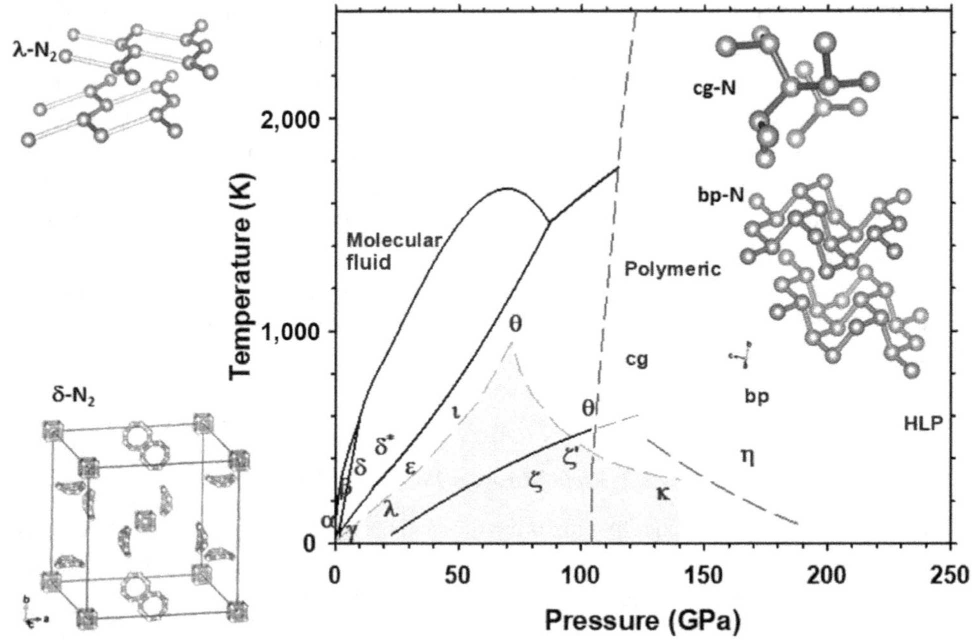

Figure 14.3 Phase diagram of nitrogen. The phase boundaries and phase occurrence domains are from Bini et al. (2000); Goncharov et al. (2000); Eremets et al. (2001), 2004); Gregoryanz et al. (2001, 2002, 2007); Tomasino et al. (2014a, 2014b); Frost et al. (2016); Weck et al. (2017); Turnbull et al. (2018); Laniel et al. (2019, 2020); Ji et al. (2020). The light gray area corresponds to the stability field of λ-N_2 (Frost et al., 2016). The insets show the structural motifs of δ-N_2, λ-N_2, cg-N, and bp-N.

motion: sphere- and disc-like (Mills et al., 1986; Hanfland et al., 1998). It was reported to transform to a tetragonally distorted δ^*-N_2 (Stinton et al., 2009) at 11 GPa (see also Scheerboom and Schouten, 1993; Hellwig et al., 2001). At 17 GPa, δ^*-N_2 transforms into a rhombohedral ε-N_2 (Mills et al., 1986). This sequence of the phase transformations (see also Nosé and Klein, 1983; Goncharov and Gregoryanz, 2005) is consistent with a stepwise orientational ordering, which includes different molecular crystallographic sites (Hellwig et al., 2001). Upon further compression, other distorted phases emerge, as evidenced by splitting of some peaks in powder diffraction patterns, interpreted as transformations to orthorhombic ζ-N_2 and monoclinic κ-N_2 (Gregoryanz et al., 2007). However, the structures of these phases were not determined. Above 150 GPa, κ-N_2 transforms to an amorphous nonmolecular phase η-N as evidenced by vibrational spectroscopy (Goncharov et al., 2000; Eremets et al., 2001) and XRD (Gregoryanz et al., 2007).

It is instructive to mention that none of the molecular structures that are observed experimentally at room temperature in the sequence $\delta\leftrightarrow\varepsilon\leftrightarrow\zeta\leftrightarrow\kappa$ (the transitions are reversible) is predicted to be the stable phase according to first-principles theoretical calculations (Pickard and Needs, 2009; Turnbull et al., 2018). However, numerous experiments have shown these structures' stability and reversibility at room and low temperatures (Bini et al., 2000; Goncharov et al., 2001). The only exception is some hysteretic behavior for the $\varepsilon\leftrightarrow\zeta$

transition that has been reported at low temperatures (Bini et al., 2000). The latter experiments were performed upon cooling down along quasi-isobars up to 41 GPa, while those reported by (Goncharov et al., 2001) were performed along the pressure ramp up at low temperatures (15–30 K) from 17.5–44 GPa, yielding different results at low temperatures.

It is the discovery of new classes of structures, when performing the experiments at various *P–T* pathways (Gregoryanz et al., 2002; Frost et al., 2016), which makes the phase diagram of nitrogen so intriguing and yet not fully understood. The Raman experiments in resistively heated DAC at high temperatures up to 850 K (Gregoryanz et al., 2002) discovered two new phases, ι and θ, at 70 and 95 GPa, respectively (cf. Tomasino et al., 2014a). These phases can be quenched to room and low temperature, and they are even stable upon the pressure release, but they transform back to ε-N_2 at 23 and 30 GPa, respectively. Most surprisingly, yet another molecular phase has been synthesized while compressing α- and β–N_2 at temperatures below 90 K (Frost et al., 2016). The new phase λ appears at as low as 2 GPa – the *P–T* conditions thought to be in the stability region of ε(δ,γ)–N_2. Once synthesized at low temperatures, λ–N_2 is stable over an exceptionally wide *P–T* range from 32–140 GPa at room temperature and below a steeply increasing *P–T* line (where it transforms to ε–N_2 and θ–N_2), which is some 50–200 K below the δ↔ε phase line; the latter ends in the triple point with the fluid (Goncharov et al., 2008). λ–N_2 transforms to an amorphous nonmolecular nitrogen above 140 GPa at room temperature, quite similar to κ–N_2. Moreover, λ–N_2 has been shown to exist down to 0.3 GPa at 10 K. This shows that this phase is metastable over the region of *P–T* space, where the ε, ζ, κ, ι, γ, and θ molecular phases (and polymeric phases, as discussed later in this chapter) of nitrogen are also observable, thus underscoring the difficulties in establishing the stability fields in the phase diagram of nitrogen.

These new ι, θ, and λ phases have very distinct Raman spectra, characterized by only one very strong Raman vibron mode, unlike δ– and ε–N_2, where there are two major vibrons due the existence of two very different crystallographic positions for the molecules. This was interpreted as a sign that the molecules of a single type occur in these structures (cf. Turnbull et al., 2018, for ι–N_2, as discussed later), while weaker additional Raman modes arise from the vibrational splitting due to the presence of many molecules in the unit cell, which are coupled via the intermolecular forces. Moreover, these new phases have very distinct low-frequency lattice translational and librational modes. It is interesting that the vibrational spectra of θ and λ phases are very similar (several weaker extra Raman bands are observed in θ–N_2) (Frost et al., 2016), which suggests that the structures are similar. Perhaps one structure is characterized by the space group, which is a subgroup of another and an extended unit cell. It is very surprising that these phases are synthesized at very different *P–T* conditions; also, λ–N_2 transforms to θ–N_2, at 70 GPa and 750 K, and the latter one can be quenched to room temperature (Frost et al., 2016). Finally, θ– and λ–N_2 demonstrate a weakened intramolecular strength at high pressures (compared to ε, ζ, κ phases), as evidenced by a lower frequency of the main Raman vibron mode (Gregoryanz et al., 2002; Frost et al., 2016), suggesting the proximity of molecular dissociation.

The direct structural characterization of high-pressure phases of nitrogen remains a challenge. Nitrogen scatters X-ray weakly, and the samples are very small at high pressures. On the other hand, the Compton scattering from diamonds is rather strong, thus detecting a powder diffraction signal is difficult. Moreover, the samples typically develop a strong texture (preferred orientation), further making structural investigations problematic as the use of a conventional Rietveld refinement technique becomes very difficult at pressures above approximately 20 GPa. This can be overcome by using the single-crystal XRD technique in the DAC as was reported for δ and δ^* (Stinton et al., 2009), and ι phases (Turnbull et al., 2018).

Structure determination of molecular $\iota{-}N_2$ at 56 GPa (Turnbull et al., 2018) is a remarkable result, demonstrating the strength of the single-crystal XRD technique. The structure features a monoclinic unit cell with a volume of 765 Å^3 that contains 48 molecules. Surprisingly, the structure refinement yields a large range of the intramolecular bond lengths from 0.88–1.13 Å, which is highly unusual and seems to reflect the uncertainties in the structure refinement procedure. Perhaps surprising is that the density functional theory (DFT) calculations show that this complex structure does not seem to represent a great enthalpy gain, so other predicted but not realized structures (e.g., $P4_12_12$; Pickard and Needs, 2009) remain substantially more energetically favorable. The DFT calculations show that $\iota{-}N_2$ is more stable than $\varepsilon{-}N_2$ above 20 GPa at room temperature, which broadly agrees with experiments.

Moving to nonmolecular (polymeric) nitrogen phases, we first notice that the discrepancy between the calculated and observed transition pressure remains not well understood (Figure 14.3). It was proposed that with the discovery of more stable molecular phase(s), the low calculated transition pressure (e.g., 50 GPa; Mailhiot et al., 1992) can be shifted up closer to the experimentally observed direct transition of amorphous nonmolecular phase η–N to cg–N at about 110 GPa and above 2,000 K (Eremets et al., 2004). Initially, this discrepancy seems to be even larger as room and low-temperature compression experiments found a transition into an amorphous semiconducting nonmolecular nitrogen at much higher pressures of 150 GPa at 300 K (Goncharov et al., 2000) and 190 GPa at 80 K (Eremets et al., 2001). The transition to nonmolecular phases is kinetically hindered because of the energy barrier, which needs to be overcome upon the synthesis. Applying high temperatures is the most natural strategy; however, photochemistry could be also a possibility (Popov, 2005). The high-temperature experiments with resistive and laser heating show that the most favorable nonmolecular nitrogen in a cubic gauche (cg) form cannot be synthesized below 110 GPa, and the synthesis occurs via η–N, which forms first on heating of molecular ζ– or $\kappa{-}N_2$ (Gregoryanz et al., 2001; Lipp et al., 2007) (Figure 14.3). These results have been recently confirmed by another experiment, where cg–N was synthesized from a fluid phase by applying pressure at high temperatures (Weck et al., 2017); the results agree with the high-pressure location of the phase line of transformation to a polymeric state. On the other hand, the work of Pickard and Needs (2009) and Turnbull et al. (2018) extensively discuss this discrepancy and note that a very large correction in enthalpy of the most stable molecular phase is needed (0.4 eV per atom) to make the theory and experiments agree. Such correction is much larger than the uncertainties due to various approximations, which are used in the theoretical calculation.

Polymorphism of polymeric nitrogen represents another interesting topic, which has recently progressed due to the development of experimental methods. Previously, only theoretical calculations have been able to address this issue. Early calculations (McMahan and LeSar, 1985; Martin and Needs, 1986), which did not have effective tools for searching for the lowest enthalpy structures, considered the structures of polymeric phosphorus (P) and arsenic (As) from the same group in the periodic table as the most plausible candidates. However, for nitrogen, which does not have *p*-electrons in its ionic core, it appears that a different structure, namely cg–N, is uniquely stable (Mailhiot et al., 1992). Numerous subsequent calculations predicted many more polymeric nitrogen structures (Kotakoski and Albe, 2008; Ma et al., 2009; Pickard and Needs, 2009; Wang et al., 2012). A layered structure of black phosphorous (BP) was thought to be a candidate for higher-pressure polymeric nitrogen >200 GPa (e.g., Kotakoski and Albe, 2008), but other calculations found more stable phases. One of these structures, a layered orthorhombic *Pba*2 polymeric nitrogen, was predicted to be stable between 188–320 GPa (Ma et al., 2009). In 2014, Tomasino et al. (2014b) reported the synthesis of a new polymeric nitrogen that has a very strong and distinct Raman spectrum; they called it LP–N. The reported XRD was indexed using the *Pba*2 structure, which was predicted theoretically (Ma et al., 2009). However, there were extra XRD peaks that were assigned to another metastable polymeric *C*2/*c* phase (Yao et al., 2008). The laser-heating experiments to approximately 3,000 K between 126–175 GPa yielded a mixture of η, cg, and LP–N, while at lower pressures only cg–N has been detected; this suggests that LP–N is stable at higher pressures than cg–N. Given the observed coexistence of both polymeric phases, it is very surprising that these modifications have very different densities deviating by about 8.2% at 112 GPa (Tomasino et al., 2014b). However, the structural determination of LP–N (Tomasino et al., 2014b) has been recently challenged when two independent groups reported that this phase has the BP structure (Ji et al., 2020; Laniel et al., 2020), which is only about 2% denser than cg–N at the same pressure. Also, the phase stability of cg–N and LP–N was recently examined by Raman spectroscopy (Cheng et al., 2020), with the conclusion that synthesis of LP–N (which appears to be bp–N) is more favorable at higher pressures and higher temperatures compared to cg–N.

Recently yet another structural modification of polymeric nitrogen has been discovered above 240 GPa (Laniel et al., 2019). It is remarkable that laser heating of amorphous η–N above 180 GPa up to this pressure did not yield any crystalline phase; it was proposed that this is because of the very close proximity in enthalpy of several phases in this pressure range. Laser heating to 3,300 K at 240 GPa results in synthesis of a crystalline phase (called HLP–N) which was identified as the previously predicted *P*42*bc* single-bonded layered polymeric nitrogen (Wang et al., 2012) based on the excellent agreement of XRD patterns and Raman spectra. This also appears to be a very surprising result, as the *Pba*2 structure of LP–N was expected to remain the most stable up to 263 GPa, while above this pressure a unique N_{10} cage-like diamondoid structure was predicted (Wang et al., 2012), which is believed to represent another level of densification based on the most efficient packing of N_{10} cages. The latter can be considered a pseudoatom due to its compactness resulting from the shape and short intracage N–N distances. This structure, which is expected to be the most stable up to 700 GPa, still needs to be discovered.

14.3.3 Dense Ices: Symmetrization of Hydrogen Bonds

The high-pressure behavior of ice is fundamental to addressing a wealth of problems in chemistry and physics under extremes. In particular, ice is the archetypal system for elucidating the properties of hydrogen bonds and hydrogen-bonded networks at high pressures, which reveal a great variety of different configurations at low pressures. This configurational richness results in the great complexity in the phase diagram of water below 2 GPa, with some 17 crystalline and three amorphous structures that have been documented (Salzmann et al., 2009; Falenty et al., 2014; del Rosso et al., 2016). However, this complexity was thought to disappear above 2 GPa, where oxygen atoms form a *bcc* cubic lattice, which nearly perfectly accommodates the optimal linear hydrogen bonds and creates two interpenetrating hydrogen-bonded networks. The locations of hydrogen atoms and hydrogen bonds in this ice VII structure is believed to be disordered, but it obeys the ice rule: Each tetrahedrally coordinated oxygen atom has two hydrogen atoms closer than the other two forming the water molecule. The questions about the geometry of the hydrogen bond, on whether oxygen atoms permanently reside in the high-symmetry positions and whether this configuration is pressure dependent, remain unclear (Nelmes et al., 1998; Guthrie et al., 2019). It is expected that under pressure the average hydrogen position drifts toward the oxygen atom of the neighboring molecule as it approaches, and at sufficiently high compression the hydrogen position becomes centered dynamically and eventually statically, forming ice X with symmetric hydrogen bonds (Kamb and Davis, 1964; Holzapfel, 1972). The DAC XRD experiments indicate that the oxygen *bcc* lattice persists up to at least 170 GPa (Hemley et al., 1987; Loubeyre et al., 1999; Sugimura et al., 2008). However, there is a great disparity concerning the hydrogen positions and their symmetrization depending on the experimental technique.

Addressing the issues of the evolution of the hydrogen atom positions and dynamics at very high pressures is extremely challenging experimentally. Hydrogen atoms are difficult to locate using "direct" techniques such as XRD because of their extremely low X-ray scattering efficiency. This can be mitigated by using single-crystal diffraction. In the work of Loubeyre et al., (1999), they applied this technique in the energy dispersive mode to ice VII and followed the intensity ratio of the (111) and (222) Bragg peaks, which is expected to be sensitive to the positions of the hydrogen atoms. The results suggest that ice X appears at as low as 30 GPa, but the data are inconclusive given the data scatter and a discrepancy for different samples. The diffraction peaks, which are used in this analysis, are weak, so their integrated intensities are subject to large uncertainties. This can be drastically improved by using neutron diffraction, especially in the case of deuterated ice; however, this technique requires much larger samples. In two works of Guthrie et al. (2013, 2019), they applied this technique to very high pressures up to 60 GPa, generated in a DAC with conical anvils with the culets of 1–1.65 mm in diameter that use a strong side support (Boehler et al., 2013). However, the results, which include the refined D–O distances, do not show that D_2O symmetrizes up to 60 GPa (Guthrie et al., 2019). These experiments remain very challenging in reaching the desired pressure with a sufficient amount of material for reliable data collection, and in the data analysis, which requires careful calibration procedures to extract the Bragg intensities.

Since the "direct" measurements are scarce, the alternative methods become very valuable; however, these require careful evaluation of their deliverables. Measurements of elasticity could provide, in principle, an independent probe of the transition into a nonmolecular state. However, accurate measurements of elasticity even for cubic ices VII and X require the use of oriented single crystals. The experiment on single crystal ice VII samples are limited to pressures below 8 GPa (Shimizu et al., 1995); investigation to higher pressures suffer from the uncertainties related to a random orientation of the crystallites (but often developing a preferred orientation, a texture). This results in rather large uncertainties because the sound velocities strongly depend on the direction of propagation.

In the work of Polian and Grimsditch (1984), they measured the longitudinal sound velocity of ice at high pressure to 67 GPa and found an anomaly at 44 GPa, which they assigned to the transition to ice X. More recently, Ahart et al. (2011) performed the experiment to 100 GPa and found no anomaly. This discrepancy was explained by a difference in the pressure scales that have been used. Additionally, Ahart et al. (2011) found no anomalies for the transverse velocity. Theory predicts that only one acoustic branch at the Brillouin zone boundary is unstable upon the transition to ice X (Caracas, 2008). This makes the coupling to the Brillouin zone center acoustic phonons unlikely, so no anomaly is expected to be observed for the transverse acoustic modes. Kuriakose et al. (2017b) applied a time-domain laser acoustic technique (time-domain Brillouin scattering), which probes crystallites of submicrometer dimensions (thus improving the accuracy statistically) to study water up to 82 GPa. The authors concluded that their experiments have not reached ice X by comparing the experimental elastic parameters (which they derived using the data of other measurements, such as the EoS and the refractive index) and the theoretically calculated ones for ice X. Obviously, much better criteria should be used to detect and identify the transition in this case, where there is no abrupt change in the bulk physical properties of the material, such as the specific volume (cf. Sugimura et al., 2008).

It is somewhat perplexing that another technique, nuclear magnetic resonance (NMR), which is expected to be sensitive to the electronic and structural environments of the nuclei, revealed largely inconsistent results. In the work of Meier et al. (2018), they performed experiments on H_2O ice up to 97 GPa and found a splitting of the proton signal and a pronounced tunneling mode above 20 GPa (cf. theoretical investigations; Benoit et al. 2002). However, there is no abrupt change in the spectra up to the highest pressure reached except a small chmage in the chemical shift at 75 GPa, and the resonance tunneling persisted, making it difficult to definitively identify ice X.

Vibrational spectroscopy is often considered an indirect tool to probe the structure. However, this is a common misconception. Each experimental technique probes certain physical quantities directly reflecting specific properties of the system. In fact, vibrational spectroscopy is extremely useful for probing lattice/molecular order/disorder, and it becomes indispensable for assessing hydrogen and other bond strengths and their dynamics. Even if the direct measure of the atomic positions cannot be obtained via vibrational spectroscopy, the information about the vibrational, intra-, and intermolecular stretching vibrational frequencies is definitive in determining the chemical structure of the molecular crystal. Moreover, an important feature of vibrational spectroscopy is that it captures the

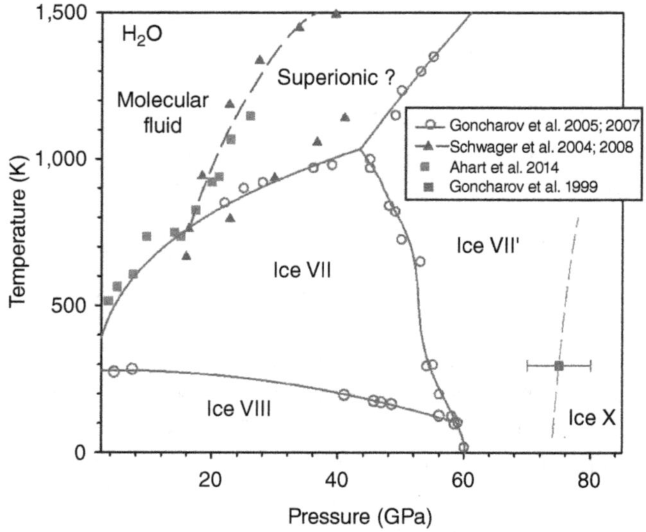

Figure 14.4 Phase diagram of water from Raman and Brillouin spectroscopy and visual observations. The phase boundaries are from Goncharov et al. (1999, 2005); Schwager et al., 2004; Goncharov and Crowhurst (2007); Schwager and Boehler (2008); Ahart et al. (2014). The location of the melt line above 18 GPa as well as the stability field of superionic water remain controversial due to the difficulties in direct identification of these states.

instantaneous molecular configurations, while XRD and NMR provide the time-averaged picture. In addition, vibrational spectroscopy is compatible with very small samples and thus can be used at very high pressures.

Hydrogen bond symmetrization is a casebook example illustrating the power of vibrational spectroscopy techniques. The principal observation is that the O–H intramolecular modes, detected by Raman and IR spectroscopy, reveal a strong softening and broadening with pressure and a new set of hardening vibrational excitations appear above the transition (Aoki et al., 1996; Goncharov et al., 1996, 1999; Goncharov and Crowhurst, 2007). Additional observations that accompany this major event include the mode coupling of the O–H stretch vibration with other fundamentals and lattice modes, asymmetric line shapes, and the existence of a pressure range (56–75 GPa) where the system response is highly damped (peaks are tremendously broadened) and is barely changing. These results clearly point to the scenario proposed theoretically (e.g., Benoit et al., 2002), where dynamical translational proton disorder occurs as an intermediate stage on the way to the formation of an atomic solid with symmetric hydrogen bonds, called ice VII′ (Goncharov and Crowhurst, 2007) (Figure 14.4). This intermediate state is characterized by the presence of ionic defects, which results in a violation of the ice rule. This state is similar to superionic water (Cavazzoni et al., 1999) in that the oxygen atoms continue forming an ordered lattice, but it is different in the proton dynamics, which shows a complete dissociation of water molecules and a very rapid proton diffusion in superionic state. The states with mobile hydrogen atoms are difficult to distinguish by XRD, NMR, and other techniques, including vibrational spectroscopy. The soft O–H stretching mode is strongly overdamped and is not clearly seen by Raman in this regime, while the IR O–H stretch mode shows a very strong oscillator strength at nearly zero frequency (below the IR spectroscopy observation limit of 200 cm^{-1} in DAC experiments; Goncharov and Crowhurst, 2007).

The existence of the intermediate dynamically disordered phase (ice VII′) stems from the quantum nature of the protons, whose behavior can be qualitatively understood using a 1D two-well potential model (Holzapfel, 1972). Upon compression, the potential barrier between the two equilibrium proton positions decreases and the regime of quantum tunneling is realized. At this stage, there is a major change in the spectra of the quantum states, which experiences a large splitting of the first excited energy level (when it reaches the barrier top) and a tunneling splitting of the ground level. This likely occurs near 20 GPa and results in profound changes in vibrational and NMR spectra. When the barrier height reaches the ground level, the proton can move freely between the wells and the dynamic disordered state is realized. The proton probability distribution is bimodal at this stage, and the vibrational selection rules are largely relaxed allowing the observation of vibrations at various Brillouin zone points. The lattice is stabilized via anharmonicity, which is evidenced by a tremendous vibrational peak broadening. The proton probability distribution becomes single modal once the potential barrier is below the ground energy level. At this stage, a new system of pressure hardening optical phonons appears, of which the Raman T_{2g} mode is the most prominent (Goncharov et al., 1999), representing the O–O vibrations, while hydrogen modes are Raman inactive in the ice X structure (Figure 14.4). No volume discontinuity is expected upon this transition; only a change in slope in V(P) dependencies (Sugimura et al., 2008) and discontinuous changes in the elastic moduli (Kuriakose et al., 2017a) are anticipated. This simple scenario can be examined by performing control experiments on D_2O, where the quantum effects must be substantially suppressed. Unfortunately, very limited experimental data exist for D_2O: IR spectroscopy investigation reported the transition to ice X at 70–90 GPa (Goncharov et al., 1996; Song et al., 1999), and neutron diffraction data to 60 GPa (Guthrie et al., 2019) revealed no transition.

Ice X is expected to be stable up to 300–400 GPa, where it is theoretically predicted to transform to a *Pbcm* orthorhombic structure with a distorted oxygen hcp sublattice (Benoit et al., 1996; Caracas, 2008) due to a phonon instability at the M-point of the Brillouin zone. It is interesting that *Pbcm* ice is expected to have symmetric hydrogen bonds, while this configuration is lost in $Pmc2_1$ (800–950 GPa) and $P2_1$ (1.17 TPa) structures, but symmetric hydrogen bonds are eventually regained at 5.62 TPa in a layered *C2/m* ice, which is metallic (McMahon, 2011) (see also Ji et al., 2011) (Figure 14.5). However, Pickard et al. (2013) reported that the H_2O composition becomes thermodynamically unstable at nearly 5 TPa, decomposing to H_2O_2 and a hydrogen-rich phase preventing metallization of H_2O system.

14.3.4 Squeezing Hydrogen into Exotic States

Hydrogen as element number one is the simplest element and the most abundant material in the universe. However, the high-pressure behavior and phase diagram of hydrogen are quite complex and have been the subject of continuous investigation. Experiments are very

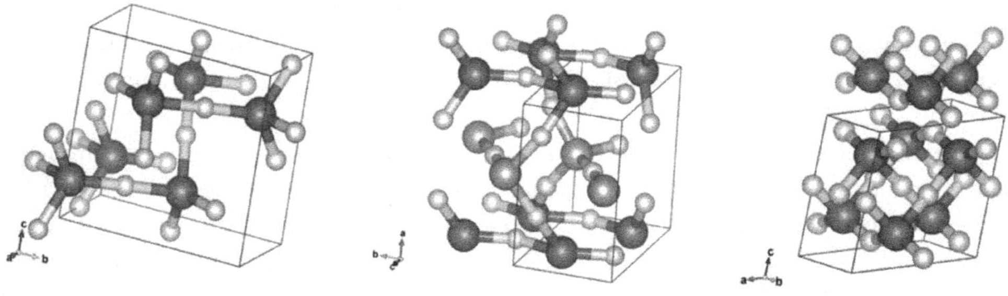

***Pbcm*, 0.4 TPa** ***Pmc*2$_1$, 1 TPa** ***C*2/*m*, 5 TPa**

Figure 14.5 Theoretically predicted phases of H_2O ice at very high pressure (after McMahon et al., 2012). *Pbcm* ice has symmetric hydrogen bonds. *Pmc*2_1 structure has oxygen atoms in two different crystallographic positions (shown by dark and light gray spheres), and the hydrogen bonds are distorted. The highest-pressure metallic *C2/m* ice is layered, and it regains symmetric hydrogen bonds within the layer.

challenging, as hydrogen is difficult to contain, and very high pressures are needed to reach the conditions for transformation from a diatomic molecular to a monatomic metallic state.

It is the simplicity of the electronic and molecular structure that makes the study of hydrogen at very high pressures fundamental for our understanding of chemical bonding and physical properties in materials under extremes (Mao and Hemley, 1994; McMahon et al., 2012; Goncharov et al., 2013; Goncharov, 2020; Gregoryanz et al., 2020). Of particular interest is metallization of hydrogen and, related to this, possible superconductivity and superfluidity (Ashcroft, 1968; Babaev et al., 2004). Hydrogen was predicted to metallize in the solid state at 25 GPa by Wigner and Huntington in 1935 (Wigner & Huntington, 1935), and until now this topic remains one of the most exciting ones in high-pressure research. Major efforts have been invested in establishing the transformation pathway and the intermediate phases into a monoatomic metal state (which undoubtedly exists), but these remain puzzling for both theory and experiment. Performing high-pressure experiments that can yield information about the physical and chemical structure and electronic properties is a necessary first step to address this issue.

Dave Mao pioneered high-pressure static compression work on hydrogen at the Geophysical Laboratory (Mao and Bell, 1979). After establishing cryogenic loading technique (Mao and Bell, 1979), the next step was to apply the most powerful techniques to investigate the high-pressure behavior of hydrogen. In 1985, Raman spectroscopy measurements to 147 GPa revealed a turnover of the internal H–H stretching mode (vibron), which was interpreted as a "harbinger" of molecular dissociation (Ashcroft, 1990). Infrared synchrotron spectroscopy at the National Synchrotron Light Source (NSLS) (Hanfland et al., 1992) was yet another important pillar of this extremely challenging research program, which has continued up to now (Zha et al., 2014; Goncharov et al., 2019). In 1988, Hemley and Mao (1988) discovered a new phase of solid hydrogen at 150 GPa and 77 K, phase III, and this finding established the high-pressure phase diagram of hydrogen for the next 24 years until a new mixed-molecular phase IV was reported in 2012 at 220 GPa and 300 K (Howie et al., 2012) (Figure 14.6). Phase III revealed an

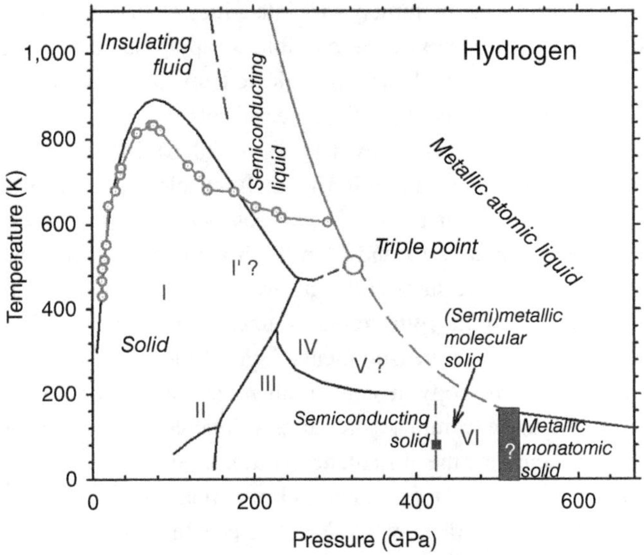

Figure 14.6 Phase diagram of hydrogen at high pressure and various temperatures. The solid black lines indicate the phase boundaries between the solid molecular phases and the melt line from Goncharov et al. (2011); Howie et al. (2015); Eremets et al. (2019), and the dashed line are the extrapolation of the melt line to higher pressures. An alternative and substantially different set of measurements of the melt line by Zha et al. (2017) is depicted by open gray circles and solid gray lines. A solid dark square corresponds to an IR bandgap closure reported by Loubeyre et al. (2020). A vertical dashed dark line is a proposed associated phase line between a semiconducting and semi(metallic) molecular phase also probed via the electrical conductivity by Eremets et al. (2019). A hypothetical transition to an atomic metallic phase is shown by a dark gray box. At higher pressure, theory predicts an atomic metallic phase with a declining melt line shown by a solid line (Chen et al., 2013). At high temperature, the experiments and theory show two almost parallel boundaries corresponding to a transition into a semiconducting state (dashed line) and insulator-metal transition (solid line) (see Goncharov, 2020, for more details).

unusually strong IR activity (Hanfland et al., 1993), suggesting spontaneous molecular polarization (Edwards and Ashcroft, 1997), which was later understood as a dynamic effect that appears when the atoms in the molecule have asymmetric crystallographic positions as in phase III (Pickard and Needs, 2007). Recent IR spectroscopy experiments indicate that phase III is stable to 420 GPa (Loubeyre et al., 2020).

The principal development in high-pressure investigations of hydrogen would be an ability to obtain high-quality structural data at very high pressures, where we currently infer the structure from theoretical calculations (Pickard and Needs, 2007; Drummond et al., 2015; Monserrat et al., 2016, 2018) and vibrational spectroscopy. For a long time, the structural information was only available for phase I of H_2, which crystallizes in an hcp structure (Hazen et al., 1987), where the molecules can freely rotate (plastic phase) (Figure 14.6). Dave Mao's contributions into the development of synchrotron X-ray methods were tremendous, resulting in technical breakthroughs that enabled XRD measurements in hydrogen to megabar pressures. The EoS of phase I of hydrogen was extended up to 120 GPa using a single-crystal approach (Loubeyre et al., 1996). These experiments are very difficult because of the extremely low intensities of the reflections with respect to the background. This makes it next to impossible to apply standard single-crystal XRD methodology, where the full set of data is collected upon a continuous rotation of the sample, followed by the analysis extracting the subsets corresponding to the crystal of interest and solving the structure (Dubrovinsky et al., 2020). Instead, one searches for the strongest Bragg peaks and collects them individually at selected geometries. This restricts this research to the simplest structural solutions, as demonstrated in recently published structural investigations of phases III and IV of hydrogen (Akahama et al., 2017; Ji et al., 2019). For the same reason, the structure of a low-temperature phase II of hydrogen and deuterium (<160 GPa) (Figure 14.6) remains uncertain (Goncharenko and Loubeyre, 2005), where neutron and XRD measurements suggest the existence of incommensurate structures. Vibrational spectroscopy investigations remain an important tool to address the structural properties of these complex structures (Goncharov et al., 1996, 1998, 2019). Raman and IR spectra capture the instantaneous atomic configurations by observing the characteristic bands (e.g., of vibron modes ν_1 and ν_2 in phase IV of hydrogen (Howie et al., 2012; Goncharov et al., 2013, 2019), while XRD detects the averaged over time electronic density, which results in a puzzling conclusion about phases I, III, and IV being isostructural (Ji et al., 2019) (Figure 14.7).

Finally, I address the widely debated question about metallic hydrogen at very high pressure (see also recent review papers by Goncharov, 2020; Gregoryanz et al., 2020). It is

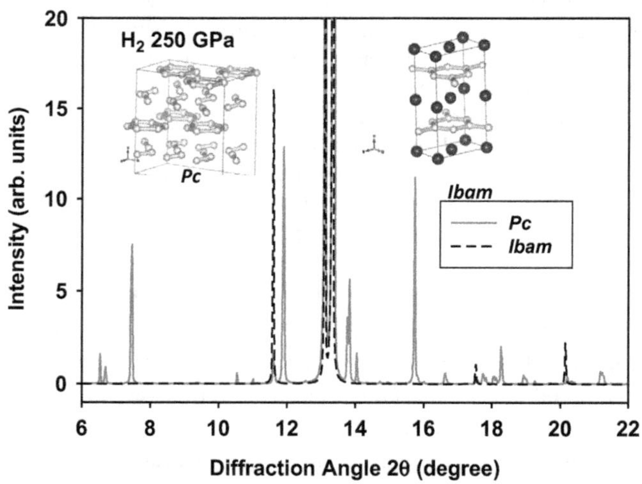

Figure 14.7 High-pressure structures and XRD of phase IV of hydrogen. The modeled XRD patterns (X-ray wavelength is 0.2952 Å) of the theoretically predicted ordered *Pc* hydrogen (Pickard et al., 2012) at 250 GPa (gray line and left inset) and a model structure, where the fluxional behavior of hydrogen atoms is time-averaged, yielding an *Ibam* structure (dashed line and right inset; dark spheres correspond to the centers of mass of the short H_2 molecules). XRD of this latter structure has fewer lines than that of *Pc*, and it is almost indistinguishable from an hcp lattice of freely rotating molecules (phase I) (Goncharov et al., 2019). Future high-quality XRD experiments will examine the effect of dynamic hydrogen modeled here.

very tempting to declare that a metallic state is reached based on a suggestive set of data, as has occurred a number of times over the last two decades or so (Hemley and Mao, 1989; Eremets & Troyan, 2011; Dias and Silvera, 2017). However, even if these speculative observations are valid, these results are not instrumental in addressing the overarching questions about the metallization of molecular systems under pressure, because characterization of presumably metallic states was essentially missing in the aforementioned works (the list is not exhaustive). Moreover, these works suffer from methodological problems (e.g., pressure measurements) and lack of advanced, reliable probes (Goncharov and Struzhkin, 2017; Liu et al., 2017). Presently, we are witnessing a time of tremendous progress in DAC technologies and in technical development of DAC probes that have enabled what would have been impossible in the previous experiments. Recent publications by Eremets et al. (2019) and Loubeyre et al. (2020) present reliable data on electrical and optical conductivity of hydrogen up to 480 GPa that reveal a bandgap narrowing in phase III of hydrogen, which can be accompanied by a phase transition. These results strongly indicate an indirect bandgap closure in molecular hydrogen above 400 GPa, which is broadly consistent with the theoretical predictions. These works represent an important step in understanding of high-pressure metallic hydrogen.

14.4 Outlook

Dave Mao's contributions to DAC technology and high-pressure science cover many aspects of technical developments, experimental methodology, and training of young scientists. High-pressure physics and chemistry are now blossoming fields with many laboratories and research centers worldwide and a large scientific and technical community involved in the research of materials under extremes. Research of molecular crystals and compounds is an important and visible part of this activity, because possible implications of this research cover a wide range of disciplines such as fundamental physics and chemistry, materials synthesis, defense and energy technologies, and planetary sciences. DAC technology has not yet reached its limits in achieving extreme *P–T* and strain-rate conditions, and the community is continuing to work on optimizing anvil design (Dewaele et al., 2018; Jenei et al., 2018), refining time-domain experiments, improving micromachining tools for advanced sample preparation, and other developments. Future research will likely combine experimental static and dynamic high-pressure techniques in reaching extreme conditions, which are poorly controlled and not reliably reproducible now. These developments will enable reaching currently unattainable extreme conditions and improve our understanding of the physical and chemical processes under such extremes. Moreover, there are opportunities to integrate experiments and theory more closely to inform experimental design, modeling of *P–T* conditions, data reduction, and interpretation.

Acknowledgments

I thank Wendy Mao, John Tse, Ross Howie, and Yingwei Fei for the excellent comments on the manuscript.

References

Ahart, M., Karandikar, A., Gramsch, S., Boehler, R., Hemley, R. J. (2014). High P–T Brillouin scattering study of H_2O melting to 26 GPa. *High Pressure Research*, **34**(3), 327–336.

Ahart, M., Somayazulu, M., Gramsch, S. A., Boehler, R., Mao, H.-k., Hemley, R. J. (2011). Brillouin scattering of H_2O ice to megabar pressures. *Journal of Chemical Physics*, **134**(12), 124517.

Akahama, Y., Mizuki, Y., Nakano, S., Hirao, N., Ohishi, Y. (2017). Raman scattering and X-ray diffraction studies on phase III of solid hydrogen. *Journal of Physics: Conference Series*, **950**, 042060.

Aoki, K., Yamawaki, H., Sakashita, M., Fujihisa, H. (1996). Infrared absorption study of the hydrogen-bond symmetrization in ice to 110 GPa. *Physical Review B*, **54**(22), 15673–15677.

Ashcroft, N. W. (1968). Metallic hydrogen: a high-temperature superconductor? *Physical Review Letters*, **21**(26), 1748–1749.

Ashcroft, N. W. (1990). Pairing instabilities in dense hydrogen. *Physical Review B*, **41**(16), 10963–10971.

Babaev, E., Sudbø, A., Ashcroft, N. W. (2004). A superconductor to superfluid phase transition in liquid metallic hydrogen. *Nature*, **431**(7009), 666–668.

Baer, B. J., Nicol, M. (1990). High-pressure binary phase diagram of nitrogen–oxygen at 295 K determined by Raman spectroscopy. *Journal of Physical Chemistry*, **94**(3), 1073–1078.

Bazarkina, E. F., Chou, I.-M., Goncharov, A. F., Akinfiev, N. N. (2020). The behavior of H2 in aqueous fluids under high temperature and pressure. *Elements*, **16**(1), 33–38.

Bell, P. M., Mao, H. K. (1981). Degrees of hydrostaticity in He, Ne, and Ar pressure-transmitting media. *Carnegie Institution of Washington Yearbook*, **80**, 404–406.

Benoit, M., Bernasconi, M., Focher, P., Parrinello, M. (1996). New high-pressure phase of ice. *Physical Review Letters*, **76**(16), 2934–2936.

Benoit, M., Romero, A. H., Marx, D. (2002). Reassigning hydrogen-bond centering in dense ice. *Physical Review Letters*, **89**(14), 145501.

Bini, R., Ulivi, L., Kreutz, J., Jodl, H. J. (2000). High-pressure phases of solid nitrogen by Raman and infrared spectroscopy. *Journal of Chemical Physics*, **112**(19), 8522–8529.

Boehler, R., Guthrie, M., Molaison, J. J., et al. (2013). Large-volume diamond cells for neutron diffraction above 90 GPa. *High Pressure Research*, **33**(3), 546–554.

Brovman, E. G., Kagan, Y., Kholas, A. (1972). Structure of metallic hydrogen at zero pressure. *Soviet Journal of Experimental and Theoretical Physics*, **34**, 1300.

Caracas, R. (2008). Dynamical instabilities of ice X. *Physical Review Letters*, **101**(8), 085502.

Cavazzoni, C., Chiarotti, G. L., Scandolo, S., Tosatti, E., Bernasconi, M., Parrinello, M. (1999). Superionic and metallic states of water and ammonia at giant planet conditions. *Science*, **283**(5398), 44–46.

Ceppatelli, M., Bini, R., Schettino, V. (2009). High-pressure photodissociation of water as a tool for hydrogen synthesis and fundamental chemistry. *Proceedings of the National Academy of Sciences*, **106**(28), 11454–11459.

Chen, J., Li, X.-Z., Zhang, Q., et al. (2013). Quantum simulation of low-temperature metallic liquid hydrogen. *Nature Communications*, **4**(1), 2064.

Chen, L., Matsunami, M., Nanba, T., et al. (2004). Far-infrared spectroscopy on solids under high pressure by infrared synchrotron radiation. Paper presented at the Infrared and Millimeter Waves, Conference Digest of the 2004 Joint 29th International

Conference on 2004 and 12th International Conference on Terahertz Electronics, 2004.

Cheng, P., Yang, X., Zhang, X., Wang, Y., Jiang, S., Goncharov, A. F. (2020). Polymorphism of polymeric nitrogen at high pressures. *Journal of Chemical Physics*, **152**(24), 244502.

Connerney, J. E. P., Benn, M., Bjarno, J. B., et al. (2017). The Juno magnetic field investigation. *Space Science Reviews*, **213**(1), 39–138.

Conway, L. J., Hermann, A. (2019). High pressure hydrocarbons revisited: from van der Waals compounds to diamond. *Geosciences*, **9**, 227.

del Rosso, L., Celli, M., Ulivi, L. (2016). New porous water ice metastable at atmospheric pressure obtained by emptying a hydrogen-filled ice. *Nature Communications*, **7**(1), 13394.

Dewaele, A., Loubeyre, P., Occelli, F., Marie, O., Mezouar, M. (2018). Toroidal diamond anvil cell for detailed measurements under extreme static pressures. *Nature Communications*, **9**(1), 2913.

Dias, R. P., Silvera, I. F. (2017). Observation of the Wigner–Huntington transition to metallic hydrogen. *Science*, **355**(6326), 715–718.

Donohue, J. (1961). A refinement of the positional parameter in a-nitrogen. *Acta Crystallographica*, **14**(9), 1000–1001.

Drozdov, A. P., Eremets, M. I., Troyan, I. A., Ksenofontov, V., Shylin, S. I. (2015). Conventional superconductivity at 203 kelvin at high pressures in the sulfur hydride system. *Nature*, **525**(7567), 73–76.

Drozdov, A. P., Kong, P. P., Minkov, V. S., et al. (2019). Superconductivity at 250 K in lanthanum hydride under high pressures. *Nature*, **569**(7757), 528–531.

Drummond, N. D., Monserrat, B., Lloyd-Williams, J. H., Ríos, P. L., Pickard, C. J., Needs, R. J. (2015). Quantum Monte Carlo study of the phase diagram of solid molecular hydrogen at extreme pressures. *Nature Communications*, **6**(1), 7794.

Duan, D., Liu, Y., Tian, F., et al. (2014). Pressure-induced metallization of dense (H2S) 2H2 with high-Tc superconductivity. *Scientific Reports*, **4**(1), 6968.

Dubrovinsky, L., Dubrovinskaia, N., Katsnelson, M. I. (2020). No evidence of isostructural electronic transitions in compressed hydrogen. arXiv:1910.10772v1 [cond-mat.mtrl-sci].

Edwards, B., Ashcroft, N. W. (1997). Spontaneous polarization in dense hydrogen. *Nature*, **388**(6643), 652–655.

Efimchenko, V. S., Kuzovnikov, M. A., Fedotov, V. K., Sakharov, M. K., Simonov, S. V., Tkacz, M. (2011). New phase in the water–hydrogen system. *Journal of Alloys and Compounds*, **509**, S860–S863.

Eremets, M., Hemley, R., Mao, H.-k., Gregoryanz, E. (2001). Semiconducting non-molecular nitrogen up to 240 GPa and its low-pressure stability. *Nature*, **411**, 170–174.

Eremets, M. I., Drozdov, A. P., Kong, P. P., Wang, H. (2019). Semimetallic molecular hydrogen at pressure above 350 GPa. *Nature Physics*, **15**, 1246–1249.

Eremets, M. I., Gavriliuk, A. G., Trojan, I. A., Dzivenko, D. A., Boehler, R. (2004). Single-bonded cubic form of nitrogen. *Nature Materials*, **3**(8), 558–563.

Eremets, M. I., Troyan, I. A. (2011). Conductive dense hydrogen. *Nature Materials*, **10**(12), 927–931.

Falenty, A., Hansen, T. C., Kuhs, W. F. (2014). Formation and properties of ice XVI obtained by emptying a type sII clathrate hydrate. *Nature*, **516**(7530), 231–233.

Frost, M., Howie, R. T., Dalladay-Simpson, P., Goncharov, A. F., Gregoryanz, E. (2016). Novel high-pressure nitrogen phase formed by compression at low temperature. *Physical Review B*, **93**(2), 024113.

Goncharenko, I., Loubeyre, P. (2005). Neutron and X-ray diffraction study of the broken symmetry phase transition in solid deuterium. *Nature*, **435**(7046), 1206–1209.

Goncharov, A., Gregoryanz, E. (2005). Solid nitrogen at extreme conditions of high pressure and temperature, in M. R. Manaa, ed., *Chemistry at Extreme Conditions*. Elsevier, pp. 241–267.

Goncharov, A. F. (2012). Raman spectroscopy at high pressures. *International Journal of Spectroscopy*, **2012**, 617528.

Goncharov, A. F. (2020). Phase diagram of hydrogen at extreme pressures and temperatures; updated through 2019 (review article). *Low Temperature Physics*, **46**(2), 97–103.

Goncharov, A. F., Chuvashova, I., Ji, C., Mao, H.-k. (2019). Intermolecular coupling and fluxional behavior of hydrogen in phase IV. *PNAS*, **116**(51), 25512–25515.

Goncharov, A. F., Crowhurst, J. (2007). Proton delocalization under extreme conditions of high pressure and temperature. *Phase Transitions*, **80**(10–12), 1051–1072.

Goncharov, A. F., Crowhurst, J. C., Struzhkin, V. V., Hemley, R. J. (2008). Triple point on the melting curve and polymorphism of nitrogen at high pressure. *Physical Review Letters*, **101**(9), 095502.

Goncharov, A. F., Eggert, J. H., Mazin, I. I., Hemley, R. J., Mao, H.-k. (1996). Raman excitations and orientational ordering in deuterium at high pressure. *Physical Review B*, **54**(22), R15590–R15593.

Goncharov, A. F., Goldman, N., Fried, L. E., Crowhurst, J. C., Kuo, I. F. W., Mundy, C. J., Zaug, J. M. (2005). Dynamic ionization of water under extreme conditions. *Physical Review Letters*, **94**(12), 125508.

Goncharov, A. F., Gregoryanz, E., Mao, H.-K., Hemley, R. J. (2001). Vibrational dynamics of solid molecular nitrogen to megabar pressures. *Low Temperature Physics*, **27** (9), 866–869.

Goncharov, A. F., Gregoryanz, E., Mao, H.-k., Liu, Z., Hemley, R. J. (2000). Optical evidence for a nonmolecular phase of nitrogen above 150 GPa. *Physical Review Letters*, **85**(6), 1262–1265.

Goncharov, A. F., Hemley, R. J., Mao, H.-k. (2011). Vibron frequencies of solid H2 and D2 to 200 GPa and implications for the P–T phase diagram. *Journal of Chemical Physics*, **134**(17), 174501.

Goncharov, A. F., Hemley, R. J., Mao, H.-k., Shu, J. (1998). New high-pressure excitations in parahydrogen. *Physical Review Letters*, **80**(1), 101–104.

Goncharov, A. F., Holtgrewe, N., Qian, G., et al. (2015). Backbone NxH compounds at high pressures. *Journal of Chemical Physics*, **142**(21), 214308.

Goncharov, A. F., Howie, R. T., Gregoryanz, E. (2013). Hydrogen at extreme pressures (review article). *Low Temperature Physics*, **39**(5), 402–408.

Goncharov, A. F., Kong, L., Mao, H.-k. (2019). High-pressure integrated synchrotron infrared spectroscopy system at the Shanghai Synchrotron Radiation Facility. *Review of Scientific Instruments*, **90**, 093905.

Goncharov, A. F., Lobanov, S. S., Prakapenka, V. B., Greenberg, E. (2017). Stable high-pressure phases in the H–S system determined by chemically reacting hydrogen and sulfur. *Physical Review B*, **95**(14), 140101.

Goncharov, A. F., Struzhkin, V. V. (2017). Comment on "observation of the Wigner–Huntington transition to metallic hydrogen." *Science*, **357**(6353), eaam9736.

Goncharov, A. F., Struzhkin, V. V., Hemley, R. J., Mao, H.-k., Liu, Z. (2000). New techniques for optical spectroscopy at ultrahigh pressures, in M. Manghnani, W. J. Nellis, M. F. Nicol, eds. Science and Technology of High Pressure: Proceedings of AIRAPT-17. Universities Press, pp. 90–95.

Goncharov, A. F., Struzhkin, V. V., Mao, H. K., Hemley, R. J. (1999). Raman spectroscopy of dense H_2O and the transition to symmetric hydrogen bonds. *Physical Review Letters*, **83**(10), 1998–2001.

Goncharov, A. F., Struzhkin, V. V., Somayazulu, M. S., Hemley, R. J., Mao, H. K. (1996). Compression of ice to 210 gigapascals: infrared evidence for a symmetric hydrogen-bonded phase. *Science*, **273**(5272), 218–220.

Goncharov, A. F., Tse, J. S., Wang, H., et al. (2013). Bonding, structures, and band gap closure of hydrogen at high pressures. *Physical Review B*, **87**(2), 024101.

Gregoryanz, E., Goncharov, A. F., Hemley, R. J., Mao, H.-k. (2001). High-pressure amorphous nitrogen. *Physical Review B*, **64**(5), 052103.

Gregoryanz, E., Goncharov, A. F., Hemley, R. J., Mao, H.-k., Somayazulu, M., Shen, G. (2002). Raman, infrared, and X-ray evidence for new phases of nitrogen at high pressures and temperatures. *Physical Review B*, **66**(22), 224108.

Gregoryanz, E., Goncharov, A. F., Sanloup, C., Somayazulu, M., Mao, H.-k., Hemley, R. J. (2007). High P–T transformations of nitrogen to 170 GPa. *Journal of Chemical Physics*, **126**(18), 184505.

Gregoryanz, E., Ji, C., Dalladay-Simpson, P., Li, B., Howie, R. T., Mao, H.-K. (2020). Everything you always wanted to know about metallic hydrogen but were afraid to ask. *Matter and Radiation at Extremes*, **5**(3), 038101.

Grin, Y., Wagner, F. R., Armbrüster, M., et al. (2006). $CuAl_2$ revisited: composition, crystal structure, chemical bonding, compressibility and Raman spectroscopy. *Journal of Solid State Chemistry*, **179**(6), 1707–1719.

Guillot, T. (2005). The interiors of giant planets: models and outstanding questions. *Annual Review of Earth and Planetary Sciences*, **33**(1), 493–530.

Guthrie, M., Boehler, R., Molaison, J. J., Haberl, B., dos Santos, A. M., Tulk, C. (2019). Structure and disorder in ice VII on the approach to hydrogen-bond symmetrization. *Physical Review B*, **99**(18), 184112.

Guthrie, M., Boehler, R., Tulk, C. A., et al. (2013). Neutron diffraction observations of interstitial protons in dense ice. *Proceedings of the National Academy of Sciences*, **110**(26), 10552–10556.

Hanfland, M., Hemley, R. J., Mao, H.-k. (1993). Novel infrared vibron absorption in solid hydrogen at megabar pressures. *Physical Review Letters*, **70**(24), 3760–3763.

Hanfland, M., Hemley, R. J., Mao, H. K., Williams, G. P. (1992). Synchrotron infrared spectroscopy at megabar pressures: vibrational dynamics of hydrogen to 180 GPa. *Physical Review Letters*, **69**(7), 1129–1132.

Hanfland, M., Lorenzen, M., Wassilew-Reul, C., Zontone, F. (1998). Structures of molecular nitrogen at high pressures. *Review of High Pressure Science and Technology*, **7**, 787–789.

Hazen, R. M., Mao, H. K., Finger, L. W., Hemley, R. J. (1987). Single-crystal X-ray diffraction of nH_2 at high pressure. *Physical Review B*, **36**(7), 3944–3947.

Hellwig, H., Daniels, W. B., Hemley, R. J., Mao, H.-k., Gregoryanz, E., Yu, Z. (2001). Coherent anti-Stokes Raman scattering spectroscopy of solid nitrogen to 22 GPa. *Journal of Chemical Physics*, **115**(23), 10876–10882.

Hemley, R. J. (2000). Effects of high pressure on molecules. *Annual Review of Physical Chemistry*, **51**(1), 763–800.

Hemley, R. J., Bell, P. M., Mao, H. K. (1987). Laser techniques in high-pressure geophysics. *Science*, **237**(4815), 605–612.

Hemley, R. J., Goncharov, A. F., Lu, R., Struzhkin, V. V., Li, M., Mao, H. K. (1998). High-pressure synchrotron infrared spectroscopy at the national synchrotron light source. *Il Nuovo Cimento D*, **20**(4), 539–551.

Hemley, R. J., Jephcoat, A. P., Mao, H. K., Zha, C. S., Finger, L. W., Cox, D. E. (1987). Static compression of H_2O-ice to 128 GPa (1.28 Mbar). *Nature*, **330**(6150), 737–740.

Hemley, R. J., Mao, H. K. (1988). Phase transition in solid molecular hydrogen at ultrahigh pressures. *Physical Review Letters*, **61**(7), 857–860.

Hemley, R. J., Mao, H. K. (1989). Optical studies of hydrogen above 200 gigapascals: evidence for metallization by band overlap. *Science*, **244**(4911), 1462–1465.

Hemley, R. J., Mao, H.-k., Struzhkin, V. V. (2005). Synchrotron radiation and high pressure: new light on materials under extreme conditions. *Journal of Synchrotron Radiation*, **12**(2), 135–154.

Holtgrewe, N., Greenberg, E., Prescher, C., Prakapenka, V. B., Goncharov, A. F. (2019). Advanced integrated optical spectroscopy system for diamond anvil cell studies at GSECARS. *High Pressure Research*, **39**(3), 457–470.

Holtgrewe, N., Lobanov, S. S., Mahmood, M. F., Goncharov, A. F. (2016). Photochemistry within compressed sodium azide. *Journal of Physical Chemistry C*, **120**(49), 28176–28185.

Holzapfel, W. B. (1972). Symmetry of hydrogen-bonds in ice-vii. *Journal of Chemical Physics*, **56**(2), 712–715.

Howie, R. T., Dalladay-Simpson, P., Gregoryanz, E. (2015). Raman spectroscopy of hot hydrogen above 200 GPa. *Nature Materials*, **14**(5), 495–499.

Howie, R. T., Guillaume, C. L., Scheler, T., Goncharov, A. F., Gregoryanz, E. (2012). Mixed molecular and atomic phase of dense hydrogen. *Physical Review Letters*, **108** (12), 125501.

Hubbard, W. B. (1981). Interiors of the giant planets. *Science*, **214**(4517), 145–149.

Jenei, Z., O'Bannon, E. F., Weir, S. T., Cynn, H., Lipp, M. J., Evans, W. J. (2018). Single crystal toroidal diamond anvils for high pressure experiments beyond 5 megabar. *Nature Communications*, **9**(1), 3563.

Ji, C., Adeleke, A. A., Yang, L., et al. (2020). Nitrogen in black phosphorus structure. *Science Advances*, **6**(23), eaba9206.

Ji, C., Goncharov, A. F., Shukla, V., et al. (2017). Stability of $Ar(H_2)_2$ to 358 GPa. *Proceedings of the National Academy of Sciences*, **114**(14), 3596–3600.

Ji, C., Li, B., Liu, W., et al. (2019). Ultrahigh-pressure isostructural electronic transitions in hydrogen. *Nature*, **573**(7775), 558–562.

Ji, M., Umemoto, K., Wang, C.-Z., Ho, K.-M., Wentzcovitch, R. M. (2011). Ultrahigh-pressure phases of H_2O ice predicted using an adaptive genetic algorithm. *Physical Review B*, **84**(22), 220105.

Kamb, B., Davis, B. L. (1964). Ice VII, the densest form of ice. *Proceedings of the National Academy of Sciences*, **52**(6), 1433–1439.

Koch-Müller, M., Jahn, S., Birkholz, N., Ritter, E., Schade, U. (2016). Phase transitions in the system $CaCO_3$ at high P and T determined by in situ vibrational spectroscopy in diamond anvil cells and first-principles simulations. *Physics and Chemistry of Minerals*, **43**(8), 545–561.

Kong, P. P., Minkov, V. S., Kuzovnikov, M. A., et al. (2019). Superconductivity up to 243 K in yttrium hydrides under high pressure. arXiv:1909.10482v1 [cond-mat.supr-con].

Kotakoski, J., Albe, K. (2008). First-principles calculations on solid nitrogen: a comparative study of high-pressure phases. *Physical Review B*, **77**(14), 144109.

Kuriakose, M., Raetz, S., Hu, Q. M., et al. (2017a). Longitudinal sound velocities, elastic anisotropy, and phase transition of high-pressure cubic H_2O ice to 82 GPa. *Physical Review B*, **96**(13), 134122.

Kuriakose, M., Raetz, S., Hu, Q. M., et al. (2017b). Longitudinal sound velocities, elastic anisotropy, and phase transition of high-pressure cubic H_2O ice to 82 GPa. *Physical Review B*, **96**(13), 134122.

Laniel, D., Geneste, G., Weck, G., Mezouar, M., Loubeyre, P. (2019). Hexagonal layered polymeric nitrogen phase synthesized near 250 GPa. *Physical Review Letters*, **122**(6), 066001.

Laniel, D., Svitlyk, V., Weck, G., Loubeyre, P. (2018). Pressure-induced chemical reactions in the $N_2(H_2)_2$ compound: from the N_2 and H_2 species to ammonia and back down into hydrazine. *Physical Chemistry Chemical Physics*, **20**(6), 4050–4057. 10.1039/C7CP07989C.

Laniel, D., Winkler, B., Fedotenko, T., et al. (2020). High-pressure polymeric nitrogen allotrope with the black phosphorus structure. *Physical Review Letters*, **124**(21), 216001.

Li, B., Ji, C., Yang, W., et al. (2018). Diamond anvil cell behavior up to 4 Mbar. *Proceedings of the National Academy of Sciences*, **115**(8), 1713–1717.

Li, Y., Feng, X., Liu, H., et al. (2018). Route to high-energy density polymeric nitrogen t–N via He–N compounds. *Nature Communications*, **9**(1), 722.

Liang, Y., Tse, J. S. (2009). Mechanisms for the formation of H_2 and O_2 from X-ray irradiated dense ice. *Physical Review B*, **79**(10), 104105.

Lipp, M. J., Klepeis, J. P., Baer, B. J., et al. (2007). Transformation of molecular nitrogen to nonmolecular phases at megabar pressures by direct laser heating. *Physical Review B*, **76**(1), 014113.

Liu, H., Naumov, I. I., Hoffmann, R., Ashcroft, N. W., Hemley, R. J. (2017). Potential high-Tc superconducting lanthanum and yttrium hydrides at high pressure. *Proceedings of the National Academy of Sciences*, **114**(27), 6990–6995.

Liu, X.-D., Dalladay-Simpson, P., Howie, R. T., Li, B., Gregoryanz, E. (2017). Comment on "observation of the Wigner–Huntington transition to metallic hydrogen." *Science*, **357**(6353), eaan2286.

Liu, Y., Duan, D., Tian, F., et al. (2014). Crystal structures and properties of the CH_4H_2 compound under high pressure. *RSC Advances*, **4**(71), 37569–37574. 10.1039/C4RA05263C.

Lokshin, K. A., Zhao, Y., He, D., et al. (2004). Structure and dynamics of hydrogen molecules in the novel clathrate hydrate by high pressure neutron diffraction. *Physical Review Letters*, **93**(12), 125503.

Londono, D., Kuhs, W. F., Finney, J. L. (1988). Enclathration of helium in ice II: the first helium hydrate. *Nature*, **332**, 141.

Loubeyre, P., LeToullec, R. (1995). Stability of O_2/H_2 mixtures at high pressure. *Nature*, **378**(6552), 44–46.

Loubeyre, P., LeToullec, R., Hausermann, D., et al. (1996). X-ray diffraction and equation of state of hydrogen at megabar pressures. *Nature*, **383**(6602), 702–704.

Loubeyre, P., LeToullec, R., Pinceaux, J. P., Mao, H. K., Hu, J., Hemley, R. J. (1993). Equation of state and phase diagram of solid ^{4}He from single-crystal X-ray diffraction over a large P–T domain. *Physical Review Letters*, **71**(14), 2272–2275.

Loubeyre, P., LeToullec, R., Wolanin, E., Hanfland, M., Hausermann, D. (1999). Modulated phases and proton centring in ice observed by X-ray diffraction up to 170 GPa. *Nature*, **397**(6719), 503–506.

Loubeyre, P., Occelli, F., Dumas, P. (2020). Synchrotron infrared spectroscopic evidence of the probable transition to metal hydrogen. *Nature*, **577**(7792), 631–635.

Ma, Y., Oganov, A. R., Li, Z., Xie, Y., Kotakoski, J. (2009). Novel high pressure structures of polymeric nitrogen. *Physical Review Letters*, **102**(6), 065501.

Machida, S.-i., Hirai, H., Kawamura, T., Yamamoto, Y., Yagi, T. (2008). Structural changes of filled ice Ic structure for hydrogen hydrate under high pressure. *Journal of Chemical Physics*, **129**(22), 224505.

Mailhiot, C., Yang, L. H., McMahan, A. K. (1992). Polymeric nitrogen. *Physical Review B*, **46**(22), 14419–14435.

Mak, T. C. W., McMullan, R. K. (1965). Polyhedral clathrate hydrates. X. Structure of the double hydrate of tetrahydrofuran and hydrogen sulfide. *Journal of Chemical Physics*, **42**(8), 2732–2737.

Mao, H. K., Bell, P. M. (1977). Technique of operating the diamond-window pressure cell: considerations of the design and functions of the diamond anvils. *Carnegie Institute of Washington Yearbook*, **76**, 646–650.

Mao, H. K., Hemley, R. J. (1994). Ultrahigh-pressure transitions in solid hydrogen. *Reviews of Modern Physics*, **66**(2), 671–692.

Mao, H. K., Bell, P. M. (1979). Observations of hydrogen at room temperature (25°C) and high pressure (to 500 kilobars). *Science*, **203**(4384), 1004–1006.

Mao, H. K., Bell, P. M., Shaner, J. W., Steinberg, D. J. (1978). Specific volume measurements of Cu, Mo, Pd, and Ag and calibration of the ruby R1 fluorescence pressure gauge from 0.06 to 1 Mbar. *Journal of Applied Physics*, **49**(6), 3276–3283.

Mao, H. K., Chen, X.-J., Ding, Y., Li, B., Wang, L. (2018). Solids, liquids, and gases under high pressure. *Reviews of Modern Physics*, **90**(1), 015007.

Mao, H. K., Jephcoat, A. P., Hemly, R. J., et al. (1988). Synchrotron X-ray diffraction measurements of single-crystal hydrogen to 26.5 gigapascals. *Science*, **239**(4844), 1131–1134.

Mao, H. K., Shirley, E. L., Ding, Y., et al. (2010). Electronic structure of crystalline ^{4}He at high pressures. *Physical Review Letters*, **105**(18), 186404.

Mao, W. L., Mao, H.-k., Goncharov, A. F., et al. (2002). Hydrogen clusters in clathrate hydrate. *Science*, **297**(5590), 2247–2249.

Mao, W. L., Mao, H.-k., Meng, Y., et al. (2006). X-ray induced dissociation of H_2O and formation of an O_2–H_2 alloy at high pressure. *Science*, **314**(5799), 636–638.

Martin, R. M., Needs, R. J. (1986). Theoretical study of the molecular-to-nonmolecular transformation of nitrogen at high pressures. *Physical Review B*, **34**(8), 5082–5092.

McMahan, A. K., LeSar, R. (1985). Pressure dissociation of solid nitrogen under 1 Mbar. *Physical Review Letters*, **54**(17), 1929–1932.

McMahon, J. M. (2011). Ground-state structures of ice at high pressures from ab initio random structure searching. *Physical Review B*, **84**(22), 220104.

McMahon, J. M., Morales, M. A., Pierleoni, C., Ceperley, D. M. (2012). The properties of hydrogen and helium under extreme conditions. *Reviews of Modern Physics*, **84**(4), 1607–1653.

Meier, T., Petitgirard, S., Khandarkhaeva, S., Dubrovinsky, L. (2018). Observation of nuclear quantum effects and hydrogen bond symmetrisation in high pressure ice. *Nature Communications*, **9**(1), 2766.

Meng, Y., Von Dreele, R. B., Toby, B. H., et al. (2006). Hard X-ray radiation induced dissociation of N_2 and O_2 molecules and the formation of ionic nitrogen oxide phases under pressure. *Physical Review B*, **74**(21), 214107.

Mills, R. L., Liebenberg, D. H., Bronson, J. C., Schmidt, L. C. (1980). Procedure for loading diamond cells with high-pressure gas. *Review of Scientific Instruments*, **51**(7), 891–895.

Mills, R. L., Olinger, B., Cromer, D. T. (1986). Structures and phase diagrams of N_2 and CO to 13 GPa by X-ray diffraction. *Journal of Chemical Physics*, **84**(5), 2837–2845.

Monserrat, B., Drummond, N. D., Dalladay-Simpson, P., et al. (2018). Structure and metallicity of phase V of hydrogen. *Physical Review Letters*, **120**(25), 255701.

Monserrat, B., Needs, R. J., Gregoryanz, E., Pickard, C. J. (2016). Hexagonal structure of phase III of solid hydrogen. *Physical Review B*, **94**(13), 134101.

Murray, M. J., Sanders, J. V. (1980). Close-packed structures of spheres of two different sizes II. The packing densities of likely arrangements. *Philosophical Magazine A*, **42**(6), 721–740.

Naumova, A. S., Lepeshkin, S. V., Oganov, A. R. (2019). Hydrocarbons under pressure: phase diagrams and surprising new compounds in the C–H system. *Journal of Physical Chemistry C*, **123**(33), 20497–20501.

Nellis, W. J., Holmes, N. C., Mitchell, A. C., van Thiel, M. (1984). Phase transition in fluid nitrogen at high densities and temperatures. *Physical Review Letters*, **53**(17), 1661–1664.

Nelmes, R. J., Loveday, J. S., Marshall, W. G., Hamel, G., Besson, J. M., Klotz, S. (1998). Multisite disordered structure of ice VII to 20 GPa. *Physical Review Letters*, **81**(13), 2719–2722.

Ness, N. F. (1979). The magnetic fields of Mercury, Mars, and Moon. *Annual Review of Earth and Planetary Sciences*, **7**(1), 249–288.

Nettelmann, N., Becker, A., Holst, B., Redmer, R. (2012). Jupiter models with improved ab initio hydrogen equation of state (H-REOS.2). *Astrophysical Journal*, **750**(1), 52.

Ninet, S., Weck, G., Loubeyre, P., Datchi, F. (2011). Structural and vibrational properties of the van der Waals compound $(N_2)_{11}He$ up to 135 GPa. *Physical Review B*, **83**(13), 134107.

Nosé, S., Klein, M. L. (1983). Structural transformations in solid nitrogen at high pressure. *Physical Review Letters*, **50**(16), 1207–1210.

Oganov, A. R., Pickard, C. J., Zhu, Q., Needs, R. J. (2019). Structure prediction drives materials discovery. *Nature Reviews Materials*, **4**(5), 331–348.

Pace, E. J., Binns, J., Alvarez, M. P., Dalladay-Simpson, P., Gregoryanz, E., Howie, R. T. (2017). Synthesis and stability of hydrogen selenide compounds at high pressure. *Journal of Chemical Physics*, **147**(18), 184303.

Pace, E. J., Liu, X.-D., Dalladay-Simpson, P., et al. (2020). Properties and phase diagram of $(H_2S)_2H_2$. *Physical Review B*, **101**(17), 174511.

Peng, F., Sun, Y., Pickard, C. J., Needs, R. J., Wu, Q., Ma, Y. (2017). Hydrogen clathrate structures in rare earth hydrides at high pressures: possible route to room-temperature superconductivity. *Physical Review Letters*, **119**(10), 107001.

Pickard, C. J., Martinez-Canales, M., Needs, R. J. (2012). Density functional theory study of phase IV of solid hydrogen. *Physical Review B*, **85**(21), 214114.

Pickard, C. J., Martinez-Canales, M., Needs, R. J. (2013). Decomposition and terapascal phases of water ice. *Physical Review Letters*, **110**(24), 245701.

Pickard, C. J., Needs, R. J. (2007). Structure of phase III of solid hydrogen. *Nature Physics*, **3**(7), 473–476.

Pickard, C. J., Needs, R. J. (2009). High-pressure phases of nitrogen. *Physical Review Letters*, **102**(12), 125702.

Plisson, T., Weck, G., Loubeyre, P. (2014). N_6Ne_7: a high pressure van der Waals insertion compound. *Physical Review Letters*, **113**(2), 025702.

Polian, A., Grimsditch, M. (1984). New high-pressure phase of H_2O: ice X. *Physical Review Letters*, **52**(15), 1312–1314.

Popov, M. (2005). Raman and IR study of high-pressure atomic phase of nitrogen. *Physics Letters A*, **334**(4), 317–325.

Qian, G.-R., Lyakhov, A. O., Zhu, Q., Oganov, A. R., Dong, X. (2014). Novel hydrogen hydrate structures under pressure. *Scientific Reports*, **4**, 5606.

Ramsey, S. B., Pena-Alvarez, M., Ackland, G. J. (2020). Localization effects on the vibron shifts in helium–hydrogen mixtures. *Physical Review B*, **101**(21), 214306.

Saleh, G., Oganov, A. R. (2016). Novel stable compounds in the C–H–O ternary system at high pressure. *Scientific Reports*, **6**(1), 32486.

Salzmann, C. G., Radaelli, P. G., Mayer, E., Finney, J. L. (2009). Ice XV: a new thermodynamically stable phase of ice. *Physical Review Letters*, **103**(10), 105701.

Scheerboom, M. I. M., Schouten, J. A. (1993). Anomalous behavior of the vibrational spectrum of the high-pressure delta phase of nitrogen: a second-order transition. *Physical Review Letters*, **71**(14), 2252–2255.

Schuch, A. F., Mills, R. L. (1970). Crystal structures of the three modifications of nitrogen 14 and nitrogen 15 at high pressure. *Journal of Chemical Physics*, **52**(12), 6000–6008.

Schwager, B., Boehler, R. (2008). H_2O: another ice phase and its melting curve. *High Pressure Research*, **28**(3), 431–433.

Schwager, B., Chudinovskikh, L., Gavriliuk, A., Boehler, R. (2004). Melting curve of H2O to 90 GPa measured in a laser-heated diamond cell. *Journal of Physics: Condensed Matter*, **16**(14), S1177.

Seitz, A. E., Hippauf, F., Kremer, W., Kaskel, S., Scheer, M. (2018). Facile storage and release of white phosphorus and yellow arsenic. *Nature Communications*, **9**(1), 361.

Shen, G., Mao, H. K. (2016). High-pressure studies with X-rays using diamond anvil cells. *Reports on Progress in Physics*, **80**(1), 016101.

Shimizu, H., Ohnishi, M., Sasaki, S., Ishibashi, Y. (1995). Cauchy relation in dense H_2O ice VII. *Physical Review Letters*, **74**(14), 2820–2823.

Sihachakr, D., Loubeyre, P. (2004). O_2–N_2 mixtures under pressure: a structural study of the binary phase diagram at 295 K. *Physical Review B*, **70**(13), 134105.

Sloan, D. E. J., Koh, C. A. (2007). *Clathrate Hydrates of Natural Gases*. CRC Press.

Snider, E., Dasenbrock-Gammon, N., McBride, R., et al. (2020). Room-temperature superconductivity in a carbonaceous sulfur hydride. *Nature*, **586**(7829), 373–377.

Somayazulu, M., Ahart, M., Mishra, A. K., et al. (2019). Evidence for superconductivity above 260 K in lanthanum superhydride at megabar pressures. *Physical Review Letters*, **122**(2), 027001.

Somayazulu, M. S., Finger, L. W., Hemley, R. J., Mao, H. K. (1996). High-pressure compounds in methane–hydrogen mixtures. *Science*, **271**(5254), 1400–1402.

Song, M., Yamawaki, H., Fujihisa, H., Sakashita, M., Aoki, K. (1999). Infrared absorption study of Fermi resonance and hydrogen-bond symmetrization of ice up to 141 GPa. *Physical Review B*, **60**(18), 12644–12650.

Spaulding, D. K., Weck, G., Loubeyre, P., Datchi, F., Dumas, P., Hanfland, M. (2014). Pressure-induced chemistry in a nitrogen-hydrogen host–guest structure. *Nature Communications*, **5**(1), 5739.

Stanley, S., Bloxham, J. (2004). Convective-region geometry as the cause of Uranus' and Neptune's unusual magnetic fields. *Nature*, **428**(6979), 151–153.

Stinton, G. W., Loa, I., Lundegaard, L. F., McMahon, M. I. (2009). The crystal structures of δ and δ^* nitrogen. *Journal of Chemical Physics*, **131**(10), 104511.

Strobel, T. A., Ganesh, P., Somayazulu, M., Kent, P. R. C., Hemley, R. J. (2011). Novel cooperative interactions and structural ordering in H_2S–H_2. *Physical Review Letters*, **107**(25), 255503.

Strobel, T. A., Somayazulu, M., Sinogeikin, S. V., Dera, P., Hemley, R. J. (2016). Hydrogen-stuffed, quartz-like water ice. *Journal of the American Chemical Society*, **138**(42), 13786–13789.

Struzhkin, V., Li, B., Ji, C., et al. (2020). Superconductivity in La and Y hydrides: remaining questions to experiment and theory. *Matter and Radiation at Extremes*, **5** (2), 028201.

Sugimura, E., Iitaka, T., Hirose, K., Kawamura, K., Sata, N., Ohishi, Y. (2008). Compression of H_2O ice to 126 GPa and implications for hydrogen-bond symmetrization: synchrotron X-ray diffraction measurements and density-functional calculations. *Physical Review B*, **77**(21), 214103.

Tomasino, D., Jenei, Z., Evans, W., Yoo, C.-S. (2014a). Melting and phase transitions of nitrogen under high pressures and temperatures. *Journal of Chemical Physics*, **140** (24), 244510.

Tomasino, D., Kim, M., Smith, J., Yoo, C.-S. (2014b). Pressure-Induced symmetry-lowering transition in dense nitrogen to layered polymeric nitrogen (LP–N) with colossal Raman intensity. *Physical Review Letters*, **113**(20), 205502.

Turnbull, R., Donnelly, M.-E., Wang, M., et al. (2018). Reactivity of hydrogen–helium and hydrogen–nitrogen mixtures at high pressures. *Physical Review Letters*, **121**(19), 195702.

Turnbull, R., Hanfland, M., Binns, J., et al. (2018). Unusually complex phase of dense nitrogen at extreme conditions. *Nature Communications*, **9**(1), 4717.

Vos, W. L., Finger, L. W., Hemley, R. J., Hu, J. Z., Mao, H. K., Schouten, J. A. (1992). A high-pressure van der Waals compound in solid nitrogen–helium mixtures. *Nature*, **358**(6381), 46–48.

Vos, W. L., Finger, L. W., Hemley, R. J., Mao, H.-k. (1993). Novel H_2-H_2O clathrates at high pressures. *Physical Review Letters*, **71**(19), 3150–3153.

Wang, H., Eremets, M. I., Troyan, I., Liu, H., Ma, Y., Vereecken, L. (2015). Nitrogen backbone oligomers. *Scientific Reports*, **5**(1), 13239.

Wang, X., Wang, Y., Miao, M., et al. (2012). Cagelike diamondoid nitrogen at high pressures. *Physical Review Letters*, **109**(17), 175502.

Wang, Y., Glazyrin, K., Roizen, V., et al. (2020). Novel hydrogen clathrate hydrate. *Physical Review Letters*, **125**(25), 255702.

Wang, Y., Zhang, X., Jiang, S., et al. (2019). Helium-hydrogen immiscibility at high pressures. *Journal of Chemical Physics*, **150**(11), 114504.

Weck, G., Datchi, F., Garbarino, G., et al. (2017). Melting curve and liquid structure of nitrogen probed by X-ray diffraction to 120 GPa. *Physical Review Letters*, **119**(23), 235701.

Wigner, E., Huntington, H. B. (1935). On the possibility of a metallic modification of hydrogen. *Journal of Chemical Physics*, **3**(12), 764–770.

Yao, Y., Tse, J. S., Tanaka, K. (2008). Metastable high-pressure single-bonded phases of nitrogen predicted via genetic algorithm. *Physical Review B*, **77**(5), 052103.

Zha, C.-s., Cohen, R. E., Mao, H.-k., Hemley, R. J. (2014). Raman measurements of phase transitions in dense solid hydrogen and deuterium to 325 GPa. *Proceedings of the National Academy of Sciences*, **111**(13), 4792–4797.

Zha, C.-s., Liu, H., Tse, J. S., Hemley, R. J. (2017). Melting and high P–T transitions of hydrogen up to 300 GPa. *Physical Review Letters*, **119**(7), 075302.

Zhang, X., Xu, W., Wang, Y., et al. (2018). Synthesis and properties of selenium trihydride at high pressures. *Physical Review B*, **97**(6), 064107.

15

Superconductivity at High Pressure

MIKHAIL I. EREMETS

Room temperature superconductivity (RTSC) is one the most challenging and longstanding problems in solid-state physics. The Bardeen–Cooper–Schrieffer (BCS) theory (1956) explained superconductivity but could not predict high critical temperatures (T_c). Extension of the BCS theory allowed RTSC in principle; however, estimations for realistic materials gave low T_c, with the only exception being metallic hydrogen. Therefore, conventional superconductors were not considered potential RTSCs. This tendency strengthened after the experimental discovery of superconductivity in cuprates with very high T_c, up to 133 K. Later, other families of nonconventional superconductivity appeared, notably, iron-based superconductors with T_c reaching 100 K. However, the mechanism of superconductivity in these materials is still not understood, and there has been no progress for many years in increasing T_c. Unexpectedly, conventional superconductors recently showed a clear prospect to be the first RTSCs: T_c = 203 K was discovered in H_3S, and then T_c = 250 K in LaH_{10}. This breakthrough resulted from a combination of factors, including the general idea to consider hydrogen-dominant materials, the appearance of ab initio predictions of structures for searching new materials, and advances in synthesis and characterization of new superconductors at megabar pressures. There is a clear prospect to achieve higher T_c in other binary or ternary hydrides. At ambient pressures, there is also a distinct possibility for substantial superconductivity, likely in materials with strong covalent bonding.

15.1 Introduction

There have been tremendous advances in the field of superconductivity at high pressures in the last several decades. The present contribution is not meant as a review of superconductivity at high pressures, a subject on which recent and excellent reviews exist (Schilling, 2007, Gor'kov and Kresin, 2018, Flores-Livas et al., 2020, Pickard et al., 2020, Pickett and Eremets, 2020). Instead, it provides my personal views on the progress in high-temperature superconductivity, one of the many directions of the modern high-pressure science that were triggered by Dave Mao. Dave played a significant role in my scientific life. I was fascinated by his prolific career already at the time when I was working in Russia. I was very excited to move in 1998 to the Geophysical Lab (GL) to work with Dave Mao and Rus Hemley on metallic hydrogen. Dave Mao opened the megabar era (Mao, 1978), started the practical search for metallic hydrogen, and obtained a huge amount of experimental results, summarized in (Mao and Hemley, 1994). Arguably, the GL was the best place at the time for high-pressure investigations, with a creative atmosphere and intense research. I remember those good times, jogging in the beautiful neighborhood of the GL and imagining how metallic hydrogen should look. Rapidly, it turned out that the way was long and tortuous. In the GL, I worked on different subjects but always kept hydrogen in my mind. We studied xenon (Eremets et al., 2000), which we considered as a model substance for metallic hydrogen. It is isoelectronic to CsI and is an archetype of a material with a dielectric–metal transition. The metallization of xenon is closely related to that of hydrogen and occurs by the closure of the bandgap and appearance of a semimetallic state at ~150 GPa. We worked on nitrogen as well and discovered, on the way, that nitrogen behaves very differently from hydrogen – instead of dissociation and metallization, it creates polymeric nitrogen (Eremets et al., 2001a). The strongly bonded nitrogen molecules dissociate at much lower pressures, ~100 GPa, and form a covalently bonded amorphous solid. Later we found that it can crystalize in a perfect cubic lattice and in other symmetric forms. This work sharply deviated from the goal of metallic hydrogen but produced significant results and gave us excellent practice in optical, electrical, and X-ray studies at megabar pressures.

Metallic hydrogen remained unattainable, and we tried to find high-temperature superconductivity by searching the light elements, a recipe prescribed by the Bardeen–Cooper–Schrieffer (BCS) theory. We reported the transformation in boron from a semiconductor to a metallic and superconducting state at $P > 160$ GPa, and the critical temperature (T_c) increased from 4 K to 11 K at 250 GPa (Eremets et al., 2001b). Lithium represents a dramatic example of the influence of pressure in materials and is a textbook "simple metal." It has a T_c of only ~20 mK at ambient pressure, but the emerging structural and electronic complexity under pressure led to the appearance of 9 K superconductivity at 23 GPa, which increases to 16 K at 80 GPa (Struzhkin et al., 2002). These values of T_c were considered high at the time but were still far away from room temperature superconductivity.

15.2 Searching for Room Temperature Superconductors

Is it possible then to create superconductors that operate at room temperature? If yes, the world would change no less than with the appearance of electricity two centuries ago. The ability to transfer electrical current without losses obviously will save a huge amount of energy. Even more essential, new applications will be enabled by the phenomenon that flow in a superconducting ring does not stop when the current source is switched off, which is a way of storing energy that can replace accumulators. Superconducting coils create strong magnetic fields, and therefore superconducting magnets will become widespread and can be used for levitating trains, and in motors, MRI scanners, accelerators, and nuclear fusion reactors.

This is a dream that is now getting closer to reality. For a long time, superconductivity was considered a low-temperature phenomenon. The first superconductor was Hg, which

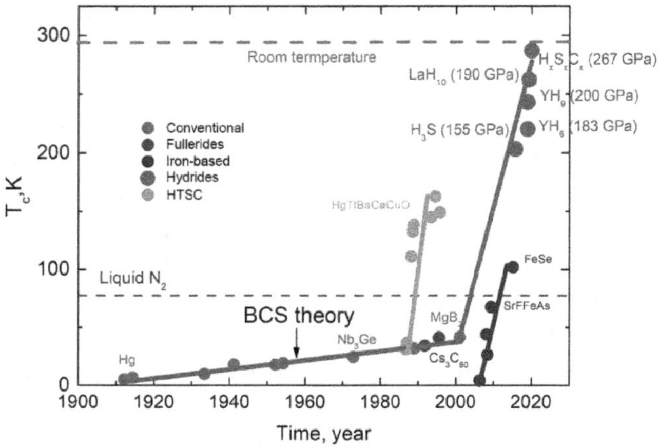

Figure 15.1 The evolution of the superconducting transition temperature for representative materials.

was discovered by Kamerlingh Onnes. Resistance sharply dropped to zero when Hg was cooled below a T_c of 4.2 K. Gradually other superconductors with higher T_c were found. Figure 15.1 shows the slow growth record of T_c values with time resulting from an empirical search, with experimentalists accumulating facts on the superconductors. One of the decisive points was the isotope effect: T_c depends on the mass of the atoms of the lattice, meaning that not only electrons but somehow also the lattice is involved in the superconductivity. Only in 1957 was a full theory of superconductivity built – the BCS theory (Bardeen et al., 1957). At the heart of this theory is the idea that electrons pair with the aid of lattice vibrations (phonons). This theory successfully explained the known properties of superconductors. It predicted that higher T_c should be achievable in materials with high frequencies of phonons and strong coupling of electrons and phonons. In particular, materials with light elements are favorable. However, the BCS theory does not predict high values of T_c. According to the theory, T_c saturates around 20 K when the electron–phonon coupling increases. Extensions of the BCS theory, such as the Migdal–Eliashberg theory (Migdal, 1958, Eliashberg, 1960), covered the strong-coupling case, and Allen and Dynes (1975) realized that in this case T_c can be proportional to the electron–phonon coupling such that there is no formal bound. However, these theoretical advances were not reflected in the experimental search: The maximum critical temperature continued to increase only very slowly, at a rate of ≈ 0.3 K/year, until a maximum T_c of 23.2 K was measured in Nb_3Ge in 1973 (Gavaler, 1973). Higher values of T_c seemed impossible to achieve, which led Cohen and Anderson to postulate in 1972 (Cohen and Anderson, 1972) that a hard limit of 25 K must exist for the T_c of conventional superconductors. This idea was based on several misconceptions.

A grand discovery came in 1986 by K. Alex Müller and Georg Bednorz (Bednorz and Mueller, 1986), when cuprates broke all the old prejudices. They found superconductivity with $T_c = 35$ K in lanthanum–copper–oxide crystals doped by barium, which was unbelievable at the time. Soon the critical temperature was raised above 77 K (the temperature of liquid nitrogen) – superconductors working above this temperature are called high-temperature superconductors (HTSCs). It seemed that room temperature superconductivity would soon be achieved in curprates, but the record T_c of 133 K at ambient pressure (Schilling et al., 1993) and of 164 K under high pressure (Gao et al., 1994) remains unmoved for more than a quarter of a century. Further progress was suspended, presumably because the mechanism of superconductivity in cuprates is not understood. It is still debated, but apparently it is not conventional superconductivity based on the phonon-mediated coupling of electrons.

Another breakthrough came from H. Hosono in 2006 when superconductivity was serendipitously discovered by looking for new magnetic semiconductors (Kamihara et al., 2006), which was surprising because long-range magnetism and superconductivity are typically incompatible. The original Fe-based superconductors had T_c of 26 K, but soon it was increased to 56 K in iron pnictides, and above 65 K in electron-doped monolayer FeSe on $SrTiO_3$ (Si et al., 2016). The simplest among all Fe-based superconductors, FeSe, did not initially attract attention because of the relatively low T_c of 8 K. However, application of a moderate pressure of 10 GPa increased T_c to 40 K (Medvedev et al., 2009). This remarkable enhancement hinted that pressure strongly enhanced spin fluctuations. The mechanism for pairing is almost certainly electronic since the bulk electron–phonon interaction is rather weak.

The discovery of superconductivity in cuprates and iron-based materials strengthened the old belief that superconductivity with high T_c should not be searched for in conventional phonon-mediated superconductors. Even the finding, in 2001, of superconductivity with $T_c = 39$ K in MgB_2 (Nagamatsu et al., 2001) and at somewhat lower temperatures of 33 K in fullerides Cs_3C_{60} (Tanigaki et al., 1991), did not shake this view, and room temperature superconductivity remained elusive.

15.3 Metallic Hydrogen and Superconductivity

The recent story of high-temperature conventional superconductivity started from metallic hydrogen. In fact, this exotic material is the most common substance in our solar system, as the main constituent of the gas giants Jupiter and Saturn. This fact alone would be enough to attempt to reach this state. But N. Ashcroft in 1968 (Ashcroft, 1968) gave qualitative arguments that metallic hydrogen can be a superconductor with high critical temperature: the lightness of hydrogen provides high phonon frequencies, and the absence of core electrons in hydrogen promotes strong coupling between electrons and phonons. According to the BCS theory, these factors favor a high T_c. Ashcroft is said to have not given an estimate of the T_c, but at lectures he mentioned $T_c = 50$ K – an extraordinary value at that time – and K. A. Mueller, attending a conference, learned then that T_c can be that high, which was psychologically important. Later, many calculations showed that metallic atomic hydrogen can have T_c ~200–700 K (McMahon et al., 2012). The discovery of metallic hydrogen, which should be a room temperature superconductor and also have other fascinating properties, became a "Holy Grail" for many researchers.

This is a difficult theoretical and experimental problem. Suffice to say that one should typically reach pressures of ~500 GPa and measure electrical and other properties of hydrogen. This goal stimulated enormous progress in recent years, both in theory and experiment. Infrared, Raman, and electrical measurements were done at pressures near 500 GPa. In our study (Eremets et al., 2019), we concluded that hydrogen first transforms into the semimetallic state through the closure of the bandgap. A semimetal is a poor metal and does not exhibit superconductivity, which should appear at higher pressures. Theoretical studies supported the transformation to semimetallic hydrogen at pressures >360 GPa and predicted that it should be a good metal and superconductor at higher pressures, closer to 500 GPa. Loubeyre et al. (2020) reported on the transition to molecular metallic hydrogen at ~420 GPa. According to the recent calculations the next dramatic transition should happen at a pressure of ~500 GPa, when molecules of hydrogen dissociate and hydrogen transforms into an atomic state. Wigner and Huntington (1935) predicted this transformation in 1935. In this state, hydrogen should be a good metal and superconductor.

15.4 Metallic Hydrogen Alloys and Superconductivity

To circumvent the problem of extremely high pressure, Ashcroft (2004) proposed to consider alloys of metallic hydrogen, for example, dense group-IVa hydrides (CH_4, SiH_4, etc.). Metallization of the hydrides should happen at much lower pressures in comparison with pure hydrogen, within the capabilities of DACs. These hydrogen-dominant materials are similar to pure hydrogen in that they have two critical ingredients for good superconductivity: high phonon frequencies and strong electron–phonon coupling.

That was a remarkable idea. The reality, however, turned out to be more complex. Methane, for instance, dissociates to ethane and butane and then to diamond and hydrogen ($P > 287$ GPa) (Gao et al., 2010). Other hydrides of the IVa group (SiH_4, GeH_4, etc.) also decompose under pressure. Fortunately, N. Ashcroft proposed this idea at the right time, just as methods of structure prediction emerged, and many other hydrogen-rich materials could be analyzed. It became possible to predict ab initio how atoms pack crystal structures from knowing just the properties of the atom in the lattice – a prospect that seemed unbelievable. However, the calculations reproduced, for instance, allotropes of carbon, diamond, and graphite and found several unknown structures that carbon atoms can form. Structures could be searched at any density, and therefore ab initio methods are ideally suited for looking for different phases at high pressure. The most developed methods of these calculations are AIRSS (Pickard and Needs, 2011), USPEX (Glass et al., 2006), and CALYPSO (Wang et al., 2010). There was, and still exists, skepticism about the reliability of the predictions. A drawback of the calculations is that there is no benchmark; the search continues to exhaustion or saturation, but there remains a probability that the real, most stable structure was missed. However, most times the calculations predict correct structures at least for simple elemental or binary materials. Methods of the calculations are developing, with predictions becoming more reliable and extending to more complex compounds. I had a shocking experience when finding a new phase in AlH_3 with *Pn–3m* structure at $P > 100$ GPa (Goncharenko et al., 2008). Later we found that Pickard and his colleagues predicted the same structure completely independently (Pickard and Needs, 2007). The same happened with the prediction of a new structure in SiH_4 (Pickard and Needs, 2006). A fantastic experience was to find transparent sodium (Ma et al., 2009) at 190 GPa, predicted by Y. Ma.

The computational prediction of structures opened a new chapter in the search for new superconductors. As soon as a structure is found, phonon and electron spectra of the solid can be calculated. The next step, the evaluation of T_c, is relatively easy; it is calculated from the McMillan equation of the BCS theory or more precisely from Eliashberg equations. In this way, the structure of many hydrides was predicted, and T_c calculated, and some of them showed very high, unbelievable values of T_c. For instance, a T_c of 220 K was predicted for CaH_6 (2012) (Wang et al., 2012). These calculations were met with skepticism in the superconducting community. Hirsch (2009) even wrote: "... and even to this day theorists continue 'predicting' that metal hydrides will yield high-temperature superconductors."

At the time of the development of the theory, we made dramatic progress in the development of experimental techniques by working on hydrogen. We regularly achieved pressures above 300 GPa while being able to perform electrical measurements – the most important method for testing new superconductors. Our first experiments seemed not to support the calculations. We did not find any superconductivity in available substances such as metallic AlH_3 (Goncharenko et al., 2008), and found only low $T_c < 17$ K in SiH_4 (Eremets et al., 2008). Despite many examples of calculations, it was challenging to find a substance suitable for experiments. Many hydrides were challenging to synthesize; CaH_6, for instance, has been synthesized only very recently.

15.5 Tale of Superconductivity of Sulfur Hydride at High Pressure

One substance attracted our attention – H_2S. Li et al. (2014) explored energetically stable crystal structures under pressure and found metallic phases of H_2S at pressures above 100 GPa, with superconducting temperatures on the order of ~80 K at 160 GPa. H_2S seemed not to be the best choice for testing, as the hydrogen content was not high and the predicted T_c ~80 K was lower than in many other predictions. But this gas is available commercially. Another reason for its choice was that the calculations were done in the Y. Ma group, who had made correct predictions for several substances, including transparent sodium, on which we worked together.

We experimented together with A. Drozdov, whose thesis involved such work. An idea was to condense the H_2S in a cryostat and clamp the liquid. We placed a DAC into a cryostat and cooled down to 191–213 K (which is within the temperature range of liquid H_2S) and then flowed the H_2S gas in the chamber through a capillary into the rim around the diamond anvil, where it liquified. We monitored the filling of the chamber visually and identified the sample by measuring Raman spectra. After clamping the DAC, we heated the system to 220 K to evaporate the remaining H_2S and subsequently increased the pressure, typically at this temperature. We also had to overcome other obstacles, such as the aggressive nature of H_2S – it spoiled the sealing, the DAC gasket, and other parts. The smell of H_2S was one of the significant obstacles in this procedure, as the sealing was not perfect in the beginning, and even small leaks resulted in foul smells – the smell of a skunk, or at its most extreme, the odor of volcanoes and geysers, like the "Smell of Hell" at the basins of Yellowstone. On the positive side, I learned that mice exposed to nontoxic levels of H_2S are characterized by superior memory and learning profiles (Pizzochero et al., 2015). We were active mostly at night with uninterrupted work hours, and with a lot of work, H_2S leaks were gradually fixed.

We found that at 90–100 GPa, the sample turns into a metal, and above 100 GPa into a superconductor: the resistance dropped abruptly to zero. A further increase in pressure resulted in a T_c of ~50–70 K. That was already a great success, and the result agreed well with the theoretical prediction (Li et al., 2014). However, we "spoiled" this agreement by continuing experimenting. Soon, we discovered that instead of declining, the T_c sharply increased, with a further increase in pressure, up to 150 K (see Figure 15.2). Later we found superconductivity at 190 K, but this value scattered in fact from run to run. Even though there was an apparent signature of superconductivity – the drop of resistance goes to zero – we did not publish our first result but spent about eight months complicating our experiments to prove and understand this enormously high T_c. We hypothesized that the abrupt emergence of T_c (Figure 15.2) might be because of the increase in temperature, as the pressure did not change much. Indeed, the samples pressurized to ~150 GPa at low temperatures (<100 K) but warmed to room temperature showed a tremendous T_c, on the order of 200 K, exceeding by far the record T_c of 164 K in cuprates. The value of the critical temperature was at odds with the original prediction for H_2S. We came up with the

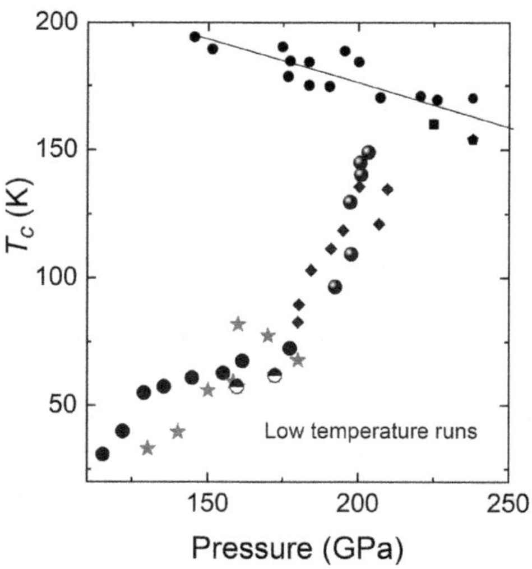

Figure 15.2 The pressure dependence of the critical temperature T_c. The points with increasing trend: T_c measured on an H_2S sample loaded and pressurized at low temperatures (100–250 K). The Gray stars: BCS calculations. The points forming a negative correlation: T_c measured from samples annealed at room temperature.

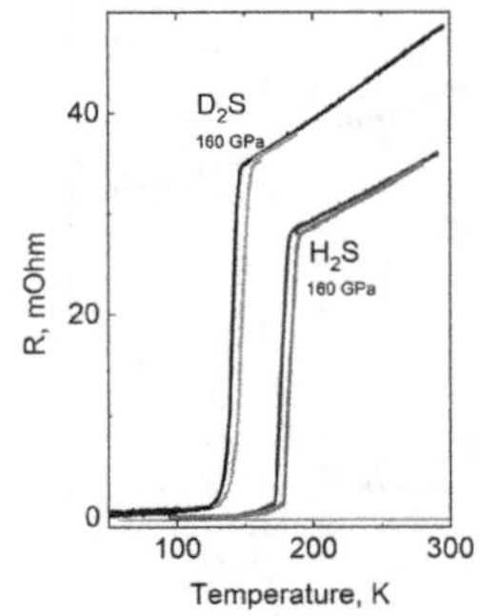

Figure 15.3 Superconducting transitions in hydrogen and deuterium sulphides at 160 GPa. The dark and gray lines were measured at cooling and warming, respectively.

guess that sulphur, known for its supervalency, could in principle have bound more hydrogen atoms than in H_2S, and presumably our samples had transformed into another hydride (H_4S or H_6S). For that to happen, H_2S should decompose with the deposition of elemental sulphur. Indeed, we found traces of sulfur in the Raman spectra.

We also found a huge shift of the superconductivity temperature by substituting hydrogen by deuterium, to get D_2S – an isotope effect and a crucial test supporting superconductivity (Figure 15.3). Also, the T_c shifted to lower temperatures in a magnetic field, in agreement with the behavior of a superconductor. We prepared a manuscript but continued experimenting, as we understood the importance of the result and knew that the community is plagued by many claims of high-T_c materials, most of them of doubtful nature. Submission of our paper was accelerated by the appearance of Duan's paper (Duan et al., 2014), where theoretically another system $(H_2S)_2H_2$, different from pure H_2S, was considered. We immediately uploaded our work to arXiv (Drozdov et al., 2014), sticking with our original interpretation. Duan's paper did not impact our work in any aspect. The almost simultaneous announcement of experimental and theoretical work tremendously sped up the understanding of the T_c in this compound, especially because the calculated T_c amazingly coincided with our result. This perfect agreement with the experiment was, in fact, accidental: As it will be revealed later, the corrections above the used DFT approximation are significant, ~50 K, but they mutually cancel. The coincidence of the predicted and experimental T_c led computational and theoretical scientists to the associate the superconductivity with the predicted *Im3m* structure. It was not apparent at that time, however, as we did not have X-ray data, which followed later.

To guarantee that superconductivity occurs, the most crucial signature that must be present is the expulsion of the magnetic field below T_c (the Meissner effect). Without these measurements, *Science* and *Nature* rejected our paper even if we had already substantial shreds of evidence of superconductivity (zero resistance, proper shift of T_c in a magnetic field, and isotope effect). We needed a quick solution. One way was to build a setup designed by Timofeev (1992) where the change of a magnetic moment is detected by a system of coils around the anvils. It was tested for several superconductors, for instance for sulphur at pressures of up to 170 GPa (Struzhkin et al., 1997). However, this is a complicated method, and it only detects the superconducting transition itself without further characterization of the superconductivity.

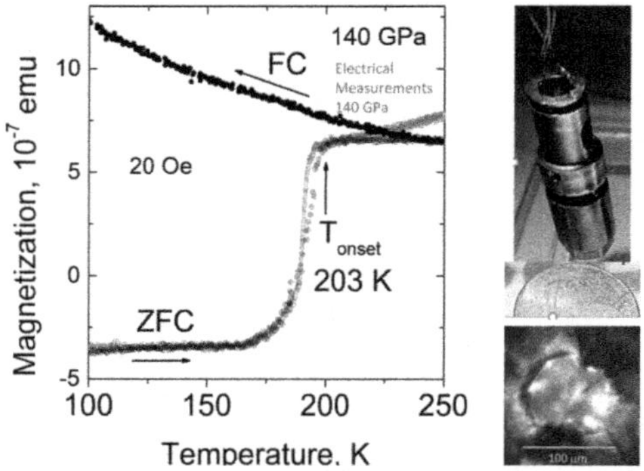

Figure 15.4 Left: Temperature dependence of the magnetization of sulfur hydride at a pressure of 155 GPa in zero-field-cooled (ZFC) and 20 Oe field-cooled (FC) modes (black circles). For comparison, the superconducting step obtained for sulfur hydride from electrical measurements at 145 GPa is shown (red circles). Right: Tiny DAC and view of the metallic hydrogen sulphide sample in the DAC at 150 GPa. For the color version, refer to the plate section.

We picked a way of direct measurement of magnetic susceptibility with the most sensitive quantum magnetometer (SQUID). At that time, DACs were developed for measurements of up to ~15 GPa. Our tiny samples of about 50–100 mm in diameter and of a few micrometers thickness were already difficult to detect with SQUID, especially inside the DAC, which adds a strong background. The DAC should be made from nonmagnetic Cu:Be or Cu:Ti alloys, but these materials are not strong or particularly suitable for building a DAC. A severe limitation was also the size: The DAC should be of diameter <9 mm in order to fit the SQUID magnetometer. Based on my previous experience, when I built a tiny cell of 10 mm diameter just for curiosity, I designed an extremely simple yet functional cell (Figure 15.4). Based on that design, our workshop quickly built these DACs made of Cu:Ti alloy within a few weeks. In a short time, we reached the required pressure of ~150 GPa and made a series of measurements. We obtained a nice superconducting transition (Figure 15.4) with the sample of diameter ~100 μm by measuring, using Quantum Design's Magnetic Property Measurement System (MPMS), magnetic susceptibility as a function of temperature for sulfur hydride. The zero-field-cooled (ZFC) curve demonstrates a rather sharp transition from the diamagnetic to the paramagnetic state, which is classified as a superconducting one. The onset temperature is 203 K, in good agreement with the resistance measurements (Drozdov et al., 2015b). We confirmed the superconducting transition by magnetization measurements, $M(H)$, at different temperatures. The magnetization curves show hysteresis, indicating a type II superconductor.

It was essential to verify the superconductivity in other laboratories. K. Shimizu invited me to test our samples at Osaka University and at the SPring-8 synchrotron. Professor Shimizu has substantial experience in studying superconductors under pressure – he is famous for building a "periodic table of superconductors." He has built a unique setup, which combines X-ray diffraction measurements at low temperatures with electrical measurements, so that the structure can be obtained at low temperatures in the superconducting

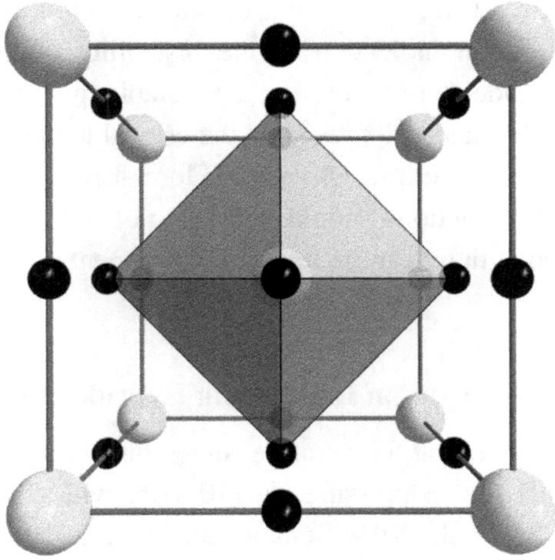

Figure 15.5 The structure of the superconducting phase of H_3S.

state. It was impossible to find a better place for testing our samples. We checked five cells with clamped samples. The first cell showed no superconducting transition – the sample was not well synthesized. We loaded another cell and went for lunch during the cooling. We were back at the time when the cell was cooled to about 210 K. After some minutes of waiting, the resistance started to decrease at the expected temperature of ~195 K and soon dropped to zero. This was amazing to observe such a high T_c in comparison with typical conventional superconductors, which have a T_c in the range of ~10 K. But it was superconductivity: The zero resistance was carefully double-checked with a sensitive bridge. The superconducting transition showed nearly no temperature hysteresis, as expected for superconductivity. At SPring-8, we obtained a significant new result: the structure of the superconducting phase (Einaga et al., 2016). It turned out to be an *Im3m* structure (Figure 15.5), as found theoretically (Duan et al., 2014) and later used in many theoretical works. In addition to the *Im3m* structure, which relates to H_3S, we observed pure sulphur, in full agreement with our hypothesis that H_2S dissociates into sulphur and H_3S. The calculations confirmed this way of dissociation (Bernstein et al., 2015). Later we characterized the superconducting state by measuring infrared spectra (Capitani et al., 2017), and we estimated the superconducting gap to be ~80 meV. We also measured electronic Raman spectra and determined the superconducting gap, which was in agreement with the infrared data.

Meanwhile, theoretical calculations helped to understand why H_3S can provide such a high T_c. It emerged that it has a phonon spectrum that is close to ideal (Nicol and Carbotte, 2015); a peak in the density of states (van Hove singularity) nearly coincides with the Fermi level (Quan and Pickett, 2016). Hydrogen in the H_3S structure is covalently bonded with two neighboring atoms of sulphur – this is a covalent metal, similar to MgB_2.

We attempted to find superconductors with higher T_c in materials similar to hydrogen sulphide. By analogy with H_2S, we tried PH_3 (phosphine) and found that it also turns to metal at 100 GPa, with a T_c of 100 K at pressures above 200 GPa. We posted preliminary results on arXiv (Drozdov et al., 2015b) but did not mention that annealing leads to a dramatic decrease of T_c – an indication of the instability of the phase. Interestingly, subsequent calculations indeed did not find any stable phase of PH_3. Estimates for metastable phases (PH, PH_2, and PH_3) revealed the critical temperatures, which reproduce to a good approximation the experimental values (Flores-Livas et al., 2016). This example indicates that theoretical predictions should consider not only stable phases but analyze metastable phases that potentially can be realized experimentally.

15.6 Superconductivity in Lanthanum Hydride at High Pressure

The next big step toward room temperature superconductivity was the discovery of superconductivity in LaH_{10} at a pressure of 170 GPa with T_c~250 K – nearly room temperature (RT) (Drozdov et al., 2019; Somayazulu et al., 2019). This work was, in fact, initiated by 2012 predictions for CaH_6 (Wang et al., 2012). This compound has a remarkable symmetric cage-like structure (Figure 15.6) with Ca ions surrounded by hydrogen atoms. It differs entirely from H_3S. Whereas hydrogen atoms in H_3S are strongly covalently bonded with sulphur atoms, hydrogen atoms in clathrate-like superconductors are covalently bonded to one another and form a network that interacts with the host atom, predominantly ionically. This hydrogen sublattice can be considered as the closest analog to atomic metallic hydrogen, which is doped and stabilized by the host atom (Ca). A T_c of 220–235 K was calculated for CaH_6, which should be even higher than the T_c later found in H_3S – the story of high-temperature conventional superconductivity would be different were it not for the difficulties to synthesize CaH_6.

Only recently, a cage structure was realized. Geballe et al. (2018) synthesized lanthanum hydride directly from lanthanum and hydrogen with the aid of laser heating to 1,300 K in the DAC at 170 GPa. The synthesis is tricky, as hot hydrogen can destroy the stressed

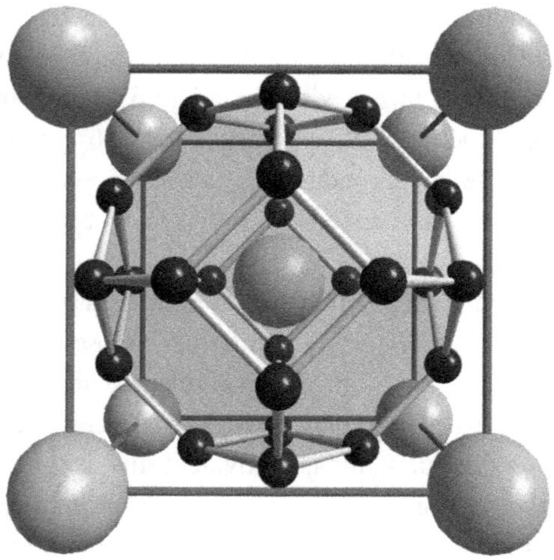

Figure 15.6 The cage structure of CaH_6.

diamonds. Therefore, pulsed laser heating was used, which is an important advance in high-pressure techniques (Goncharov et al., 2010). The structure of the synthesized sample exactly corresponded to the calculated structure of LaH_{10}, which was predicted to have superconductivity near RT (Liu et al., 2017, Peng, Sun et al., 2017).

The next push for us was a step in *R*(*T*) presented by R. Hemley in 2018 at a conference in Madrid (International Symposium on "Superconductivity and Pressure: A Fruitful Relationship on the Road to Room Temperature Superconductivity" by ARECES Foundation, 21-22 May 2018). Even though there was not a representative step around 260 K, and the resistance did not drop to zero as expected for a superconductor, we sped up searching superconductivity in lanthanum hydride. That was a troublesome problem, as the sample should be synthesized at very high pressures. The synthesized sample should contact all four electrodes (Figure 15.7), and the gasket should be insulating. We overcame these and many other difficulties and synthesized and systematically studied superconductivity in LaH_{10} with the aid of electrical measurements (zero resistance), the dependence of the superconductivity on pressure, the magnetic field, and the substitution of hydrogen with deuterium (the isotope effect) (Drozdov et al., 2019). We have done our work nearly simultaneously with Somayazulu et al. (2019). It was great that two groups obtained very similar results with different techniques: ~260 K was claimed in Somayazulu et al. (2019) and superconductivity at 250 K in our work (Drozdov et al., 2019). We synthesized samples directly from lanthanum and hydrogen, whereas Somayazulu et al. (2019) worked

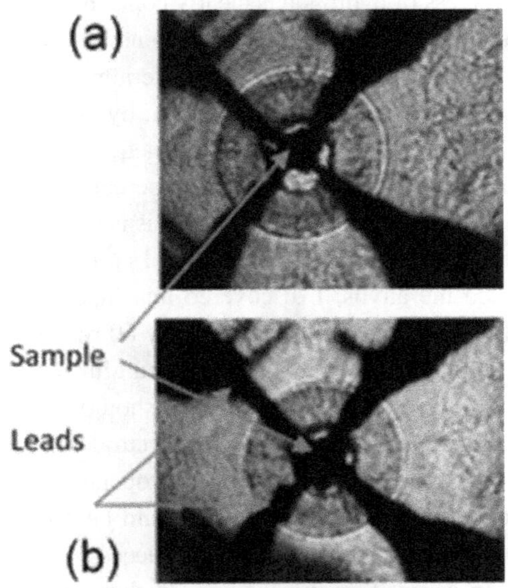

Figure 15.7 A sample synthesized at 159 GPa from a mixture of La and an excess of H_2. View of the sample inside the DAC with the attached four electrodes at transmission illumination (a) before and (b) after laser heating. As the result of heating, the sample significantly increased in volume and nearly filled the entire sample space but is still surrounded by hydrogen.

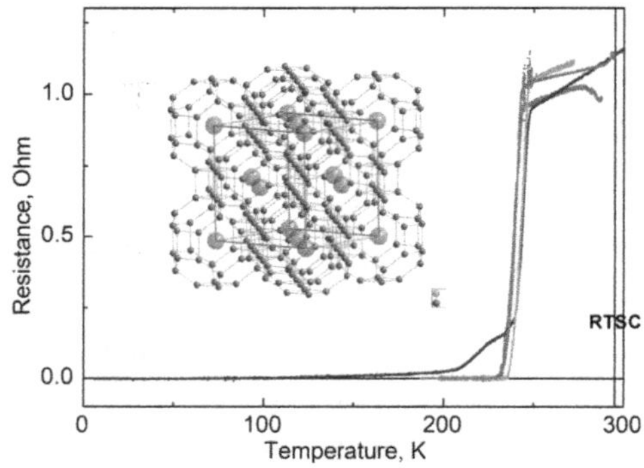

Figure 15.8 Superconducting transitions in lanthanum superhydride LaH_{10} measured in different samples shown with different colors. Below 250 K (–23°C), the resistance drops to zero, indicating a transition to a superconducting state. A vertical line drawn at 290 K marks the room temperature limit. The insert shows the structure of LaH_{10}: Blue balls indicate positions of La atoms, and gray small balls represent hydrogen atoms. Apparently, this structure can be considered metallic atomic hydrogen doped with lanthanum.

using ammonia borane as a precursor. Using BH_3NH_3 was an excellent idea – the experiment, especially electrical measurements, are greatly simplified. It works surprisingly well, but there are many uncertainties, such as whether there is an excess of hydrogen required for the synthesis, and if the sample is contaminated or doped with boron or nitrogen. In our direct synthsis of the hydride lanthanum, the changes in the sample hydrogenation can be clearly detected and an excess of hydrogen is clearly controlled (Figure 15.7). Compared to our colleagues (Somayazulu et al., 2019), we obtained zero resistance (Figure 15.8), determined an isotope effect via substitution of deuterium for hydrogen, and found that the deuterium samples adopt structures different from hydrogen samples. We also measured the dependence of superconductivity on the magnetic field, providing critical supporting evidence that the transitions are indeed due to superconductivity.

Our straightforward but accurate method of synthesis is more complicated, and we spent a lot of effort on it. I found an interesting statistic. As I polished all diamond anvils for the experiments and recorded the anvils, I discovered that input to the experiment included more than 100 pairs of diamond anvils; approximately 60 runs were done for hydride and 40 for deuterium. As output, we had about 20 meaningful runs. This low success ratio underlines the difficulty of the experiment. The DAC should be properly clamped, and the sample should be in the center and touching all four electrodes. In reality, in many cases the sample either did not touch all the electrodes or, if it touched all electrodes, the resistance was too high, and so on. In these cases, the cell should be opened and reloaded and the anvils repolished. Of course, there were many unsuccessful cases (not counted in the statistics) when hydrogen escaped during loading. In these cases, if the pressure did not reach ~20 GPa, the cell could be safely unloaded and used once again. Sometimes electrodes were damaged and should have been sputtered again. The rate of successful runs increases with experience. Nonetheless, the result was not bad: Approximately 70% of the clamped and laser-heated samples revealed superconductivity. For the best samples, with clearly measured T_c, we measured X-ray diffraction and the dependence on magnetic field.

15.7 Other Hydrides for RTSC on the Horizon

Achieving room temperature superconductivity at high pressure is already very close to being achieved. The next candidate is apparently yttrium hydride, which is very similar to LaH, possessing the same cage structure (Figure 15.6). According to calculations (Liu et al., 2017; Peng et al., 2017), superconducting *fcc*–YH_{10} should have an extremely high T_c of ~300–330 K in the 250–400 GPa pressure range. In addition to YH_{10}, there are other phases predicted to display very high values of T_c as well and to be stable at lower and more accessible pressures. In our experiments, we found the phases YH_6 and YH_9, in agreement with the predicted structures (Kong et al., 2021), but unfortunately, T_c was significantly lower than predicted, by ~30 K: The highest T_c was 243 K in YH_9. These results indicate that YH_{10} is also likely to have a lower T_c, below room temperature. Nevertheless, we tried to synthesize the YH_{10} phase. However, we found *P6₃/mmc* YH_9 as the most stable phase in the pressure–temperature domain up to 410 GPa and 2,250 K in contradiction to the ab initio calculations (Peng et al., 2017), which pointed to stabilization of *Fm–3m* YH_{10} phase at these conditions.

Theoretical calculations continue to produce interesting predictions. Both H_3S and LaH_{10} have cubic high-symmetry structures, which seemed obviously preferable. However, an unexpected *hexagonal* "penta-graphene-like" structure was found in ScH_{10} (Peng et al., 2017). For the similar structure in HfH_{10}, a very high T_c, up to ~230 K at pressures of ~250 GPa, was predicted (Xie et al., 2020).

Recently superconductivity in the room temperature range at 287.8 K and pressure of 267 GPa was claimed in a carbonaceous sulphur hydride (Snider et al., 2020). This material, however, is poorly characterized, and no X-ray diffraction data are presented. One can speculate that this is a ternary S–C–H compound or a sulphur hydride doped with carbon. At present, two years later after the announced room temperature superconductivity (Snider et al., 2020), this work has not been convincingly supported theoretically or confirmed experimentally by our and other groups despite extensive efforts.

15.8 Perspectives

Over the past five years, following the discovery of high T_c in H_3S (Drozdov et al., 2015b), we have witnessed a dramatic development of conventional superconductivity both in theory and experiment. The T_c values increased to 250–260 K. This hints toward the imminent achievement of room temperature (300 K) superconductivity in the coming years, most likely in hydrides at high pressures. Already many binary hydrides (more than 100) have been explored, mostly theoretically. The search is not exhausted yet, as there are many more, about 300 binary hydrides. Even more, about 1,770, stable ternary hydride compounds exist, and >30,000 quaternary hydrides also can be synthesized. How to find a good superconductor in this mass? Clearly, computation and theory should support experimental searches. For ternary hydrides with three elements, the challenges for calculations increase enormously. However, one of the first attempts brought an outstanding result: superconductivity at 473 K (200°C) at a pressure of 250 GPa for the ternary hydride Li_2MgH_{16} (Sun et al., 2019). This compound can be thought of as the binary hydride, MgH_{16}, doped with Li, an electron donor; this doping stabilizes the cage structure similar to LaH_{10}. The experiment should answer whether this is the right material and, hopefully, a superconductor with T_c above 300 K.

Another way to increase T_c of known superconductors or creating new superconductors is doping: inserting into a crystal (substitutionally or interstitially) extra atoms. This is a versatile and powerful method. For example, Ge et al. (2016) proposed doping H_3S with elements from neighboring groups in the periodic table and concluded that a T_c as high as 280 K could be reached at 250 GPa in $S_{0.925}P_{0.075}H_3$. Another amazing example is due to Flores-Livas et al., 2017, 2018), who showed that doped H_2O and polyethylene might exhibit superconductivity with a T_c of about 60 K at moderate pressures. Given the vast number of hydrides that are covalently bonded but remain insulating up to rather high pressures, this approach represents a very promising route to hunt for high-T_c hydrides.

Practical applications require *ambient* conditions of pressure and temperature. Presently the leading candidates for superconductivity at room temperature and pressure are unconventional superconductors, such as cuprates or iron pnictides and chalcogenides, which currently hold the record for the highest T_c at ambient pressure. These systems are still heavily investigated and might hold many surprises. However, a clear advantage of conventional superconductors is that theoretical and computational methods can be used to guide experimental discovery. The same can hardly be said about materials where exotic pairing mechanisms mediate superconductivity, as, in these cases, a full ab initio description of the superconducting pairing is missing.

The direct transfer of the successful experience of achieving high T_c in superhydrides at high pressure to ambient pressure is unlikely, because it is difficult for hydrogen at ambient pressure to be stable in any other chemical configuration than a fully saturated covalent bond due to its strong ionic potential and high mobility. However, given its small atomic size, hydrogen can be incorporated into existing materials and could act indirectly to induce or enhance superconductivity. This indirect key role of hydrogen is clearly seen in graphane (graphene saturated with hydrogen) and nanotubes (Sanna et al., 2018). When graphene is doped with holes, superconductivity has been predicted to occur at temperatures as high as 100 K. Following the same line of thought, and using an MgB_2 monolayer as a template material, Bekaert et al.. 2019) predicted a T_c increase in MgB_2 from 39–67 K.

A radical way is to look for nonhydrogen-related substances. Many other materials also have the required high phonon frequencies and electron–phonon coupling strengths. They include, in particular, carbon-based materials such as diamond (high Debye frequency ~1,860 K), graphite, or fullerenes. These are intrinsically insulators, semiconductors, or semimetals, but can be made metallic and superconducting by doping. For example, Si is a semiconductor, but becomes a metal and a superconductor with $T_c = 0.35$ K when doped at 5–8% with boron, which replaces carbon atoms and produces free carriers (holes)

(Bustarret et al., 2006). The case of carbon is even more impressive, as dielectric diamond turns to a metal upon 3% boron doping, and it is a superconductor with T_c up to 11 K (Ekimov et al., 2004), whereas amorphous carbon can be heavily doped with boron and reaches at 30% a T_c of 55 K (Bhaumik et al., 2017). Carbon-based clathrates are another prospective superconductor. The compound FC_{34} is predicted to have a T_c of up to 77 K (Zipoli et al., 2006). The T_c of sodalite-like NaC_6 is predicted to be above 100 K (Lu et al., 2016).

Finally, not only phonon-mediated superconductors but unconventional superconductors with other "glue" for electrons might be encountered in high- and ambient-pressure experiments as well. Unfortunately, predictive computational methods are currently not helpful in this case. However, should quantitative theories emerge for unconventional superconducting states, we could anticipate the same fruitful symbiosis among theory, computation, and experiment that proved so successful for the superconducting hydrides.

References

Allen, P. B., Dynes, R. C. (1975). Transition temperature of strong-coupled superconductors reanalyzed. *Physical Review B*, **12**, 905–922.

Ashcroft, N. W. (1968). Metallic hydrogen: a high-temperature superconductor? *Physical Review Letters*, **21**, 1748–1750.

Ashcroft, N. W. (2004). Hydrogen dominant metallic alloys: high temperature superconductors? *Physical Review Letters,* **92**, 187002.

Bardeen, J., Cooper, L. N., Schrieffer, J. R. (1957). Theory of superconductivity. *Physical Review*. **108**, 1175–1204.

Bednorz, J. G., Mueller, K. A. (1986). Possible high T superconductivity in the Ba– La–Cu–0 system. *Z. Phys. B – Condensed Matter*, **64**, 189–193.

Bekaert, J., Petrov, M., Aperis, A., Oppeneer, P. M., Milošević, M. V. (2019). Hydrogen-induced high-temperature superconductivity in two-dimensional materials: the example of hydrogenated monolayer MgB_2. *Physical Review Letters,* **123**, 077001.

Bernstein, N., Hellberg, C. S., Johannes, M. D., Mazin, I. I., Mehl, M. J. (2015). What superconducts in sulfur hydrides under pressure and why. *Physical Review B*, **91**, 060511(R).

Bhaumik, A., Sachan, R., Gupta, S., Narayan, J. (2017). Discovery of high-temperature superconductivity (Tc = 55 K) in B-doped Q-carbon. *ACS Nano,* **11**, 11915–11922.

Bustarret, E., Marcenat, C., Achatz, P., et al. (2006). Superconductivity in doped cubic silicon. *Nature,* **444**, 465.

Capitani, F., Langerome, B., Brubach, J.-B., et al. (2017). Spectroscopy of H3S: evidence of a new energy scale for superconductivity. *Nature Physics*, **13**, 859–863.

Cohen, M. L., Anderson, P. W. (1972). Comments on the maximum superconducting transition temperature. *AIP Conference Proceedings*, **4**, 17–27.

Drozdov, A. P., Eremets, M. I., Troyan, I. A. (2014). Conventional superconductivity at 190 K at high pressures. arXiv:1412.0460.

Drozdov, A. P., Eremets, M. I., Troyan, I. A. (2015a). Superconductivity above 100 K in PH3 at high pressures. arXiv:1508.06224.

Drozdov, A. P., Eremets, M. I., Troyan, I. A., Ksenofontov, V., Shylin, S. I. (2015b). Conventional superconductivity at 203 K at high pressures. *Nature,* **525**, 73.

Drozdov, A. P., Kong, P. P., Minkov, V. S., et al. (2019). Superconductivity at 250 K in lanthanum hydride under high pressures. *Nature*, **569**, 528.

Duan, D., Liu, Y., Tian, F., et al. (2014). Pressure-induced metallization of dense $(H_2S)_2H_2$ with high-Tc superconductivity. *Scientific Reports*, **4**, 6968.

Einaga, M., Sakata, M., Ishikawa, T., et al. (2016). Crystal structure of 200 K-superconducting phase of sulfur hydride. *Nature Physics*, **12**, 835–838.

Ekimov, E. A., Sidorov, V. A., Bauer, E. D., et al. (2004). Superconductivity in diamond. *Nature*, **428**, 542.

Eliashberg, G. M. (1960). Interactions between electrons and lattice vibrations in a superconductor. *Soviet Physics, JETP*, **11**(3), 696–702.

Eremets, M. I., Drozdov, A. P., Kong, P. P., Wang, H. (2019). Semimetallic molecular hydrogen at pressure above 350 GPa. *Nature Physics*, **15**, 1246.

Eremets, M. I., Gregoryanz, E., Mao, H. K., Hemley, R. J., Mulders, N., Zimmerman, N. (2000). Electrical conductivity of Xe at megabar pressures. *Physical Review Letters*, **85**(13), 2797–2800.

Eremets, M. I., Hemley, R. J., Mao, H. K., Gregoryanz, E. (2001a). Semiconducting non-molecular nitrogen up to 240 GPa and its low-pressure stability. *Nature*, **411**, 170–174.

Eremets, M. I., Struzhkin, V. V., Mao, H. K., Hemley, R. J. (2001b). Superconductivity in boron. *Science*, **293**, 272–274.

Eremets, M. I., Trojan, I. A., Medvedev, S. A., Tse, J. S., Yao, Y. (2008). Superconductivity in hydrogen dominant materials: silane. *Science*, **319**, 1506–1509.

Flores-Livas, J. A., Amsler, M., Heil, C., et al. (2016). Superconductivity in metastable phases of phosphorus–hydride compounds under high pressure. *Physical Review B*, **93**, 020508(R).

Flores-Livas, J. A., Boeri, L., Sanna, A., Profeta, G., Arita, R., Eremets, M. (2020). A perspective on conventional high-temperature superconductors at high pressure: methods and materials. *Physics Reports*, **856**, 1–78.

Flores-Livas, J. A., Graužinyte, M., Boeri, L., Profeta, G., Sanna, A. (2018). Superconductivity in doped polyethylene at high pressure. *European Physics Journal B*, **91**, 176.

Flores-Livas, J. A., Sanna, A., Graužinytė, M., Davydov, A., Goedecker, S., Marques, M. A. L. (2017). Emergence of superconductivity in doped H_2O ice at high pressure. *Scientific Reports*, **7**, 6825.

Gao, G., Oganov, A. R., Ma, Y., et al. (2010). Dissociation of methane under high pressure. *Journal of Chemical Physics*. **133**, 144508.

Gao, L., Xue, Y. Y., Chen, F., et al. (1994). Superconductivity up to 164 K in HgBa2Cam−1CumO2m+2+δ (m=1, 2, and 3) under quasihydrostatic pressures. *Physical Review B*, **50**, 4260–4263.

Gavaler, J. R. (1973). Superconductivity in Nb-Ge films above 22 K. *Applied Physics Letters*, **23**, 480.

Ge, Y., Zhang, F., Yao, Y. (2016). First-principles demonstration of superconductivity at 280 K in hydrogen sulfide with low phosphorus substitution. *Physical Review B,* **93**, 224513.

Geballe, Z. M., Liu, H., Mishra, A. K., et al. (2018). Synthesis and stability of lanthanum superhydrides. *Angewandte Chemie, International Edition*, **57**, 688.

Glass, C. W., Oganov, A. R., Hansen, N. (2006). USPEX – evolutionary crystal structure prediction. *Computer Physics Communications*, **175**, 713–720.

Goncharenko, I., Eremets, M. I., Hanfland, M., et al. (2008). Pressure-induced hydrogen-dominant metallic state in aluminum hydride. *Physical Review Letters,* **100**, 045504.

Goncharov, A. F., Prakapenka, V. B., Struzhkin, V. V., Kantor, I., Rivers, M. L., Dalton, D. A. (2010). X-ray diffraction in the pulsed laser heated diamond anvil cell. Review of Scientific Instruments, **81**, 113902.

Gor'kov, L. P., Kresin, V. Z. (2018). High pressure and road to room temperature superconductivity. *Review of Modern Physics*, **90**, 011001.

Hirsch, J. E. (2009). BCS theory of superconductivity: it is time to question its validity. *Physica Scripta*, **80**, 03570.

Kamihara, Y., Hiramatsu, H., Hirano, M., et al. (2006). Iron-based layered superconductor: LaFeP. *Journal of the American Chemical Society*, **128**, 10012–10013.

Kong, P. P., Minkov, V. S., Kuzovnikov, M. A., et al. (2021). Superconductivity up to 243 K in yttrium hydrides under high pressure. Nature Communications, **12**, 5075.

Li, Y., Hao, J., Li, Y., Ma, Y. (2014). The metallization and superconductivity of dense hydrogen sulfide. *Journal of Chemical Physics*, **140**, 174712.

Liu, H., Naumov, I. I., Hoffmann, R., Ashcroft, N. W., Hemley, R. J. (2017). Potential high-Tc superconducting lanthanum and yttrium hydrides at high pressure. *PNAS*, **114**, 6990.

Loubeyre, P., Occelli, F., Dumas, P. (2020). Synchrotron infrared spectroscopic evidence of the probable transition to metal hydrogen. *Nature*, **577**, 631.

Lu, S., Liu, H., Naumov, I. I., et al. (2016). Superconductivity in dense carbon-based materials. *Physical Review B*, **93**, 104509.

Ma, Y., Eremets, M., Oganov, A. R., et al. (2009). Transparent dense sodium. *Nature*, **458**, 182–185.

Mao, H. (1978). High-pressure physics: sustained static generation of 1.36 to 1.72 megabars. *Science*, **200**, 1145–1147.

Mao, H. K., Hemley, R. J. (1994). Ultrahigh pressure transitions in solid hydrogen. *Review of Modern Physics*, **66**, 671–692.

McMahon, J. M., Morales, M. A., Pierleoni, C., Ceperley, D. M. (2012). The properties of hydrogen and helium under extreme conditions. *Review of Modern Physics*, **84**, 1607–1653.

Medvedev, S., McQueen, T. M., Troyan, I. A., et al. (2009). Electronic and magnetic phase diagram of β–Fe1.01Se with superconductivity at 36.7 K under pressure. *Nature Materials*, **8**, 630–633.

Migdal, A. B. (1958). Interaction between electrons and lattice vibrations in a normal metal. *Soviet Physics, JETP*, **7**, 996.

Nagamatsu, J., Nakagawa, N., Muranaka, T., Zenitani, Y., Akimitsu, J. (2001). Superconductivity at 39 K in magnesium diboride. *Nature*, **410**, 63.

Nicol, E. J., Carbotte, J. P. (2015). Comparison of pressurized sulfur hydride with conventional superconductors. *Physical Review B,* **91**, 220507(R).

Peng, F., Sun, Y., Pickard, C. J., Needs, R. J., Wu, Q., Ma, Y. (2017). Hydrogen clathrate structures in rare earth hydrides at high pressures: possible route to room-temperature superconductivity. *Physical Review Letters*, **119**, 107001.

Pickard, C. J., Errea, I., Eremets, M. I. (2020). Superconducting hydrides under pressure. *Annual Review of Condensed Matter Physics*, **11**, 57–76.

Pickard, C. J., Needs, R. J. (2006). High-pressure phases of silane. *Physical Review Letters*, **97**, 045504.

Pickard, C. J., Needs, R. J. (2007). Metallization of aluminum hydride at high pressures: a first-principles study. *Physical Review Letters*. **76**, 144114.

Pickard, C. J., Needs, R. J. (2011). Ab initio random structure searching. *Journal of Physics: Condensed Matter*, **23**, 053201

Pickett, W., Eremets, M. I. (2020). The quest for room-temperature superconductivity in hydrides. *Physics Today*, **72**(5), 52–58.

Pizzochero, M., Leenaerts, O., Partoens, B., Martinazzo, R., Peeters, F. M. (2015). Hydrogen adsorption on nitrogen and boron doped graphene. *Journal of Physics: Condensed Matter*, **27**, 425502.

Quan, Y., Pickett, W. E. (2016). Van Hove singularities and spectral smearing in high-temperature superconducting H3S. *Physical Review B,* **93**, 04526.

Sanna, A., Davydov, A., Dewhurst, J. K., Sharma, S., Flores-Livas, J. A. (2018). Superconductivity in hydrogenated carbon nanostructures. *European Physics Journal B,* **91**, 177.

Schilling, A., Cantoni, M., Guo, J. D., Ott, H. R. (1993). Superconductivity above 130 K in the Hg–Ba–Ca–Cu–O system. *Nature*, **363**, 56–58.

Schilling, J. S. (2007). Studies in superconductivity at extreme pressures. *Physica C*, **460–462**, 182–185.

Si, Q., Yu, R., Abrahams, E. (2016). High-temperature superconductivity in iron pnictides and chalcogenides. *Nature Reviews Materials*, **1**, 16017.

Snider, E., Dasenbrock-Gammon, N., McBride, R., et al. (2020). Room-temperature superconductivity in a carbonaceous sulfur hydride. *Nature*, **586**, 373–377.

Somayazulu, M., Ahart, M., Mishra, A. K., et al. (2019). Evidence for superconductivity above 260 K in lanthanum superhydride at megabar pressures. *Physical Review Letters*, **122**, 027001.

Struzhkin, V. V., Eremets, M. I., Gan, W., Mao, H. K., Hemley, R. J. (2002). Superconductivity in dense lithium. *Science*, **298**, 1213-1215.

Struzhkin, V. V., Hemley, R., Mao, H. K., Timofeev, Y. A. (1997). Superconductivity at 10–17 K in compressed sulphur. *Nature*, **390**, 382.

Sun, Y., Lv, J., Xie, Y., Liu, H., Ma, Y. (2019). Route to a superconducting phase above room temperature in electron-doped hydride compounds under high pressure. *Physical Review Letters*, **123**, 097001.

Tanigaki, K., Ebbesen, T. W., Saito, S., et al. (1991). Superconductivity at 33 K in CsxRbyC60. *Nature,* **352** 222–223.

Timofeev, Y. A. (1992). Detection of superconductivity in high-pressure diamond anvil cell by magnetic susceptibility technique. *Prib. Tekh. Eksper.*, **5**, 186–189.

Wang, H., Tse, J. S., Tanaka, K., Iitaka, T. i., Ma, Y. (2012). Superconductive sodalite-like clathrate calcium hydride at high pressures. *PNAS*, **109**, 6463–6466.

Wigner, E. and H. B. Huntington (1935). "On the possibility of a metallic modification of hydrogen." J. Chem. Phys. **3**: 764-770.

Xie, H., et al. (2020). Hydrogen "penta-graphene-like" structure stabilized via hafnium: a high-temperature conventional superconductor. arXiv:2001.04076.

Wang, Y., Lv, J., Zhu, L., Ma, Y. (2010). Crystal structure prediction via particle-swarm optimization. *Physical Review B*, **82**, 094116.

Zipoli, F., Bernasconi, M., Benedek, G. (2006). Electron–phonon coupling in halogen-doped carbon clathrates from first principles. *Physical Review B*, **74**, 205408.

16

Thermochemistry of High-Pressure Phases

ALEXANDRA NAVROTSKY

In honor of H. K. (David) Mao and our interactions over half a century, this chapter focuses on the techniques and application of calorimetry to high-pressure research. The chapter reviews thermodynamic concepts and calorimetric methodology. It summarizes a large body of work over many years, with emphasis on mantle mineralogy, and also discusses recent developments in a broader context, including calorimetric studies of hydrous phases, nonoxides, and nanophase materials.

16.1 The Power and Utility of Thermodynamics

At the risk of being pedagogical but hopefully not pedantic, I start this review by addressing some fundamental thermodynamic relationships. There are two major goals of high-pressure research applied to Earth and planetary science. The first is to determine what phases form at equilibrium under high-pressure and high-temperature conditions in systems of petrologic interest and, conversely, to deduce the pressure, temperature, and composition (P, T, X) conditions that allow the formation of a given phase assemblage in a complex natural system. With growing interest in and information about planetary systems, both in our Solar System and the wealth of exoplanets beyond, the P, T, X conditions of interest are greatly expanded. The second goal is to determine, if one knows or can approximate the P, T, X conditions under which a given rock formed, whether that material attained and preserved its original equilibrium state, and if not, what it was originally and how it has been altered by its subsequent history. This second question is especially important for high-pressure phases, as they are very far from equilibrium under ambient conditions, and in many cases will not persist on release of pressure and temperature. Thus, thermodynamics is the starting point for kinetics, leading to understanding the present, past, and future states of planetary systems.

The Gibbs free energy at constant P and T can be broken into two terms, arising from enthalpy and entropy: G = H − TS. The enthalpy dominates at low temperature and is amenable to theoretical calculation using density functional theory (DFT) and other methods. The entropy term becomes more important as temperature increases. It contains terms arising from lattice vibrations, magnetic and electronic interactions, and configurational disorder, and is harder to calculate theoretically. At a given pressure, the balance between ΔH and TΔS (the enthalpy and entropy for a phase change) determines the temperature at which the transition (melting, vaporization, change in crystal structure) occurs.

Thermodynamics requires that the enthalpy and entropy of the high-temperature phase assemblage be higher than that of the low-temperature assemblage, i.e., ΔH of the transition on heating must be endothermic. This offers an immediate opportunity for experimental thermochemistry to be a "policeman" of equilibrium. If a phase transforms exothermically to another phase on heating at constant pressure (e.g., in a differential scanning calorimetry [DSC] experiment), that is incontrovertible evidence that the initial phase was metastable below the observed transformation temperature.

The temperature at which the exothermic transformation to the stable phase is observed is kinetically controlled, and on cooling the transformed material below that temperature, the stable phase is thermodynamically forbidden to transform back to the original metastable form.

While entropy increase is the driving force for phase transition with increasing temperature, with $(d\Delta G/dT)_{P,X} = -\Delta S$, volume decrease is the driving force for phase transition as a function of pressure, with $(d\Delta G/dP)_{T,X} = \Delta V$. Thus changes in entropy and volume determine the pressure–temperature slope of an equilibrium phase transition through the Clausius–Clapeyron equation: $(dP/dT)_{equilibrium} = \Delta S/\Delta V$. Commonly, ΔS and ΔV have the same sign; the denser phase is of lower entropy, and dP/dT is positive. Cases of negative P–T slopes are interesting from a structural and geophysical point of view, as discussed later for the magnesium silicate system.

For solid–solid transitions, it is often sufficient to assume that ΔS and ΔV are independent of P and T, producing a linear P–T boundary. In my opinion, there are relatively few high-pressure experiments that warrant the inclusion of curvature in the boundary when one considers uncertainties in P and T determination and the tightness (or lack thereof) of reversal brackets on the equilibrium.

There are a number of thermodynamic relations among the correction terms arising from compressibility (bulk modulus), thermal expansivity, and heat capacity, and these must be included at the same level of complexity in a self-consistent manner to obtain meaningful constraints on the curvature of a phase boundary. For transitions involving melting or vaporization, the P and T dependence of enthalpy, entropy, and volume changes becomes much more important.

To be widely useful, the thermochemical data must be incorporated in databases and made internally consistent in format and values. There are several complementary approaches, often distinct in the geochemical and materials community. Detailed discussion of databases is outside the scope of this chapter. Geochemical and materials science approaches to databases and phase equilibrium calculations have developed independently, with emphasis on different systems, temperature and pressure ranges, and parameters. Taking advantage of modern computational capabilities and data science, one has the opportunity to unify these diverse methodologies to provide much more complete, up-to-date, versatile, and user-friendly databases.

An equally important use of thermochemical data is to advance molecular-level understanding of structure and bonding, addressing the fundamental question of solid-state chemistry: "What structures can form at a given composition and why?" Here one needs strong synergy among thermochemical, structural, and computational approaches. Rapid advancement in all three areas makes this a very exciting time for linking microscopic and macroscopic aspects of mineral thermodynamics. Theorists and experimentalists are in a good position to integrate their approaches and keep each other honest.

Calorimetry is the science and art of measuring heat effects and can be grouped into two broad categories. The first is the measurement of heat capacities (C_p) or enthalpy increments ($H_{T2} - H_{T1}$) as a function of temperature on a sample that usually does not undergo chemical change. For materials that are fully ordered at low temperature, the integral of $(C_p/T)dT$ from a temperature close to absolute zero to the temperature of interest gives the standard entropy $S^o(T)$ at that temperature. The difference in standard entropy of two phases then provides the entropy of transformation at a given temperature. Cryogenic heat capacity measurements take specialized instrumentation and expertise, as discussed later. Heat capacities at and above room temperature are generally measured by DSC, which, though using commercial equipment, must be done carefully with proper attention to calibrations and other protocols. The second group of calorimetric methods involves intentional and controlled chemical reaction during calorimetry. Solution calorimetry, in which reactants and products are each dissolved separately in a suitable solvent, obtains the enthalpy of reaction through a thermochemical cycle, as the difference in enthalpies of solution. Dissolution in molten oxide solvents in specialized calorimeters at 700–800°C is now an established and generally applicable methodology (as discussed later in this chapter). For less refractory materials, solution calorimetry in aqueous solvents (e.g., HCl, NaOH, HF) can be used. Combustion calorimetry in oxygen or fluorine gas gives heats of formation of compounds from elements.

Once the enthalpy of reaction, ΔH, is obtained, it can be combined with the entropy of reaction, ΔS, from heat capacity measurements to provide the Gibbs free energy of reaction, $\Delta G^o = \Delta H^o - T\Delta S^o$ (at a given temperature, T, and ambient pressure, indicated by the superscript zero). The Gibbs free energy change at high pressure can then be calculated if the volume change of reaction is known. It is important that all these calculations be done consistently and at the same level of approximation for the temperature and pressure dependence of transition enthalpy, entropy, and volume terms. These calculations, relying on measured enthalpy and entropy, are sometimes called "Third Law" calculations because they rely on the assumption of zero entropy at 0 K. This method has been applied to the coesite–stishovite transition (Akaogi et al., 2011), obtaining satisfactory agreement between the purely calorimetric approach and measured phase equilibria.

If some disorder (usually in cation distributions) is present in a mineral, a calculated configurational entropy term can be added to the entropy of reaction, but this must be done with due caution because such a term rests on many assumptions.

A second approach, sometimes called a "Second Law" calculation, uses the measured enthalpy of reaction and the constraint that, at the equilibrium transition P and T, $\Delta G (P,T) = 0$. Then ΔS can be calculated from ΔG and ΔH. This method produces an entropy value containing all contributions, including those from disorder, at the given P and T. This approach is often used for high-pressure equilibria since standard entropies of high-pressure phases are not often known from heat capacity measurements. It is advantageous to take a P, T point, preferably in the middle of the range of *P–T* measurements, that is tightly bracketed or reversed. The value of ΔS can be compared to that obtained using the observed *P–T* slope of transition through the Clausius–Clapeyron equation and can help identify whether equilibrium has been attained in the *P–T* slope measurements, especially at lower temperature. Analysis of the coesite–stishovite transition (Liu et al., 1996) is an example of such a study in which care was taken in sample preparation and calorimetry, phase boundary bracketing, and calculations using available thermophysical parameters. The Second Law approach is used more extensively for high-pressure phases than the Third Law approach because of the challenge in obtaining sufficient sample for cryogenic heat capacity measurements.

Thus, calorimetric measurements complement direct high *P–T* experiments. The goal is to combine both types of studies into an internally consistent set of data to describe high-pressure phase equilibria.

16.2 Advances in Calorimetric Methodology

16.2.1 Low-Temperature Heat Capacity Measurements

Cryogenic heat capacity measurements have been performed for almost a century using large custom-built adiabatic calorimeters. These experiments were the domain of specialists and required grams of material, making them out of reach for high-pressure materials. Quantum Design Inc. has for some time developed and sold the Physical Property Measurement System (PPMS) that has a heat capacity option utilizing the thermal relaxation method. The PPMS heat capacity option is highly automated and can generate acceptable Third Law quality heat capacity data in the temperature range of 2–300 K using much smaller samples. Woodfield's group at Brigham Young University has developed sample mounting techniques that have eliminated errors for poor thermally conducting ceramic and mineral samples (Rosen and Woodfield, 2020). These methods are now more generally accessible, and reliable heat capacities and entropies can be obtained on relatively small samples (10–100 mg) in a few days of measurements.

Custom built "calorimeters-on-a-chip" have been used to measure heat capacities of microgram samples, especially of thin films. An exploratory study several years ago showed the feasibility of applying this method to high-pressure silicates by performing measurements on a sample of Fe_2SiO_4 spinel (Navrotsky et al., 2007) (Figure 16.1). However, the lack of a user-friendly and generally available apparatus is a hindrance to application in mineral physics. The next frontier for the PPMS is to develop sample platforms accessible to the general scientist that push sample sizes down to the 100 μg level while maintaining acceptable accuracy and resolution.

Woodfield has also recently developed methods for measuring the heat capacity of samples under gas pressures of less than 100 psi, which is useful for looking at the energetics of molecular confinement, especially of CO_2 (Rosen et al., 2020). Further developments and applications are expected. The emphasis in these studies is on changes

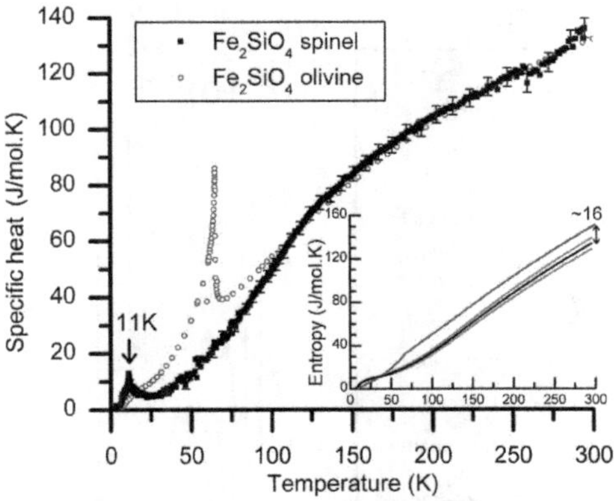

Figure 16.1 Example: Heat capacities and entropies of Fe2SiO4: olivine and spinel polymorphs. The data for the spinel were obtained by a "calorimeter-on-a-chip," using a sample weighing less than a milligram. Note that the main difference in room temperature entropies arises from heat capacity differences below 100 K. The peaks in heat capacity represent magnetic transitions (Navrotsky et al., 2007).

in lattice dynamics, especially in porous materials, triggered by gas sorption. Though the materials science applications currently driving this work are in catalysis and separations, similar sorption phenomena are likely encountered in Earth and planetary environments, and their thermodynamics will influence the solubility and transport of H_2O, CO_2, CH_4, and other volatiles. This control of water content is especially important when dealing with minerals with variable degrees of hydration.

16.2.2 High-Temperature Solution and Reaction Calorimetry

Because many minerals and ceramics are not readily soluble in room temperature aqueous solvents, oxide melt solution calorimetry has been the method of choice for obtaining their enthalpies of formation since the 1960s. The signature instruments are Calvet high-temperature calorimeters (Figure 16.2) used for oxide melt (or other solvent) solution calorimetry and for heats of mixing in molten systems at 700–800°C. The basic calorimeter and especially the calorimetric methodology have undergone extensive improvements in the design, electronics, operation, and range of materials that can be studied over the past several decades (Navrotsky, 1977, 1997, 2014). This apparatus, initially custom built, primarily in Navrotsky's laboratory, has recently been commercialized as the Setaram AlexSYS calorimeter and is now installed in a number of laboratories (e.g., Notre Dame University, Washington State University, Los Alamos National Laboratory, Pacific Northwest National Laboratory, Clemson University, NASA Glenn Research Center, University of California Davis, Arizona State University, and Karlsruhe Institute of Technology). This availability represents a major expansion in capabilities and active projects. These new efforts are linked through an informal Thermodynamics Consortium,

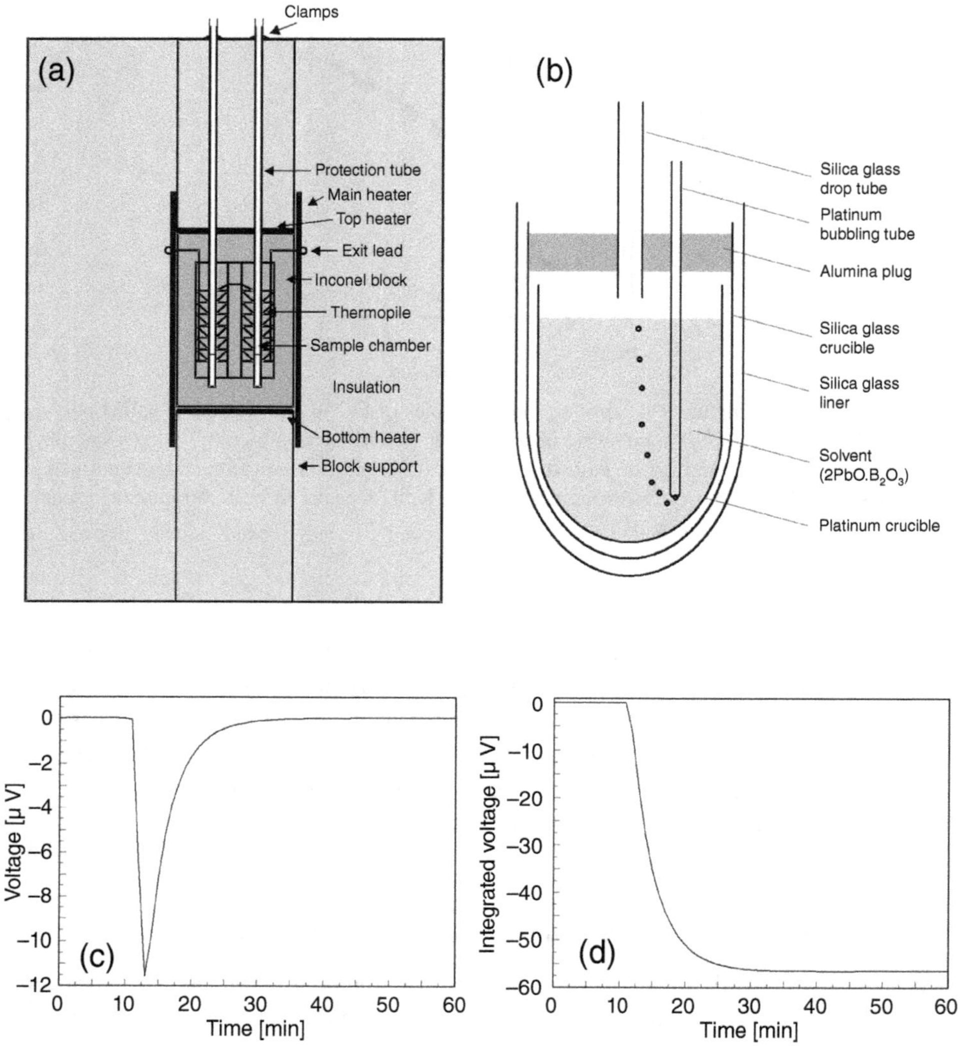

Figure 16.2 (a) Schematic cross section of a high-temperature solution calorimeter. The actual height of the instrument is about 1.5 m. (b) Detail of the sample chamber of the calorimeter. (c) Voltage generated in thermopile of the calorimeter as a function of duration of the experiment. Note that the signal (baseline) approaches 0 after about 40 minutes, indicating the end of the reaction. (d) Integrated voltage versus time curve (proportional to measured heat effect).

which brings together investigators for collaborations and workshops and is expected to grow and diversify.

Oxide melt solution calorimetry has been used extensively for materials containing H_2O and CO_2, which are evolved as gases upon dissolution of the sample in the oxide melt (Navrotsky, 1997; Navrotsky et al., 2007). Of interest to high-pressure research is older work on katoite garnet (Schoenitz and Navrotsky, 1999) and $Ca(OH)_2$ (Schoenitz et al.,

2000) and more recent work on hydrous stishovite (Spektor et al., 2011, 2016). Recent advances in solution calorimetry methodology include better control of oxygen fugacity and sample oxidation state (Birkner and Navrotsky, 2012; Lilova et al., 2012, 2014; Sahu et al., 2015), applications to carbides (Niu et al., 2017), carbon nanotubes (Levchenko et al., 2010), carbon in silicates (Navrotsky et al., 2020), nitrides (Liang et al., 1999; Tessier et al., 2000; Navrotsky, 2001; Zhang et al., 2006), chalcogenides (Deore and Navrotsky, 2006; Deore et al., 2008; Navrotsky and Xu, 2010; Tavakoli and Navrotsky, 2013), and metals (Sharma et al., 2016). A sample holding and dropping mechanism suitable for highly air-sensitive, toxic, or radioactive samples has been developed (Guo et al., 2019). Direct combustion calorimetry of small samples of metals, alloys, and organic and hybrid organic–inorganic solids such as metal organic frameworks (MOFs) is being developed. At present, sample sizes down to about 1 mg per experiment can be used, making the methodology compatible with materials synthesized in a multi-anvil apparatus.

Figure 16.2 shows the high-temperature Calvet-type reaction calorimeter. Using a thermopile (PtRh thermocouples in series), the calorimeter measures heat flow between the sample chamber, where a reaction occurs, and a large metal block heat sink; thus the surroundings are constant (isoperibol conditions). The calorimeter is twinned (two sample chambers with their thermopiles in opposition), providing greater signal stability, but each chamber can be used sequentially for measuring reactions.

The typical solution calorimetric measurement, called a drop-solution experiment, consists of dropping a loosely pressed sample pellet between 1–5 mg in weight from room temperature into the molten salt solvent in the calorimeter. Solvents commonly used are lead borate ($2PbO–B_2O_3$) and sodium molybdate ($3Na_2O–4MoO_3$). The former is more suitable for silicates, the latter is generally used for oxides with large trivalent, tetravalent, and pentavalent ions. Oxidation state can be controlled by bubbling gas through the solvent, which also provides efficient stirring. Calorimeter calibration is provided by dropping a sample of known heat content, typically corundum. In general, heat effects can be measured with an accuracy of about 1%.

Another commercial calorimeter of modified Calvet design, the Setaram HT96, can be used up to 1,500°C. Direct calorimetric measurements of enthalpies of solution of solid oxides in silicate melts have been performed successfully, although the accuracy is somewhat less than at 700°C or 800°C (Linard et al., 2008; Koryttseva and Navrotsky, 2016; Costa et al., 2019).

Solution calorimetry near room temperature in aqueous acid, base, or organic solvents (Wu and Navrotsky, 2015; Sharma et al., 2016; Wu et al., 2019) is useful for materials that dissolve readily. There have been improvements in calorimetric techniques aimed at using smaller samples. Dissolution in hydrofluoric acid is also useful for specific materials. It has been used to study organic–inorganic interactions in SiO_2, such as those between an organic "structure-directing agent" and a growing silica phase (Trofymluk et al., 2012). Such studies may be extended to questions of biomineralization and the occurrence and interactions of organics with minerals at modest and high pressures.

DSC, often coupled with thermogravimetric analysis (TGA), is well established, with excellent commercial instrumentation and standard protocols for temperatures up to about

800°C. A smaller selection of instruments is available for work to about 1,600°C, but accuracy is often compromised by reactions between the sample and its container (ceramic or noble metal) and/or the calorimetric detector. A commercial instrument is available for differential thermal analysis (DTA) to about 2,500°C, but the quantification of enthalpies of phase transitions such as melting requires the use of enthalpy standards. Until recently, the highest-temperature oxide thermal analysis standard was alumina, with a melting point near 2,000°C. To overcome this limitation, and to avoid increasingly acute containment problems, a miniaturized drop calorimeter was designed and tested in which laser-heated levitated samples of about 100 mg are equilibrated at high temperatures and then dropped into a small room temperature calorimeter to measure their heat content as a function of temperature (Kapush et al., 2017; Ushakov et al., 2018). This "drop-n-catch" calorimeter (see Figure 16.3) has been used to measure heats of fusion of high melting oxides of the rare earths and related materials (Navrotsky and Ushakov, 2017; Ushakov et al., 2018). These measurements have established yttria as a high-temperature DSC standard. The measured heats of fusion have been benchmarked against first-principles calculations with good agreement (Kapush et al., 2017). Thermochemical data on the melting of refractory oxides are essential to build a database applicable to petrologic processes at both ambient and high pressures. Future work is expected to refine the accuracy of this new instrument and to solve problems of temperature gradients needed to enable the calculation of heat capacities in addition to enthalpies of phase transitions at temperatures up to about 3,000°C.

16.3 Specific Applications

16.3.1 Silicate Spinels, Perovskites, and Related Phases

Early calorimetric work from Navrotsky's laboratory focused on the calorimetry of phenacite, olivine, spinelloid, and spinel phases in the silicates and germanates of Mg, Fe, Co, Ni, and Zn (Navrotsky, 1980, 1981, 1987, 1998: Akaogi et al., 1984, 1989; Navrotsky and Akaogi, 1984). A triumph of high-pressure synthesis and oxide melt solution calorimetry about three decades ago was the determination of the enthalpy of $MgSiO_3$ ilmenite (Ito and Navrotsky, 1985) and then of $MgSiO_3$ perovskite, the latter using one 2 mg sample synthesized in a multi-anvil apparatus (Ito et al., 1990). The enthalpies of transition from ilmenite to perovskite and from spinel to perovskite assemblages gave direct confirmation of negative *P–T* slopes for the transformations and suggested that perovskite $MgSiO_3$ has a higher vibrational entropy than $MgSiO_3$ ilmenite, despite its higher density. Spectroscopic, computational, and lattice dynamic studies (Akaogi et al., 1984; Ito et al., 1990; Kojitani et al., 2012) suggested features in the vibrational density of states of the perovskite that concentrate vibrational modes at low frequencies and are probably related to the longer than optimal Mg–O bond length in the eight-coordinate site of the perovskite. This work pioneered the value of combining thermochemical, spectroscopic, and computational approaches to high-pressure phase transitions. Later work (Kojitani et al., 2012) refined the calorimetric data with better samples but did not change the fundamental picture. Once more, material could be synthesized, cryogenic heat capacity measurements became possible, and the entropies of a number of high-pressure magnesium silicates were determined, further

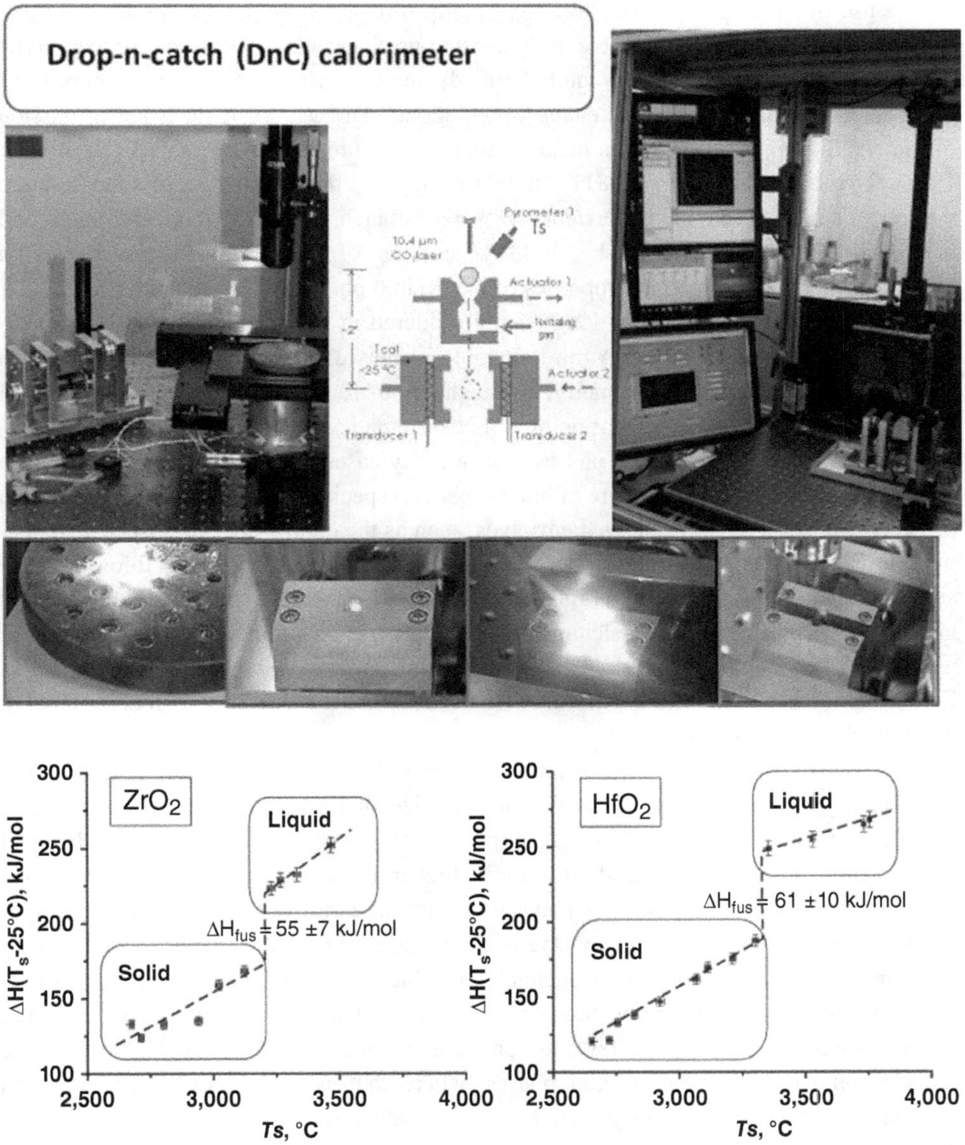

Figure 16.3 Top: Drop-n-catch calorimeter for measurements to 3,000°C. Bottom: Data obtained for melting of zirconia and hafnia. Fusion enthalpy from drop-n-catch calorimetry on HfO_2 (a) and ZrO_2 (b) in argon flow. T_s – surface temperature measured by FAR spectro-pyrometer. For the color version, refer to the plate section.
(Hong et al., 2018).

constraining phase diagrams and *P–T* slopes (Kojitani et al., 2012). Such work continues to the present day, mainly in Akaogi's laboratory in Japan, where synthesis, characterization, and calorimetry are performed in-house on the same materials, such as postspinel phases (Kojitani et al., 2016).

The use of "analog materials," different compositions having analogous phase transitions under more easily attainable *P–T* conditions, is a well-established strategy. The systematics of crystal chemistry and thermodynamics across a range of compositions enhances our understanding and enables predictions. An early example is the discussion of the stability of rocksalt versus nickel arsenide structures in binary oxides (Navrotsky, 1981; Navrotsky and Davies, 1981). Another example is the systematics of solid-solution thermodynamics based on the extent of size mismatch in binary oxides, spinels and perovskites (Davies and Navrotsky, 1983). The extent of solid solubility, governed both by the thermodynamic mixing properties of the desired phase and the competition of other possible phase assemblages, has long been considered in the geochemistry of multicomponent mineral solid solutions. A third example is the well-documented relation between perovskite tolerance factor, symmetry, and enthalpy of formation (Takayama-Muromachi and Navrotsky, 1988). The salient point is that systematics provides predictive power.

Spinels are moderately dense phases, but in many cases they are somewhat less dense than other polymorphs or a mixture of binary oxides, especially when the latter can undergo phase transitions to denser phases themselves, such as the quartz–coesite–stishovite transitions in silica (Akaogi and Navrotsky, 1984). There is an inherent competition between single-phase high-pressure polymorphs and multiphase mixtures. Two possible postspinel high-pressure phases are the calcium titanate (CT) and calcium ferrite (CF) structures, which have been studied extensively in vanadate and chromite systems (Ishii et al., 2018). Crystal chemical systematics suggest that the CF structure is more flexible in accommodating a much larger range of cations.

The discovery of postperovskite phases in $MgSiO_3$ and related materials has been a game changer in mineral physics. We now realize there are likely whole families of complex structures, slightly denser than perovskite, where small adjustments in local coordination and symmetry lead to slightly higher density. Thus, highly symmetrical phases such as cubic perovskite or rocksalt are not necessarily the limiting case in high-pressure experiments. Most of the postperovskite phases of petrologic interest occur at very high pressure and appear to be unquenchable to ambient conditions. Therefore, they are not accessible to calorimetric study; however, a number of analog materials having similar structures exist at lower pressures and can be maintained to ambient conditions. These include some rhodium and iridium phases, where the energetic relationships between perovskite and postperovskite phases has been studied by calorimetry (Kojitani et al., 2007). The high-temperature calorimetric methodology needed to be modified to take advantage of, instead of being compromised by, the easy reduction of Rh and Ir ions to the metallic state. This was done by dropping samples into a calorimeter without any solvent present where the final state was shown to be CaO and metallic Ir or Rh.

16.3.2 Chalcogenides, Nitrides, and Carbides

Chalcogenides, carbides, and nitrides are of interest to materials science and mineral physics. Both their low-pressure and high-pressure chemistry and thermodynamics are far less explored than those of oxides. Slight modifications of the high-temperature oxide melt solution calorimetric techniques enable measurement of formation enthalpies of these materials. Chalcogenides (sulfides selenides, tellurides) are dissolved readily in molten sodium molybdate solvent at 700°C via an oxidative reaction to produce dissolved metal ions and sulfate, selenate, or tellurate anions. The method was vetted using a number of binary chalcogenides (Deore and Navrotsky, 2006; Deore et al., 2008) and then applied pyrrhotite (Navrotsky and Xu, 2010) and to the Fe–Ni–S system (Tavakoli and Navrotsky, 2013). The enthalpy of formation of the spinel form of Fe_3S_4 (greigite) and its high-pressure polymorph have been measured recently (Subramani et al., 2020). Sulfates can also be dissolved directly in the molten molybdate solvent to determine their heats of formation (Majzlan et al., 2005; Navrotsky et al., 2005).

Under an oxidizing atmosphere, nitrides and oxynitrides dissolve as oxides in either lead borate or sodium molybdate solvent. Nitrogen gas is quantitatively evolved during dissolution. The enthalpies of formation of a number of crystalline nitrides and oxynitrides and of a series of rare earth containing oxynitride glasses have been determined by this technique (Navrotsky, 2001). An example of potential geological importance is the iron nitrides (Tessier et al., 2000). An interesting case is an oxynitride perovskite, where a relatively stable intermediate in the oxidation process appears to contain N_2 molecules trapped in interstitial sites (Tessier et al., 2005). One must ask whether similar intermediates with trapped gas molecules are possible at high pressure. The enthalpies of formation of atmospheric-pressure polymorphs and of the high-pressure spinel form of silicon nitride have also been determined (Liang et al., 1999; Zhang et al., 2006).

Carbon phases, carbides, and oxycarbides can also be studied, with oxidative drop-solution calorimetry producing dissolved oxides and evolved CO_2. Silicon oxycarbide "polymer-derived ceramics" have been studied extensively for their thermal stability and interesting mechanical properties (Morcos et al., 2008; Widgeon et al., 2012; Mera et al., 2013; Niu et al., 2017). These materials are amorphous at the bulk scale but contain local intertwined carbon-rich and silica-rich nanodomains. Their somewhat surprising energetic stability has led us to hypothesize that carbon can be incorporated into silicates, including geologic and planetary materials at high pressure, by forming "mixed tetrahedra," where carbon substitutes for oxygen (Sen et al., 2013; Navrotsky et al., 2020). Calorimetric study of synthetic high-pressure carbon-containing materials is a growing field.

16.3.3 Water and Defects in High-Pressure Phases

Nominally anhydrous phases may contain significant amounts of H_2O in their crystal structures when synthesized in a hydrous environment. At a phase boundary, the activity (but not the concentration) of water must be equal in the phases in equilibrium. If the system is water saturated, that activity is given by the fugacity of water, which is moderately well known at up to several gigapascals but less constrained at higher pressure. In general, the solubility of H_2O in a given phase increases with pressure but decreases with temperature, so synthesis of materials with high water content is usually done at relatively low temperature and high pressure. Stishovite appears to dissolve more water than quartz or coesite, and the energetics of hydrous stishovite has been studied by oxide melt solution calorimetry (Spektor et al., 2011, 2016). Higher water solubility is suggestive of greater thermodynamic stabilization of that phase in a hydrous environment. Thus, the stability field of the more hydrous phase is expected to increase at the expense of the less hydrous phase, especially in a water-saturated environment.

Deviations from ideal stoichiometry and incorporation of altervalent ions (especially Al^{3+} and Fe^{3+}) in silicates are an important complication in high-pressure equilibria, especially in multicomponent natural systems. The substitution of trivalent ions can occur by several distinct mechanisms. Ferric iron can substitute for ferrous iron (representing oxidation) by the creation of one cationic vacancy for every two iron atoms oxidized. Spinels in the system $Fe_3O_4 - \gamma\ Fe_2O_3$ (magnetite–maghemite) are the classic example of this substitution, and their thermochemistry has recently been refined by detailed calorimetric studies (Lilova et al., 2012). Quantitative thermodynamic study of high-pressure iron-containing silicates is challenging because of the difficulty of defining and maintaining a well-defined oxygen fugacity in high *P–T* experiments, followed by appropriate characterization of the run product(s). An oxide melt solution calorimetric study of materials with known compositions and ferric/ferrous ratios should be possible if appropriate time and care are invested. Based on recent calorimetric experiments on iron oxides (Lilova et al., 2012, 2014), earlier problems of controlling the oxidation state and avoiding parasitic side reactions during calorimetry have been brought under control. Experiments under oxidizing conditions, which ensure the dissolved state of iron in the calorimetric solvent is completely ferric, are accurate and reliable.

Aluminum can substitute for magnesium in octahedral sites by an analogous mechanism, creating cationic vacancies. Because aluminum can also substitute for silicon in tetrahedral sites, there are additional defect mechanisms, including the charge-balanced simultaneous substitution of aluminum for silicon and for magnesium. Systematic calorimetric study of the energetics of these substitutions in high-pressure phases are feasible, and some preliminary studies have been reported (Navrotsky et al., 2002, 2003). In distinction from olivines and spinels, where anion vacancies appear to be minor or nonexistent, perovskites are well known to harbor anion vacancies with relative ease, and the substitution of aluminum into $MgSiO_3$ perovskite may follow multiple mechanisms (Navrotsky et al., 2003; Stebbins et al., 2003). Thus, the variety of possible defect equilibria in silicate perovskites is potentially greater than that in lower-pressure phases.

16.3.4 Nanoscale Effects

It has been recognized for over two decades now that nanoscale materials, whether separate particles in air or in water or as solid nanocomposites, have different thermodynamic properties from bulk materials, arising mainly from the additional energy of their surfaces and interfaces (Navrotsky, 2016).

Because such surfaces and interfaces can have different energies for different structures (even at the same overall composition), the thermodynamics of phase transitions will be

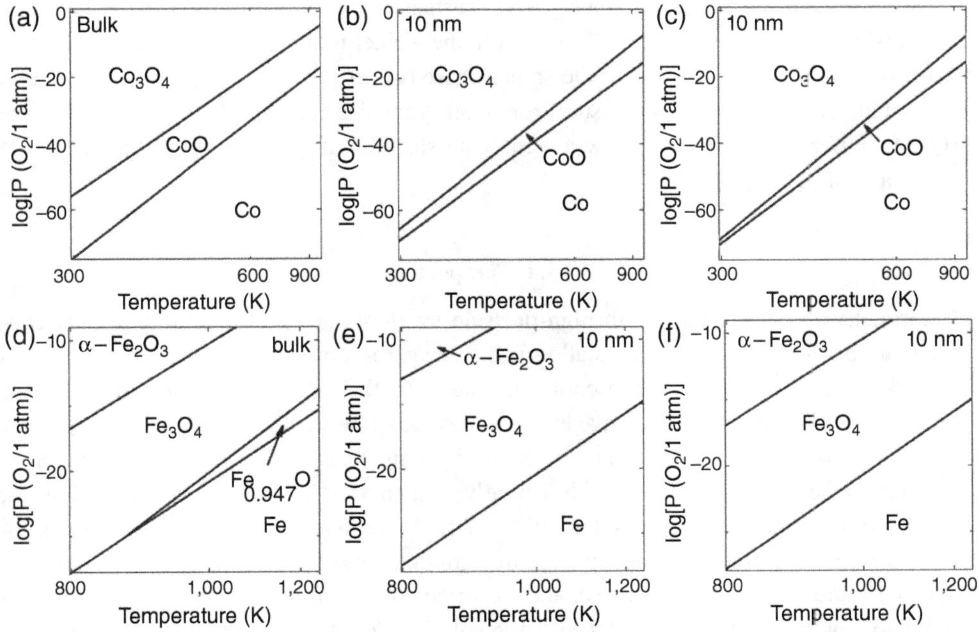

Figure 16.4 Calculated Fe–O phase diagram for bulk ((a) and (d)) and 10 nm particles ((b), (c), (e), and (f)). Note expansion of the spinel (magnetite) field and disappearance of the rocksalt (wustite) field, reflecting the low and high surface energies of these two phases, respectively
(Navrotsky et al., 2010).

affected, with the phase(s) of lower surface or interface energy expanding their stability fields at the expense of those with higher energies. These effects can lead to the thermodynamic stabilization at the nanoscale of phases metastable in the bulk and to significant changes in dehydration and redox equilibria (see Figure 16.4) (Navrotsky et al., 2008, 2010; Birkner and Navrotsky, 2012; Navrotsky, 2011, 2016).

While isolated nanoparticles are probably not common at high pressure, their aggregates, dense nanocomposites with small individual crystallites commonly occur, joined together by a large number of grain boundaries and other interfaces. In efforts to get faster reactions and approach to equilibrium, especially at low temperatures, in high-pressure experiments, one often uses amorphous, nanophase, or sol–gel starting materials. There have been discrepant versions of the quartz – coesite and coesite–stishovite phase boundaries (see, for example, Akaogi and Navrotsky, 1984; Liu et al., 1996). Though difficulties in attaining equilibrium reversals below about 800°C are certainly part of the problem, the use of nanoparticulate starting materials under *P–T* conditions that do not lead to grain growth may also be contributing complications. Other transitions affected by particle size are those between monoclinic, tetragonal, and cubic phases in zirconia and hafnia (Sharma et al., 2018) and between anatase, rutile, α−PbO_2, and fluorite phases in TiO_2.

Because spinels generally have lower surface energies than olivines (Chen and Navrotsky, 2010; Lilova et al., 2018), the olivine–spinel transition in silicates and germanates is expected to shift to favor the spinel phase (lower pressure at a given temperature, lower temperature at a given pressure) for small grain size. The quartz–fayalite–magnetite (QFM) buffer is predicted to show a significant shift favoring magnetite at the nanoscale (Lilova et al., 2018).

16.4 Perspective

Calorimetry in combination with high-pressure synthesis and phase equilibria must continue to be integrated with crystallographic, spectroscopic, and theoretical studies to provide macroscopic and microscopic insights into the structure and dynamics of the interior of the Earth and other planetary bodies. Rapidly evolving information on other planets, moons, and asteroids in our Solar System, and on the staggering wealth of exoplanets being discovered weekly, greatly extends the range of relevant pressures, temperatures, and compositions for study of high-pressure materials. For example, an Earth-like planet with a shallower geothermal gradient might contain hitherto undiscovered hydrous minerals; the hot, corrosive, and CO_2-rich conditions on Venus-like planets might stabilize new carbonates and sulfates; while cold icy worlds may contain a wealth of new inorganic, organic, and hybrid organic–inorganic minerals. Such materials, potentially attainable in the laboratory, would require structural and thermodynamic characterization, including calorimetric studies. At the same time, new materials for technology, many of them far from equilibrium as nanomaterials, composites, and thin films, are being created at a great pace. Their stability or metastability must be characterized to optimize their synthesis and to understand the driving forces for their degradation in use. Both nature and technology access a dense energy landscape of possible structures and transformations; understanding the similarities and differences in these two domains brings materials science and Earth and planetary sciences closer together.

It is difficult to predict advances in calorimetric technology. More and more, instruments are developed commercially, with decreasing interest in the "home built." Faster, more stable, more accurate, more computerized, more convenient, and cheaper are the desiderata and current buzzwords. At the same time, one must be wary of instruments good enough for routine work but not sufficiently accurate or flexible for the highest-quality measurements on demanding materials. One hopes for stronger collaboration between research scientists and instrument developers.

The computer and data revolution is upon us. The interplay among first-principles calculations, data science, and machine learning is becoming more intense. New databases must be constructed to be flexible, user-friendly, easily upgraded, and reliable. In all such work, accurate experimental thermochemical data continue to be the cornerstone and arbiter; the relation between experiment and computation must be a synergy, not a competition. Computation, whether of fundamental physical parameters or of phase diagrams, will identify problems and inconsistencies, which will provide focal points for additional experiments. Calorimetry will continue to be essential.

Acknowledgments

Calorimetric work over several decades has been supported by a series of grants from the US National Science Foundation and the Sloan Deep Carbon Observatory. I thank many colleagues here and abroad for continuing interest and encouragement, discussions, and samples of high-pressure phases.

References

Akaogi, M., Ito, E., Navrotsky, A. (1989). Olivine-modified spinel–spinel transitions in the system Mg_2SiO_4–Fe_2SiO_4: calorimetric measurements, thermochemical calculations, and geophysical application. *Journal of Geophysical Research*, 94(**B11**), 15671–15685.

Akaogi, M., Navrotsky, A. (1984). The quartz–coesite–stishovite transformations: new calorimetric measurements and calculation of phase diagrams. *Physics of the Earth and Planetary Interiors*, **36**(2), 124–134.

Akaogi, M., Oohata, M., Kojitani, H., Kawaji, H. (2011). Thermodynamic properties of stishovite by low-temperature heat capacity measurements and the coesite–stishovite transition boundary. *American Mineralogist*, **96**(8–9), 1325–1330.

Akaogi, M., Ross, N. L., McMillan, P., Navrotsky, A. (1984). The Mg_2SiO_4 polymorphs (olivine, modified spinel, and spinel) – thermodynamic properties from oxide melt calorimetry, phase relations, and models of lattice vibrations. *American Mineralogist*, **69**(5–6), 499–512.

Birkner, N., Navrotsky, A. (2012). Thermodynamics of manganese oxides: effects of particle size and hydration on oxidation-reduction equilibria among hausmannite, bixbyite, and pyrolusite. *American Mineralogist*, **97**(8), 1291–1298.

Chen, S., Navrotsky, A. (2010). Calorimetric study of the surface energy of forsterite. *American Mineralogist*, **95**(1), 112–117.

Costa, G. C. C., Harder, B. J., Weisner, V. L., et al. (2019). Thermodynamics of reaction between gas-turbine ceramic coatings and ingested CMAS corrodents. *Journal of the American Ceramic Society*, **102**(5), 2948–2964.

Davies, P. K., Navrotsky, A. (1983). Quantitative correlations of deviations from ideality in binary and pseudo-binary solid solutions. *Journal of Solid State Chemistry*, **46**(1), 1–22.

Deore, S., Navrotsky, A. (2006). Oxide melt solution calorimetry of sulfides: enthalpy of formation of sphalerite, galena, greenockite and hawleyite. *American Mineralogist*, **91**(2), 400–403.

Deore, S., Xu, F., Navrotsky, A. (2008). Oxide-melt solution calorimetry of selenides: enthalpy of formation of zinc, cadmium and lead selenide. *American Mineralogist*, **93** (5), 779–783.

Guo, X., Boukhalfa, H., Mitchell, J., et al. (2019). Sample seal-and-drop device and methodology for high temperature oxide melt solution calorimetric measurements of PuO_2. *Review of Scientific Instruments*, **90**(4), 044101–044107.

Hong, Q.-J., Ushakov, S. V., Kapush, D., et al. (2018). Combined computational and experimental investigation of high temperature thermodynamics and structure of cubic ZrO_2 and HfO_2. *Scientific Reports*, **8**(1), 14962.

Ishii, T., Sakai, T., Kojitani, H., et al. (2018). High-pressure phase relations and crystal structures of postspinel phases in MgV_2O_4, FeV_2O_4, and $MnCr_2O_4$: crystal chemistry of AB_2O_4 postspinel compounds. *Inorganic Chemistry*, **57**(11), 6648–6657.

Ito, E., Akaogi, M., Topor, L., Navrotsky, A. (1990). Negative pressure-temperature slopes for reactions forming $MgSiO_3$ perovskite from calorimetry. *Science*, **249**(4974), 1275–1278.

Ito, E., Navrotsky, A. (1985). $MgSiO_3$ Ilmenite: calorimetry, phase equilibria, and decomposition at atmospheric pressure. *American Mineralogist*, **70**(9–10), 1020–1026.

Kapush, D. Ushakov, S. V., Navrotsky, A., et al. (2017). A combined experimental and theoretical study of enthalpy of phase transition and fusion of yttria above 2000°C using "drop-n-catch" calorimetry and first-principles calculations. *Acta Materialia*, **124**, 204–209.

Kojitani, H., Furukawa, A., Akaogi, M. (2007). Thermochemistry and high-pressure equilibria of the post-perovskite phase transition in $CaIrO_3$. *American Mineralogist*, **92**(1), 229–232.

Kojitani, H., Inoue, T., Akaogi, M. (2016). Precise measurements of enthalpy of postspinel transition in Mg_2SiO_4 and application to the phase boundary calculation. *Journal of Geophysical Research: Solid Earth*, **121** (2), 729–742.

Kojitani, H., Oohata, M., Inoue, T., Akaogi, M. (2012). Redetermination of high-temperature heat capacity of Mg_2SiO_4 ringwoodite: measurement and lattice vibrational model calculation. *American Mineralogist*, **97**(8), 1314–1319.

Koryttseva, A., Navrotsky, A. (2016). High temperature calorimetric study of oxide component dissolution in a $CaO–MgO–Al_2O_3–SiO_2$ slag at 1450°C. *Journal of the American Ceramic Society*, **100**(3), 1172–1177.

Levchenko, A. A., Kolesnikov, A., Trofymluk, O., Navrotsky, A. (2010). Energetics of single-wall carbon nanotubes as revealed by calorimetry and neutron scattering. *Carbon*, **49**(3), 949–954.

Liang, J.-J., Topor, L., Navrotsky, A., Mitomo, M. (1999). Silicon nitride: enthalpy of formation of the α- and β- polymorphs and the effect of C and O impurities. *Journal of Materials Research*, **14**(5), 1959–1968.

Lilova, K. I., DeAngelis, M., Anovitz, L. M., Navrotsky, A. (2018). Surface energy of fayalite and its effect on Fe–Si–O oxygen buffers and the olivine–spinel transition. *American Mineralogist*, **103**(10), 1599–1603.

Lilova, K. I., Pearce, C. I., Rosso, K. M., Navrotsky, A. (2014). Energetics of spinels in the Fe–Ti–O system at the nanoscale. *ChemPhysChem*, **15**(16), 3655–3662.

Lilova, K. I., Xu, F., Rosso, K. M., et al. (2012). Oxide melt solution calorimetry of Fe(II)–bearing oxides and application to the magnetite – maghemite ($Fe_3O_4–Fe_{8/3}O_4$) system. *American Mineralogist*, **97**(1), 164–175.

Linard, Y., Wilding, M. C., Navrotsky, A. (2008). High temperature calorimetric studies of heat of solution of NiO, CuO, La_2O_3, TiO_2, HfO_2 in sodium silicate liquids. *Geochimica et Cosmochimica Acta*, **72**(2), 590–601.

Liu, J., Topor, L., Zhang, J., Navrotsky, A., Liebermann, R. (1996). Calorimetric study of the coesite–stishovite transformation and calculation of the phase boundary. *Physics and Chemistry of Minerals*, **23**, 11–16.

Majzlan, J., Navrotsky, A., Stevens, R., et al. (2005). Thermodynamics of monoclinic $Fe_2(SO_4)_3$. *Journal of Chemical Thermodynamics*, **37**(8), 802–809.

Mera, G., Navrotsky, A., Sen, S., Kleebe, H.-J., Riedel, R. (2013). Polymer-derived SiCN and SiOC ceramics – structure and energetics at the nanoscale. *Journal of Materials Chemistry A*, **1**(12), 3826–3836.

Morcos, R. M., Navrotsky, A., Varga, T., et al. (2008). Thermodynamically stable $Si_wC_xN_yO_z$ polymer–like, amorphous ceramics made from organic precursors. *Journal of the American Chemical Society*, **91**(7), 2391–2393.

Navrotsky, A. (1977). Progress and new directions in high temperature calorimetry. *Physics and Chemistry of Minerals*, **2**, 89–104.

Navrotsky, A. (1980). Lower mantle phase transitions may generally have negative pressure – temperature slopes. *Geophysics Research Letters*, **7**(9), 709–711.

Navrotsky, A. (1981). Energetics of phase transitions in AX, ABO_3, and AB_2O_4 compounds, in M. O'Keeffe, A. Navrotsky, eds., *Structure and Bonding in Crystals*, Vol. II. Academic Press, pp. 71–93.

Navrotsky, A. (1987). Silicate and germanate garnets, ilmenites, and perovskites: thermochemistry, lattice vibrations, and spectroscopy, in M. Manghnani, Y. Syono, eds., *High Pressure Research in Mineral Physics*. Terra Publications, pp. 261–268.

Navrotsky, A. (1997). Progress and new directions in high temperature calorimetry revisited. *Physics and Chemistry of Minerals*, **24**(3), 222–241.

Navrotsky, A. (1998). Thermodynamics of high pressure phases, in R. J. Hemley, ed., *Ultrahigh-Pressure Mineralogy: Physics and Chemistry of the Earth's Deep Interior, Reviews in Mineralogy*. Mineralogical Society of America, 37, pp. 319–341.

Navrotsky, A. (2001). Thermochemical studies of nitrides and oxynitrides by oxidative oxide melt calorimetry. *Journal of Alloys and Compounds*, **321**(2), 300–306.

Navrotsky, A. (2011). Nanoscale effects on thermodynamics and phase equilibria in oxide systems. *ChemPhysChem*, **12**(12), 2207–2215.

Navrotsky, A. (2014). Progress and new directions in calorimetry: a 2014 perspective. *Journal of the American Ceramic Society*, **97**(11), 3349–3359.

Navrotsky, A. (2016). Energetics at the nanoscale: impacts for geochemistry, the environment, and materials. *MRS Bulletin*, **41**(2), 139–145.

Navrotsky, A., Akaogi, M. (1984). The α, β, γ phase relations in Fe_2SiO_4–Mg_2SiO_4 and Co_2SiO_4–Mg_2SiO_4: calculation from thermochemical data and geophysical applications. *Journal of Geophysical Research*, **89**(B12), 10135–10140.

Navrotsky, A., Davies, P. K. (1981). Cesium chloride versus nickel arsenide as possible structures for (Me,Fe)O in the lower mantle. *Journal of Geophysical Research*, **86** (B5), 3689–3694.

Navrotsky, A., Dorogova, M., Hellman, et al. (2007). Application of calorimetry on a chip to high pressure materials. *Proceedings of the National Academy of Sciences*, **104** (22), 9187–9191.

Navrotsky, A., Forray, F. L. Drouet, C. (2005). Jarosite stability on Mars. *Icarus*, **176**(1), 250–253.

Navrotsky, A., Ma, C., Lilova, K., Birkner, N. (2010). Nanophase transition metal oxides show large thermodynamically driven shifts in oxidation-reduction equilibria. *Science*, **330**(6001), 199–201.

Navrotsky, A., Mazeina, L., Majzlan, J. (2008). Size-driven structural and thermodynamic complexity in iron oxides. *Science*, **319**(5870), 1635–1638.

Navrotsky, A., Percival, J., Dobrzhinetskaya, L., et al. (2020). A geologic Si–O–C pathway to incorporate carbon in silicates, in C. E. Manning, ed., *Carbon in Planetary Interiors*, American Geophysical Union Monograph, pp. 47–54.

Navrotsky, A., Schoenitz, M., Kojitani, H., et al. (2002). Aluminum in magnesium silicate perovskite: synthesis and energetics of defect solid solutions. *Materials Research Society Symposium Proceedings*, **718**(D2.2), 103–108.

Navrotsky, A., Schoenitz, M., Kojitani, H., et al. (2003). Aluminum in magnesium silicate perovskite: formation, structure, and energetics of magnesium-rich defect solid solutions. *Journal of Geophysical Research B*, **108**(B7), 2330.

Navrotsky, A., Ushakov, S. V. (2017). Hot matters – experimental methods for high-temperature property measurement. *American Ceramic Society Bulletin*, **96**(2), 22–28.

Navrotsky, A., Xu, F. (2010). Enthalpies of formation of pyrrhotite $Fe_{1-0.125x}$ S ($0 \leq x \leq 1$) solid solutions. American Mineralogist, **95**(5), 717–723.

Niu, M., Wang, H., Chen, J., et al. (2017). Structure and energetics of SiOC and SiOC-modified carbon-bonded carbon fiber composites. *Journal of the American Ceramic Society*, **100**(8), 3693–3702.

Rosen, P. F., Dickson, M. S., Calvin, J. J., et al. (2020). Thermodynamic evidence of structural transformations in CO_2-loaded metal-organic framework $Zn(MeIm)_2$ from heat capacity measurements. *Journal of the American Chemical Society*, **142**(10), 4833–4841.

Rosen, P. F., Woodfield, B. F. (2020). Standard methods for heat capacity measurements on a quantum design physical property measurement system. *Journal of Chemical Thermodynamics*, **141**(C), 105974.

Sahu, S. K., Huang, B., Lilova, K. I., Woodfield, B. F., Navrotsky, A. (2015). Thermodynamics of Fe_3O_4–Co_3O_4 and Fe_3O_4–Mn_3O_4 spinel solid solutions at bulk and nanoscale. *Physical Chemistry Chemical Physics*, **17**(34), 22286–22295.

Schoenitz, M., Navrotsky, A. (1999). Enthalpy of formation of katoite $Ca_3Al_2[(OH)_4]_3$: energetics of the hydrogarnet substitution. *American Mineralogist*, **84**(3), 389–391.

Schoenitz, M., Navrotsky, A., Leinenweber, K. (2000). The enthalpy of transformation of $Ca(OH)_2$-I (portlandite) to $Ca(OH)_2$-II (EuI_2 structure) by low-temperature DSC. *Physics and Chemistry of Minerals*, **27**(9), 604–9.

Sen, S., Widgeon, S. J., Navrotsky, A., et al. (2013). Carbon substitution for oxygen in silicates in planetary interiors. *Proceedings of the National Academy of Sciences*, **110** (40), 15904–15977.

Sharma, G., Naguib, M., Feng, D., Gogotsi, Y., Navrotsky, A. (2016). Calorimetric determination of thermodynamic stability of MAX and MXene phases. *Journal of Physical Chemistry C*, **120**(49), 28131–28137.

Sharma, G., Ushakov, S., Navrotsky, A. (2018). Size-driven thermodynamic crossovers in phase stability in zirconia and hafnia. *Journal of the American Ceramic Society*, **101** (1), 31–35.

Spektor, K., Nylen, J., Mathew, R., et al. (2016). Formation of hydrous stishovite from coesite in high-pressure hydrothermal environments. *American Mineralogist*, **101** (11), 2514–2524.

Spektor, K., Nylen, J., Stoyanov, E., et al. (2011). Ultrahydrous stishovite from high-pressure hydrothermal treatment of SiO_2. *Proceedings of the National Academy of Sciences*, **108**(52), 20918–20922.

Stebbins, J. F., Kojitani, H., Akaogi, M., Navrotsky, A. (2003). Aluminum substitution in $MgSiO_3$ perovskite: investigation of multiple mechanisms by ^{27}Al NMR. *American Mineralogist*, **88**(7), 1161–1164.

Subramani, T., Lilova, K., Abramchuk, M., Leinenweber, K., Navrotsky, A. (2020). Greigite (Fe_3S_4) is thermodynamically stable: implications for its terrestrial and planetary occurrence. *Proceedings of the National Academy of Sciences*, **117**(46), 28645–28648.

Takayama-Muromachi, E., Navrotsky, A. (1988). Energetics of compounds ($A^{3+}B^{4+}O_3$) with the perovskite structure. *Journal of Solid State Chemistry*, **72**(2), 244–256.

Tavakoli, A. H., Navrotsky, A. (2013). Enthalpies of formation of Fe–Ni monosulfide solid solutions. *American Mineralogist*, **98**(8), 1508–1515.

Tessier, F., Le Gendre, L., Chevirė, F., Marchand, R., Navrotsky, A. (2005). Thermochemistry of a new class of materials containing dinitrogen pairs in an oxide matrix. *Chemistry of Materials*, **17**(13), 3570–3574.

Tessier, F., Navrotsky, A., Niewa, R., et al. (2000). Energetics of binary iron nitrides. *Solid State Sciences*, **2**(4), 457–462.

Trofymluk, O., Levchenko, A. A., Navrotsky, A. (2012). Mesoporous silica synthesis: energetics of interaction between framework and structure directing agent. *Microporous and Mesoporous Materials*, **149**(1), 119–125.

Ushakov, S. V., Maram, P. S., Kapush, D., et al. (2018). Phase transformations in oxides above 2000 °C: experimental technique development. *Advanced Applied Ceramics*, **117**(sup1), s82–s89.

Widgeon, S., Mera, G., Gao, Y., et al. (2012). Nanostructure and energetics of carbon-rich SiCN ceramics derived from polysilylcarbodiimides: role of the nanodomain interfaces. *Chemistry of Materials*, **24**(6), 1181–1191.

Wu, D., Navrotsky, A. (2015). Thermodynamics of metal-organic frameworks. *Journal of Solid State Chemistry*, **223**(C), 53–58.

Wu, L., Li, L., Evans, S. F., et al. (2019). Lithium aluminum layered double hydroxide chlorides (LDH): formation enthalpies and energetics of lithium ion capture. *Journal of the American Ceramic Society*, **102**(5), 2398–2404.

Zhang, Y., Navrotsky, A., Sekine, T. (2006). Energetics of cubic Si_3N_4. *Journal of Materials Research*, **21**(1), 41–44.

Index

www.ingramcontent.com/pod-product-compliance
Lightning Source LLC
LaVergne TN
LVHW062200200926
848484LV00006B/1290
* 9 7 8 1 1 0 8 4 7 9 7 5 2 *